Edible Flowers: Source of Phytonutrients, Valorization and Technological Advancements

Amrita Poonia • Parmjit S. Panesar
Maricarmen Iñiguez Moreno
Editors

Edible Flowers: Source of Phytonutrients, Valorization and Technological Advancements

Springer

Editors
Amrita Poonia
Department of Dairy Science and Food Technology, Institute of Agricultural Sciences
Banaras Hindu University
Varanasi, Uttar Pradesh, India

Parmjit S. Panesar
Department of Food Engineering & Technology
S.L. Institute of Engineering & Technology
Longowal, Punjab, India

Maricarmen Iñiguez Moreno
School of Engineering and Technology, Department of Physics and Mathematics
Universidad de Monterrey
San Pedro Garza García, NL, Mexico

ISBN 978-3-032-00666-0 ISBN 978-3-032-00667-7 (eBook)
https://doi.org/10.1007/978-3-032-00667-7

This Springer imprint is published by the registered company Springer Nature Switzerland AG
The registered company address is: Gewerbestrasse 11, 6330 Cham, Switzerland

This book is dedicated to ***Shri Kashi Vishwanath Ji*** *located on the Western bank of the holy river* ***Ganga****, Varanasi, Uttar Pradesh, India, and is one of the twelve Jyotirlingas*

The main deity is known by the names Shri Vishwanath and Vishweshwara, literally meaning ***Lord of the Universe****. Varanasi city was called Kashi in ancient times, and hence the temple is popularly called Kashi Vishwanath Temple.*

&

Bharat Ratna Mahamana Pandit Madan Mohan MaiaviyaJi*, Founder of Banaras Hindu University, the largest residential university in Asia and one of the largest in the world.*

Preface

Flowers are not only used for decoration of savory dishes and desserts but also provide a unique combination of sensations, enhancing the nutritional profile of various foods. Owing to increasing knowledge about the composition, antioxidant, and biological properties of edible flowers and availability of metabolomics to identify their chemical components, there are many possibilities for development of nutraceuticals and functional foods using edible flowers. Since edible flowers have antioxidants, fiber, and even some proteins, they can be used in various medicines and therapeutic meal plans. The most common phytochemical compounds found in edible flowers are carotenoids, flavonoids, phenolic acids, and anthocyanins, which are well known for health benefits. This can be an emerging business as far as food processing and marketing is concerned. Edible flowers have a high nutritional potential due to their rich content of bioactive compounds, which can lead to significant economic and social impacts in the global food economy. Food industry could benefit by fulfilling the market demand of functional foods through the development of edible flower-based foods, therefore, allowing the valorization of unexplored or underexplored species of these flowers.

The diverse range of colors and flavors associated with edible flowers could help to substitute artificial colors and flavors. In addition to the antioxidant properties, researches have proved that these flowers have numerous significant biological activities such as antidiabetic, anti-inflammatory, antidiabetic, antitumor, gastroprotective, and antimicrobial activities, which make them more special along with their nutritional advantages. In recent years, the edible flower supply chain increased owing to their high demand as new food ingredients. Use of edible flowers has potentials in alleviation of malnutrition, alternative source of income as well as rescue of wild edible flower species.

This book comprises 15 chapters and focuses on the current information on edible flower species, consumer acceptance, nutritional and functional properties, as well as about viewpoints on processing, post-harvest management and preservation, biological activity, health benefits, formulation of functional foods, nanoparticle synthesis, safety issues, toxicological effects, cultural uses, and current market trends and possibilities. Diversification and sustainable production of edible flowers from different countries contribute to food security by providing a variety of flavours, nutrients, and health benefits.

Unique Features of the Book

- Bioactive compounds, antioxidant potential, nanoparticle synthesis, current trends and innovations
- Fulfilling the market demand of functional foods through the utilization of edible flower-based foods and food formulations using unexplored species of these flowers
- Explores potential industrial applications of edible flowers as alternative source of income to support world economy

Varanasi, Uttar Pradesh, India — Amrita Poonia
Longowal, Punjab, India — Parmjit S. Panesar
San Pedro Garza García, NL, Mexico — Maricarmen Iñiguez Moreno

Contents

List of Figures

List of Tables

Editors and Contributors

About the Editors

Amrita Poonia is working as an Assistant Professor in the Department of Dairy Science and Food Technology, Institute of Agricultural Sciences, Banaras Hindu University, Varanasi, Uttar Pradesh, India. She has published 2 books with Springer, 2 laboratory manuals, and more than 26 book chapters at national and international level. She has conducted/offered a Massive Open Online Course (MOOC) under agMOOCs on Functional Foods: Concept, Technology and Health Benefits from October 8, 2018, to November 19, 2018, and was awarded with the Certificate of Appreciation for offering a Massive Open Online Course (MOOC) under agMOOCs, by Commonwealth of Learning (COL) Canada. She has published more than 78 publications in esteemed national and international journals. She has supervised 3 Ph.D. scholars and 42 M.Sc. students. She has 16 years of teaching experience at UG, PG, and Ph.D. levels. She has received maximum Research Incentive Grant under Institutions of Eminence (IoE Scheme) of Banaras Hindu University. She is Co-PI in the project titled "Morpho-Molecular characterization, and utilization of land races of finger millet (*Eleusine coracana*) germplasms" funded by Uttar Pradeh Council of Agricultural Research (UPCAR). She is a Life Member of the Association of Food Scientists & Technologists, India, and Indian Dairy Association.

Parmjit S. Panesar is currently working as Professor, Food Biotechnology Research lab, Sant Longowal Institute of Engineering and Technology (SLIET) (Deemed to be University: Established by Govt. of India), Longowal, Punjab. He has more than 26 years of teaching and research experience and has also served in administrative positions like Dean (Research & Consultancy), Dean (Planning & Development), and Head, Department of Food Engineering & Technology, SLIET, Longowal. In 2005, he has been awarded BOYSCAST fellowship by Department of Science & Technology (DST), Govt. of India, to carry out advance research at Chembiotech labs, University of Birmingham Research Park, UK. He was also involved in US and Australian sponsored projects during his stay at the UK. In 1999, he was awarded Young Scientist Fellowship by Punjab State Council for Science & Technology, Punjab. He has been selected in Top 2% Scientist list of Stanford University in 2021, 2022, and 2023. He has successfully completed nine research projects funded by Department of Biotechnology (DBT), Council of Scientific and Industrial Research (CSIR), Ministry of Human Research Development (MHRD), All India Council For Technical Education (AICTE), New Delhi, and one project is in progress. He has published more than 200 international/national scientific papers, 50 book reviews in peer-reviewed journals, and 60 chapters and has authored/edited 15 books.

Maricarmen Iñiguez Moreno is a professional specialized in food biotechnology and microbiology, particularly in the valorization of agro-industrial waste for the development of value-added products, as well as in the development and validation of green alternatives for food preservation integrating principles of circular economy and sustainability. Her thesis projects in M.Sc. and Ph.D. were about the formation and dynamics of microbial populations and control of biofilms developed by spoilage and foodborne pathogenic bacteria on inert surfaces under conditions that mimic the food industry. She has completed two-year postdoctoral stay at the Tecnológico Nacional de México/Instituto Tecnológico de Tepic with focus on the development of coatings added with biocontrol agents or bioactive compounds into nanostructures to enhance the postharvest preservation of tropical fruits.

Contributors

Muhammad Afzaal Department of Food Science, Government College University Faisalabad, Faisalabad, Pakistan

Aftab Ahmed Department of Nutritional Sciences, GC University Faisalabad, Faisalabad, Pakistan

Oscar Cruz Álvarez Facultad de Ciencias Agrotecnológicas, Universidad Autónoma de Chihuahua, Chihuahua, Mexico

V. S. Anjana Department of Biotechnology, Pondicherry University, Puducherry, India

Laxmikant S. Badwaik Department of Food Engineering and Technology, School of Engineering, Tezpur University, Napaam, Assam, India

Damaris Leopoldina Ojeda Barrios Facultad de Ciencias Agrotecnológicas, Universidad Autónoma de Chihuahua, Chihuahua, Mexico

Aneeba Busharat Department of Food Science, Government College University Faisalabad, Faisalabad, Pakistan

Muhammad Usman Butt Department of Food Science, Government College University Faisalabad, Faisalabad, Pakistan

Indrani Chetia Department of Food Engineering and Technology, School of Engineering, Tezpur University, Napaam, Assam, India

Wee Sim Choo School of Science, Monash University Malaysia, Selangor, Malaysia

E. J. Delgado-Núñez Departamento de Agronomía, Facultad de Ciencias Agropecuarias y Ambientales de la Universidad Autónoma de Guerrero, Iguala, Guerrero, Mexico

Maryam Ehsan Department of Food Science, Government College University Faisalabad, Faisalabad, Pakistan

Felipe Ascencio-Valle Felipe SECIHTI-Centro de Investigaciones Biológicas del Noroeste, Instituto Politécnico, La Paz, Baja California Sur, Mexico

A. Francisco-Palemon Departamento de Agronomía, Facultad de Ciencias Agropecuarias y Ambientales de la Universidad Autónoma de Guerrero, Iguala, Guerrero, Mexico

Saikat Gantait Crop Research Unit (Genetics and Plant Breeding), Bidhan Chandra Krishi Viswavidyalaya, Mohanpur, Nadia, West Bengal, India

Fernando González-Chávez Universidad Politécnica del Estado de Nayarit, Tepic, Mexico

Lizet Aguirre Güitrón Universidad Tecnológica de Nayarit, Carretera Federal, Xalisco, Nayarit, Mexico

Debdip Hazra Department of Dairy Science and Food Technology, Institute of Agricultural Sciences, Banaras Hindu University, Varanasi, Uttar Pradesh, India

Muhammad Bilal Hussain Department of Food Science, Government College University Faisalabad, Faisalabad, Pakistan

Haseeba Khalil Janjua Department of Human Nutrition and Dietetics, Mirpur University of Science and Technology Mirpur, Mirpur, Pakistan

D. E. Jimenez-Sánchez Centro Regional de Educación Superior de la Zona Norte, Campus Taxco el viejo/Universidad Autónoma de Guerrero, Taxco, Guerrero, Mexico

Marisela Calderón Jurado Facultad de Ciencias Agrotecnológicas, Universidad Autónoma de Chihuahua, Chihuahua, Mexico

M. Karthik Department of Biotechnology, Pondicherry University, Puducherry, India

Mahendra Kumar Department of Food Processing Technology, College of Dairy and Food Technology, Bassi (Rajasthan University of Veterinary and Animal Sciences), Bikaner, Rajasthan, India

Shankar Lal Department of Dairy Science & Food Technology, Institute of Agricultural Sciences, Banaras Hindu University, Varanasi, Uttar Pradesh, India

M. Luna-Flores Departamento de Ingeniería en Procesos Bioalimentarios, Universidad Tecnológica del Centro de Veracruz, Veracruz, Mexico

Bimlesh Mann Indian Council of Agricultural Research, New Delhi, India

Rubí A. Martínez-Camacho Instituto Tecnológico de La Paz, Tecnológico Nacional de México, La Paz, Baja California Sur, Mexico

Adalberto Benavides Mendoza Departamento de Horticultura, Universidad Agraria Antonio Narro, Saltillo, Mexico

Anwesha Mukherjee Department of Biotechnology, Pondicherry University, Puducherry, India

Tuba Mustafa Department of Food Science, Government College University Faisalabad, Faisalabad, Pakistan

Iqra Nawaz Department of Food Science, Government College University Faisalabad, Faisalabad, Pakistan

S. A. Ortega-Acosta Departamento de Agronomía, Facultad de Ciencias Agropecuarias y Ambientales de la Universidad Autónoma de Guerrero, Iguala, Guerrero, Mexico

Parmjit S. Panesar Department of Food Engineering & Technology, Sant Longowal Institute of Engineering and Technology, (Deemed to be University, Estd. by Govt of India, Longowal, Punjab, India

Chinmay Patil Department of Dairy Science and Food Technology, Institute of Agricultural Sciences, Banaras Hindu University, Varanasi, Uttar Pradesh, India

Amrita Poonia Department of Dairy Science and Food Technology, Institute of Agricultural Sciences, Banaras Hindu University, Varanasi, Uttar Pradesh, India

Priyanka Department of Food and Nutrition, I.C College of Community Science, Chaudhary Charan Singh Haryana Agricultural University, Hisar, Haryana, India

Graciela Dolores Ávila Quezada Facultad de Ciencias Agrotecnológicas, Universidad Autónoma de Chihuahua, Chihuahua, Mexico

Saranga Rajesh Department of Biotechnology, Pondicherry University, Puducherry, India

Diana Ramírez-Gamboa Tecnologico de Monterrey, Institute of Advanced Materials for Sustainable Manufacturing, Monterrey, Mexico

Bisma Rasheed Department of Food Science, Government College University Faisalabad, Faisalabad, Pakistan

Ana Gisela Reyes-Alvarado SECIHTI-Centro de Investigaciones Biológicas del Noroeste, Instituto Politécnico, La Paz, Baja California Sur, Mexico

Tomás Rivas-García SECIHTI-Universidad Autonoma Chapingo, Texcoco, Estado de Mexico, Mexico

Farhan Saeed Department of Food Science, Government College University Faisalabad, Faisalabad, Pakistan

Muhammad Aqib Saeed Department of Human Nutrition and Dietetics, Mirpur University of Science and Technology Mirpur, Mirpur, Pakistan

Muhammad Awais Saleem Department of Human Nutrition and Dietetics, Mirpur University of Science and Technology Mirpur, Mirpur, Pakistan

Jayoti Majumder Sarkar Department of Floriculture and Landscape Architecture, Bidhan Chandra Krishi Viswavidyalaya, Mohanpur, Nadia, West Bengal, India

Awanish Singh Department of Food Engineering and Technology, Tezpur University, Tezpur, Assam, India

Nandan Sit Department of Food Engineering and Technology, Tezpur University, Tezpur, Assam, India

Pan Nu Thwe School of Science, Monash University Malaysia, Selangor, Malaysia

Hannah R. Vasanthi Department of Biotechnology, Pondicherry University, Puducherry, India

Marwa Waheed Riphah International University Faisalabad, Faisalabad, Pakistan

Arati Yadav Department of Dairy Science and Food Technology, Institute of Agricultural Sciences, Banaras Hindu University, Varanasi, Uttar Pradesh, India

Rekha Yadav Department of Food and Nutrition, I.C College of Community Science, Chaudhary Charan Singh Haryana Agricultural University, Hisar, Haryana, India

Abbreviations

AAE	Ascorbic acid equivalents
AuNPs	Gold nanoparticles
Ca	Calcium
CA	Controlled atmosphere
CAGR	Compound annual growth rate
CAP	Controlled atmospheric packaging
CAT	Catalase
Cd	Cadmium
CEA	Controlled environment agriculture
Co	Cobalt
CO_2	Carbon dioxide
CPE	*Chimonanthus praecox* extract
Cu	Copper
Cyn.H	Cyanidin hexoside
DLS	Dynamic light scattering
DM	Dry matter
DNA	Deoxyribonucleic acid
DPPH	2, 2-diphenyl-1-picrylhydrazyl
DPPH	2,2-diphenyl-1-picrylhydrazyl
DSC	Differential scanning calorimeter
ECM	Extracellular matrix
EFs	Edible flowers
EFSA	European Food Safety Authority
ENMs	Engineered nanomaterials
FAO	Agriculture Organisation of the United Nations
FDA	Food and Drug Administration
FRAP	Ferric-reducing antioxidant potential
FRAP	Ferric reducing antioxidant power
FRAP	Ferric-reducing antioxidant potential
FSMA	Food Safety Modernization Act
FW	Fresh weight

GABA	Gamma-aminobutyric acid
GAE	Gallic acid equivalents
GAP	Good agricultural practices
GMP	Good manufacturing practices
GPx	Glutathione peroxidase
H_2O_2	Hydrogen peroxide
HACCP	Hazard Analysis and Critical Control Points
HepG2 cells	Human liver cancer cell line
HG	Mercury
HHP	High hydrostatic pressure
HOAP	High oxygen atmospheric packaging
HOAP	High-oxygen atmospheric packaging
HPP	High hydrostatic pressure
INS-1 cell line	Rat pancreatic β-cell line
IoT	Internet of things
IPM	Integrated pest management
IPM	Integrated pest management
K562 cells	Human leukemia cell line
LPS	Lipopolysaccharide
MAE	Microwave-assisted extraction
MAP	Modified atmospheric packaging
MCF-7 cells	Human breast cancer cell line
MDA	Malondialdehyde
MI	Myocardial infarction
MRLs	Maximum residue levels
MTD	Maximally tolerated dose
N2a cells	Mouse neuroblastoma cell line (ATCC® CCL-131™)
Ni	Nickel
NK	Nattokinase
NMR	Nuclear magnetic resonance
NPs	Nanoparticles
OAS	Oral allergic syndrome
OECD	Organization for Economic Co-operation and Development
ORAC	Oxygen radical absorbance capacity
Pb	Lead
PEF	Pulsed electric field technology
PLE	Pressurized liquid extraction
PPO	Polyphenol oxidase
RASFF	Rapid Alert System for Food and Feed
RAW 264.7 macrophages	Murine macrophage cell line
RH	Relative humidity
ROS	Reactive oxygen species
S	Sulphur
SEM	Scanning electron microscopy

SFE	Supercritical fluid extraction
SOD	Superoxide dismutase
SOD	Superoxide dismutase
TAE	Tannic acid equivalents
TBARS	Thiobarbituric acid reactive substances
TE	Trolox equivalent
TEM	Transmission electron microscopy
TPC	Total phenolic content
TRAP	Tartrate-resistant acid phosphatase
UAB	Ultraviolet B
UAE	Ultrasound-assisted extraction
UNESCO	United Nations Educational, Scientific, and Cultural Organization
UV	Ultraviolet
V	Vanadium
WHC	Water holding capacity
WHO	World Health Organisation
WTO	World Trade Organization
XRD	X-ray diffraction

Chapter 1
Edible Flowers: An Overview

Amrita Poonia, Shankar Lal, Mahendra Kumar, and Parmjit S. Panesar

1.1 Introduction

The ancient history of edible flowers is extensive and crosses many continents and cultures. These flowers have been utilised for millennia not only for their aesthetic value but also for their culinary and therapeutic qualities (Lokare and Pandya 2024). Ancient societies such as the Romans, Greeks, and Egyptians used flowers in their food and medicine because they had special health advantages, such as aiding in digestion or curing illnesses (Goswami and Ram 2017). Edible flowers are an excellent source of antioxidants, vitamins, and minerals. The market for nutraceuticals is driven primarily by increased consumer awareness, improved purchasing power, the popularisation of the beneficial effects of phytochemicals on the human body, consumer readiness to purchase functional foods, and a strong desire to maintain a healthy lifestyle. It is anticipated that dairy-based beverages will propel the market expansion for functional foods (Rao and Poonia 2024a). These flowers have long been prized for their capacity to improve the taste and appearance of food and also a natural source of nutrients, contains vitamins, minerals, antioxidants and various bioactive compounds that promote general health. These flowers have been

A. Poonia (✉) · S. Lal
Department of Dairy Science & Food Technology, Institute of Agricultural Sciences, Banaras Hindu University, Varanasi, UP, India
e-mail: amrita12@bhu.ac.in

M. Kumar
Department of Food Processing Technology, College of Dairy and Food Technology, Bassi (Rajasthan University of Veterinary and Animal Sciences), Bikaner, Rajasthan, India

P. S. Panesar
Department of Food Engineering & Technology, Sant Longowal Institute of Engineering and Technology (Deemed to be University, Estd. by Govt of India), Longowal, Punjab, India

A. Poonia et al. (eds.), *Edible Flowers: Source of Phytonutrients, Valorization and Technological Advancements*, https://doi.org/10.1007/978-3-032-00667-7_1

rediscovered for their health advantages and adaptability, making them a healthy and fascinating addition to our gardens and kitchens as interest in organic and sustainable foods develops. In addition to their possible medical uses in herbal treatments, they can be added to salads, drinks, desserts, and garnishes, among other culinary applications.

These flowers are present worldwide in around 180 species, 100 genera, and 97 families, yet only a limited number have been extensively studied (Kumari and Bhargava 2021). There is a highly need for more comprehensive research to recognise the complexities of this natural flowers. Such study is crucial to enhancing the acceptance of These flowers as viable food ingredients while mitigating potential risks (Benvenuti and Mazzoncini 2021). Some of the flowers possess the necessary criteria to be deemed safe for human consumption-such as being non-toxic, innocuous, and nutritionally beneficial (Duggirala et al. 2024). These flowers are gaining popularities because of their taste, colour, flavour, medicinal properties, natural and minimal processed food can be packaged without preservatives, and low in sugar, fat and salt content and can be immersed in most of food products without any side effects. These flower also possess several biological properties such as anti-hyperglycaemic and anticholinergic activity (Nowicka and Wojdyło 2019), improvement of ulcerative colitis (Meurer et al. 2020), anticancer (Nguyen et al. 2019), protection against oxidative effects in erythrocyte (Yang 2019). Recent studies shows that they have medicinal properties, such as antidiabetic (Janarny et al. 2021), anticancer (Fakhri et al. 2022), anti-anxiety (Skrajda-Brdak et al. 2020), immunomodulatory, anti-inflammatory, antimicrobial (Faisal et al. 2022). Although some flower species are considered as non-edible, because they include compounds that are harmful or antinutritional, like alkaloids, cyanogenic glycosides, oxalic acid, trypsin inhibitors, or hemagglutinins (Purohit et al. 2021), Because of that the production of innovative food products with therapeutic benefits using these flowers becomes a crucial challenge for Agro-food industry by maintaining food safety, quality and security, Because ornamental flowers and these flowers are almost similar in appearance and difficult to differentiate them by chemical and biological parameters. Some decorative flowers are also toxic and can be fatal in some cases (Coyago-Cruz et al. 2023).

The consumer acceptance of these flowers is a crucial and important for their marketing, commercialization and post-harvest performances (Malakar et al. 2023). The flowers are highly perishable (suffer oxidation) products and can be deteriorate by insects, so it's difficult to maintain their physical and nutritional quality like odour, taste, appearance and nutraceutical components (Majerska et al. 2019). Mostly they are packaged in small plastic containers. Edible coating and Modified Atmospheric Packaging (MAP) are also important techniques to preserve the perishable flowers by maintaining quality, extension self-life and reducing microbial load (Jadhav et al. 2023). Some other post-harvest techniques such as drying, refrigeration, preservation in distillates and sugar canning also helpful in preservation but they bring some undesirable nutritional and biochemical changes that impacts the overall quality. Food irradiation is an economical and viable technique to preserve the food, maintain quality and preserve phytochemicals and bioactive characteristics (Verma and Joshi 2000). In this chapter, we explore the availability various kinds of flowers, their historical and cultural significance, health benefits, and creative ways to incorporate them into culinary creations and also their sensory, nutritional, and functional qualities (Table 1.1). Opinions on processing, contents, biological activity, cultural usage, and possible future markets were also to be discussed.

1.2 Nutritional Composition of Edible Flowers

These flowers have different amounts of carbohydrates, minerals, and dietary fibre, although they are mostly made up of water (more than 80%) and are typically low in fat and protein (Mlcek and Rop 2011). Nutritional composition of different flowers has been mentioned in Table 1.2. Although many research has been done on the biological characteristics of the flowers, there is still a lack of comprehensive nutritional information for many species, especially for widely utilised types. The context of sustainable flower production may make this knowledge gap more significant, especially for species like sunflowers (*Helianthus* annuus L.). The sunflower "flower" is actually a composite structure made up of hundreds of individual ray florets, which are long, yellow structures that resemble petals, and disc florets, which are the tiny, brownish florets in the centre of the "flower." In 2016 alone, 47.3 million tonnes of sunflower seeds were processed for oil production, demonstrating the enormous scale of sunflower production (FAOSTAT 2019). Even though they are a rich source of important nutrients, the blossoms that are left over after the oil is extracted are sometimes thrown away as waste. Some studies have suggested that these blossoms could be used as a supplement for athletes because of their high amount of essential amino acids (3.09–3.21% dry weight), which includes notable levels of leucine, isoleucine, and valine (Liang et al. 2013). With values ranging from 1983.83 to 2006, sunflower blooms are exceptionally rich in minerals, especially potassium, calcium, and phosphorus. 127.81–335.67 mg/100 g dry weight, 622.81–748.26 mg/100 g, and 76 mg/100 g, in that order (Liang et al. 2013). These flowers may also help those with anaemia because of their high iron content (12.84–34.81 mg/100 g dry weight) (Müseler 2005). Furthermore, according to (Liang et al. 2013), they have significant concentrations of vitamins C (43.65–68.08 mg/100 g dry weight) and E (around 0.5%).

The chives (*Allium schoenoprasum* L.), which are native to Europe and Asia, have unsaturated fatty acids like linoleic and oleic acid, are abundant in these exquisite pale purple blossoms (about 13%) (Grzeszczuk et al. 2011). Furthermore, it has been demonstrated that blooms from plants including sunflower, primula, and marigold (*Calendula officinalis*) (Pires et al. 2017; Chensom et al. 2019) contain important fatty acids. Marigold's remarkable 59.3% unsaturated fatty acid content makes it stand out (Takahashi et al. 2020). These flowers contain less fatty acids than foods like chicken breast, but they nevertheless have more fatty acids than many plant-based foods including cauliflower (USDA 2018d), broccoli (USDA 2018c), bananas

Table 1.1 Common species of edible flowers with their bioactive compounds

Common name	Scientific name	Place of origin	Bioactive compounds	Used in functional food	References
Lavender	*Lavandula angustifolia*	Mediterranean	Linalool, tannins, flavonoids	Teas, desserts, functional beverages (stress relief)	Kmail (2024)
Chamomile	*Matricaria chamomilla*	Europe, Western Asia	Apigenin, bisabolol, flavonoids	Teas, sleep aids, digestive health products	Yuca and Karakaya (2022)
Hibiscus	*Hibiscus sabdariffa*	West Africa, Tropical Regions	Anthocyanins, vitamin C, flavonoids	Teas, juices, weight loss and blood pressure products	Kouakou et al. (2024)
Rose	*Rosa* spp.	Asia, Europe, North America	Anthocyanins, flavonoids, vitamin C	Teas, syrups, functional beauty drinks (skin health)	Takahashi (2025)
Calendula	*Calendula officinalis*	Europe, North Africa	Flavonoids, carotenoids, saponins	Soothing teas, anti-inflammatory and digestive	Matejić et al. (2024)
Elderflower	*Sambucus nigra*	Europe, Asia	Flavonoids, antioxidants, phenolic acids	Syrups, immunity-boosting products	Mercola (2021)
Borage	*Borago officinalis*	Mediterranean	Gamma-linolenic acid, flavonoids	Salads, teas, skin health (anti-inflammatory)	Asadi-Samani et al. (2014)
Sunflower	*Helianthus annuus*	North America	Flavonoids, carotenoids	Salads, oils, heart health, and antioxidant products	Adeleke and Babalola (2020)
Tuberose	*Polianthes tuberosa*	Mexico	Linalool, flavonoids, saponins	Perfumes, syrups, relaxation and wellness products	Chadha et al. (2014)
Bee Balm	*Monarda didyma*	North America	Thymol, carvacrol, flavonoids	Teas, anti-microbial and digestive health supplements	Gontar et al. (2022)
Fennel Blossom	*Foeniculum vulgare*	Mediterranean	Anethole, flavonoids, essential oils	Digestive aids, anti-inflammatory products	Rafieian et al. (2024)
Zucchini Blossom	*Cucurbita pepo*	Americas	Carotenoids, flavonoids, vitamin C	Stuffed, soups, anti-aging (rich in antioxidants)	Sadia et al. (2024)

(continued)

Table 1.1 (continued)

Common name	Scientific name	Place of origin	Bioactive compounds	Used in functional food	References
Dandelion	*Taraxacum officinale*	Europe, Asia, North America	Flavonoids, carotenoids, polyphenols	Teas, detox drinks, liver support products	Bjørklund et al. (2024)
Marigold	*Tagetes erecta*	Mexico, Central America	Lutein, zeaxanthin, flavonoids	Soups, salads, eye health (rich in lutein)	Chitrakar et al. (2019)
Jasmine	*Jasminum* spp.	Tropical and Subtropical Asia	Flavonoids, alkaloids, benzyl alcohol	Teas, syrups, relaxation and sleep aids	Chadha et al. (2014)
Fuchsia	*Fuchsia* spp.	Central and South America	Anthocyanins, flavonoids	Teas, antioxidant-rich beverages	Alighadri et al. (2023)
Linden Blossom	*Tilia* spp.	Europe, North America	Flavonoids, tannins, essential oils	Teas, calming beverages, anti-inflammatory	Hritcu and Cioanca (2016)
Lemon Balm	*Melissa officinalis*	Europe, Asia	Rosmarinic acid, flavonoids, phenolic acids	Teas, relaxation, digestive health products	Sharifi-Rad et al. (2021)
Pansy	*Viola tricolour*	Europe, Asia	Flavonoids, saponins	Garnish, antioxidant, anti-inflammatory applications	Mittu et al. (2024)
Orange Blossom	*Citrus sinensis*	Southeast Asia	Flavonoids, essential oils	Jams, syrups, calming drinks for sleep and relaxation	Agarwal et al. (2022)

(USDA 2018b). Some of these flowers stand out for their high protein content. For example, the dry weight protein content of banana flowers is 12.5% (Bhaskar et al. 2012), while sunflower flowers have notable protein levels (Liang et al. 2013). Despite having lower protein content than animal-based foods like chicken, other species such as cornflower (*Centaurea cyanus*), dahlia (*Dahlia mignon*), pot marigold (*Calendula officinalis*), and rose (*Rosa damascene*) also exhibit intriguing protein concentrations (de São Pedro Pires 2020). Their protein level, however, is higher than that of plant-based alternatives such as cauliflower (USDA 2018d), broccoli (USDA 2018c), and bananas (USDA 2018b) (Ijabadeniyi 2023). Many These flowers, such as Agave salmiana and Aloe vera, were examined by (Sotelo et al. 2007). They discovered that the moisture content ranged from 860 to 932 g/kg, the crude protein ranged from 113–275 g/kg dry matter (DM), and the crude fibre ranged from 104–177 g/kg DM. These flowers have nutritional profiles that are comparable to those of other plants, according to (Navarro-Gonzalez et al. 2015). They are high in fibre and water content but low in protein and fat. The nutritional makeup of petals and infusions from flowers like cornflower, calendula, rose, and

Table 1.2 Nutritional composition of edible flowers (Wet basis)

Common name	Energy (kcal/100 g)	Water (%)	Protein (g/100 g)	Fat (g/100 g)	Carbohydrates (g/100 g)	Fiber (g/100 g)	References
Nasturtium	45	87	2.0	0.5	8.0	1.3	Ek et al. (2018)
Chamomile	25	91	1.2	0.3	5.0	1.2	Askari-Khorasgani et al. (2017)
Hibiscus	37	89	1.0	0.4	8.0	1.3	Salami and Afolayan (2021)
Rose	42	90	1.0	0.2	9.6	1.0	Prata et al. (2017)
Calendula	45	87	1.8	0.4	8.3	2.1	Karimi et al. (2020)
Borage	44	86	1.4	0.5	8.0	1.0	Montaner et al. (2021)
Dandelion	45	87	2.7	2.7	9.2	3.5	Biel et al. (2017)
Marigold	38	89	1.5	0.3	7.4	1.8	Ahmed et al. (2017)
Zucchini Blossom	19	92	1.3	0.1	4.3	1.4	Sood et al. (2017)
Bee Balm	30	89	1.2	0.3	6.0	1.5	Jeet et al. (2024)
Clover	44	85	3.0	0.5	6.2	2.0	Gounden et al. (2018)
Lemon Balm	30	90	1.0	0.2	6.0	1.7	Son et al. (2021)
Orange Blossom	40	2.0	0.9	0.1	9.6	2.0	Labadie et al. (2015)
Tuberose	50	85	1.6	0.4	9.0	1.2	Iftikhar et al. (2023)
Jasmine	40	88	1.4	0.3	8.2	1.5	Chopde et al. (2017)
Lavender	49	85	1.9	0.3	9.0	2.3	Mavandi et al. (2021)

dahlia was investigated by (Pires et al. 2017). They discovered that the most prevalent macronutrients were ash, proteins, and carbs. Particularly abundant in organic acids such as quinic and malic acids were infusions made from calendula and rose petals. The carpel and petals of *Yucca elephantipes* flowers have a high fibre content and unsaturated fatty acids including palmitic, linoleic, and oleic acids (Juarez-Trujillo et al. 2018). When Fernandes et al. (2019) examined pansy (Viola × Wittrockiana) flowers at various stages of bloom, they discovered that fully open flowers had higher protein (2.0 g/100 g) and carbohydrate (8.0 g/100 g) contents per 100 g fresh weight. The study of some edible flowers shows that Dianthus chinensis had the highest levels of vitamin C, polyphenols, anthocyanins, and antioxidant activity, while Monarda didyma had the highest levels of dry matter, ash, fibre, and protein (Stefaniak and Grzeszczuk 2019). Important minerals including calcium, magnesium, phosphorus, and iron can be found naturally in edible flowers (Table 1.3). They are also a great source of vitamins A and C.

1.3 Phytochemicals Present in Edible Flowers

Phytochemicals are naturally occurring biologically active compounds found in plants that are not essential for nutrition but offer various health benefits when consumed (Leitzmann 2016). These compounds protect plants from diseases and environmental stress, while also contributing to the plant's colour, scent, and taste (Paul et al. 2022). Carotenoids, flavonoids, phenolic acids and anthocyanins are the main bioactive compounds present in these flowers. (Pires Jr et al. 2021). The main ingredients that give flowers their vivid colours are carotenoids and flavonoids, which also have antioxidant qualities. In marigolds, compounds like lutein and flavonoids are linked to anti-inflammatory effects, the ability to neutralize free radicals, and cytotoxic activity against certain cancer cells (Fakhri et al. 2022) (Fig. 1.1).

1.3.1 Anthocyanin

Anthocyanins are belongs to the flavonoid group, and are primarily responsible for the vivid colours seen in flowers. The exact colour of these pigments depends on factors such as pH, metal ions, and the presence of co-pigments. Research has demonstrated that anthocyanins possess antioxidant properties, which neutralize free radicals, and have a crucial role in protecting against diseases such as cancer, cardiovascular conditions, obesity, and diabetes (Mattioli et al. 2020). Flowers are particularly rich in anthocyanins (Łysiak 2022). In studies of different rose varieties, Cyanidin 3,5-di-O-glucoside in red and pink flowers and pelargonidin 3,5-di-O-glucoside in orange types were found to be the main anthocyanins (Kumari et al. 2022). Additionally, other anthocyanins such as peonidin 3-O-sophoroside and peonidin 3-O-glucoside have been detected in rose petals (Cendrowski et al.

Table 1.3 Microelement content in selected flowers of ornamental plant species (g/kg dry matter)

Plant species and cultivar	N (Nitrogen)	P (Phosphorus)	K (Potassium)	Na (Sodium)	Ca (Calcium)	Mg (Magnesium)	References
Mimulus × hybridus L. (Magic Yellow)	12–15	8.11	53–55	1.24	1.62	2.50	Grzeszczuk et al. (2018)
Mimulus × hybridus L. (Magic Red)	13.40	7.84	43–46	0.56	2.87	2.52	Stefaniak and Grzeszczuk (2019)
Antirrhinum majus L. (Cavalier)	10.13	7.70	24–27	0.37	1.11	1.32	Del Castillo-Munera (2015)
Dianthus chinensis L. (Chianti)	14–16	7.85	20	0.24	0.42	1.57	Chunwichit et al. (2024)
Hemerocallis × hybrida Hort.	18–22	8.40	26.80	0.38	0.58	1.0	Fernandes et al. (2017)
Paeonia officinalis L. (Sarah Bernardt)	5–7	6.58	8.13	0.05	1.47	1.44	Suksathan et al. (2021)
Paeonia officinalis L. (Dr Aleksander Fleming)	7.60	6.54	26–29	0.05	0.84	1.12	Rachkeeree et al. (2018)
Paeonia officinalis L. (Karl Rosenfield)	8–12	6.00	27	0.08	1.05	1.43	Stefaniak and Grzeszczuk (2019)
Monarda didyma L.	22–26	9.15	38–42	0.2	16.77	3.99	Pires Jr et al. (2021)
Monarda fistulosa L.	24	9.11	32–36	0.2	17.4	3.27	Chensom et al. (2019)
Monarda citriodora subsp. austromontana Cerv. ex Lag. (Bees Favourite)	19.40	7.32	24.60	0.10	2.62	4.76	Araújo et al. (2019)

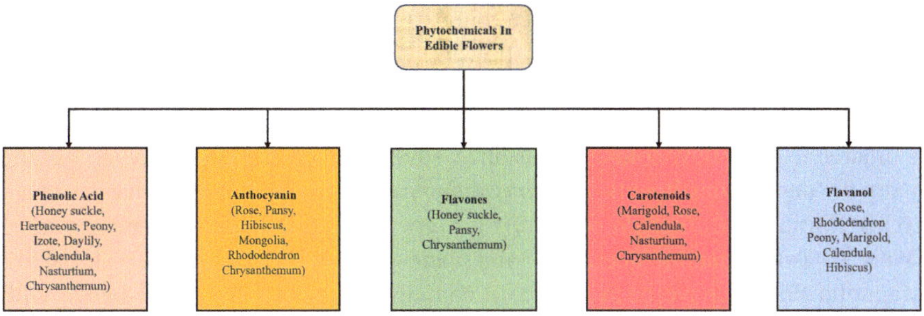

Fig. 1.1 Different phytochemicals found in edible flowers

2017). In hibiscus, delphinidin-3-O-sambubioside and cyanidin-3-O-sambubioside were found to be the predominant anthocyanins (Marquez-Rodriguez et al. 2021).

1.3.2 Carotenoids

The natural pigments known as carotenoids, which belong to the isoprenoid class, give plants their yellow, orange, and red colours. (Zia-Ul-Haq 2021). The carotenoid composition in flowers varies significantly across different plant species. These pigments play a crucial role in human and animal nutrition, as they cannot be synthesized de novo and must be obtained from dietary sources or supplements (Tan and Norhaizan 2019). Various investigations have highlighted marigold flowers as a promising source of lutein and its derivatives, which are key carotenoid compounds (Kumari and Bhargava 2021). Other carotenoids such as zeaxanthin, α-carotene, β-carotene, and violaxanthin were also detected, albeit in trace amounts. Research on calendula flowers revealed the presence of lutein, flavoxanthin,. Additionally, a study noted that *Calendula officinalis* flowers contain the highest concentration of β-carotene, measured at 637 μg/100 g (Bocho-Janiszewska et al. 2016). Rose petals, on the other hand, are found to contain carotenoid pigments such as luteoxanthin, zeaxanthin, β-carotene, lutein and neoxanthin (Madurangi and Maduwanthi 2024).

1.3.3 Flavanols

Flavanols is a subclass of flavonoids, include compounds such as myricetin, quercetin, isorhamnetin, and kaempferol, which are abundantly present in these flowers (Shen et al. 2022). These flavonols appear in various forms across a range of flowers, including roses, chrysanthemums, day lilies, Chinese roses, roselle, magnolia, and honeysuckle. For example, numerous flavonols have been identified in *Paeonia suffruticosa*, with significant varieties found in the petals of *Paeonia lactiflora* (Sun et al. 2020). In dried honeysuckle flowers, key flavonols include quercetin and kaempferol (Gołba et al. 2020) which found in rose petals (Wan et al. 2019). In many These flowers, flavanols tend to be present in relatively higher concentrations compared to other flavonoids. For instance, studies have highlighted flavonols as the main flavonoid constituents in *Chrysanthemum indicum* flowers (Sharma et al. 2023). Similarly, *Calendula officinalis* flowers have been found to contain notable levels of isorhamnetin, querceṭin, and their glycosides (Tavallali et al. 2024). Magnolia flowers have been shown to contain a diverse array of flavonols, with kaempferol, quercetin, and isorhamnetin being the most prevalent, along with their glycosidic forms (Cristea et al. 2024). These findings support the idea that flavonols are widely distributed across various These flowers.

1.3.4 Flavones

Flavones are present in a variety of forms in edible flowers such as acacetin, chrysoeriol and their respective glucosides (Kandylis 2022). These compounds are commonly found in numerous These flower species. Luteolin-7-glucoside is particularly abundant in *Chrysanthemum morifolium* flowers (Chen et al. 2021). Luteolin has been extensively reported in honeysuckle flowers, with several studies confirming its presence (Becker and Szakiel 2019) It has also been detected in extracts from cornflower and Yucca. Additionally, apigenin glycosides have been recognized as the predominant flavones in pansy flowers. These findings highlight the widespread occurrence of flavones across a range of These flowers.

1.3.5 Phenolic Acids

These flowers are often rich in phenolic acids, an important family of phytonutrients. Trans-cinnamic acid, caffeic acid and 4-hydroxybenzoic acid are among the phenolic acids that have been found in izote (Yucca elephantipes) flowers (Mulík and Ozuna 2020). The main phenolic acids found in honeysuckle blossoms are caffeic acid, 3,5-dicaffeoylquinic acid, chlorogenic acid, and 4,5-dicaffeoylquinic acid (Cheng et al. 2023) which are present in mutant variants of Chrysanthemum morifolium (Pandey et al. 2022). The variety of phenolic acids found in various These flowers is highlighted by these findings.

1.4 Health Benefits of Edible Flowers

These flowers have long been prized for their remarkable therapeutic and medicinal qualities, as well as their vivid beauty and culinary diversity. These flowers are a great complement to a balanced diet and provide a host of health advantages, because they are abundant in a variety of vital nutrients, antioxidants, vitamins, and minerals. Numerous flowers, such as nasturtiums, lavender, and chamomile, are rich in potent antioxidants like flavonoids, polyphenols, and carotenoids that aid in the body's defence against dangerous free radicals (Madurangi and Maduwanthi 2024). By reducing oxidative stress, this antioxidant effect lowers the risk of chronic illnesses like cancer, heart disease, and neurological diseases. To support their function in promoting long-term health, blue cornflowers and violets are rich in anthocyanins, which are powerful antioxidants with anti-inflammatory and anticancer qualities (Amrouche et al. 2022). The anti-inflammatory qualities of many These flowers are well known, in addition to their antioxidant advantages. Calendula (marigold) and chamomile are two flowers that have been used for generations to boost the immune system, lessen pain, and reduce inflammation. Due to its calming properties, chamomile is very useful for reducing bloating, indigestion, and gastrointestinal pain. Indirectly improving digestion by lowering stress-related digestive problems, it is also known to encourage relaxation and better sleep. To treat burns, rashes, and skin irritation, calendula is frequently administered topically. Its strong anti-inflammatory properties make it a popular treatment for minor skin irritations and wounds because they reduce inflammation, hasten the healing process, and guard against infections (Ullah et al. 2023). A natural source of vital nutrients that support a variety of body functions, These flowers are also high in vitamins and minerals. Nasturtiums are a great source of vitamin C and iron, which are necessary for immunological strength and healthy blood circulation, while dandelion blossoms are rich in vitamins A and C, which are important for skin health and immune function (Tiwari et al. 2018). Minerals like potassium, calcium, and magnesium, which are present in these flowers, support healthy bones, strong muscles, and appropriate cellular activity, all of which improve general vitality. A lot of These flowers are likewise valued for their calming and mood-boosting properties. One of the most well-known medicinal flowers, lavender has been used extensively for its capacity to lower anxiety, relieve tension, and enhance the quality of sleep, Because of its relaxing qualities, it is frequently ingested as a tea, added to desserts, or infused into oils. (Batiha et al. 2023). Another flower having anxiolytic (anxiety-reducing) properties is passionflower, which has also been researched for its capacity to enhance sleep and encourage relaxation, making it a useful treatment for nervous tension and insomnia. By relaxing the body and mind, these flowers support emotional well-being, which is crucial for lowering stress and creating a peaceful atmosphere (Janda et al. 2020). Many of these flowers have the ability to cure the skin, which adds to their therapeutic value. Because of their calming and anti-inflammatory properties, lavender, chamomile, and rose petals are frequently included in skincare products. These flowers shield the skin from harm, encourage healing, and lessen skin irritation (Aswandi et al. 2024). In particular, calendula is well known for its capacity to prevent infections, lessen scarring, and promote wound healing. It is frequently applied externally as oils, ointments, or lotions to minor burns, wounds, bruises, and even skin disorders like eczema. By lowering inflammation and promoting tissue regeneration, their use in skincare regimens enhances both the appearance of the skin and its general health. Additionally important to detoxifying are several these flowers such as dandelion. The capacity of dandelion blossoms to help digestion, encourage the removal of toxins from the body, and stimulate liver function is well documented. By acting as a natural diuretic, dandelion aids in the body's overall cleansing by removing waste materials and extra fluid. Because of its purifying qualities, dandelion is a useful flower for kidney and liver support, as well as for enhancing general health and energy (Yan et al. 2024). Certain flowers, such as hibiscus, are also linked to cardiovascular health. Flowers of the hibiscus plant are very good for lowering blood pressure, increasing circulation, and lowering cholesterol. Hibiscus, which is high in anthocyanins, lowers the risk of cardiovascular disorders by enhancing blood flow and supporting healthy blood vessels. Hibiscus tea is a powerful ally in the prevention of hypertension and other cardiovascular problems because regular use has been associated with better heart health (Ellis et al. 2022). Moreover, These flowers can help with weight control. Flowers like hibiscus and lavender have been researched for their ability to support a healthy weight and regulate metabolism. While lavender's relaxing properties aid in reducing stress-induced overeating, which helps improve weight control, hibiscus in particular has demonstrated promise in avoiding the growth of fat cells and controlling fat metabolism. These flowers can help preserve a healthy weight indirectly by promoting digestion and lowering stress (Moltaji 2020). It has been demonstrated that some flowers, such as sage and rosemary, improve cognitive performance. Compounds in both herbs increase blood flow to the brain, improving concentration, memory, and focus. Particularly, rosemary has been researched for its ability to promote general brain health and shield the brain from age-related cognitive loss. They provide important support for mental clarity and long-term cognitive health by improving cognitive function and guarding against neurodegenerative illnesses like Alzheimer's (Kosmopoulou et al. 2024) (Fig. 1.2).

1.5 Advanced Techniques for Extraction of Bioactive Compounds from Edible Flowers

Extraction of bioactive compounds from edible flowers has garnered considerable interest due to their rich phytochemical profile, which contributes to their diverse applications in food, cosmetics, and pharmaceuticals (Sarode et al. 2024). Recent advancements in extraction methods aim to improve yield, purity, and sustainability while preserving the bioactive integrity of these delicate compounds. Key

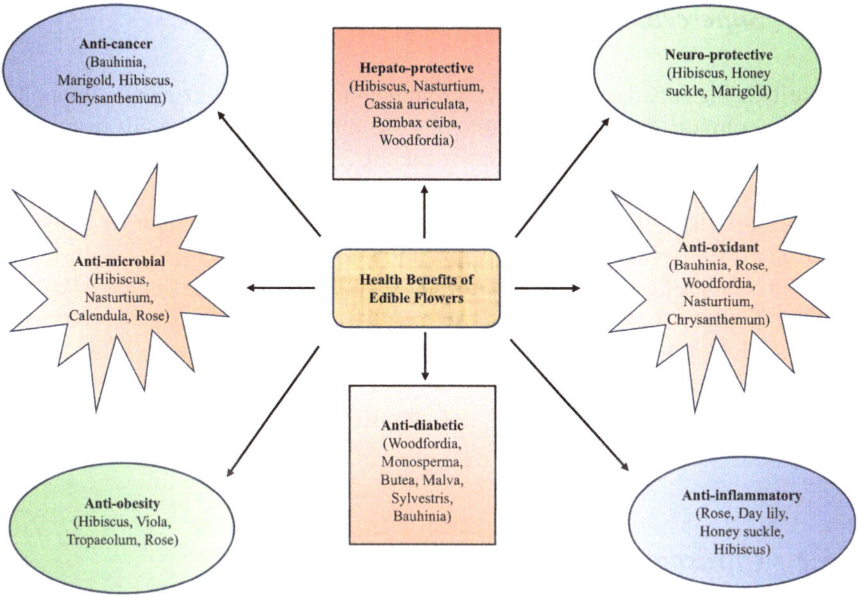

Fig. 1.2 Health benefits of edible flowers

techniques such as Supercritical Fluid Extraction (SFE) (Granone et al. 2022), Ultrasound-Assisted Extraction (UAE) (Benimana et al. 2024), Microwave-Assisted Extraction (MAE) (Cerdá-Bernad et al. 2022), Pressurized Liquid Extraction (PLE) (Zhao et al. 2019), Enzyme-Assisted Extraction (EAE) (Nadar et al. 2018), and green solvents like Ionic Liquids (ILs) (Hashemi et al. 2022), and Deep Eutectic Solvents (DES) (Ivanović et al. 2020), have been optimized to increase efficiency and selectivity in the extraction of valuable compounds from These flowers. Advanced extraction techniques such as SFE, UAE, MAE, PLE, EAE, and the use of green solvents have revolutionized the way bioactive compounds are extracted from These flowers. Research has shown that methods like SFE and UAE can optimize the extraction of oils and antioxidants from flowers like lavender, hibiscus, and chamomile (Khalid et al. 2024), while MAE and PLE provide rapid and efficient extraction of a broad spectrum of bioactive compounds (Jha and Sit 2022). Additionally, the use of enzyme-assisted extraction and green solvents offers new opportunities for sustainable and environmentally friendly processing. As research in this field continues to evolve, the refinement of these techniques is expected to lead to more sustainable, cost-effective, and high-quality extracts from These flowers, paving the way for their broader use in various industries, including food, cosmetics, and pharmaceuticals. The full potential of these flowers as significant sources of bioactive chemicals with a variety of functional applications will surely be unlocked by these developments (Krishnamoorthi et al. 2023).

1.5.1 Supercritical Extraction of Fluids (SFE)

Primarily using supercritical CO_2 is a prominent technique for extracting lipophilic bioactive compounds such as essential oils, flavonoids, and carotenoids from flowers like lavender and chamomile (Zorić et al. 2022). The unique properties of supercritical CO_2, such as its ability to dissolve a wide range of compounds and its tuneable polarity with pressure and temperature adjustments, allow for selective extractions with minimal thermal degradation (Hiraga et al. 2018). For instance, SFE extract high-purity essential oils from lavender (*Lavandula angustifolia*) and jasmine (*Jasminum sambac*) (Makeri and Salihu 2023), Preserving their aromatic profile and bioactive components like linalool and linalyl acetate. This method offers significant advantages over traditional steam distillation, such as higher extraction yields and better retention of sensitive compounds (Pourmortazavi and Hajimirsadeghi 2007).

1.5.2 Ultrasound-Assisted Extraction (UAE)

Utilizes high-frequency sound waves to generate cavitation bubbles within the solvent, improving mass transfer and enhancing the penetration of the solvent into plant cells (Shen et al. 2023). This process accelerates the release of compounds such as polyphenols, and anthocyanins. UAE has been effectively employed for extracting antioxidant-rich compounds from hibiscus (*Hibiscus sabdariffa*) (Mohamad et al. 2023). and elderflower (*Sambucus nigra*) (Bljajić et al. 2021), known for their high concentrations of anthocyanins, flavonoids, and phenolic acids, which have well-documented anti-inflammatory and anti-cancer properties. UAE is favoured for its energy efficiency, reduced solvent use, and shorter extraction times (Kumar et al. 2021).

1.5.3 Microwave-Assisted Extraction (MAE)

Accelerates the extraction process by using microwave radiation to heat both the solvent and plant material simultaneously. This results in faster extractions while preserving sensitive compounds that might degrade under prolonged heat (Ekezie et al. 2017). MAE has shown significant success in extracting essential oils and antioxidants from flowers such as roses (*Rosa* spp.) and jasmine (Chemat et al. 2020), where the technique produces extracts with higher concentrations of bioactive compounds. Research indicates that MAE offers superior extraction yields compared to conventional methods, making it an attractive option for both small-scale and industrial applications (Ekezie et al. 2017).

1.5.4 Pressurized Liquid Extraction (PLE)

Accelerated Solvent Extraction (ASE) uses high temperature and pressure in a closed system to improve the solubility of compounds in solvents. This method can extract both polar and non-polar compounds efficiently. PLE has been applied to flowers like calendula (*Calendula officinalis*) and chamomile to extract a broad range of flavonoids, carotenoids, and saponins (Tomasella et al. 2022). The advantage of PLE lies in its ability to operate at high pressures and temperatures, which shortens extraction times and improves yields while minimizing solvent use (Geow et al. 2021).

1.5.5 Enzyme-Assisted Extraction (EAE)

EAE involves breaking down plant cell walls with certain enzymes to release bioactive substances, especially water-soluble molecules like proteins, polysaccharides, and phenolics (Das et al. 2021). This technique has been applied to flowers like chrysanthemums (*Chrysanthemum morifolium*) and lotus (*Nelumbo nucifera*) to extract bioactive flavonoids like luteolin and apigenin, which possess significant antioxidant, anti-inflammatory, and anticancer properties (Hu et al. 2022). EAE offers several benefits, including milder extraction conditions, higher selectivity, and the preservation of sensitive bioactive compounds. These solvents are biodegradable, non-toxic, and can offer high selectivity in extracting bioactive compounds while reducing the environmental impact typically associated with conventional organic solvents. ILs and DES have been employed in the extraction of flavonoids, polyphenols, and essential oils from a variety of these flowers, including borage (*Borago officinalis*) and geranium (*Pelargonium graveolens*). The unique properties of these solvents allow for efficient and selective extraction while also providing a more sustainable approach to bioactive compound isolation (Greche et al. 2009).

1.6 Role of Edible Flowers as Natural Pesticides and Biocides

Growing awareness of the harmful effects of synthetic pesticides on the environment and human health has spurred interest in investigating natural alternatives, such as plant-based biocides. Among these, edible flowers are known for their rich phytochemical profiles, have garnered attention for their potential as natural pesticides and biocides. Alkaloids, flavonoids, terpenoids, and phenolics, exhibit significant pesticidal and antimicrobial properties, making them promising candidates for sustainable pest management. Flowers such as marigold (*Tagetes* spp.) and chrysanthemum (*Chrysanthemum* spp.) are particularly well-known for their pesticidal properties (Walia et al. 2017). Marigold flowers contain thiophenes and other sulfur-containing compounds, which are effective in repelling nematodes, aphids, and certain beetles. These compounds act as both deterrents and toxins to insect pests, providing an eco-friendly alternative to synthetic chemical pesticides (Sharma et al. 2022). Chrysanthemum, on the other hand, contains pyrethrins, a class of natural insecticides that interfere with the nervous system of pests (Gupta et al. 2021). Pyrethrins are widely used in both organic farming and household pest control due to their low toxicity to humans and animals. Another notable example is lavender (*Lavandula angustifolia*), which is revered for its potent repellent properties against a range of insect pests, including moths, fleas, and mosquitoes. The essential oils in lavender, primarily linalool and linalyl acetate, exhibit insecticidal activity by disrupting the nervous systems of insects. These oils also possess antimicrobial and antifungal properties, further contributing to the flower's utility as a biocide in pest management. Lavender's non-toxic nature, combined with its pleasant fragrance, makes it an attractive alternative for use in both agricultural and domestic settings (Sil and Barman 2020).

The calendula (*Calendula officinalis*) flower, commonly known for its medicinal properties, also exhibits antifungal, antibacterial, and insecticidal activities (Pandey and Despande 2022). Calendula extracts have demonstrated effectiveness in combating fungal infections in plants, such as powdery mildew, as well as acting as a deterrent against insect pests like aphids and whiteflies. The antimicrobial properties of calendula, attributed to its high flavonoid and carotenoid content, further enhance its value as a natural biocide. Incorporating These flowers as natural pesticides and biocides into integrated pest management (IPM) systems holds substantial promise for reducing the reliance on synthetic chemicals, fostering sustainable agriculture, and promoting ecological balance. Their biodegradability, minimal toxicity, and multifaceted properties make them ideal candidates for use in organic farming and eco-friendly pest control strategies. With further research and development, These flowers may offer a potent solution to the global challenges posed by pesticide resistance and environmental degradation (Khouchlaa et al. 2023).

1.7 Safety, Toxicity, and Regulatory Concerns of Edible Flowers

The consumption of some flowers may possess serious safety, toxicological, and regulatory issues despite the nutritional and aesthetic advantages. Many flowers such as lavender (*Lavandula angustifolia*), chamomile (*Matricaria chamomilla*), and roses (*Rosa* spp.), are generally regarded as safe for human eating; nevertheless, those that contain natural toxins can be quite hazardous. Examples of flowers that contain dangerous cardiac glycosides that can result in serious poisoning if consumed include oleander (*Nerium oleander*) and lily of the valley (*Convallaria majalis*) (Patil et al. 2023). Even though they are non-toxic, some flowers can cause allergic reactions in people who are already sensitive. For instance, because of its affinity for the botanical family, chamomile can cause severe allergic reactions in those who are allergic to ragweed. The possibility of pesticide contamination is another urgent issue because flowers that are cultivated conventionally may have dangerous chemical residues that are extremely dangerous to human health (Preda et al. 2024). To reduce exposure to these poisons, it is crucial to purchase flowers from reliable, pesticide-free or organic sources. These flowers have the potential to be used in medicine and cooking, but they are still not adequately controlled, and there are no common standards to distinguish between edible and potentially hazardous types. In order to close this regulatory vacuum, stronger regulations for the safe production, labelling, and sale of These flowers must be established, together with stricter enforcement of safety regulations and increased consumer education. Given these factors, These flowers have potential uses in both food and health.

1.8 Marketing Trends of Edible Flowers

Due to rising consumer demand for sustainable, natural, and visually appealing products in the food, beverage, and cosmetics sectors, the global market for edible flowers has grown significantly in recent years. These flowers are increasingly being used in gourmet cooking, particularly in fine dining and desserts, where they are valued for their vibrant colours, delicate flavours, and health benefits. Additionally, flowers like lavender, chamomile, hibiscus, and roses are gaining popularity in herbal teas, infused oils, and natural skincare products. The export and import rates of these flowers have steadily risen, with Europe and North America being major importers of these flowers, while countries in Asia, particularly India and China, as well as regions in Africa and South America, serve as key exporters. Nanotechnology represents an emerging field of scientific exploration focused on the examination and control of matter at the smallest scales, encompassing molecules and atoms (Rao and Poonia 2024b). The quantity of These flowers traded varies by species, with hibiscus and chamomile dominating the market, followed by lavender and rose petals. According to recent trade data, the global export value of These flowers was estimated at over USD 500 million in 2022, with a projected annual growth rate of 5–7% (Worku et al. 2023). Growing interest in eco-friendly beauty products and the growing popularity of plant-based and organic food products are the main drivers of this developing business. As consumer preferences shift toward wellness-oriented products and experiences, These flowers are anticipated to play an increasingly central role in the health, culinary, and cosmetic sectors, further boosting global trade and consumption.

1.9 Conclusion

These flowers present a fascinating intersection of nature's beauty and nutritional benefits. Their rich array of vitamins, minerals, and antioxidants offers promising health benefits, enhancing the nutritional profile of our diets in a variety of forms-from fresh garnishes to infused teas and culinary delights. Moreover, the presence of bioactive compounds, such as flavonoids, alkaloids, and essential oils, underscores their potential in supporting human health and combating oxidative stress and inflammation. However, the increasing popularity of these flowers in the global food market also brings forward the need for informed consumption. As consumers seek to incorporate these vibrant ingredients into their meals, awareness regarding their toxicity is crucial. While many These flowers are safe and nutritious, others may contain compounds that can be harmful if consumed in excessive amounts or without proper identification. The market for These flowers continues to grow, driven by trends in sustainable eating, functional foods, and culinary experimentation. This growing demand offers opportunities for innovation, but it also underscores the importance of proper education, standardization, and regulation to ensure consumer safety. Ultimately, These flowers have the potential to be a valuable addition to modern diets, provided that they are chosen wisely and enjoyed in moderation. Their diverse and colourful presence in the culinary and health spheres demonstrates their broad appeal, but it is essential to approach their use with both excitement and caution to fully unlock their benefits.

References

Adeleke BS, Babalola OO (2020) Oilseed crop sunflower (Helianthus annuus) as a source of food: nutritional and health benefits. Food Sci Nutr 8(9):4666–4684

Agarwal P, Sebghatollahi Z, Kamal M, Dhyani A, Shrivastava A, Singh KK et al (2022) Citrus essential oils in aromatherapy: therapeutic effects and mechanisms. Antioxidants 11(12):2374

Ahmed R, Hussain MJ, Ahmed S, Karim MR, Siddiky MA (2017) Effect of nitrogen, phosphorus and potassium fertilizers on yield and yield attributes of marigold (Tagetes patula L.). Agriculturists 15(1):101–109

Alighadri T, Tabibiazar M, Mohammadi M, Jahanban-Esfahlan A, Amarowicz R (2023) Nasunin, an amazing chemical constituent in eggplants (Solanum melongena L.): a review of its major properties and health-promoting effects. ACS Food Sci Technol 4(1):16–35

Amrouche TA, Yang X, Capanoglu E, Huang W, Chen Q, Wu L et al (2022) Contribution of edible flowers to the Mediterranean diet: phytonutrients, bioactivity evaluation and applications. Food Front 3(4):592–630

Araújo S, Matos C, Correia E, Antunes MC (2019) Evaluation of phytochemicals content, antioxidant activity and mineral composition of selected edible flowers. Qual Assur Saf Crops Foods 11(5):471–478

Asadi-Samani M, Bahmani M, Rafieian-Kopaei M (2014) The chemical composition, botanical characteristic and biological activities of Borago officinalis: a review. Asian Pac J Trop Med 7:S22–S28

Askari-Khorasgani O, Mortazaeinezhad F, Rafiee P (2017) Variations in the growth, oil quantity and quality, and mineral nutrients of chamomile genotypes under salinity stress. J Cent Eur Agric 18(1):150–169

Aswandi A, Kholibrina CR, Kuspradini H (2024) Essential oils for cosmetics application. In: Biomass-based cosmetics: research trends and future outlook. Springer, Singapore, pp 151–173

Batiha GES, Teibo JO, Wasef L, Shaheen HM, Akomolafe AP, Teibo TKA et al (2023) A review of the bioactive components and pharmacological properties of Lavandula species. Naunyn Schmiedeberg's Arch Pharmacol 396(5):877–900

Becker R, Szakiel A (2019) Phytochemical characteristics and potential therapeutic properties of blue honeysuckle Lonicera caerulea L. (Caprifoliaceae). J Herb Med 16:100237

Benimana F, Kawata K, Kong F, Roy A, Mohan A (2024) Role of ultrasound application in the production, processing, and preservation of edible flowers: a review. J Food Process Eng 47(12):e70012

Benvenuti S, Mazzoncini M (2021) The biodiversity of edible flowers: discovering new tastes and new health benefits. Front Plant Sci 11:569499

Bhaskar JJ, Chilkunda ND, Salimath PV (2012) Banana (Musa sp. var. elakki bale) flower and pseudostem: dietary fiber and associated antioxidant capacity. J Agric Food Chem 60(1):427–432

Biel W, Jaroszewska A, Łysoń E, Telesiński A (2017) The chemical composition and antioxidant properties of common dandelion leaves compared with sea buckthorn. Can J Plant Sci 97(6):1165–1174

Bjørklund G, Cruz-Martins N, Goh BH, Mykhailenko O, Lysiuk R, Shanaida M et al (2024) Medicinal plant-derived phytochemicals in detoxification. Curr Pharm Des 30(13):988–1015

Bljajić K, Brajković A, Čačić A, Vujić L, Jablan J, Saraiva de Carvalho I, Zovko Končić M (2021) Chemical composition, antioxidant, and α-glucosidase-inhibiting activity of aqueous and hydroethanolic extracts of traditional antidiabetics from Croatian ethnomedicine. Horticulturae 7(2):15

Bocho-Janiszewska A, Ogorzałek M, Wasilewski T (2016) VII. Technology and innovation products and processes. In: Commodity science in a changing world, p 532

Cendrowski A, Ścibisz I, Mitek M, Kieliszek M, Kolniak-Ostek J (2017) Profile of the phenolic compounds of Rosa rugosa petals. J Food Qual 2017(1):7941347

Cerdá-Bernad D, Baixinho JP, Fernández N, Frutos MJ (2022) Evaluation of microwave-assisted extraction as a potential green technology for the isolation of bioactive compounds from saffron (Crocus sativus L.) floral by-products. Foods 11(15):2335

Chadha KL, Singh SK, Srivastav M, & Behera TK (2014) Book of abstracts (Poster Papers). In: 6th Indian Horticulture Congress-an international event on horticulture for inclusive growth, Tamil Nadu Agricultural University, pp 6–9

Chemat F, Vian MA, Fabiano-Tixier AS, Nutrizio M, Jambrak AR, Munekata PE et al (2020) A review of sustainable and intensified techniques for extraction of food and natural products. Green Chem 22(8):2325–2353

Chen S, Liu J, Dong G, Zhang X, Liu Y, Sun W, Liu A (2021) Flavonoids and caffeoylquinic acids in Chrysanthemum morifolium Ramat flowers: A potentially rich source of bioactive compounds. Food Chem 344:128733

Cheng Z, Bao Y, Li Z, Wang J, Wang M, Wang S et al (2023) Lonicera caerulea (Haskap berries): a review of development traceability, functional value, product development status, future opportunities, and challenges. Crit Rev Food Sci Nutr 63(27):8992–9016

Chensom S, Okumura H, Mishima T (2019) Primary screening of antioxidant activity, total polyphenol content, carotenoid content, and nutritional composition of 13 edible flowers from Japan. Prev Nutr Food Sci 24(2):171

Chitrakar B, Zhang M, Bhandari B (2019) Edible flowers with the common name "marigold": their therapeutic values and processing. Trends Food Sci Technol 89:76–87

Chopde N, Patil S, Kuchanwar O, Raut VU (2017) Growth, yield and quality of jasmine vary by integrated plant nutrition. J Pharmacogn Phytochem 6(6):1201–1203

Chunwichit S, Phusantisampan T, Thongchai A, Taeprayoon P, Pechampai N, Kubola J et al (2024) Influence of soil amendments on phytostabilization, localization and distribution of zinc and cadmium by marigold varieties. Sci Total Environ 919:170791

Coyago-Cruz E, Moya M, Méndez G, Villacís M, Rojas-Silva P, Corell M et al (2023) Exploring plants with flowers: from therapeutic nutritional benefits to innovative sustainable uses. Foods 12(22):4066

Cristea RM, Sava C, Căpățână C, Kanellou A (2024) Phytochemical analysis and specific activities of bark and flower extracts from four Magnolia Plant species. Horticulturae 10(2):141

Das S, Nadar SS, Rathod VK (2021) Integrated strategies for enzyme assisted extraction of bioactive molecules: A review. Int J Biol Macromol 191:899–917

de São Pedro Pires TC (2020) Development of new food products of high nutritional and functional value using flowers, fruits and plant stems (Doctoral dissertation, Universidad de Salamanca)

Del Castillo-Munera J (2015) Characterization, population genetics and management of Pythium spp. from floriculture crops in Michigan. Dissertation, Michigan State University

Duggirala K, Mummaleti G, Kong F, Roy A, Mohan A (2024) Edible flowers: a sustainable source of natural food ingredient. J Food Bioactives 27:1–14

Ek P, Araujo AC, Oliveira SM, Ramos IN, Brandao TR, Silva CL (2018) Assessment of nutritional quality and color parameters of convective dried watercress (Nasturtium officinale). J Food Process Preserv 42(2):e13459

Ekezie FGC, Sun DW, Cheng JH (2017) Acceleration of microwave-assisted extraction processes of food components by integrating technologies and applying emerging solvents: A review of latest developments. Trends Food Sci Technol 67:160–172

Ellis LR, Zulfiqar S, Holmes M, Marshall L, Dye L, Boesch C (2022) A systematic review and meta-analysis of the effects of Hibiscus sabdariffa on blood pressure and cardiometabolic markers. Nutr Rev 80(6):1723–1737

Faisal Z, Saeed F, Afzaal M, Akram N, Shah YA, Islam F, Ateeq H (2022) Phytochemical profile and food applications of edible flowers: A comprehensive treatise. J Food Process Preserv 46(11):e17061

Fakhri S, Tomas M, Capanoglu E, Hussain Y, Abbaszadeh F, Lu B et al (2022) Antioxidant and anticancer potentials of edible flowers: where do we stand? Crit Rev Food Sci Nutr 62(31):8589–8645

Fernandes L, Casal S, Pereira JA, Saraiva JA, Ramalhosa E (2017) Edible flowers: A review of the nutritional, antioxidant, antimicrobial properties and effects on human health. J Food Compos Anal 60:38–50

Fernandes L, Saraiva JA, Pereira JA, Casal S, Ramalhosa E (2019) Post-harvest technologies applied to edible flowers: a review: edible flowers preservation. Food Rev Int 35(2):132–154

Geow CH, Tan MC, Yeap SP, Chin NL (2021) A review on extraction techniques and its future applications in industry. Eur J Lipid Sci Technol 123(4):2000302

Gołba M, Sokół-Łętowska A, Kucharska AZ (2020) Health properties and composition of honeysuckle berry Lonicera caerulea L. An update on recent studies. Molecules 25(3):749

Gontar Ł, Geszprych A, Przybył J, Buła M, Osińska E (2022) Chemical variability of lemon beebalm (Monarda citriodora Cerv. Ex lag.) during plant phenology. JAppl Res Med Aromat Plants 31:100433

Goswami HK, Ram HK (2017) Ancient food habits dictate that food can be medicine but medicine cannot be "food"!! —"food is the substance digested by an individual to suppress the basic vital instinct of hunger"(HKG). Medicine 4(4):82

Gounden T, Moodley R, Jonnalagadda SB (2018) Elemental analysis and nutritional value of edible Trifolium (clover) species. J Environ Sci Health B 53(8):487–492

Granone LI, Hegel PE, Pereda S (2022) Citrus fruit processing by pressure intensified technologies: A review. J Supercrit Fluids 188:105646

Greche H, Belkheir K, Hajjaji N, Boukir A and Loukili A (2009) Les huiles essentielles «bio» du Maroc. In: Hassan Greche and Abdeslam Ennabili (Eds.) Recherche sur les plantes médicinales et aromatiques, Acte du Congrès International, pp. 130–40

Grzeszczuk M, Wesolowska A, Jadczak D, Jakubowska B (2011) Nutritional value of chive edible flowers. Acta Sci Pol Hortorum Cultus 10(2):85–94
Grzeszczuk M, Stefaniak A, Meller E, Wysocka G (2018) Mineral composition of some edible flowers. J Elem 23(1):151–162
Gupta R, Gupta A, Jain S, Singh D, Verma N (2021) Chrysanthemum production, viral diseases and their management. In: Virus Diseases of Ornamental Plants: Characterization, Identification, Diagnosis and Management. Springer Singapore, Singapore, pp 261–275
Hashemi B, Shiri F, Švec F, Nováková L (2022) Green solvents and approaches recently applied for extraction of natural bioactive compounds. TrAC Trends Anal Chem 157:116732
Hiraga Y, Duereh A, Smith RL Jr (2018) Aspects of solvent polarity and solvent properties in developing efficient systems for processing biomass with ionic liquid mixtures and supercritical CO2. J Supercrit Fluids 134:12–20
Hritcu L, Cioanca O (2016) Prevalence of use of herbal medicines and complementary and alternative medicine in Europe. In: Herbal Medicine in Depression: Traditional Medicine to Innovative Drug Delivery. Springer, Cham, pp 135–181
Hu Y, Chen X, Hu M, Zhang D, Yuan S, Li P, Feng L (2022) Medicinal and edible plants in the treatment of dyslipidemia: advances and prospects. Chin Med 17(1):113
Iftikhar T, Majeed H, Waheed M, Zahra SS, Niaz M, Bilal B, Riaz M (2023) Tuberose. In: Essentials of medicinal and aromatic crops. Springer, Cham, pp 373–397
Ijabadeniyi OA (ed) (2023) Food science and technology: fundamentals and innovation. Walter de Gruyter GmbH & Co KG, Berlin
Ivanović M, Islamčević Razboršek M, Kolar M (2020) Innovative extraction techniques using deep eutectic solvents and analytical methods for the isolation and characterization of natural bioactive compounds from plant material. Plants 9(11):1428
Jadhav HB, Badwaik LS, Annapure U, Casanova F, Alaskar K (2023) A review on the journey of edible flowers from farm to consumer's plate. Appl Food Res 3(2):100312
Janarny G, Gunathilake KDPP, Ranaweera KKDS (2021) Nutraceutical potential of dietary phytochemicals in edible flowers – a review. J Food Biochem 45(4):e13642
Janda K, Wojtkowska K, Jakubczyk K, Antoniewicz J, Skonieczna-Żydecka K (2020) Passiflora incarnata in neuropsychiatric disorders—A systematic review. Nutrients 12(12):3894
Jeet S, Verma R, Yadav GK, Bhagat S, Tabassum S, Kumar K, Bhanwaria R (2024) Morphological and phenological characteristic, oil yield, quality and economics of lemon bee balm (Monarda citriodora Cerv. ex-Lag) at diverse environment condition, India. J Essent Oil Bear Plants 27(4):1079–1101
Jha AK, Sit N (2022) Extraction of bioactive compounds from plant materials using combination of various novel methods: a review. Trends Food Sci Technol 119:579–591
Juarez-Trujillo N, Luis Monribot-Villanueva J, Manuel Jimenez-Fernandez V, Suárez-Montaño R, Sahid Aguilar-Colorado A, Antonio Guerrero-Analco J, Jiménez M (2018) Phytochemical characterization of Izote (Yucca elephantipes) flowers. *J Appl Bot Food Qual* 91:202–210
Kandylis P (2022) Phytochemicals and antioxidant properties of edible flowers. Appl Sci 12(19):9937
Karimi E, Shirmardi M, Dehestani Ardakani M, Gholamnezhad J, Zarebanadkouki M (2020) The effect of humic acid and biochar on growth and nutrients uptake of calendula (Calendula officinalis L.). Commun Soil Sci Plant Anal 51(12):1658–1669
Khalid S, Chaudhary K, Amin S, Raana S, Zahid M, Naeem M et al (2024) Recent advances in the implementation of ultrasound technology for the extraction of essential oils from terrestrial plant materials: A comprehensive review. Ultrason Sonochem 107:106914
Khouchlaa A, El Baaboua A, El Moudden H, Lakhdar F, Bakrim S, El Menyiy N et al (2023) Traditional uses, bioactive compounds, and pharmacological investigations of Calendula arvensis L.: a comprehensive review. Adv Pharmacol Pharm Sci 2023(1):2482544
Kmail A (2024) Mitigating digestive disorders: action mechanisms of Mediterranean herbal active compounds. Open Life Sci 19(1):20220857

Kosmopoulou D, Lafara MP, Adamantidi T, Ofrydopoulou A, Grabrucker AM, Tsoupras A (2024) Neuroprotective benefits of Rosmarinus officinalis and its bioactives against Alzheimer's and Parkinson's diseases. Appl Sci 14(15):6417

Kouakou TH, Kouadio KS, Yao KB, Yapo SES, Kouakou NGD, Kone M (2024) The Bissap, a beverage of Hibiscus sabdariffa: production, composition, biological activities, and Main uses. In: Natural products in beverages: botany, phytochemistry, pharmacology and processing. Springer, Cham, pp 341–363

Krishnamoorthi A, Minz V, Behera SD, Singh SK, Prasad L, Jain S (2023) Bioactive compounds from horticulture crops and their utilization: a comprehensive review. Int J Plant Soil Sci 35(23):268–277

Kumar K, Srivastav S, Sharanagat VS (2021) Ultrasound assisted extraction (UAE) of bioactive compounds from fruit and vegetable processing by-products: A review. Ultrason Sonochem 70:105325

Kumari P, Bhargava B (2021) Phytochemicals from edible flowers: opening a new arena for healthy lifestyle. J Funct Foods 78:104375

Kumari P, Raju DVS, Prasad KV, Saha S, Panwar S, Paul S et al (2022) Characterization of anthocyanins and their antioxidant activities in Indian rose varieties (Rosa× hybrida) using HPLC. Antioxidants 11(10):2032

Labadie C, Ginies C, Guinebretiere MH, Renard CM, Cerutti C, Carlin F (2015) Hydrosols of orange blossom (Citrus aurantium), and rose flower (Rosa damascena and Rosa centifolia) support the growth of a heterogeneous spoilage microbiota. Food Res Int 76:576–586

Leitzmann C (2016) Characteristics and health benefits of phytochemicals. Forschende Komplementärmedizin 23(2):69–74

Liang Q, Cui J, Li H, Liu J, Zhao G (2013) Florets of sunflower (Helianthus annuus L.): potential new sources of dietary fiber and phenolic acids. J Agric Food Chem 61(14):3435–3442

Lokare PD, Pandya JB (2024) Entrepreneurial botany. Book Saga Publications, Rahuri

Łysiak GP (2022) Ornamental flowers grown in human surroundings as a source of anthocyanins with high anti-inflammatory properties. Foods 11(7):948

Madurangi A, Maduwanthi S (2024) Edible flowers in herbal beverage development. Asian Food Sci J 23:1–21

Majerska J, Michalska A, Figiel A (2019) A review of new directions in managing fruit and vegetable processing by-products. Trends Food Sci Technol 88:207–219

Makeri M, Salihu A (2023) Jasmine essential oil: production, extraction, characterization, and applications. In:Gulzar AN, Mohammad JA (eds), Essential oils. Academic Press, pp 147–177. https://doi.org/10.1016/B978-0-323-91740-7.00013-X

Malakar M, Paiva PDDO, Beruto M, Cunha Neto ARD (2023) Review of recent advances in postharvest techniques for tropical cut flowers and future prospects: Heliconia as a case-study. Front Plant Sci 14:1221346

Marquez-Rodriguez AS, Guimarães M, Mateus N, De Freitas V, Ballinas-Casarrubias ML, Fuentes-Montero ME et al (2021) Disaccharide anthocyanin delphinidin 3-O-sambubioside from Hibiscus sabdariffa L.: Candida antarctica lipase B-catalyzed fatty acid acylation and study of its color properties. Food Chem 344:128603

Matejić JS, Dragićević AV, Jovanović MS, Žarković LD, Džamić AM, Hinić SS, Pavlović DR (2024) Plant products for musculoskeletal, respiratory, circulatory, and genitourinary disorders in Eastern and South-Eastern Serbia–Folk uses comparison with official recommendations. Rec Nat Prod 18(1):1–52

Mattioli R, Francioso A, Mosca L, Silva P (2020) Anthocyanins: a comprehensive review of their chemical properties and health effects on cardiovascular and neurodegenerative diseases. Molecules 25(17):3809

Mavandi P, Abbaszadeh B, Emami Bistgani Z, Barker AV, Hashemi M (2021) Biomass, nutrient concentration and the essential oil composition of lavender (Lavandula angustifolia Mill.) grown with organic fertilizers. J Plant Nutr 44(20):3061–3071

Mercola J (2021) Upgrade your immunity with herbs: herbal tonics, broths, brews, and elixirs to supercharge your immune system. Hay House, Inc., Carlsbad
Meurer K, Barron J, Chenu C, Coucheney E, Fielding M, Hallett P et al (2020) A framework for modelling soil structure dynamics induced by biological activity. Glob Chang Biol 26(10):5382–5403
Mittu B, Chaturvedi A, Sharma R, Singh M, Begum Z (2024) Viola tricolor and Viola odorata. In: Edible flowers. Elsevier, Amsterdam, pp 325–354
Mlcek J, Rop O (2011) Fresh edible flowers of ornamental plants – a new source of nutraceutical foods. Trends Food Sci Technol 22(10):561–569
Mohamad NR, Borhan FP, Razali MS, Wahab RA (2023) Optimizing the protocol for extraction of bioactive components from Hibiscus sabdariffa with potent antioxidant activity. J Angiother 7(1):1–9
Moltaji R (2020) The effects of auditory and olfactory cues on visual attention toward food images under acute stress. Dissertation, University of Arkansas
Montaner C, Zufiaurre R, Movila M, Mallor C (2021) Evaluation of borage (Borago officinalis L.) genotypes for nutraceutical value based on leaves fatty acids composition. Foods 11(1):16
Mulík S, Ozuna C (2020) Mexican edible flowers: cultural background, traditional culinary uses, and potential health benefits. Int J Gastron Food Sci 21:100235
Müseler DL (2005) Evaluation of the quality characteristics of the marama bean (Tylosema esculentum), an underutilized grain and tuber producing legume in southern Africa (Doctoral Dissertation, University of Namibia)
Nadar SS, Rao P, Rathod VK (2018) Enzyme assisted extraction of biomolecules as an approach to novel extraction technology: a review. Food Res Int 108:309–330
Navarro-Gonzalez N, Porrero MC, Mentaberre G, Serrano E, Mateos A, Cabal A et al (2015) Escherichia coli O157: H7 in wild boars (Sus scrofa) and Iberian ibex (Capra pyrenaica) sharing pastures with free-ranging livestock in a natural environment in Spain. Vet Q 35(2):102–106
Nguyen NT, Wereley ST, Shaegh SAM (2019) Fundamentals and applications of microfluidics. Artech house, Norwood
Nowicka P, Wojdyło A (2019) Anti-hyperglycemic and anticholinergic effects of natural antioxidant contents in edible flowers. Antioxidants 8(8):308
Pandey P, Despande B (2022) Pharmacological activities of Calendula officinalis. Int J Res Anal Rev 9(1):460–468
Pandey J, Bastola T, Dhakal B, Poudel A, Devkota HP (2022) Chrysanthemum morifolium ramat.: a medicinal plant with diverse traditional uses, bioactive constituents, and pharmacological activities. In: Medicinal plants of the asteraceae family: traditional uses, phytochemistry and pharmacological activities. Springer Nature Singapore, Singapore, pp 125–143
Patil A, Paikrao HM, Patil S (2023) The chemistry and biology of the plant poisons and their forensic significance. Stud Nat Prod Chem 78:255–321
Paul I, Poddar Sarkar M, Bhadoria PBS (2022) Floral secondary metabolites in context of biotic and abiotic stress factors. Chemoecology 32(2):49–68
Pires EDO Jr, Di Gioia F, Rouphael Y, Ferreira IC, Caleja C, Barros L, Petropoulos SA (2021) The compositional aspects of edible flowers as an emerging horticultural product. Molecules 26(22):6940
Pires TC, Dias MI, Barros L, Ferreira IC (2017) Nutritional and chemical characterization of edible petals and corresponding infusions: valorization as new food ingredients. Food Chem 220:337–343
Pourmortazavi SM, Hajimirsadeghi SS (2007) Supercritical fluid extraction in plant essential and volatile oil analysis. J Chromatogr A 1163(1–2):2–24
Prata GGB, DSK O, MMA L, Oliveira LS, FAS A, Alves RE, Silva SM (2017) Nutritional characterization, bioactive compounds and antioxidant activity of brazilian roses (Rosa spp). J Agric Sci Technol 19(3):929–941
Preda M, Smolinska S, Popescu FD (2024) Diagnostic workup in IgE-mediated allergy to Asteraceae weed pollen and herbal medicine products in Europe. Medicina 60(9):1494

Purohit SR, Rana SS, Idrishi R, Sharma V, Ghosh P (2021) A review on nutritional, bioactive, toxicological properties and preservation of edible flowers. Future Foods 4:100078

Rachkeeree A, Kantadoung K, Suksathan R, Puangpradab R, Page PA, Sommano SR (2018) Nutritional compositions and phytochemical properties of the edible flowers from selected Zingiberaceae found in Thailand. Front Nutr 5:3

Rafieian F, Amani R, Rezaei A, Karaça AC, Jafari SM (2024) Exploring fennel (Foeniculum vulgare): composition, functional properties, potential health benefits, and safety. Crit Rev Food Sci Nutr 64(20):6924–6941

Rao V, Poonia A (2024a) Bioactive compounds, nanoparticle synthesis and potential utilization of edible flowers for development of functional dairy products: a review. J Food Sci Technol 61(6):1053–1068. https://doi.org/10.1007/s13197-023-05853-6

Rao V, Poonia A (2024b) Microbial biosynthesis of selenium nanoparticles using probiotic strain and its characterization. J Food Meas Charact 18(7):5477–5488. https://doi.org/10.1007/s11694-024-02581-z

Sadia H, Malik K, Naqvi SAM, Hassan A, Abbas S (2024) Botany, ethnomedicine, phytochemistry and pharmacology of musk cucumber (Sicana odorifera) – a review. Ethnobot Res Appl 27:1–38

Salami SO, Afolayan AJ (2021) Evaluation of nutritional and elemental compositions of green and red cultivars of roselle: Hibiscus sabdariffa L. Sci Rep 11(1):1030

Sarode DK, Pagariya MC, Jadhav PR, Patil SA, Devarumath RM, Shingote PR et al (2024) Edible flowers: biotechnological interventions for improving bioactives of food and health significance. J Food Compos Anal 134:106506

Sharifi-Rad J, Quispe C, Herrera-Bravo J, Akram M, Abbaass W, Semwal P et al (2021) [Retracted] Phytochemical constituents, biological activities, and health-promoting effects of the Melissa officinalis. Oxidative Med Cell Longev 2021(1):6584693

Sharma G, Rajhansa KC, Sharma P, Singh A, Sharma A, Sahu MK et al (2022) Marigold (Tagetes spp.): a diverse crop with multipurpose value for health and environment: a review. Agric Rev 45(4):618–626

Sharma N, Kumar M, Kumari N, Puri S, Rais N, Natta S et al (2023) Phytochemicals, therapeutic benefits and applications of chrysanthemum flower: a review. Heliyon 9(10):e20232

Shen N, Wang T, Gan Q, Liu S, Wang L, Jin B (2022) Plant flavonoids: classification, distribution, biosynthesis, and antioxidant activity. Food Chem 383:132531

Shen L, Pang S, Zhong M, Sun Y, Qayum A, Liu Y et al (2023) A comprehensive review of ultrasonic assisted extraction (UAE) for bioactive components: principles, advantages, equipment, and combined technologies. Ultrason Sonochem 101:106646

Sil A, Barman M (2020) Chapter-3 green pesticides: an effective pest control approach. Recent Trends 115:45

Skrajda-Brdak M, Dąbrowski G, Konopka I (2020) Edible flowers, a source of valuable phytonutrients and their pro-healthy effects – a review. Trends Food Sci Technol 103:179–199

Son YJ, Park JE, Kim J, Yoo G, Nho CW (2021) The changes in growth parameters, qualities, and chemical constituents of lemon balm (Melissa officinalis L.) cultivated in three different hydroponic systems. Ind Crop Prod 163:113313

Sood R, Gupta S, Rana MK (2017) Zucchini. In: Vegetable crop science. CRC Press, Boca Rotan, pp 573–582

Sotelo A, López-García S, Basurto-Peña F (2007) Content of nutrient and antinutrient in edible flowers of wild plants in Mexico. Plant Foods Hum Nutr 62:133–138

Stefaniak A, Grzeszczuk ME (2019) Nutritional and biological value of five edible flower species. Notulae Botanicae Horti Agrobotanici Cluj-Napoca 47(1):128–134

Suksathan R, Rachkeeree A, Puangpradab R, Kantadoung K, Sommano SR (2021) Phytochemical and nutritional compositions and antioxidants properties of wild edible flowers as sources of new tea formulations. NFS J 24:15–25

Sun J, Wu Y, Shi M, Zhao D, Tao J (2020) Isolation of PlANS and PlDFR genes from herbaceous peony (Paeonia lactiflora Pall.) and its functional characterization in Arabidopsis and tobacco. Plant Cell Tissue Organ Cult (PCTOC) 141:435–445

Takahashi N (2025) Rose (Rosa sp.) more than just beautiful: exploring the therapeutic properties of the rose species. In: Advances in medicinal and aromatic plants. Apple Academic Press, Palm Bay, p 882. vol 2-263
Takahashi JA, Rezende FAGG, Moura MAF, Dominguete LCB, Sande D (2020) Edible flowers: bioactive profile and its potential to be used in food development. Food Res Int 129:108868
Tan BL, Norhaizan ME (2019) Carotenoids: how effective are they to prevent age-related diseases? Molecules 24(9):1801
Tavallali V, Rahmati S, Bahmanzadegan A, Lasibi MJM (2024) Phenolic profile andevaluation of antimicrobial and anticancer activities of Calendula officinalis L. using exogenous polyamines application. Ind Crop Prod 214:118571
Tiwari R, Latheef SK, Ahmed I, Iqbal HM, Bule MH, Dhama K et al (2018) Herbal immunomodulators-a remedial panacea for designing and developing effective drugs and medicines: current scenario and future prospects. Curr Drug Metab 19(3):264–301
Tomasella C, Binello A, Guccione S, Cravotto G (2022) Therapeutically valuable bioactive extracts from Mediterranean plants: green technologies and molecular modelling for a viable endeavour. In: Green chemistry and computational chemistry. Elsevier, Amsterdam, pp 425–448
Ullah MA, Hassan A, Hamza A (2023) Calendula (Calendula officinalis) marigold as medicinal plant. Orthop Case Rep 2(1):1–9
Verma, L. R., & Joshi, V. K. (2000). Post-harvest technology of fruits and vegetables: handling, processing, fermentation and waste management. Volume 2: Technology, Indus Publishing Company, New Delhi, 1(1), 1–76
Walia S, Saha S, Tripathi V, Sharma KK (2017) Phytochemical biopesticides: some recent developments. Phytochem Rev 16:989–1007
Wan H, Yu C, Han Y, Guo X, Luo L, Pan H et al (2019) Determination of flavonoids and carotenoids and their contributions to various colors of rose cultivars (Rosa spp.). Front Plant Sci 10:123
Worku MA, Debela KL, Mudde HL (2023) Competitive performance of the Ethiopian flower industry from a pre-to post COVID-19 pandemic era (2003-2022): A comparative study. Int J Organ Leadersh 12(first special issue 2023):91–114
Yan Q, Xing Q, Liu Z, Zou Y, Liu X, Xia H (2024) The phytochemical and pharmacological profile of dandelion. Biomed Pharmacother 179:117334
Yang Z (2019) XLNet: generalized autoregressive pretraining for language understanding. arXiv preprint arXiv:1906.08237
Yuca H, Karakaya S (2022) Matricaria chamomilla L. In: Novel drug targets with traditional herbal medicines: scientific and clinical evidence. Springer International Publishing, Cham, pp 387–400
Zhao L, Fan H, Zhang M, Chitrakar B, Bhandari B, Wang B (2019) Edible flowers: review of flower processing and extraction of bioactive compounds by novel technologies. Food Res Int 126:108660
Zia-Ul-Haq M (2021) Historical and introductory aspects of carotenoids. In: Carotenoids: structure and function in the human body. Springer International Publishing, Cham, pp 1–42
Zorić M, Banožić M, Aladić K, Vladimir-Knežević S, Jokić S (2022) Supercritical CO2 extracts in cosmetic industry: current status and future perspectives. Sustain Chem Pharm 27:100688

Chapter 2
Postharvest Processing Methods of Edible Flowers for Their Extended Preservation

Jayoti Majumder Sarkar and Saikat Gantait

2.1 Introduction

Herb flowers, known as edible flowers (EFs), possess a deep significant past in phytotherapy, valued for their nutraceutical and medicinal values for centuries. Traditionally, fresh flowers were primarily used for decorative and aromatic purposes, enhancing the sensory appeal of spaces and dishes (Matyjaszczyk and Śmiechowska 2019). However, a contemporary shift has emerged in the culinary world, where fresh flowers are increasingly appreciated for their nutritional and therapeutic benefits (Mlcek and Rop 2011; Fernandes et al. 2019a). This trend highlights their role as functional foods, contributing pharmacologically active compounds such as flavonoids, polyphenols, and essential oils, recognized for their antioxidant, anti-inflammatory, and antibacterial uses. Their vibrant colors, diverse flavors, and health-promoting attributes have made EFs a significant addition to modern gastronomy, merging aesthetic appeal with wellness (Chen and Wei 2017; Pires et al. 2018; Fernandes et al. 2019a). Bioactive compounds in EFs vary significantly depending on plant anatomical parts, family, and environmental conditions (Yang et al. 2018). Carbohydrates are the predominant constituents, providing energy and structural support, along with vitamins and minerals. Flowers have a high content of natural antioxidants such as anthocyanin, flavonoids and various other phenolic components (Rao and Poonia 2024a). Phenolic compounds,

J. M. Sarkar (✉)
Department of Floriculture and Landscape Architecture, Bidhan Chandra Krishi Viswavidyalaya, Mohanpur, Nadia, West Bengal, India
e-mail: majumderr.jayoti@bckv.edu.in

S. Gantait
Crop Research Unit (Genetics and Plant Breeding), Bidhan Chandra Krishi Viswavidyalaya, Mohanpur, Nadia, West Bengal, India

A. Poonia et al. (eds.), *Edible Flowers: Source of Phytonutrients, Valorization and Technological Advancements*, https://doi.org/10.1007/978-3-032-00667-7_2

particularly abundant in the petals, are the primary antioxidants, offering protection against oxidative stress. These phenolics, including flavonoids, carotenoids, and polyphenols, play a crucial role in promoting health by reducing inflammation, neutralizing free radicals, and supporting immune functions. The variation in bioactive content underscores the importance of species selection, environmental factors, and flower anatomy in maximizing the health benefits of EFs (Grzeszczuk et al. 2018; Pires et al. 2018; Takahashi et al. 2020).

These flowers can be consumed fresh, as seen with marigold or nasturtium flowers in salads, or incorporated into savory dishes with meat, fish, and soups. EFs also elevate beverages, including wine and beer, and add flair to desserts, jellies, and sweets. Beyond fresh consumption, they serve as spices and natural dyes, enriching the sensory experience of food. Dried forms are popular in infusions or desserts featuring dried rose petals, while powdered forms (Chen and Wei 2017) are used for seasoning or coloring. Innovations in molecular gastronomy have introduced crystallized flowers and flower-based foams, broadening their culinary applications (Fernandes et al. 2019a). Nanotechnology represents an emerging field of scientific exploration focused on the examination and control of matter at the smallest scales, encompassing molecules and atoms (Rao and Poonia 2024b). Despite their growing popularity, the market for EFs remains relatively underdeveloped compared to more established sectors like fruits and vegetables (Fernandes et al. 2019b). This is primarily due to limited production, positioning EFs as a niche product. Factors such as the delicate nature of flowers, short shelf life, and specialized cultivation requirements contribute to their lower availability. Additionally, consumer awareness and demand for EFs are still emerging, often restricted to gourmet dining, upscale markets, and health-conscious segments (Marchioni et al. 2021). Expanding this market will require increased production, better postharvest handling techniques, and broader consumer education on the culinary and nutritional potential of EFs (Benvenuti et al. 2016; Simoni et al. 2018).

Postharvest technologies are widely employed for ornamental and aromatic flowers due to their high economic value and perishability. Optimizing postharvest technologies for EFs can enhance their market potential by ensuring longer shelf life, maintaining phytonutrient content, and reducing postharvest losses (Fernandes et al. 2020). Their high antioxidant content, mainly polyphenols, carotenoids, and ascorbic acid, act in scavenging free radicals and extending shelf life during postharvest storage (Cavaiuolo et al. 2013). Maintaining flowers at low temperatures (5–10 °C) effectively delays senescence and preserves marketability for up to 8 days (Menegaes et al. 2019; Landi et al. 2017). However, the ascorbic acid trend during cold storage varies by species. For example, it decreases in *Cucurbita pepo*, *Tulbaghia cominsii*, and *Tropaeolum majus*, increases in *Acmella oleracea*, and remains stable in certain *Salvia* species (Landi et al. 2017; Aquino-Bolaños et al. 2013). Additionally, fragrance retention is vital for consumer appeal, though limited studies exist on EFs aroma profiling (Fernandes et al. 2019a; Marchioni et al. 2020).

EFs, are valued for rich nutritional components and bioactive compounds, contributing to their vibrant colors, distinct flavors, and overall palatability. Their rise in popularity aligns with the growing trend of wholesome eating, making them an emerging dietary fad. However, despite their strong sensory appeal, EFs remain less commonly used than fruits and vegetables, largely due to their short shelf life.

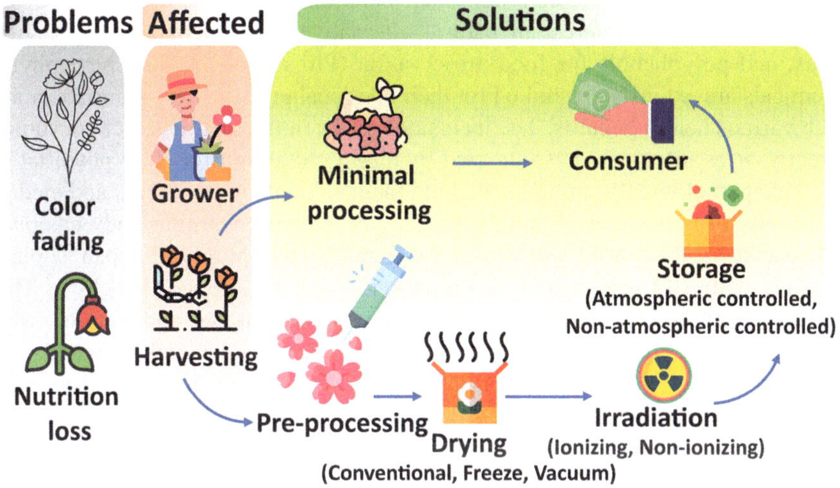

Fig. 2.1 Postharvest management strategies of edible flowers. (Source: Diagram prepared by the authors) (icon source: https://www.flaticon.com)

Various postharvest processing technologies have been developed to overcome this limitation and extend their market viability. Innovative methods such as edible coatings, high hydrostatic pressure (HHP), radiation exposure, controlled atmosphere (CA) packaging, and modified atmosphere packaging (MAP) have proven effective in maintaining freshness and delaying senescence (Fig. 2.1). Additionally, advanced drying techniques, including freeze-drying and vacuum drying, are employed to preserve the flowers' sensory qualities, nutritional value, and bioactive compounds during long-term storage. These advancements strive to enhance the quality and extend the shelf life of EFs, thereby promoting their incorporation into mainstream markets and broadening their commercial applications within the food industry. The growing integration of EFs into various food products highlights the need for advanced technologies to enhance their distribution and marketability, whether as fresh or minimally processed products. Extending their post-harvest life while maintaining quality is key to industrial growth and consumer satisfaction. As a result, significant investments are being made in innovative preservation techniques, including traditional and emerging non-thermal methods, as well as improved packaging solutions. Specific EFs like artichoke, broccoli, and cauliflower have also been discussed due to their unique consumer perceptions.

2.2 Nutritional Value of EFs

The vibrant colors of EFs offer therapeutic benefits to the body, mind, and soul. With rising chronic diseases linked to industrialized food and sedentary lifestyles, there's a growing need for healthier, functional food options. Although synthetic antioxidants are common, their potential mutagenic and carcinogenic effects have driven interest in natural alternatives like tocopherols, phenolic acids, and polyphenols for food preservation (Pires et al. 2021). These phytochemicals are extensively studied for their chemical profiles and their association with various health benefits. The inclusion of EFs in the diet, whether consumed directly or as infusions, has attracted increasing attention for their potential to alleviate pathological symptoms, making them valuable in functional and medicinal foods (Takahashi et al. 2020). To date, no global governing body has published an official list of EFs. However, Lu et al. (2016) reported EFs span 97 plant families with 100 genera and encompass 180 species around the globe. This diversity highlights their potential in culinary and industrial applications, warranting further research and standardization. Phenolic acids and flavonoids are key contributors to antioxidant activity, aiding in the neutralization of free radicals and the reduction of oxidative stress. EFs have been used historically in various cultures as ingredients in various dishes (Mlcek and Rop 2011). Today, their popularity is rising due to their aromatic and health-promoting roles (Chen and Wei 2017). EFs are visually appealing because of their vibrant colors (Kelley et al. 2002) and offer diverse flavors, from sweet to spicy (Fernandes et al. 2018a). Nutritionally, EFs are considered safe due to organic cultivation and are rich in beneficial compounds such as amino acids (Sotelo et al. 2007), polyphenols (Lu et al. 2016), polyunsaturated fatty acids (Fernandes et al. 2018a), minerals (Rop et al. 2012), xanthophylls (Fernandes et al. 2018a), and vitamin E (Pires et al. 2018). Many of these components exhibit, anti-irritant, and antimicrobial (Lu et al. 2016).

EFs are increasingly valued for their bioactive compounds with therapeutic potential, supporting their role in health-conscious diets and nutraceuticals. For example, roses have been utilized to relieve menstrual discomfort, enhance circulation, and show promise in inhibiting cancer cell growth, making them significant in preventive healthcare and functional foods (Yang and Shin 2017). Several studies highlight the role of EFs in combating bacteria and fungi (Młynarczyk et al. 2018) and their potential in treating critical issues like Alzheimer's (Hamdy et al. 2018) and diabetes (Morittu et al. 2019). *Calendula officinalis* exhibits gastroprotective and hepatoprotective impacts (Kumari et al. 2021), attributed to its bioactive components. Antioxidant-rich diets are linked to reduced risks of cardiovascular, diabetes, neurological, and respiratory diseases, with various EFs showing in vivo antioxidant activity (Bao et al. 2018; Marchelak et al. 2019). Cultures cell assays, including DPPH scavenging (Pires et al. 2018; Morales-Luna et al. 2019), FRAP (Zhang et al. 2019), ABTS (Bao et al. 2018; Morales-Luna et al. 2019), β-carotene inhibition (Lockowandt et al. 2019), etc. confirm the antioxidant potential of EFs.

2.3 EFs as Fresh or Processed Food

A few key factors influence food choice: convenience, familiarity, health, mood, content, cost, sensory appeal, obesity management, and moral considerations (Pereira et al. 2020). The acceptance of new foods is closely tied to cultural perceptions and consumption patterns (Rodrigues et al. 2017). EFs are more likely to be consumed by individuals who are well-informed about their chemical composition (Chen and Wei 2017). Aroma is a key factor in attracting attention and generating interest in flowers (Chen and Wei 2017). Understanding food consumption factors is essential for marketing, allowing professionals to highlight curiosity, aroma, and health benefits. Petals are the most commonly consumed flower parts (Fernandes et al. 2017), while pollen, which is high in lipids, proteins, and flavonoids, may cause allergies (Mlcek and Rop 2011). Nectar is rich in alkaloids, amino acids, and sugars (Mlcek and Rop 2011). Key quality factors for consumer acceptance include color, aroma, size, flavor, and shape, with preferences for bright colors such as white, yellow, orange, and blue (Kelley et al. 2001). Each flower species' flavor and aroma make them suitable for specific dishes or drinks (Fernandes et al. 2017) (Table 2.1).

EFs are widely used for decorative purposes in cakes, desserts, and ice creams. They are utilized in the preparation of drinks and brews, enhancing both visual appeal and flavor (Fernandes et al. 2017). They have immense potential for various processed foods due to their unique properties. They can be incorporated into the making of candies, confectionery, jams, marmalades, and others, offering both flavor enhancement and visual appeal (Jadhav et al. 2023). Hussain et al. (2019) compared the sensory attributes and bioactive profiles of EFs' tea infusions. Their study focused on osmanthus, jasmine, and rosebud flowers. The results indicated that *Osmanthus* flower had the highest polyphenolic and caffeine content compared to the other two, highlighting its potential as a nutrient-rich tea infusion. Matyjaszczyk and Śmiechowska (2019) surveyed consumer acceptance of EFs-enriched foods, revealing a generally optimistic attitude. EFs are popular as processed foods due to

Table 2.1 Edible Flower Species Consumed as Fresh and Processed Products

Scientific name	Edible part	Used in food	References
Calendula officinalis	Petals	Salad	Mlcek and Rop (2011)
Viola tricolor	Petals	Salad	Takahashi et al. (2020)
Antigonon leptopus, *Bougainvillea glabra*, *Cosmos sulphureus*, *Tagetes erecta*	Flowers	Salad and tea	Barani et al. (2022)
Agave spp., *Cucurbita* sp., *Dahlia* spp., *Erytrhrina americana*, *Ferocactus* sp., *Plumeria rubra*, *Yucca* spp., *Opuntia*, *Myrtillocactus geometrizans*, *Euphorbia pulcherrima*, and *Phaseolus vulgaris*	Whole inflorescence	Hot dishes, desserts, and beverages	Mulík and Ozuna (2020)
Rosa spp.	Petals	Sweetener	Mlcek and Rop (2011)

their health benefits. Studies show that color, taste, aroma, and appearance significantly influence consumer choices, with yellow, blue, and orange flowers being the most preferred (Kelley et al. 2001). Matyjaszczyk and Śmiechowska (2019) noted that odor plays a crucial role, as different species of roses can have varying scent intensities. Despite regional differences, consumers generally prefer health-promoting products containing EFs over similar products without them (Rodrigues et al. 2017). Roses and pansies are known for their sweet flavor, while a spicy and peppery taste is offered by nasturtium (Takahashi et al. 2020). Consumers appreciate EFs for their vibrant colors and unique flavors. For instance, roses are often used in salads and jams, while marigolds are preferred for adding flavor to savory dishes (Zhang et al. 2019).

2.4 Morphophysiological Changes During Postharvest of EFs

Morphological changes, like EF quality loss in stems, leaves, flowers, or seeds, can lead to market rejection. This may result from yellowing, wilting, petal abscission, and bending of scapes or stems. Yellowing is often linked to the end of the flower's display life, influenced by environmental factors. Flower shattering can occur due to ethylene activity and postharvest stress. Postharvest techniques are essential for extending the shelf life of edible ornamental flowers due to their high commercial value (Fernandes et al. 2020). These flowers are extensively perishable, with a limited shelf life of 2–5 days after harvest, requiring storage at temperatures between 5 and 10 °C to slow down senescence (Menegaes et al. 2019). Antioxidants in EFs, such as carotenoids, flavonoids, polyphenols, and ascorbic acid, help to neutralize free radicals (Mlcek and Rop 2011). High antioxidant levels during postharvest storage can extend shelf life, as evidenced by flowers that maintained their marketability, polyphenols, and anthocyanins content up to 8 days after harvest (Landi et al. 2017). Ascorbic acid content typically decreases during cold storage, with varying trends depending on the plant species. For example, decreasing ascorbic acid metabolism was observed in *Cucurbita pepo* (Aquino Bolaños et al. 2013) and *Tropaeolum majus* (Landi et al. 2017). An increase in vitamin C was recorded in *Acmella oleracea* after 2 days storage (Landi et al. 2017), however, no post-storage changes were observed in two *Salvia* spp. (Landi et al. 2017).

Chlorophyll gives green color, carotenoids produce yellow, orange, and red hues, and anthocyanins provide blue, purple, red, and colors, which are influenced by the pH of the flower sap. In acidic conditions, anthocyanins appear red, while they shift to blue in alkaline environments. This shift, known as 'blueing' in roses, occurs with aging due to sugar depletion and protein breakdown (Takahashi et al. 2020). White flowers reflect all visible light, often from aerated tissues. Variegated leaves in some EFs may lack chloroplasts, resulting in white, orange, or yellow colors. Over time, petals may be discolored, wilt, and blackened.

2.5 Postharvest Processing of EFs

Advancements in production systems are crucial for boosting bioactive compound synthesis in EFs. Key factors include selecting superior cultivars, optimizing water, temperature, and light conditions, and ensuring adequate mineral nutrition through micro- and macronutrients and growth regulators that can stimulate bioactive production. Additionally, implementing optimal harvesting techniques and advanced pre-processing methods enhances quality, minimizes postharvest losses, and meets consumer safety standards, supporting the floriculture industry's ability to meet rising demands for EFs.

2.5.1 Harvesting

The farming of flowers specifically for edible purposes represents a shift in the floriculture sector, driven by growing consumer interest. Traditionally, flowers were cultivated primarily for beautification, but changing preferences have prompted countries to adopt practices for cultivating EFs. Harvesting is a vital step in ensuring quality and usability, with the timing and method significantly influencing flavor, texture, and bioactive content. EFs are known for their delicate nature and are very perishable, with shelf life typically ranging between 2 and 5 days after harvest (Fernandes et al. 2019a). This short postharvest window is attributed to various quality-degrading factors, including early senescence. These issues significantly limit the usability and appeal of EFs for both consumers and businesses.

Compared to other cut ornamental flowers utilized for decorative purposes, EFs are particularly vulnerable. Their foliage is often short, reducing their ability to draw moisture, and they are mostly stored without supplemental moisture. This results in faster degradation compared to ornamental flowers, which are handled with techniques aimed at extending their visual freshness (Kou et al. 2012).

While some findings suggest that EFs should be harvested and consumed on the same day to ensure optimal freshness, such a recommendation imposes severe limitations on their commercial potential. This practice is impractical for large-scale distribution and retail, where extended storage periods and shipping times are necessary. Addressing these challenges requires innovative postharvest handling techniques and storage strategies, such as controlled temperature environments, specialized packaging materials, and drying, to increase the shelf life and maintain quality. Such advancements would not only improve the commercial viability of EFs but also enable them to reach broader markets, ensuring accessibility for a wider range of consumers and culinary applications (Lu et al. 2016).

EFs are best harvested at full maturity, judged by factors such as color vibrancy, flower size, and full bloom stage. Morning hours are ideal for harvesting because the flowers are fresh and have their peak natural flavors. However, flowers should not be harvested when dew is present, as it can cause petals to stick together, leading to damage during transport. Once harvested, parts like stamens, pistils, and sepals are often removed due to their unappealing flavors (Netam 2021).

Research underscores the impact of harvesting conditions on EF quality. For example, pansy flowers cultivated with optimal irrigation (100% water replenishment) and no mycorrhizal inoculation were found to have higher levels of bioactive molecules like phenolics, flavonoids, and carotenoids (González-Barrio et al. 2018). Peony flowers, when harvested at full bloom, exhibited superior edible quality due to enhanced nutrient and bioactive profiles (Jadhav et al. 2021). Proper cultivation methods, including organic growth without synthetic inputs, are critical to ensuring safety and avoiding toxic residues during harvest (Mlcek and Rop 2011).

2.5.2 *Minimal Processing for Short-Duration Storage*

EFs are delicate and perishable, prone to quality issues such as wilting, discoloration, damaged petals, and undesirable aromas. To address these challenges, processing techniques must focus on preserving their aesthetic and sensory qualities while extending postharvest life. Methods of storage in low temperatures, appropriate packaging, and using gentle preservatives are employed to maintain freshness. To date, no report has been found in EFs in this aspect. Whereas, most supermarkets sell the EFs in fresh. Proper harvesting, handling, and storage conditions are critical to minimizing quality deterioration and ensuring EFs remain appealing and suitable for culinary uses. Sorting, sectioning, and cutting are included in minimal processing (Fig. 2.2). It also needs innovative packaging strategies designed to maintain the flowers' sensory attributes and extend their shelf life. This approach aims to address the perishability challenges of EFs.

2.5.3 *Pre-processing*

Post harvest, the EFs are highly perishable due to their high water content, making pre-processing and storage crucial in maintaining the quality. Immediate handling and appropriate storage methods can significantly extend shelf life and preserve bioactive properties. EFs are highly perishable, with their quality starting to degrade soon after harvest. This makes pre-processing or postharvest management a crucial task in maintaining their freshness, nutritional value, and visual appeal. The methods used for pre-processing often vary depending on the species, as the physical robustness, chemical composition, and intended use of the flowers dictate the optimal approach.

Pre-processing techniques, such as coating with egg white powder where, freshly harvested petals are lightly coated by pasteurized powder of egg white, which acts as a protective barrier. This coating helps preserve the petals' quality by preventing contamination from foreign materials or disease-causing microorganisms

Fig. 2.2 Minimally processed EFs in the supermarket. (Source: https://kinarino.jp/cat4/12766)

(Shanthamma et al. 2021). Proper pasteurization of the egg white powder is essential to ensure food safety. Another important aspect is temperature control which is critical for the quality preservation of EFs. Studies have shown that nasturtium, pansy, and viola exhibit good storability around −2.5 °C for up to 14 days when stored in polyethylene bags, species like scarlet and borage showed a noticeable quality decline after 7 days under the same conditions (Kelley et al. 2001). Proper packaging materials such as breathable bags or containers can prevent dehydration and preserve flower integrity.

However, chlorine dioxide, a powerful oxidizing agent, enhances the shelf life of EFs by lowering microbial growth. Ran et al. (2019) reported that chlorine dioxide could be applied alone or combined with other food-grade preservatives to improve storage outcomes. Species with fragile petals require lower storage temperatures with shorter storage durations to preserve their quality. Flowers with a robust physical structure can endure slightly longer storage times without significant loss of quality.

Proper handling techniques during pre-processing ensure that EFs maintain their quality until they are ready to be processed into high-quality food products. EFs, with enhanced moisture content, are more susceptible to discoloration, wilting, and microbial spoilage. They must be stored in optimal conditions to prevent such issues. The presence of dew during harvesting, for instance, can lead to petal damage during transportation if not dried properly. Flowers that are bruised or damaged during handling lose their commercial and culinary appeal. Innovative pre-processing methods, tailored to specific flower species, are needed to extend shelf life and improve commercial viability. Research into alternative natural preservatives, optimized packaging technologies, and precise temperature management can help ensure that EFs retain their visual and nutritional qualities, enabling wider market reach and consumer satisfaction.

2.5.4 Drying Techniques for Long-Duration Storage

Drying EFs extends their shelf life, facilitates transportation, packaging, and usage, and can preserve bioactive compounds under optimal conditions. Studies highlight the importance of methodology, with low-temperature drying being favorable for retaining biologically active compounds and minimizing degradation (Aboltins and Kic 2016). These findings emphasize the innovative potential of exploring and optimizing drying techniques for EFs (Fig. 2.3).

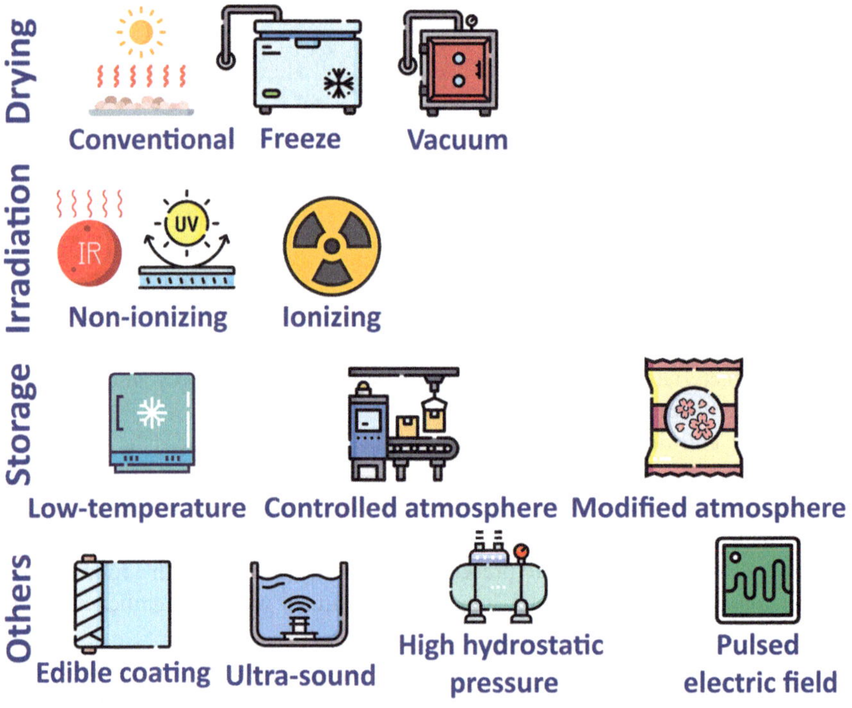

Fig. 2.3 Various processing techniques of EFs. (Source: Diagram prepared by the authors) (icon source: https://www.flaticon.com)

2.5.4.1 Conventional Drying

Traditional methods of preserving EFs, such as sun drying or hot air oven drying often result in undesirable changes in color, texture, sensory attributes, and nutritional quality due to prolonged exposure to high temperatures (Zevallos-Concha et al. 2016). Hot air drying has been studied primarily in broccoli inflorescences, where at 48 °C and 3 h of treatment resulted in delayed senescence, preserved pigments, higher sugar, protein, and antioxidant levels at 20 °C. This process ensures better storage quality by preserving key attributes such as plant pigment and antioxidants (Costa et al. 2005). Consumers seek fresh-like products with higher nutritional values, which traditional methods often cannot provide. High hydrostatic pressure and irradiation offer advanced solutions by using mild temperatures to keep the sensory and nutraceutical properties of EFs during preservation and storage (Jadhav et al. 2023).

2.5.4.2 Freeze Drying

Freeze drying, or lyophilization, is a preservation method that dehydrates flowers by freezing them and by reducing the surrounding pressure causing frozen water to sublimate directly to form vapor from solid. This method is ideal for retaining the structural integrity, color, fragrance, and bioactive compounds of EFs (Zheng et al. 2015). A review by Shanthamma et al. (2021) explained the different effects of storage temperatures on artichoke (*Cynara scolymus*) flowers, showing that phenolic content peaked at 5 °C, surpassing levels in freshly harvested flowers. Chromogenic content was highest in internal bracts stored between 0–7 °C but diminished at 10 °C and −10 °C, emphasizing the importance of precise temperature control during storage. Following up the research by Gębczyński and Kmiecik (2007) evaluated the nutraceutical and sensory attributes of broccoli and two cauliflower varieties (white and green) under traditional and modified freezing methods. In the study, cauliflowers were either blanched and cooked or treated with citric acid before cooking, then stored at −20 °C or −30 °C for up to 12 months. The modified freezing process showed better vitamin C retention (18% higher) and antioxidant activity (13% greater) compared to traditional methods. Despite vitamin C reductions of 27–37% and antioxidant losses of up to 42% during prolonged storage, the green cauliflower variety demonstrated superior sensory attributes. Fernandes et al. (2020) explored freezing techniques for EFs, including direct freezing in ice cubes applied to white borage (*Borago officinalis*), heartsease (*Viola tricolor*), yellow kalanchoe (*Kalanchoe blossfeldiana*), and dandelion (*Taraxacum officinale*). Ice cube freezing preserved fresh-like appearances, whereas direct freezing maintained and occasionally increased antioxidant activity after 1 month. However, bioactive compound losses were observed over longer storage durations. Innovative freezing methods, such as combining freezing with non-thermal techniques like ultrasound processing, high-pressure technology, and osmotic dehydration, have demonstrated benefits like enhanced ice nucleation and improved crystal distribution, which protect product quality during storage (Fan et al. 2019; Zhu et al. 2020).

2.5.4.3 Vacuum Drying

Dehydration, a time-tested preservation method, involves removing moisture to lower water activity, thereby prolonging shelf life. Traditional sun drying, still widely practised in tribal regions for EFs, is slow and often leads to undesirable changes, such as loss of texture, color, and nutritional quality. Modern advancements have introduced efficient techniques, such as vacuum drying. Vacuum drying, studied by Ahluwalia et al. (2014), excels in preserving β-carotene and quality while being faster than traditional methods. However, its high cost makes it less viable for widespread use. As a more economical option, this method was recommended for EFs like Pusa Basanti and Pusa Narangi marigold varieties through Siriamornpun et al. (2012) along with infrared-hot air drying for *Tagetes erecta* (marigold) outperformed other methods in retaining antioxidants, bioactive compounds, and vibrant petal color. House-hold microwave was used to dry edible flower petals' of common daisy (*Bellis perennis* L.), pot marigold (*Calendula officinalis* L.), and lavender (*Lavandula angustifolia* L.) found to retain higher phenolic content in flowers (Dorozko et al. 2019). Future research is essential to refine these techniques for specific EFs, facilitating their commercial and industrial preservation. Developing species-specific protocols will address challenges in maintaining quality while scaling up for broader market applications. Such advancements promise to support both sustainability and innovation in EFs preservation.

2.5.5 Irradiation

Irradiation of food is a cost-effective method to enhance the well-being and post-harvest/processing life of perishable foods, including EFs (Koike et al. 2019) (Fig. 2.3). It delays cell death and alters biochemical properties by accumulating oxidative stress and reactive oxygen species (ROS). Gamma radiation has similarly been shown to preserve quality and bioactivity (Koike et al. 2015b). Overall, irradiation can enhance antioxidants and phenols while maintaining product integrity. To encourage market adoption, consumer education on the benefits of irradiation in food preservation is essential. Food irradiation is a cost-effective mechanism (Koike et al. 2019). To address oxidation and insect attacks, Villavicencio et al. (2018) applied irradiation through an electron beam to EFs at doses of 0.5, 0.8, and 1 kGy. The treatment extended shelf life without significant changes in nutritional composition or cytotoxic effects. However, there was a slight reduction in anti-inflammatory activity. At 0.5 kGy, phenolic compound levels increased, indicating a potential benefit of low-dose irradiation for preserving certain bioactive properties while enhancing the flowers' stability and usability. The technology ensures minimal qualitative damage while enhancing antioxidants and phenolic compounds.

2.5.5.1 Nonionizing Radiation

Recent studies have explored the use of sub-lethal dosages of ultraviolet (UV) radiation for potential treatment for EFs at postharvest, according to the hormesis concept. This concept suggests that low doses of stress-inducing agents like UV light can have beneficial effects on plants. UV radiation is categorized into three types: UV-A (320–400 nm), UV-B (280–320 nm), and UV-C (200–280 nm) (Guerrero-Beltrán and Barbosa-Cánovas 2004). While UV treatments have been investigated primarily on other crops like broccoli, studies indicate that UV-B and UV-C radiation can slow the degradation of chlorophyll, tissue damage, and retain antioxidant content in broccoli florets (Guerrero-Beltrán and Barbosa-Cánovas 2004; Costa et al. 2006). Additionally, UV-C treatment in broccoli increased phenolic and ascorbic acid contents, antioxidant activity, and soluble sugars while reducing microbial populations (Lahir 2023). These effects are likely due to the enhanced activity of phenylalanine ammonia-lyase, an enzyme involved in phenolic metabolism (Costa et al. 2006). However, it is imperative to note that the impact of UV treatments does vary among flower species, and thus, similar studies need to be conducted on EFs to assess their specific responses to UV radiation and determine optimal doses.

2.5.5.2 Ionizing Radiation

Ionizing radiation, regulated by strict guidelines, is used in food processing to delay ripening, reduce microbial contamination, and extend shelf life. This process, including gamma rays and electron beams, is considered safe by organizations such as the WHO and FAO. It has been successfully applied in various food products to improve hygiene, eliminate pests, and ensure food quality (Farkas 2006). However, its use is subject to legal constraints to ensure safety and effectiveness. Food irradiation, especially through gamma rays and electron beams, is effective for increasing shelf life, ensuring hygiene, and food disinfectant (Koike et al. 2015a). Research has shown that phenolic content enhancement and increased antioxidant activity in *Viola tricolor* and *Bauhinia variegata* flowers were due to irradiation (Koike et al. 2015a, b; Villavicencio et al. 2018). Irradiation may reduce anthocyanin content but improve the overall bioactive composition due to atmospheric changes during the process (Koike et al. 2015a, b). Additionally, the preservation of carotenoids in *Rosa chinensis* and *Tagetes patula* has been documented with gamma irradiation (Koike et al. 2021). Currently, irradiation is widely used for preserving spices, herbs, and dried vegetable seasonings. However, there are limited studies on its effects on EFs. Koike et al. (2011) investigated the tolerance of certain EFs to gamma rays and found that *Borago officinalis* could not withstand a 0.3 kGy dose, which is required to eradicate insects. In disparity to *Borago officinalis*, other flowers such as *Dianthus chinensis*, *V. tricolor*, *V. odorata*, and *Lobularia maritima* exhibited lenience to gamma radiation doses up to 1 kGy. This suggests that flower tolerance to ionizing irradiation varies by species. However, it was noted that high doses of gamma radiation could cause damage to the flowers, including petal withering, browning, and injury (Koike et al. 2011). In a study, irradiation of nasturtium flowers at 1 kGy showed that antioxidant activity did not decrease significantly, with some irradiated samples even exhibiting higher antioxidant levels compared to untreated controls (Koike et al. 2015b). The second study on *V. tricolor* found that gamma irradiation, regardless of the dose, resulted in higher phenolic content and enhanced antioxidant activity (0 kGy). This suggests that ionizing radiation may boost the accumulation of bioactive molecules in EFs, with potential benefits for their nutritional quality (Koike et al. 2015a). Although no direct explanation was provided for the increased antioxidant activity in irradiated *V. tricolor*, it is hypothesized that irradiation may stimulate secondary metabolism. However, as this research is limited to one species, more extensive studies are required to explore the effects on other EFs. In conclusion, ionizing irradiation, when applied at appropriate doses, appears to maintain visual appearance and bioactive compounds, though additional research on its role in disinfestation and microbial load reduction is necessary.

2.5.6 *Storage*

The EFs need a specialized storage system and are more likely to use other conventional and non-conventional flowers. The methods are described in Fig. 2.3.

2.5.6.1 Low-Temperature Storage without Atmospheric Control

Temperature is always a critical parameter affecting the shelf life of EFs (Watada and Qi 1999). Cold storage, including refrigerating and freezing, is the primary method for preserving EFs. By lowering the temperature, the process slows down respiration, enzyme activity, and ethylene production while reducing water loss and microbial growth, thus delaying senescence and quality deterioration (Hettiarachchi and Balas 2004). For example, EFs preserved at −2.5 °C, 2.5 °C, and 5 °C temperatures can be preserved for up to 2 weeks. However, optimal temperature conditions may vary by species (Fernandes et al. 2020).

2.5.6.2 Controlled Atmospheric (CA) Storage

CA storage is a popular preservation method that regulates temperature, moisture, and the concentrations of gases like CO_2, O_2, and N_2 throughout the storage period (Fernandes et al. 2018b). Unlike modified atmospheric packaging, which changes the gas levels inside the packaging, CA storage allows for continuous control of these parameters, providing a more stable environment for preserving the quality of stored products over an extended time. CA storage modifies the gas environment to lower respiration rates, reduce enzymatic activity, and delay senescence (Aquino Bolaños et al. 2013). It makes CA an effective method of storage for preserving EFs in their normal state. For example, Cefola et al. (2016) explained the CA storage of *C. pepo* flower under 4 t atmospheric conditions at 5 °C, demonstrating the potential of CA storage to retain quality and enhanced shelf life in delicate flowers. In CA storage, flowers showed better features and marketability for up to 9 days, outlasting those stored in ambient air. The optimal combination of 97% N_2 and 3% O_2 was found to be very effective, minimizing loss of weight and preserving vitamin C. Additionally, a study under four different gas composition combinations on squash flowers further confirmed the advantages of CA storage in better the shelf life and maintain nutritional components (Aquino Bolaños et al. 2013; Cefola et al. 2016).

2.5.6.3 Modified Atmosphere Packaging (MAP)

Simply, packaging has an importance in preserving the quality of food products. Over time, packaging technology has advanced significantly, leading to the development of new methods. MAP is one kind of technique aimed at shelf life extension of food by adjusting the gaseous combination within the packaging. This modification helps retain the food's freshness and quality (Waghmare and Annapure 2015). This alteration of the gas environment in MAP influences several factors, such as color development, ethylene production, respiration rate, volatile compound deposition, and microbial growth in fresh products. Wang et al. (2020) explored how high oxygen atmospheric packaging (HOAP) impacted hydrogen peroxide (H_2O_2) production with oxidative stress in *Brassica campestris* L. Wang et al. (2020) found that increasing oxygen levels in HOAP altered the quality of EFs by regulating oxidative stress pathways. The higher oxygen concentrations may affect reactive oxygen species (ROS) levels through ROS-related protein metabolism. Additionally, HOAP inhibits antioxidant enzyme activity, potentially modifying cell wall content. These findings highlight the biochemical effects of MAP on flower quality and storage potential by influencing stress responses at the cellular level. Alden et al. (2019) examined the postharvest qualities of cauliflower packed in MAP (92% N_2, 5% CO_2, and 3% O_2) and compared it to usual air storage. They found that cauliflower in polyethylene pouches under MAP had a maximum shelf life of 30 days, with a potential extension to 50 days through color analysis via artificial neural networks. Although MAP shows promise for extending shelf life, developing infrastructure for bulk handling remains a challenge. Recent advancements in IoT and cloud-based systems are aiding the packaging, traceability, and transportation of EFs.

2.5.7 Other Methods of Preservation

2.5.7.1 Edible Coatings

Edible films and coatings are used on food products to enhance shelf life and maintain quality. Films are thin layers formed and applied separately from the food, while coatings are directly applied as emulsions or suspensions. Both are primarily composed of biopolymers like proteins, lipids, and polysaccharides. They serve as barriers against gas exchange, moisture loss, and microbial growth, reducing metabolic activity and deterioration in fruits and vegetables, thus enhancing freshness (Souza et al. 2010; Ansorena et al. 2011; Maria Leena et al. 2020). The food industry is increasingly focusing on eco-friendly edible coatings and cling wraps made from bio-materials such as complex carbohydrates, proteins, and fats (Vimala Bharathi et al. 2020). Methods like dipping, spraying, and electro-spraying are used to coat, ensuring uniform coverage of the product's surface (Khan et al. 2014). These innovations aim to improve product shelf life while being sustainable. Limited studies have explored the likely edible coatings on flowers like artichokes and brassicas. Alginate, a commonly used biopolymer, has shown positive results. For instance, flowers with alginate coating remained fresh for 4 days longer than control. Fernandes et al. (2018b) with pansies indicated that coatings of alginate could effectively preserve flower freshness. The flowers were coated by immersing them in a 0.5% sodium alginate solution and crosslinked with 1% $CaCl_2$. When stored at 4 °C, this treatment extended their shelf life from 3 to 4 days to up to 14 days by significantly delaying degradation. Ben-Fadhel et al. (2017) explored an innovative preservation method for broccoli that combined an edible coating with gamma rays. An alginate-based edible coating with essential oils and natamycin, combined with gamma irradiation, effectively extended broccoli's shelf life at 4 °C, demonstrating enhanced preservation potential.

2.5.7.2 Ultrasound Process

Another nonthermal method is ultrasound, which disrupts microbial cells by damaging their membranes. It also generates free radicals, which release intracellular components, effectively inactivating microorganisms while preserving product quality (Zhu et al. 2020). Ultrasonication induces acoustic cavitation in liquid samples, involving bubble formation, growth, and collapse. This process generates free radicals that chemically disrupt amino acids in enzymes and compromise the structural integrity of microbial cell membranes, leading to microorganism inactivation. Mechanical effects of ultrasonication include shear stresses, turbulence, and collapse pressures. These forces contribute to the breakdown of microbial cells and enhance the processing efficiency of liquid samples (Lv et al. 2019). Numerous findings, such as Moses et al. (2017), have explored the significance of ultrasound on natural food colorants. Ultrasound has also been effectively employed in combination with added systems to improve the quality of EFs, showcasing its versatility in enhancing bioactive compounds and preserving natural pigments. Xin et al. (2013) explored ultrasound-assisted osmotic dehydration in broccoli, reducing processing time from 2 h to 30–40 min. The increased dehydration rate was due to acoustic cavitation, which enhances degassing, promotes micro-streaming, and creates microscopic pores in cell membranes. These effects improve mass transfer and moisture loss, enhancing the overall dehydration process. The application of ultrasound increased moisture and mass transfer by breaking down cell structures, improving the material's permeability (Xin et al. 2014). This structural disruption facilitates improved dehydration efficiency, making the process faster and more effective. Similarly, Ansorena et al. (2011) adjusted the collective use of ultrasound with mild heating (50 °C for 3 min) on broccoli, followed by 7.5 min of ultrasonic processing. This approach preserved the green color, enhanced nutraceutical value, maintained microbial safety, and extended shelf life. Such combined treatments improve quality and nutritional retention in minimally processed broccoli.

2.5.7.3 High Hydrostatic Pressure (HHP)

HHP is becoming a popular method in the preservation of food due to its capacity to maintain nutritional and functional properties. Operating between 200–800 MPa, HHP ensures consistent results across different product sizes, reduces the need for chemical preservatives, and is eco-friendly. It is especially effective for inactivating enzymes and microbes in fresh foods, making it a valuable tool for preserving organic and preservative-free products (Zhao et al. 2019). HHP extends product shelf life while preserving freshness, stability, and nutrient content. This technology is widely used for both animal-origin products and plant-based products (Komora et al. 2018). It ensures microbial and enzymatic inactivation without relying on chemical additives, making it an effective preservation method while maintaining functional quality. Fernandes et al. (2017) investigated the effects of HHP on EFs, revealing significant variation in pressure tolerance based on flower type. Pansies demonstrated better tolerance than camellia, borage, and *Centaurea*. Despite *Centaurea*'s pressure resistance, pansies exhibited a longer shelf life, attributed to their superhydrophobic papillae. While low-pressure treatments were insufficient to immediately reduce microbial loads in pansies, microbial counts diminished by storage's end. The study endorses HHP for pansies due to their resilience and suitability for preservation. In another study, Westphal et al. (2017) revealed that HHP treatment on broccoli buds at 100–600 MPa for 3 min at 30 °C led to decreased glucosinolate levels, browning, and physical degradation. For EFs, fewer treatments are suggested to prevent damage, while moderate to intense pressures can still be effective for processed products. The Juicy Line-Fruity Line® in Holland uses HHP technology to create smoothies from EFs, including cauliflower and broccoli. This approach helps preserve the nutritional qualities and freshness of the flowers while maintaining the quality of the final product (Fernandes et al. 2017). HHP in mild treatments can preserve the sensory qualities of EFs but may not fully inactivate microbes and enzymes, risking spoilage. Combining HHP with bioactive compounds enhances antimicrobial effects. For example, cauliflower by-products showed quicker reductions of *Salmonella enterica*. This hybrid approach reduces treatment intensity while preserving freshness—under HHP (Sanz-Puig et al. 2016).

2.5.7.4 Pulsed Electric Field Technology (PEF)

PEF is a non-thermal food-preserving technology using electric fields to disrupt microbial cell membranes, causing cell death. It operates over short treatment times and its effectiveness depends on parameters like field strength, frequency, and pulse energy. Aguiló-Aguayo et al. (2014) experimented with the PEF on broccoli inflorescences, focusing on the strength of the electric field and its duration. They found PEF to be effective for microbial inactivation while maintaining glucosinolate content, thereby preserving product quality. Glucosinolates are important phytochemicals in broccoli that contribute to its health benefits. Aguiló-Aguayo et al. (2014) demonstrated that PEF enhanced the recovery of glucosinolate from broccoli during extraction. This increase was linked to elevated activity of the myrosinase enzyme, which hydrolyzes glucosinolates into bioactive compounds like isothiocyanates with health-promoting potential. Sánchez-Vega et al. (2014) found that PEF increased lutein and beta-carotene levels in broccoli by stimulating biosynthesis and reducing oxidative enzymes. However, at 35 kV/cm and 500–2000 μs, vitamin degradation occurred, likely due to heavy metals accelerating enzymatic breakdown, emphasizing the need for precise PEF optimization. At 25 kV/cm, PEF treatment preserved up to 96.1% of the phenolic content of broccoli (Sanz-Puig et al. 2016). Research on PEF's impact on EFs is limited, but its antimicrobial potential is highlighted in studies combining PEF with cauliflower by-products against *S. typhimurium*. Further investigation is needed to understand bacterial resistance and enzyme activity. Scaling up PEF for commercial use and assessing its role as a pre-treating for EFs are also key research areas.

2.6 Challenges Encountered

The EF industry must adopt post-harvest technologies tailored to its products to enhance shelf life and stability. Current research largely focuses on broccoli and cauliflower, underscoring the need for global studies on other EFs with varying preservation needs. Safety concerns include intrinsic toxic compounds, pathogenic microorganisms like *Salmonella*, and chemical residues. Non-thermal technologies, strict hygiene practices, accurate species identification, and stronger regulatory frameworks are crucial in addressing these issues. In addition, continued innovation is necessary to ensure quality and safety in this emerging industry. Lu et al. (2016) and Matyjaszczyk and Śmiechowska (2019) identified safety concerns for EFs, divided into intrinsic (toxic plant compounds) and extrinsic factors (e.g., *Salmonella*, insecticides), these concerns emphasize the need for proper handling and regulation to ensure EFs safety and quality. Fernandes et al. (2017) explained foodborne risks from EFs highlight the need for non-thermal processing, good manufacturing practices, and proper hygiene to reduce contamination. Accurate identification of flower species is crucial to avoid toxicological issues and ensure safety. Additionally, enhancing the regulatory framework can help mitigate the risks associated with detrimental chemicals, like insecticides or pesticides, commonly in use in agricultural operations. Strengthening these regulations would ensure the safety of EFs and minimize potential health hazards.

2.7 Conclusion

This book chapter outlines the need for optimized processing methods for EFs, emphasizing the use of hybrid techniques to preserve their fresh attributes. Key research directions include understanding species-specific effects, cost-effectiveness, scalability, and sustainability. It also highlights the potential benefits of combining processing methods for improved quality and efficiency while investigating the impact of such approaches on the nutraceutical and therapeutic properties of EFs. These developments are crucial as the market for EFs continues to expand and gain consumer acceptance.

Conflict of Interests The authors declare that they have no conflict of interest.

References

Aboltins A, Kic P (2016) Research in some medical plant drying process. In: 15th international scientific conference engineering for rural development. Latvia University of Agriculture, Jelgava, pp 1145–1150

Aguiló-Aguayo I, Suarez M, Plaza L, Mohammad B, Brunton N, Lyng G, Rai DK (2014) Optimization of pulsed electric field pre-treatments to enhance health-promoting glucosinolates in broccoli flowers and stalk. J Sci Food Agric 95:1868–1875. https://doi.org/10.1002/jsfa.6891

Ahluwalia P, Kaur A, Dhillon GK (2014) Effect of dried marigold flower powder as a source of natural color on rheological properties of flour. Int J Food Nutri Saf 5:63–73

Alden KM, Omid M, Rajabipour A, Tajeddin B, Firouz MS (2019) Quality and shelf-life prediction of cauliflower under modified atmosphere packaging by using artificial neural networks and image processing. Comput Electron Agric 163:104861. https://doi.org/10.1016/j.compag.2019.104861

Ansorena MR, Marcovich NE, Roura SI (2011) Impact of edible coatings and mild heat shocks on quality of minimally processed broccoli (*Brassica oleracea* L.) during refrigerated storage. Postharvest Biol Technol 59:53–63. https://doi.org/10.1016/j.postharvbio.2010.08.011

Aquino-Bolaños EN, Urrutia-Hernández TA, López Del Castillo-Lozano M, Chávez-Servia JL, Verdalet-Guzmán I (2013) Physicochemical parameters and antioxidant compounds in edible squash (*Cucurbita pepo*) flower stored under controlled atmospheres. J Food Qual 36:302–308. https://doi.org/10.1111/jfq.12053

Bao Y, Qu Y, Li J, Li Y, Ren X, Maffucci KG, Li R, Wang R, Zeng R (2018) *In vitro* and *in vivo* antioxidant activities of the flowers and leaves from *Paeonia rockii* and identification of their antioxidant constituents by UHPLC-ESI-HRMSn via pre-column DPPH reaction. Molecules 23:392. https://doi.org/10.3390/molecules23020392

Barani YH, Zhang M, Mujumdar AS, Chang L (2022) Preservation of color and nutrients in anthocyanin-rich edible flowers: Progress of new extraction and processing techniques. J Food Process Preserv 46:e16474. https://doi.org/10.1111/jfpp.16474

Ben-Fadhel Y, Saltaji S, Khlifi MA, Salmieri S, Dang VK, Lacroix M (2017) Active edible coating and γ-irradiation as cold combined treatments to assure the safety of broccoli florets (*Brassica oleracea* L.). Int J Food Microbiol 241:30–38. https://doi.org/10.1016/j.ijfoodmicro.2016.10.010

Benvenuti S, Bortolotti E, Maggini R (2016) Antioxidant power, anthocyanin content and organoleptic performance of edible flowers. Sci Hortic 199:170–177. https://doi.org/10.1016/j.scienta.2015.12.052

Cavaiuolo M, Cocetta G, Ferrante A (2013) The antioxidants changes in ornamental flowers during development and senescence. Antioxidants 2:132–155. https://doi.org/10.3390/antiox2030132

Cefola M, Amodio ML, Colelli G (2016) Extending postharvest life of ready-to-use zucchini flowers: effects of the atmosphere composition. Acta Hortic 1141:123–130. https://doi.org/10.17660/ActaHortic.2016.1141.13

Chen NH, Wei S (2017) Factors influencing consumers' attitudes towards the consumption of edible flowers. Food Qual Prefer 56(A):93–100. https://doi.org/10.1016/j.foodqual.2016.10.001

Costa ML, Civello PM, Chaves AR, Martínez GA (2005) Effect of hot-air treatments on senescence and quality parameters of harvested broccoli (*Brassica oleracea* L. var. Italica) heads. J Sci Food Agric 85:1154–1160. https://doi.org/10.1002/jsfa.2081

Costa L, Vicente AR, Civello PM, Chaves AR, Martínez GA (2006) UV-C treatment delays postharvest senescence in broccoli florets. Postharvest Biol Technol 39:204–210. https://doi.org/10.1016/j.postharvbio.2005.10.012

Dorozko J, Kunkulberga D, Sivicka I, Kruma Z (2019) The influence of various drying methods on the quality of edible flower petals. Foodbalt 2019:182–187. https://doi.org/10.22616/FoodBalt.2019.045

Fan K, Zhang M, Bhandari B (2019) Osmotic-ultrasound dehydration pretreatment improves moisture adsorption isotherms and water state of microwave-assisted vacuum fried purple-fleshed sweet potato slices. Food Bioprod Process 115:154–164. https://doi.org/10.1016/j.fbp.2019.03.011

Farkas J (2006) Irradiation for better foods. Trends Food Sci Technol 17:148–152. https://doi.org/10.1016/j.tifs.2005.12.003

Fernandes L, Casal S, Pereira JA, Saraiva JA, Ramalhosa E (2017) Edible flowers: A review of the nutritional, antioxidant, antimicrobial properties and effects on human health. J Food Compos Anal 60:38–50. https://doi.org/10.1016/j.jfca.2017.03.017

Fernandes L, Ramalhosa E, Pereira JA, Saraiva JA, Casal S (2018a) The unexplored potential of edible flowers lipids. Agriculture 8:146. https://doi.org/10.3390/agriculture8100146

Fernandes L, Pereira JA, Saraiva JA, Casal S, Ramalhosa E (2018b) The effect of different post-harvest treatments on the quality of borage (*Borago officinalis*) petals. Acta Sci Pol Technol Aliment 17:5–10. https://doi.org/food.actapol.net/pub/1_1_2018

Fernandes L, Casal S, Pereira JA, Saraiva JA, Ramalhosa E (2019a) An overview on the market of edible flowers. Food Rev Int 36:258–275. https://doi.org/10.1080/87559129.2019.1639727

Fernandes L, Saraiva JA, Pereira JA, Casal S, Ramalhosa E (2019b) Post-harvest technologies applied to edible flowers: A review. Food Rev Int 35:132–154. https://doi.org/10.1080/87559129.2018.1473422

Fernandes L, Casal S, Pereira JA, Saraiva JA, Ramalhosa E (2020) Freezing of edible flowers: effect on microbial and antioxidant quality during storage. J Food Sci 85:1151–1159. https://doi.org/10.1111/1750-3841.15097

Gębczyński P, Kmiecik W (2007) Effects of traditional and modified technology, in the production of frozen cauliflower, on the contents of selected antioxidative compounds. Food Chem 101:229–235. https://doi.org/10.1016/j.foodchem.2006.01.021

González-Barrio R, Periago MJ, Luna-Recio C, Garcia-Alonso FJ, Navarro-González I (2018) Chemical composition of the edible flowers, pansy (*Viola wittrockiana*) and snapdragon

(*Antirrhinum majus*) as new sources of bioactive compounds. Food Chem 252:373–380. https://doi.org/10.1016/j.foodchem.2018.01.102
Grzeszczuk M, Stefaniak A, Meller E, Wysocka G (2018) Mineral composition of some edible flowers. J Elem 23:151–162. https://doi.org/10.5601/jelem.2017.22.2.1352
Guerrero-Beltrán JA, Barbosa-Cánovas GV (2004) Advantages and limitations on processing foods by UV light. Food Sci Technol Int 10:137–148. https://doi.org/10.1177/1082013204044359
Hamdy SA, El Hefnawy HM, Azzam SM, Aboutabl EA (2018) Chemical profiling, volatile oil analysis and anticholinesterase activity of *Hydrocotyle umbellata* L. aerial parts cultivated in Egypt. S Afr J Bot 115:108–112. https://doi.org/10.1016/j.sajb.2018.01.019
Hettiarachchi MP, Balas J (2004) Effects of cold storage on post-harvest keeping quality of gloriosa (*Gloriosa superba* L.) flowering stems. Trop Agric Res Ext 7:88–94. https://doi.org/10.17660/ActaHortic.2005.683.60
Hussain N, Ishak I, Mohd Harith N, Leong Pau Kuan G (2019) Comparison of bioactive compounds and sensory evaluation on edible flowers tea infusion. Ital J Food Sci 31:264–273
Jadhav H, Annapure US, Deshmukh RR (2021) Non-thermal technologies for food processing. Front Nutr 8:1–14. https://doi.org/10.3389/fnut.2021.657090
Jadhav HB, Badwaik LS, Annapure U, Casanova F, Alaskar K (2023) A review on the journey of edible flowers from farm to consumer's plate. Appl Food Res 3:100312. https://doi.org/10.1016/j.afres.2023.100312
Kelley KM, Behe BK, Biernbaum JA, Poff KL (2001) Consumer and professional chef perceptions of three edible-flower species. Hort Sci 36:162–166. https://doi.org/10.21273/hortsci.36.1.162
Kelley K, Behe B, Biernbaum J, Poff K (2002) Consumer purchase and use of edible flowers: results of three studies. Hort Technol 12:282–287. https://doi.org/10.21273/horttech.12.2.282
Khan MKI, Cakmak H, Tavman Ş, Schutyser M, Schroën K (2014) Anti-browning and barrier properties of edible coatings prepared with electrospraying. Innov Food Sci Emerg Technol 25:9–13. https://doi.org/10.1016/j.ifset.2013.10.006
Koike ACR, Araújo MM, Costa HSF, Almeida MC, Villavicencio ALCH (2011) Tolerance of edible flowers to gamma irradiation. International Nuclear Atlantic Conference-INAC. (ISBN:978-85-99141-04-5)
Koike A, Barreira JCM, Barros L, Santos-Buelga C, Villavicencio ALCH, Ferreira ICFR (2015a) Edible flowers of *Viola tricolor* L. as a new functional food: antioxidant activity, individual phenolics and effects of gamma and electron-beam irradiation. Food Chem 179:6–14. https://doi.org/10.1016/j.foodchem.2015.01.123
Koike A, Barreira JCM, Barros L, Santos-Buelga C, Villavicencio ALCH, Ferreira ICFR (2015b) Irradiation as a novel approach to improve quality of *Tropaeolum majus* L. flowers: benefits in phenolic profiles and antioxidant activity. Innov Food Sci Emerg Technol 30:138–144. https://doi.org/10.1016/j.ifset.2015.04.009
Koike ACR, Barros L, Antoniob AL, Ferreirab ICFR, Villavicencioa ALC (2019) Antioxidant activity of *Dianthus chinensis* flowers processed by ionizing radiation. Braz J Radiat Sci 7:01–09
Koike ACR, Negrão BG, Lana JG, Araújo ES, Almeida-Muradian LB, Villavicencio ALCH (2021) Effects of electron-beam and gamma irradiation on the antioxidant compounds of edible flowers species. In: Proceedings of the INAC 2021: International Nuclear Atlantic Conference. Nuclear technology: reducing our carbon footprint and increasing quality of life, pp 1–4
Komora N, Bruschi C, Ferreira V, Maciel C, Brandão TRS, Fernandes R, Saraiva JA, Castro SM, Teixeira P (2018) The protective effect of food matrices on *Listeria* lytic bacteriophage p 100 application towards high pressure processing. Food Microbiol 76:416–425. https://doi.org/10.1016/j.fm.2018.07.002
Kou L, Turner ER, Luo Y (2012) Extending the shelf life of edible flowers with controlled release of 1-methylcyclopropene and modified atmosphere packaging. J Food Sci 77:188–193. https://doi.org/10.1111/j.1750-3841.2012.02683.x
Kumari P, Ujala, Bhargava B (2021) Phytochemicals from edible flowers: opening a new arena for healthy lifestyle. J Funct Foods 78:104375. https://doi.org/10.1016/j.jff.2021.104375

Lahir YK (2023) Non-ionizing radiations and their biochemical and biomedical impacts: a review on types of non-ionizing radiations, sources, features, applications, and their impacts. J Radiat Cancer Res 14:53–66. https://doi.org/10.4103/jrcr.jrcr_17_22

Landi M, Ruffoni B, Combournac L, Guidi L (2017) Nutraceutical value of edible flowers upon cold storage. Ital J Food Sci 30:1–18. https://doi.org/10.14674/IJFS-756

Lockowandt L, Pinela J, Roriz CL, Pereira C, Abreu RMV, Calhelha RC, Alves MJ, Barros L, Bredol M, Ferreira ICFR (2019) Chemical features and bioactivities of cornflower (*Centaurea cyanus* L.) capitula: the blue flowers and the unexplored non-edible part. Ind Crops Prod 128:496–503. https://doi.org/10.1016/j.indcrop.2018.11.059

Lu B, Li M, Yin R (2016) Phytochemical content, health benefits, and toxicology of common edible flowers: a review (2000–2015). Crit Rev Food Sci Nutr 56(Suppl 1):S130–S148. https://doi.org/10.1080/10408398.2015.1078276

Lv R, Zou M, Chantapakul T, Chen W, Muhammad AI, Zhou J, Ding T, Ye X, Liu D (2019) Effect of ultrasonication and thermal and pressure treatments, individually and combined, on inactivation of *Bacillus cereus* spores. Appl Microbiol Biotechnol 103:2329–2338. https://doi.org/10.1007/s00253-018-9559-3

Marchelak A, Owczarek A, Rutkowska M, Michel P, Kolodziejczyk-Czepas J, Nowak P, Olszewska MA (2019) New insights into antioxidant activity of *Prunus spinosa* flowers: extracts, model polyphenols and their phenolic metabolites in plasma towards multiple *in vivo*-relevant oxidants. Phytochem Lett 30:288–295. https://doi.org/10.1016/j.phytol.2019.02.011

Marchioni I, Najar B, Ruffoni B, Copetta A, Pistelli L, Pistelli L (2020) Bioactive compounds and aroma profile of some Lamiaceae edible flowers. Plants 9:691. https://doi.org/10.3390/plants9060691

Marchioni I, Dimita R, Gioè G, Pistelli L, Ruffoni B, Pistelli L, Najar B (2021) The effects of post-harvest treatments on the quality of *Agastache aurantiaca* edible flowers. Horticulturae 7:83. https://doi.org/10.3390/horticulturae7040083

Maria Leena M, Yoha KS, Moses JA, Anandharamakrishnan C (2020) Edible coating with resveratrol loaded electrospun zein nanofibers with enhanced bioaccessibility. Food Biosci 36:100669. https://doi.org/10.1016/j.fbio.2020.100669

Matyjaszczyk E, Śmiechowska M (2019) Edible flowers: benefits and risks pertaining to their consumption. Trends Food Sci Technol 91:670–674. https://doi.org/10.1016/j.tifs.2019.07.017

Menegaes JF, Nunes UR, Bellé RA, Backes FAAL (2019) Post harvesting of cut flowers and ornamental plants. Sci Agr Paran 18:313–323. https://doi.org/10.18188/sap.v18i4.21261

Mlcek J, Rop O (2011) Fresh edible flowers of ornamental plants—a new source of nutraceutical foods. Trends Food Sci Technol 22:561–569. https://doi.org/10.1016/j.tifs.2011.04.006

Młynarczyk K, Walkowiak-Tomczak D, Łysiak GP (2018) Bioactive properties of *Sambucus nigra* L. as a functional ingredient for food and pharmaceutical industry. J Funct Foods 40:377–390. https://doi.org/10.1016/j.jff.2017.11.025

Morales-Luna E, Pérez-Ramírez IF, Salgado LM, Castaño-Tostado E, Gómez Aldapa CA, Reynoso-Camacho R (2019) The main beneficial effect of roselle (*Hibiscus sabdariffa*) on obesity is not only related to its anthocyanin content. J Sci Food Agric 99:596–605. https://doi.org/10.1002/jsfa.9220

Morittu VM, Musco N, Mastellone V, Bonesi M, Britti D, Infascelli F, Lombardi P (2019) *In vitro* and *in vivo* studies of *Cucurbita pepo* L. flowers: chemical profile and bioactivity. Nat Prod Res 35(17):2905–2909. https://doi.org/10.1080/14786419.2019.1672067

Moses JA, Rajauria G, Tiwari BK (2017) Effect of ultrasound on anthocyanins. In: Ultrasound in food processing: recent advances. Wiley, Hoboken, pp 485–495

Mulík S, Ozuna C (2020) Mexican edible flowers: cultural background, traditional culinary uses, and potential health benefits. Int J Gastronomy Food Sci 21:100235. https://doi.org/10.1016/j.ijgfs.2020.100235

Netam N (2021) Edible flower cultivation: a new approach in floriculture industry. The Pharma Innovation 10(3):857–859. https://doi.org/10.22271/tpi.2021.v10.i3l.5896

Pereira A, Cruz R, Gadelha T, Silva Á, Costa F, Souto Ribeiro W (2020) Edible flowers: beauty, health, and nutrition. Res Soc Dev 9(7):e336972994. https://doi.org/10.33448/rsd-v9i7.2994

Pires EO Jr, Di Gioia F, Rouphael Y, Ferreira ICFR, Caleja C, Barros L, Petropoulos SA (2021) The compositional aspects of edible flowers as an emerging horticultural product. Molecules 26(22):6940. https://doi.org/10.3390/molecules26226940

Pires TCSP, Dias MI, Barros L, Calhelha RC, Alves MJ, Oliveira MBPP, Santos-Buelga C, Ferreira ICFR (2018) Edible flowers as sources of phenolic compounds with bioactive potential. Food Res Int 105:580–588. https://doi.org/10.1016/j.foodres.2017.11.014

Ran X-L, Zhang M, Wang Y, Adhikari B (2019) Novel technologies applied for recovery and value addition of high-value compounds from plant byproducts: a review. Crit Rev Food Sci Nutr 59(3):450–461. https://doi.org/10.1080/10408398.2017.1377149

Rao V, Poonia A (2024a) Bioactive compounds, nanoparticle synthesis and potential utilization of edible flowers for development of functional dairy products: a review. J Food Sci Technol 61(6):1053–1068. https://doi.org/10.1007/s13197-023-05853-6

Rao V, Poonia A (2024b) Microbial biosynthesis of selenium nanoparticles using probiotic strain and its characterization. J Food Meas Charact 18(7):5477–5488. https://doi.org/10.1007/s11694-024-02581-z

Rodrigues H, Cielo DP, Goméz-Corona S, Silveira AAS, Marchesan TA, Galmarini MV, Richards NSPS (2017) Eating flowers? Exploring attitudes and consumers' representation of edible flowers. Food Res Int 100(2):227–234. https://doi.org/10.1016/j.foodres.2017.08.018

Rop O, Mlcek J, Jurikova T, Neugebauerova J, Vabkova J (2012) Edible flowers – a new promising source of mineral elements in human nutrition. Molecules 17:6672–6683. https://doi.org/10.3390/molecules17066672

Sánchez-Vega R, Elez-Martínez P, Martín-Belloso O (2014) Influence of high-intensity pulsed electric field processing parameters on antioxidant compounds of broccoli juice. Innov Food Sci Emer Technol 29:70–77. https://doi.org/10.1016/j.ifset.2014.12.002

Sanz-Puig M, Santos-Carvalho L, Cunha LM, Pina-Pérez MC, Rodrigo D, Martínez-López A (2016) Antimicrobial capacity of a cauliflower by-product infusion combined with the PEF treatment against *S. Typhimurium*. In: Jarm T, Kramar P (eds) Pulsed electric fields in biology, medicine and food & environmental technologies. Proceedings, IFMBE53. Springer, Singapore. https://doi.org/10.1007/978-981-287-817-5_29

Shanthamma S, Evanjalin V, Roji W, Shubham N, Jeyan M, Anandharamakrishnan C (2021) Emerging techniques for the processing and preservation of edible flowers. Future Food 4:100094. https://doi.org/10.1016/j.fufo.2021.100094

Simoni NK, Santos FF, Andrade TA, Villavicencio ALC, Pinto-e Silva MEM (2018) The use of edible flowers in human food: sensory analysis of preparations. Int J Food Eng 4(2):140–143. https://doi.org/10.18178/ijfe.4.2.140-143

Siriamornpun S, Phuseerit O, Meeso N (2012) Changes in colour, antioxidant activities and carotenoids (lycopene, β-carotene, lutein) of marigold flower (*Tagetes erecta* L.) resulting from different drying processes. J Funct Foods 4:757–766. https://doi.org/10.1016/j.jff.2012.05.002//

Sotelo A, López-García S, Basurto-Peña F (2007) Content of nutrient and antinutrient in edible flowers of wild plants in Mexico. Plant Foods Hum Nutr 62:133–138. https://doi.org/10.1007/s11130-007-0053-9

Souza BWS, Cerqueira MA, Teixeira JA, Vicente AA (2010) The use of electric fields for edible coatings and films development and production: a review. Food Eng Rev 2:244–255. https://doi.org/10.1007/s12393-010-9029-x

Takahashi JA, Rezende FAGG, Moura MAF, Dominguete LCB, Sande D (2020) Edible flowers: Bioactive profile and its potential to be used in food development. Food Res Int 129:108868. https://doi.org/10.1016/j.foodres.2019.108868

Villavicencio ALCH, Heleno SA, Calhelha RC, Santos-Buelga C, Barros L, Ferreira ICFR (2018) The influence of electron beam radiation in the nutritional value, chemical composition and bioactivities of edible flowers of *Bauhinia variegata* L. var. *candida alba* Buch.-Ham from Brazil. Food Chem 241:163–170. https://doi.org/10.1016/j.foodchem.2017.08.093

Vimala Bharathi SK, Maria Leena M, Moses JA, Anandharamakrishnan C (2020) Zein-based antibrowning cling wraps for fresh-cut apple slices. Int J Food Sci Technol 55:1238–1245

Waghmare RB, Annapure US (2015) Integrated effect of sodium hypochlorite and modified atmosphere packaging on quality and shelf life of fresh-cut cilantro. Food Packag Shelf Life 3:62–69

Wang L, Wen M, Chen F, Luo Z, Yin J, Chen Y, Huang H (2020) High oxygen atmospheric packaging (HOAP) reduces H_2O_2 production by regulating the accumulation of oxidative stress-related proteins in Chinese flowering cabbage. Postharvest Biol Technol 165:111183. https://doi.org/10.1016/j.postharvbio.2020.111183

Watada AE, Qi L (1999) Quality of fresh-cut produce. Postharvest Biol Technol 15(3):201–205. https://doi.org/10.1016/S0925-5214(98)00085-4

Westphal A, Riedl KM, Cooperstone JL, Kamat S, Balasubramaniam VM, Schwartz SJ, Böhm V (2017) High-pressure processing of broccoli sprouts: influence on bioactivation of glucosinolates to isothiocyanates. J Agric Food Chem 65(39):8578–8585. https://doi.org/10.1021/acs.jafc.7b01380

Xin Y, Zhang M, Adhikari B (2013) Effect of trehalose and ultrasound-assisted osmotic dehydration on the state of water and glass transition temperature of broccoli (*Brassica oleracea* L. var. *botrytis* L.). J Food Eng 119(3):640–647. https://doi.org/10.1016/j.jfoodeng.2013.06.035

Xin Y, Zhang M, Adhikari B (2014) The effects of ultrasound-assisted freezing on the freezing time and quality of broccoli (*Brassica oleracea* L. var. *botrytis* L.) during immersion freezing. Int J Refrig 41:82–91. https://doi.org/10.1016/j.ijrefrig.2013.12.016

Yang H, Shin Y (2017) Antioxidant compounds and activities of edible roses (*Rosa hybrida* spp.) from different cultivars grown in Korea. Appl Biol Chem 60(2):129–136. https://doi.org/10.1007/s13765-017-0261-4

Yang L, Wen KS, Ruan X, Zhao YX, Wei F, Wang Q (2018) Response of plant secondary metabolites to environmental factors. Molecules 23:762. https://doi.org/10.3390/molecules23040762

Zevallos-Concha A, Nuñez D, Gasco M, Vasquez C, Quispe M, Gonzales GF (2016) Effect of gamma irradiation on phenol content, antioxidant activity and biological activity of black maca and red maca extracts (*Lepidium meyenii* Walp). Toxicol Mech Methods 26:67–73. https://doi.org/10.3109/15376516.2015.1090512

Zhang N, He Z, He S, Jing P (2019) Insights into the importance of dietary chrysanthemum flower (*Chrysanthemum morifolium* cv. Hangju)-wolfberry (*Lyciumbarbarum* fruit) combination in antioxidant and anti-inflammatory properties. Food Res Int 116:810–818. https://doi.org/10.1016/j.foodres.2018.09.015

Zhao L, Fan H, Zhang M, Chitrakar B, Bhandari B, Wang B (2019) Edible flowers: Review of flower processing and extraction of bioactive compounds by novel technologies. Food Res Int 126:108660. https://doi.org/10.1016/j.foodres.2019.108660

Zheng M, Xia Q, Lu S (2015) Study on drying methods and their influences on effective components of loquat flower tea. LWT-Food Sci Technol 63(1):14–20. https://doi.org/10.1016/j.lwt.2015.03.090

Zhu Z, Zhang P, Sun DW (2020) Effects of multi-frequency ultrasound on freezing rates and quality attributes of potatoes. Ultrason Sonochem 60:104733. https://doi.org/10.1016/j.ultsonch.2019.104733

Chapter 3
Exploring the Nutrients, Antioxidants, and Bioactive Compounds in Edible Flowers

Diana Ramírez-Gamboa

3.1 Introduction

Natural herbs and flowers have served an important purpose in human culture and traditions throughout history (Rodrigues and Spence 2023). While some species are toxic and can cause harmful reactions upon contact, others have been highly valued for their medicinal properties (Guiné et al. 2021; Navarro-González et al. 2015). Across cultures, certain flowers, classified as edible flowers, have been utilized not only for their healing properties but also for their aesthetic appeal, flavor, and ability to elevate culinary creations (Motti et al. 2022; Newman and Kirker 2016).

The tradition of using edible flowers traces back to ancient times, with their consumption persisting as a widespread practice worldwide in rural and more traditional settings (Dankhade et al. 2024; Valencia-Cordova et al. 2024). In America, China, Europe, and the Middle East they continue to uphold these traditions as an integral part of their cultural heritage (Newman and Kirker 2016; Rodrigues and Spence 2023). Flowers have historically contributed to the development of medicinal practices, and many species continue to be employed in modern treatments (Liu et al. 2023b). Compounds extracted from flowers and plants are used to prevent post-operative infections (Abirami et al. 2021), alleviate symptoms of atopic dermatitis (Meysami et al. 2021), treat constipation (Jabri et al. 2017), support the regulation of blood sugar in diabetes control (Nguyen et al. 2024; Venkatesh et al. 2008), they are even used for slowing skin aging (Bujak et al. 2022).

In culinary traditions, edible flowers have been prized not only for their health benefits but also for their ability to make dishes more appetizing. Historical recipes

D. Ramírez-Gamboa (✉)
Tecnologico de Monterrey, Institute of Advanced Materials for Sustainable Manufacturing, Monterrey, Mexico
e-mail: diana.ramirez.gamboa@tec.mx

A. Poonia et al. (eds.), *Edible Flowers: Source of Phytonutrients, Valorization and Technological Advancements*, https://doi.org/10.1007/978-3-032-00667-7_3

from Europe, particularly in France and Rome, uncover the use of edible flowers in salads, drinks, wines, purées, omelets, desserts, and more. Similarly, in China, flowers have been used to enhance both the flavor and visual appeal of meals (Guiné et al. 2020; Mlcek and Rop 2011; Newman and Kirker 2016). These traditions persist today, upheld by individuals who value cultural heritage (Khomdram et al. 2019) and by chefs who incorporate edible flowers into innovative culinary creations (Guiné et al. 2020).

Motti et al. (2022) provide a comprehensive review of edible flower traditions in Mediterranean countries, detailing their diverse applications, such as sweeteners, teas, juices, liqueurs, seasonings, snacks, salads, and vegetables. Similar practices are evident in other regions, such as Mexico, where UNESCO-recognized cuisine (UNESCO 2010) highlights dishes like quesadillas with *Cucurbita pepo* (squash blossoms) and beverages such as *agua de Jamaica* made from *Hibiscus sabdariffa* (Mulík and Ozuna 2020). These examples showcase the integration of edible flowers into daily life and culinary heritage.

Modern scientific advancements have further increased interest in edible flowers, particularly in understanding their composition and health benefits. Scientific findings indicate that edible flowers contain phenolic and flavonoid compounds with antioxidant properties, that could potentially prevent degenerative diseases such as diabetes and hypertension. Additionally, these compounds have demonstrated anticancer, anti-inflammatory, and antimicrobial effects, among other benefits (Prabawati et al. 2021; Rivas-García et al. 2021). Despite these findings, a significant number of edible flowers remain underexplored as potential sources of bioactive compounds (Valencia-Cordova et al. 2024).

In today's global context, dietary patterns in urbanized cities increasingly rely on processed and ultra-processed foods, which often lack nutritional value. Financial constraints, time limitations, and modern trends contribute to this reliance (Zagorsky and Smith 2017). For instance, in Korea, the proportion of ultra-processed foods in the overall energy intake grew from 23.1% during 2010–2012 to 26.1% in the next 6 years (Shim et al. 2021). Similarly, in China, processed and ultra-processed food consumption rose significantly between 2002 from 2.64% and 0.88% to 8.87% and 8.52% respectively in 2022 (Li et al. 2024). These dietary shifts have been linked to malnutrition and are considered to impact the increase in chronic illnesses, including obesity, cardiovascular disease, osteoporosis, and cancer (Balhara 2022; Begum et al. 2023; Fang et al. 2024; Pagliai et al. 2021; Vitale et al. 2024). Edible flowers, with their nutrient-dense composition and bioactive properties, present a promising and affordable alternative to combat these trends (Jadhav et al. 2023), however, this current trend seems to be only reaching wealthy and/or educated populations as a source of status.

This chapter delves into the nutrient composition and biological effects of edible flowers, focusing on their ability to promote human health and well-being. Reviews the current trends in edible flower use within modern cuisine and discusses culinary techniques that preserve their beneficial properties, ensuring that their potential is fully reached. By examining their nutritional composition and health benefits, this chapter highlights how edible flowers can contribute to contemporary approaches to nutrition and disease prevention, offering insights into their transformative potential for modern wellness practices.

3.2 Understanding the Nutritional Composition and Bioactive Compounds of Edible Flowers

Commonly edible flowers are grouped as ornamental and fruit flowers, both of which offer a wide range of health and wellbeing benefits as well as an innovative food source (Motti 2022). These flowers are nutritionally diverse, containing macronutrients (carbohydrates, proteins, fats) and micronutrients (vitamins, minerals, and phenolic compounds). These compounds make the diverse types of flowers a well-rounded food source making them an extremely versatile product with so many advantages when they are consumed (Benvenuti and Mazzoncini 2021). Edible flowers have a high content of natural antioxidants such as anthocyanin, flavonoids and various other phenolic components (Rao and Poonia 2024a).

Despite these advantages, there is a significantly big area of opportunity in understanding the complete nutritional composition of most edible flowers. Most studies focus on specific extracts or secondary metabolites of interest, rather than evaluating the entire flower. This narrow scope limits their comparison with integral food sources like vegetables and fruits, to make them more well known among the population. To fully unlock their potential as a dietary component, comprehensive studies on the macronutrient and micronutrient profiles of edible flowers are required (Zhang et al. 2023).

The following sections explore the research efforts to understand the nutritional makeup of edible flowers, examining their macronutrient and micronutrient profiles, bioactive secondary metabolites, and a comparison of their nutrient density with some well-known vegetables and fruits.

3.2.1 Macronutrient and Micronutrient Profile of Edible Flowers

In addition to essential macronutrients such as carbohydrates, proteins, and fats, edible flowers also supply important micronutrients, including vitamins and minerals. These nutrients contribute to their potential as a functional food ingredient. Vitamins, particularly vitamin C, and minerals like potassium and calcium are commonly found in these flowers, while their protein and fat content are relatively modest compared to traditional food sources. Due to these factors providing necessary nutrients while also contributing to their flavor profile, is a great way to impact consumer acceptance (Zhang et al. 2023). Table 3.1 presents a comparative analysis

Table 3.1 Comparative nutritional composition of selected edible flowers edible flowers, fruits, and vegetables

Flowers, fruit and vegetables	Moisture (%)	Dry Mater (%)	Total Ash (% DM)	Protein (% DM)	Carbo-hydrates (% DM)	Fiber (% DM)	Fat (% DM)	Vitamin C (mg/100 g FW)	Potassium (mg/kg DM)	Calcium (mg/kg DM)	References
Calendula officinalis L. 'Marigold'	89.34 ± 0.10[g]	10.66 ± 0.10[g]	6.93 ± 0.14[b]	6.43 ± 0.68[b]	81.32 ± 0.75[b]	1.59 ± 0.10[g]	5.33 ± 0.45[b]	40.00 ± 0.00[c]	3.90 ± 0.004[c]	N/A	De Lima Franzen et al. (2019), Ninama et al. (2024), Tânia C. S. P. Pires et al. (2017)
Centaurea cyanus L. 'Cornflower'	88.11 ± 0.17[h]	11.89 ± 0.17[h]	5.68 ± 0.13[b]	5.79 ± 0.1[b]	88.39 ± 0.13[b]	5.35 ± 0.05[h]	0.14 ± 0.00[b]	154.40 ± 4.16[h]	N/A	N/A	Król et al. (2024) Tânia C. S. P. Pires et al. (2017)
Cosmos bipinnatus Cav. 'Cosmos'	N/A	N/A	N/A	N/A	N/A	N/A	N/A	48.00 ± 4.47[c]	2.85 ± 0.004[c]	N/A	Ninama et al. (2024)
Cucurbita moschata Duchesne 'Pumpkin flower'	92.05 ± 0.45[e]	7.95 ± 0.45[e]	N/A	14.47 ± 1.14[e]	N/A	N/A	N/A	0.85 ± 0.05[e]	31,540 ± 760[e]	47,510 ± 2200[e]	Marchioni et al. (2024)
Cytisus striatus[a] 'Portuguese broom'	88.25 ± 1.31	11.75 ± 1.31	4.47 ± 0.67	20.56 ± 1.61	54.74 ± 3.63	18.29 ± 1.46	1.95 ± 0.32	N/A	20,094.17 ± 5076.80	4883.33 ± 0,580.42	Caramelo et al. (2024)

Flowers, fruit and vegetables	Moisture (%)	Dry Mater (%)	Total Ash (% DM)	Protein (% DM)	Carbo-hydrates (% DM)	Fiber (% DM)	Fat (% DM)	Vitamin C (mg/100 g FW)	Potassium (mg/kg DM)	Calcium (mg/kg DM)	References
Dahlia mignon 'Dahlia flower'	N/A	N/A	5.83 ± 0.04^b	5.93 ± 0.2^b	86.02 ± 0.2^b	N/A	2.23 ± 0.05^b	N/A	N/A	N/A	Tânia C. S. P. Pires et al. (2017)
Dianthus chinensis L. *'China pink'*	79.77 ± 1.94^e	20.23 ± 1.94^e	6.23 ± 1.66^f	9.69 ± 0.00^e	N/A	8.13 ± 0.72^f	N/A	24.70 ± 1.34^e	$30{,}910 \pm 450^e$	$46{,}630 \pm 2130^e$	Marchioni et al. (2024), Stefaniak and Grzeszczuk (2018)
Fuchsia regia 'fuchsia flower'	83.87 ± 6.30^e	16.13 ± 6.30^e	N/A	6.09 ± 0.27^e	N/A	N/A	N/A	8.90 ± 0.40^e	$26{,}110 \pm 540^e$	$90{,}540 \pm 1650^e$	Marchioni et al. (2024)
Pelargonium hortorum Bailey 'geranium'	N/A	N/A	N/A	N/A	N/A	N/A	N/A	42.00 ± 4.47^c	3.72 ± 0.329^c	N/A	Ninama et al. (2024)
Petunia hybrida 'petunia'	N/A	N/A	N/A	N/A	N/A	N/A	N/A	28.00 ± 4.47^c	4.48 ± 0.005^c	N/A	Ninama et al. (2024)
Rosa	84.56 ± 0.122^g	15.44 ± 0.122^g	4.29 ± 0.1^b	7.58 ± 0.84^b	86.12 ± 0.8^b	3.20 ± 0.095^g	2.01 ± 0.04^b	N/A	N/A	N/A	De Lima Franzen et al. (2019), Tânia C. S. P. Pires et al. (2017)
Viola cornuta L. 'horned pansy'	82.99 ± 1.25^e	17.01 ± 1.25^e	N/A	12.94 ± 0.01^e	N/A	N/A	N/A	42.55 ± 1.69^e	$26{,}490 \pm 730^e$	$29{,}700 \pm 1020^e$	Marchioni et al. (2024)
Viola wittrockiana 'pansy'	91.3 ± 2.1^d	8.7 ± 2.1^d	0.92 ± 0.08^d	23.34 ± 0.69^d	44.82 ± 21.83^d	17.24 ± 1.15^d	5.17 ± 0.11^d	32 ± 4.47^c	2.24 ± 0.000^c	N/A	Fernandes et al. (2019), Ninama et al. (2024)

(continued)

Table 3.1 (continued)

Flowers, fruit and vegetables	Moisture (%)	Dry Mater (%)	Total Ash (% DM)	Protein (% DM)	Carbo-hydrates (% DM)	Fiber (% DM)	Fat (% DM)	Vitamin C (mg/100 g FW)	Potassium (mg/kg DM)	Calcium (mg/kg DM)	References
Brassica oleracea L. 'Broccoli'[i]	88.20	11.80	N/A	36.44	27.12	33.90	5.08	79.00	33,644.07	4067.80	Public Health England (2015)
Malus domestica 'Apple'[i]	86.2	13.8	N/A	4.35	84.06	8.70	3.62	6	7246.38	362.32	
Citrus sinensis 'Orange'[i]	87	13	N/A	6.15	63.08	9.23	1.54	52	1220.00	240.00	
Lactuca sativa 'Lettuce'[i]	96.1	3.9	N/A	30.77	35.90	38.46	2.56	1	2220.00	240.00	
Asparagus officinalis 'Asparagus'[i]	91.4	8.6	N/A	33.72	23.26	N/A	6.98	12	2600.00	270.00	

DM Dry Matter, *FW* Fresh Weight, *N/A* Not available

[a]Averages for each flower were taken from Caramelo et al. (2024), while the standard deviations were recalculated based on reported means across locations to provide a generalized variability

[b]Data retrieved from the results of Pires et al. (2017)

[c]Data retrieved from the results of Ninama et al. (2024)

[d]Data retrieved and converted to DM based on the results of the moisture content from the results of white completely open flowers by Fernandes et al. (2019)

[e]Data retrieved and converted to % based on the results by Marchioni et al. (2024)

[f]Data retrieved and converted to DM based on the results of the moisture content from the results by Stefaniak and Grzeszczuk (2018)

[g]Data retrieved from the results of De Lima Franzen et al. (2019)

[h]Data retrieved from the results of Król et al. (2024)

[i]Data retrieved and converted to DM based on the results of the moisture content from the results by Public Health England (2015)

of the nutritional composition of various edible flowers against common vegetables and fruits. This comparison highlights the unique properties of edible flowers, offering insights into their suitability as a functional food component. Data were sourced from research articles that conducted nutritional determinations of various flowers, fruits, and vegetables. To ensure comparability, all data were converted to consistent units where necessary.

The analysis reveals that edible flowers, like their fruits and vegetables contra parts, exhibit high moisture content, often exceeding 80%. For example, pansies in total bloom and pumpkin flowers have moisture levels of 91.3% (Fernandes et al. 2019) and 92.05% (Marchioni et al. 2024), respectively, comparable to vegetables like lettuce and asparagus, which exceed 90% (Public Health England 2015).

Edible flowers exhibit varied macronutrient content, including proteins, carbohydrates, fiber, and fats. The protein content of most edible flowers is relatively low but stays in the same range as the fruits. Flowers such as *Portuguese broom*, pansy, and broccoli (considered an edible flower in this comparison) show higher protein levels of 20.5% (Caramelo et al. 2024), 23.34% (Fernandes et al. 2019), and 36.44% (Public Health England 2015) of dry matter, respectively. These values are close to those of lettuce and asparagus, which contain 30.77% and 33.72% (Public Health England 2015) of dry matter.

Carbohydrate levels in edible flowers vary widely. Flowers such as marigolds, cornflowers, dahlia, and roses exhibit carbohydrate content that exceeds 80% of dry matter, similar to fruits like apples and oranges. This may be attributed to their high sugar content, making them suitable for sweet culinary applications, as is the case with many fruits. On the contrary, flowers with lower carbohydrate content may be best used as an alternative to vegetables in dishes like salads, soups, and other savory dishes.

Most edible flowers contain less than 10% of fiber by dry matter, except for Portuguese broom (18.29%) (Caramelo et al. 2024), pansy (17.24%) (Fernandes et al. 2019), and broccoli. Broccoli stands out with a fiber content of 33.9%, only surpassed by lettuce at 38.46% (Public Health England 2015).

Fat content across edible flowers, fruits, and vegetables remains below 7% of dry matter. The highest fat content is found in asparagus (6.98%)(Public Health England 2015), followed by marigold (5.33%) (Pires et al. 2017) and pansy (5.17%) (Fernandes et al. 2019). In contrast, cornflower (0.14%) (Pires et al. 2017), orange (1.54%)(Public Health England 2015), and *Portuguese broom* (1.95%) (Caramelo et al. 2024) exhibit the lowest fat levels.

Edible flowers are also rich in essential micronutrients like vitamin C, potassium, and calcium. Edible flowers are excellent sources of vitamin C, crucial for daily intake, usually advised at 90 mg daily for men and 75 mg daily for women (Food and Nutrition Board Institute of Medicine 2000). Cornflower, broccoli, and orange demonstrate the highest vitamin C levels at 154.4 (Król et al. 2024), 79, and 59 mg/100 g of fresh weight (Public Health England 2015), respectively. Most edible flowers contain 40–50 mg of vitamin C per 100 g, comparable to oranges, which are commonly recognized as a dietary source of this nutrient.

The recommended daily intake of potassium typically falls between 2,000 and 3,000 mg per day, depending on the individual's age and gender (Stallings et al. 2019). Broccoli flowers and pumpkin flowers provide 33,644.07 mg/kg (Public Health England 2015) and 30,910 mg/kg (Marchioni et al. 2024) of dry matter in potassium, respectively, showcasing their richness in this mineral.

Daily calcium intake recommendations range from 1000 to 1300 mg, depending on the life stage (Institute of Medicine (US) 2011). Fuchsia flowers have the highest calcium content at 90,540 mg/kg of dry matter, followed by pumpkin flowers at 47.510 mg/kg (Marchioni et al. 2024).

Overall, Table 3.1 underscores the potential of edible flowers as nutrient-rich alternatives to conventional fruits and vegetables. Although this analysis represents a fraction of the diverse edible flora, it demonstrates the substantial nutritional value and cultural significance of edible flowers. Their incorporation into diets can provide a well-rounded source of essential nutrients, promoting a healthy lifestyle.

3.2.2 Secondary Metabolites in Edible Flowers

Edible flowers are abundant in secondary metabolites, also known as phytochemicals that, although not necessary for the plant's growth, are vital for its survival and play significant roles in various biological processes (Singh et al. 2023). These compounds, synthesized exclusively by plants, contribute to the plant's taste, aroma, and color. Notably, edible flowers are particularly rich in these bioactive compounds, which enhance their unique characteristics (Kandylis 2022). In addition to their nutrient composition, these secondary metabolites are integral to the plant's overall functionality and appeal.

The most common bioactive compounds contained in edible flowers are a variety of polyphenols, plant pigments like carotenoids, vitamin E compounds among other tocopherols, and aromatic terpenoids. The classification of these compounds is illustrated in Fig. 3.1.

Phytochemicals in plants, including edible flowers, are widely recognized for their bioactive properties, with antioxidant activity being one of the most significant. By neutralizing free radicals, these compounds can help reduce oxidative stress, a factor linked to the development of chronic diseases (Lu et al. 2016). Exploring the antioxidant activity of these compounds further highlights their potential to contribute to human health.

3.3 Harnessing the Antioxidant Qualities of Edible Flowers

Different foods and supplements have antioxidant properties that offer significant benefits to human health, especially by reducing the possibility of cellular damage that could be contributing to the progression of diseases, such as cardiovascular

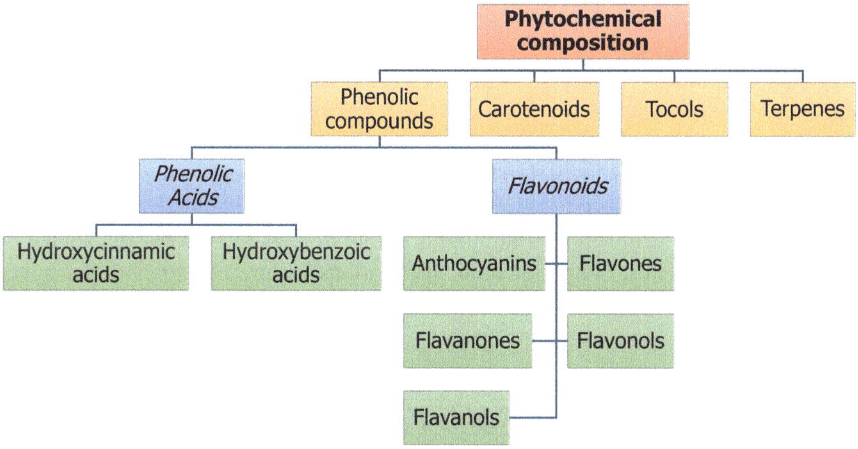

Fig. 3.1 Classification of the phytochemical composition of edible flowers. Graph created based on data from Eleomar de O. Pires et al. (2021) and Kandylis (2022)

disorders, neurodegenerative disorders, or even cancer. Below, is a discussion of the implications that oxidative stress has in the cells and how the antioxidant molecules protect the damaged against the cells by this means. This section also examines the phytochemicals found in edible flowers and their potential health benefits.

3.3.1 Mechanisms of Action: How Antioxidants Protect Human Cells

A well-balanced, antioxidant-rich food is largely considered responsible for enhancing health and mitigating the risk of chronic illnesses. But why are antioxidants so vital? The key lies in their ability to counteract free radicals and reactive oxygen species (ROS) in human cells (Lobo et al. 2010).

Free radicals are reactive molecules characterized by having one or more unpaired electrons, making them highly reactive with cellular molecules. This reactivity can disrupt cellular functions and lead to damage (Kurutas 2016). Free radicals play essential roles in low concentrations, this is why they are naturally produced in aerobic organisms as byproducts of normal biological processes. Their essential roles are in signaling pathways and immune defense (Pizzino et al. 2017). Nevertheless, genetic predispositions or external stressors can cause an overproduction of free radicals, surpassing the body's inherent defense mechanisms and leading to oxidative imbalance. This state can harm essential cellular structures, such as the DNA, membranes, and even lipids, or proteins that eventually play a role in the onset of chronic and degenerative conditions, including heart disease, diabetes, Alzheimer's, and even cancer, among others (Chaudhary et al. 2023; Hajam et al. 2022; Jena et al. 2023).

To counteract oxidative stress, cells rely on antioxidant enzymes that neutralize ROS and repair the damage caused by them (Forman and Zhang 2021). Enhancing this natural defense system is a key therapeutic strategy, often achieved through dietary intake of antioxidant-rich foods, particularly fruits, vegetables, and edible flowers. These foods contain phytochemicals, compounds with potent antioxidant properties. Phytochemicals neutralize free radicals by donating electrons or protons, thereby reducing ROS levels, and protecting cells from oxidative damage (Rui Hai Liu 2003). Recent research highlights edible flowers as a promising source of these phytochemicals, with potential applications in preventing and managing chronic illnesses (Fakhri et al. 2022).

3.3.2 Phytochemicals in Edible Flowers: A Natural Source of Antioxidant Defense

As mentioned before, edible flowers contain a great amount of antioxidant metabolites, some examples of this are ascorbic acid, also known as vitamin C, pigments or carotenoids, flavonoids, and phenolic compounds (Skrajda-Brdak et al. 2020). It has been demonstrated that polyphenols, from all of the compounds in edible flowers, have the greatest impact on antioxidant capacity. Consequently, the method often used to assess the antioxidant potential of edible flowers is total phenolic content, or TPC for abbreviation. Additional assays, such as 2,2-diphenyl-1-picrylhydrazyl (DPPH) and ferric-reducing antioxidant potential (FRAP), are also used to evaluate their antioxidant properties.

Studies have consistently demonstrated that many edible flowers possess significant antioxidant properties. However, the lack of standardization in reporting antioxidant metrics poses challenges in comparing results across studies and against other fruits and vegetables. Below are notable findings from various studies on the antioxidant capacity of edible flowers:

Ninama et al. (2024) analyzed six edible flowers for macronutrients, micronutrients, and phytochemicals, including TPC, anthocyanins, carotenoids, flavonoids, total reducing capacity, and antioxidant activity (using the DPPH assay). In short, *Geranium* had the highest TPC (107.96 ± 0.13 mg Catechol/g fw). *Dianthus* exhibited the highest anthocyanin content (1.23 ± 0.01 mg Cyn.H./g fw). *Cosmos* had the highest carotenoid content (18.82 ± 0.08 μg/g fw) and total reducing capacity (9.55 ± 0.32 mg GAE/g fw). Petunia showed the highest DPPH activity (98.3 ± 0.55% fw). Tannin content was substantial in both Dianthus and Pansy, with Dianthus presenting the highest value (36.96 ± 0.34 mg TAE/g fw).

Zhang et al. (2023) characterized 34 edible flowers from different genera. Their findings revealed that the family *Asteraceae* displayed the most abundant flavonoid content (8.53%). The family *Rosaceae* showcased superior antioxidant performance in the FRAP assay (106.33 to 662.2 mmol/100 g fw). For the DPPH assay, Rosaceae also displayed the highest free radical scavenging activity (87.98%), underscoring its strong antioxidant potential.

Mlcek et al. (2021) evaluated six edible flower species for their phytochemical concentrations. Notable results include: *Hemerocallis* exhibited the highest TPC (6.59 ± 0.23 g GAE/kg fw), flavonoid content (3.76 ± 0.23 g rutin/kg fw), and total antioxidant capacity (7.94 ± 0.26 g AAE/kg fw). *Tagetes patula* demonstrated the highest antioxidant capacity in the FRAP assay (5.62 ± 0.28 g GAE/kg fw).

These findings highlight the significant antioxidant potential of various edible flowers bridging the information of their capacity, emphasizing their promising role in reducing oxidative stress and contributing to human health when incorporated into the diet. While a high content of TPC and antioxidant action is essential for any biological impact, an important consideration is whether these compounds remain effective after digestion. Janarny et al. (2022) led a study examining the bioavailability of edible flowers and concluded that, although the phytochemical content significantly diminished in a simulated gastrointestinal system, the remaining compounds retained steady antioxidant capacity. This resilience reveals that edible flowers can continue to provide essential antioxidants, making them a promising addition to a health-focused diet or functional foods.

3.4 Exploring the Biological Activities of Edible Flowers: Implications for Health and Wellness

Edible flowers have been shown to possess an abundance of nutrients and health-promoting compounds like flavonoids, anthocyanins, and phenolic acids, which influence various biological functions. Frequent intake of these substances may enhance physical resilience and metabolic stability. Additionally, they have the potential to be utilized as adjuvant treatments, offering positive outcomes for various diseases and reducing the risk of illness.

Most of the research done to explore the health benefits of edible flowers focus on their extracts, which concentrate these bioactive compounds. Table 3.2 summarizes key findings from research examining the biological effects of these extracts, highlighting their diverse applications in improving human health.

The inflammation of cells is a natural defense mechanism prompted by different sources like pathogens, chemicals, or viruses that can become harmful when prolonged, potentially leading to the development of chronic diseases (Pahwa et al. 2023). Current research reports that bioactive compounds from edible flowers present anti-inflammatory and immunomodulatory effects that mitigate inflammation and, at the same time protect cells from damage. For example, 3-cinnamoyltribuloside from *C. nitidissima* Chi flowers has demonstrated potent anti-inflammatory activity in macrophage cells (Wang et al. 2020). Likewise, the extracts obtained using ethanol from *Chimonanthus praecox* flowers showed both anti-inflammatory and antioxidant effects in the same cell model described before (Kang et al. 2022). Additionally, saponins obtained from *Panax notoginseng* flowers presented an inhibition of the skin inflammation that was induced by UV radiation in keratinocyte

Table 3.2 Biological activities of extracts from edible flowers for human health

Edible flower	Extract Assessed	Model/System Used	Reported effects	Reference
Borago officinalis L.	Organic extraction of flowers	N2a cells (Mouse neuroblastoma cell line, ATCC® CCL-131™)	Neuroprotective and antioxidant effect	Moliner et al. (2022)
C. nitidissima Chi	3-cinnamoyltribuloside	RAW 264.7 macrophages stimulated with lipopolysaccharide (LPS)	Anti-inflammatory	Wang et al. (2020)
Calendula officinalis L.	Polyamines extracted	Both Gram-positive and Gram-negative bacteria, and foodborne fungal species	Antibacterial and antifungal activity	Tavallali et al. (2024)
Chimonanthus praecox (L.)	CPE: Ethanol-derived extract	LPS-stimulated RAW264.7 macrophages	Anti-oxidative and anti-inflammatory effects	Kang et al. (2022)
Hibiscus rosasinensis	Organic extraction	Wistar albino rats ofboth sex	Hypoglycemic effect	Venkatesh et al. (2008)
Hypericum perforatum	Powdered flower, solubilized, administered by gavage	Thirty six adult male Wistar rats weighing	Hypoglycemic effect	El Motteleb et al. (2017)
Nine wild edible herbaceous species	Extracts	Seven fungal pathogens	Antifungal activity	Gatto et al. (2011)
Panax notoginseng flower	Saponins	Human keratinocyte cells	Inhibiting inflammation in the skin	Liu et al. (2023a)
Plumeria rubra	Organic extraction of flowers	Gram-positive, Gram-negative strains and fungi trains	Antibacterial and antifungal activity	Ahaotu et al. (2020)
Prunus persica (L.)	Fractions of hot water extract of the flower	Immortalized pancreatic β-cells from the rat insulinoma INS-1 cell line	Antidiabetic	Lee et al. (2021)
Psidium cattleianum Sabine	Essential oil	5-lipoxygenase inhibitor screening assay kit, HepG2, MCF-7, and K562 cell lines	Anti-inflammatory and anticancer	Elsayed et al. (2023)

(continued)

Table 3.2 (continued)

Edible flower	Extract Assessed	Model/System Used	Reported effects	Reference
Rosa canina L.	Infusions and hydromethanolic extracts	Gram-positive and Gram-negative strains	Moderate antibacterial activity	Polumackanycz et al. (2020)
Saraca asoca (Roxb.)	Organic extract	Male and female rats from the Wistar albino lineage	Anti-hyperglycemic and anti-hyperlipidemic	Thilagam et al. (2021)

cells, presenting a promising natural alternative for cosmetic applications (Liu et al. 2023a).

Antimicrobial and antiviral properties of edible flowers are also a significant focus of research. With the increasing demand for eco-friendly alternatives to synthetic antimicrobial agents, compounds extracted from edible flowers have shown potential to combat bacteria and fungi. For instance, phenolic-rich infusions of *Rosa canina* L. demonstrated moderate antibacterial activity and high antioxidant properties (Polumackanycz et al. 2020). Similarly, polyamines extracted from *Calendula officinalis* L. enhanced antibacterial and antifungal activity in treated samples (Tavallali et al. 2024). These findings make it more evident the potential that edible flower extracts must improve food preservation and reduce the environmental impact of traditional preservatives.

Altogether, edible flowers represent a valuable source of bioactive compounds with diverse health-promoting properties. Evidence makes clear the abundant properties they have for their antioxidant, anti-inflammatory, and antimicrobial activities, among significantly more, that underline their potential not only as functional foods but also as a resource for developing rich phytochemical extracts to manage chronic and degenerative diseases effectively. By linking traditional knowledge with modern technology, we can unlock the vast capacities of edible flowers to advance preventative healthcare, promote sustainability, and enhance overall quality of life.

3.5 Innovative Culinary Uses of Edible Flowers: Enhancing Flavor, Aesthetic, and Nutrition

The integration of edible flowers into culinary practices is far from a modern trend. Ancient cultures incorporated them into foods and beverages, exploiting their vibrant appearance, delicate flavors, and nutritional benefits (Newman and Kirker 2016; Valencia-Cordova et al. 2024). This historical tradition has been revitalized in contemporary gastronomy, blending traditional wisdom with modern innovation to create unique dining experiences. The following subchapters explore the diverse applications of edible flowers and outline preparation methods that preserve their aesthetic and nutritional value.

Table 3.3 List of a few common edible flowers, their characteristics, and culinary uses

Flower	Plant name	Color	Taste	Recommended use	Reference
Fennel Flower	*Foeniculum vulgare*	Bright yellow	Anise and licorice-like notes	Pasta primavera, roasted chicken, and soups made with seasonal ingredients	Murray (2022)
Fava Bean Blossom	*Vicia faba*	White, pink, red, brown, or yellow	Light, nutty and earthy flavor	Garnish salads, sandwiches, meats, pasta, bowls, drinks, or stir-fries. Pair with citrus, herbs, nuts, feta, poultry, or fish.	Specialty produce (n.d.)
English Lavender	*Lavandula angustifolia*	Purple	Strong, herbal flavor	Best used in sweet preparations like cookies, cakes, teas, lemonades or making sugar	Murray (2022)
Rose	*Rosaceae* family	Red, white, pink, yellow, cream, and lavender	Naturally sweet or sour	As garnish, in deserts, jams, salads, ice cream, tea.	Hegde et al. (2022)
Chamomile	*Matricaria chamomilla* and *Chamaemelum nobile*	White with a yellow center	Sweet, and slightly herbal flavor	Tea, salads, cocktails, cakes, and as a garnish	Adamant (2024)

3.5.1 Innovative Applications and Trends in Edible Flower Gastronomy

Edible flowers have become a hallmark of modern gastronomy, celebrated for their ability to enhance dishes with vibrant colors, subtle aromas, and delicate flavors. Consumers are increasingly drawn to foods that deliver on taste while also offering nutritional value and visual appeal, contributing to their rising popularity (dos Santos and Reis 2021). As culinary practices progress, edible flowers are playing a vital role in the formulation of functional foods and dietary supplements, responding to the rising need for healthier and more eco-friendly choices (Takahashi et al. 2020).

Globalization has broadened the use of edible flowers, from garnishes and beverage ingredients to infusions in oils and syrups, and some recipes are stunning (Copeland 2024). Vegetables such as broccoli and cauliflower, technically edible flowers, demonstrate how floral elements are already integral to everyday diets. Table 3.3 lists a few common edible flowers and their uses.

Chefs worldwide incorporate edible flowers to elevate flavors and enhance the dining experience. For example, Chef Raymond Wong from the Seafire Steakhouse in Dubai emphasizes that flowers must elevate a dish to justify their inclusion, a sentiment echoed by Michelin-starred chefs Laëtitia Rouabah and Chris Malone

Fig. 3.2 Example of culinary usages of edible flowers in modern cuisine. (Photo by Michelle Henderson (**a**), Jovan Vasiljević (**b**), and Deepthi Clicks (**c**) via Unsplash)

(Tiwari 2021). Interestingly enough, research conducted in luxury hotels highlights how the use of edible flowers significantly enhances customer satisfaction, fosters loyalty, and reinforces brand reputation despite the added costs (Dikshit et al. 2024). Consumer acceptability was one of the main focuses of a research made to enhance the compositional bioactive and nutritional compounds of cookies. By incorporating various ingredients like edible flowers, the researchers could enrich the cookies by leveling up the contents in fiber, ash, antioxidants, and phenolic content, attracting more consumers based on the original recipe (Šťastná et al. 2021) Fig. 3.2 illustrates examples demonstrating how the incorporation adds to the aesthetic charm of dishes, providing chefs with a versatile medium to create nutritious meals featuring intricate and engaging presentations, thereby elevating the dining experience for customers.

Consumer trends and cultural influences play a pivotal role in accepting and using edible flowers. Studies show that geographic location, education level, and demographic factors (such as age, income, and gender) shape consumer knowledge and willingness to incorporate these flowers into their diets (Guiné et al. 2021; Mulík et al. 2024). These findings underscore the importance of tailoring culinary innovations to regional preferences and consumer demographics.

Innovations in edible flower applications extend beyond their use as whole flowers. Extracts and derivatives have gained prominence in the creation of advanced food products. For example, flower extracts are being increasingly used to replace synthetic anthocyanins in yogurt formulations, while powdered milk products enriched with flower extracts and essential oils serve as eco-friendly food preservatives with antimicrobial properties (Mortas 2023; Pires et al. 2018; Valencia-Cordova et al. 2024). Nanotechnology represents an emerging field of scientific exploration focused on the examination and control of matter at the smallest scales, encompassing molecules and atoms (Rao and Poonia 2024b).

These innovations not only promote sustainability but also enhance the nutritional profiles of food products, demonstrating the versatility of edible flowers in modern gastronomy.

3.5.2 Preserving Nutritional Value in Edible Flower Preparation

The preparation and cooking of edible flowers can significantly impact their nutrient and bioactive content (Mireya et al. 2021). Serving flowers raw is often the best way to retain their vitamins, antioxidants, and delicate flavor, and is a preferred method for chefs (Jürgen 2022). However, their high-water content makes them perishable, which makes it a necessity to handle them carefully (Purohit et al. 2021). Some traditional ways of preserving them are by drying them, however cooking methods like steaming or boiling retain some nutrients while ensuring food safety, often aligning with traditional recipes. However, these methods must undergo an assessment to determine the effects and the best alternative to ensure the retention of important nutrients from edible flowers.

Research on endemic Mexican edible flowers illustrates the interplay between cooking methods and nutrient preservation. To illustrate, Mulík et al. (2022) found that endemic flowers are rarely consumed raw; instead, they are prepared with complementary ingredients depending on the flower. Building on this research, studies made specifically on squash blossoms (*Cucurbita pepo L.*) revealed that cooking styles significantly alter nutritional properties. Steaming retained the highest nutrient levels, while stir-frying increased bioactive compound bioaccessibility by 41.6% of total carotenoid content (Castañeda-Rodríguez et al. 2024a).

For agave flowers, steaming preserved physicochemical properties, while stir-frying enhanced bioactive compound bioaccessibility by 52% (Castañeda-Rodríguez et al. 2023). Conversely, traditional long cooking times negatively impacted yucca flowers' composition. Instead, steaming for a maximum of 3 min proved optimal for retaining bioactive and chemical properties (Castañeda-Rodríguez et al. 2024b). Ultimately, determining the desired properties of a dish informs the best preparation method. Further research on cooking techniques could optimize the health benefits of edible flowers, advancing their applications in modern gastronomy.

3.6 Future Research Directions: Unlocking Their Nutritional and Medicinal Potential

Edible flowers hold immense promise as sustainable and nutrient-dense food sources. Comparisons between edible flowers, fruits, and vegetables reveal that edible flowers often have comparable, and in some cases even higher, levels of carbohydrates, fiber, and micronutrients such as potassium and calcium. These attributes highlight their potential to enhance diets and complement well-known food sources. Current research also emphasizes their rich phytochemical profiles and biological activities, which play a role in preventing and managing chronic diseases. Trials conducted on various cells and organisms suggest that edible flower extracts could become critical sources of bioactive compounds for the pharmaceutical industry. However, the widespread adoption and utilization of edible flowers face limitations, including low public awareness, limited accessibility, and the lack of standardized research methodologies.

3.6.1 Prospects

The integration of edible flowers into functional foods, nutraceuticals, and pharmaceuticals presents exciting opportunities. Advanced techniques such as metabolomic profiling and bioavailability studies can further elucidate their health benefits and refine current applications with cutting-edge technology. Additionally, their incorporation into sustainable agriculture and eco-friendly packaging could transform both food and environmental industries. This approach may also make edible flowers valuable for marginalized populations, as they often grow naturally in small land areas within these regions.

Given their high-water content and perishable nature, research into innovative preservation methods is essential. Techniques such as drying and edible coatings have already shown promise in extending freshness and facilitating easy storage and usage during off-seasons. Moreover, exploring sustainable cultivation practices that maximize the nutritional and nutraceutical content of metabolites, while reducing pesticide use, could enhance their value as a reliable food source.

3.6.2 Challenges and Opportunities

Maximizing the potential of edible flowers requires addressing several key challenges. Currently, their utilization as food is limited to a narrow social demographic where they are perceived as luxury items. It is essential to disseminate more information to shift public perception, making edible flowers accessible and appealing to a broader audience.

Secondly, there is a need for the development of standardized methods to evaluate the nutritional and bioactive properties of edible flowers. The reports for the macronutrient and micronutrient analysis by most authors have gained from established protocols, the determination of phytochemical content and antioxidant activity lacks consistency. Having no protocol or report standardization from this data complicates the comparison of findings across studies and interferes with the efforts to position edible flowers as mainstream dietary components alongside more familiar food sources. Establishing clear and standardized methodologies will enhance comparability, improve understanding, and facilitate global acceptance.

Lastly, their short shelf life poses another significant barrier to widespread adoption. Developing sustainable preservation techniques that extend their shelf life without compromising their nutritional and bioactive properties is essential to make edible flowers a viable addition to modern diets.

3.6.3 Conclusion

Overall, it is clear that the use of edible flowers is not merely to offer decorative elements to dishes, their nutritional and phytochemical content makes them powerful allies in promoting health and sustainability. By bridging traditional knowledge still persistent in some regions with modern science, edible flowers can become an integral part of global diets, contributing to the prevention and treatment of chronic diseases, and promoting a healthy diet and a better lifestyle. Future research must prioritize accessibility, transcending barriers of education and economic capacity, to ensure that the benefits of edible flowers are available to all.

References

Abirami S, Jabesta J, Renitta RE, Anand DA, Samrot AV (2021) Antimicrobial activity of flower extracts against wound pathogens and fungi. Curr Res Green Sustainable Chem 4:100076. https://doi.org/10.1016/j.crgsc.2021.100076

Adamant A (2024) 60+ Chamomile recipes (More than just tea!). Adamant Kitchen. https://adamantkitchen.com/chamomile-recipes/.

Ahaotu EO, Nwabueze AP, Azubuike F, Anyaegbu F (2020) Investigating the implications of antimicrobial and anti-inflammatory properties of Plumeria rubra (frangipani) in preventing and treating of diseases in livestock farms. J Bio Innov 9(5):711–724. https://doi.org/10.46344/jbino.2020.v09i05.06

Balhara M (2022) Evaluating trends in consumption of ultra-processed foods from adolescence to young adulthood, with implications for human health. Proc Nutr Soc 81(OCE4):E159. https://doi.org/10.1017/s0029665122001884

Begum RF, Singh A, Mohan S (2023) Impact of junk food on obesity and polycystic ovarian syndrome: mechanisms and management strategies. Obes Med 40:100495. https://doi.org/10.1016/j.obmed.2023.100495

Benvenuti S, Mazzoncini M (2021) The biodiversity of edible flowers: discovering new tastes and new health benefits. Front Plant Sci 11:569499. https://doi.org/10.3389/fpls.2020.569499

Bujak T, Zagórska-Dziok M, Ziemlewska A, Nizioł-łukaszewska Z, Lal K, Wasilewski T, Hordyjewicz-Baran Z (2022) Flower extracts as multifunctional dyes in the cosmetics industry. Molecules 27(3):922. https://doi.org/10.3390/molecules27030922

Caramelo D, Pitacas I, Vitória C, Gonçalves J, Gominho J, Gallardo E, Anjos O (2024) Chemical composition and nutritional value of flowers and fruits of Cytisus striatus (Hill) Rothm. Plants 13(15):2121. https://doi.org/10.3390/plants13152121

Castañeda-Rodríguez R, Larrea V, Hernando I, Ozuna C (2023) Cooking changes agave flower properties, including the bioaccessibility of bioactive compounds. Int J Gastron Food Sci 32:100749. https://doi.org/10.1016/j.ijgfs.2023.100749

Castañeda-Rodríguez R, Quiles A, Hernando I, Ozuna C (2024a) Cooking methods determine chemical composition and functional properties of squash blossoms: a study of microstruc-

tural and bioaccessibility changes. Food Res Int 180:114095. https://doi.org/10.1016/j.foodres.2024.114095
Castañeda-Rodríguez R, Quiles A, Llorca E, Ozuna C (2024b) How to cook Yucca spp. flowers? An analysis of their chemical composition, microstructure, and bioactive compound bioaccessibility. Appl Food Res 4(1):100414. https://doi.org/10.1016/j.afres.2024.100414
Chaudhary P, Janmeda P, Docea AO, Yeskaliyeva B, Razis A, Faizal A, Modu B, Calina D, Sharifi-Rad J (2023) Oxidative stress, free radicals and antioxidants: potential crosstalk in the pathophysiology of human diseases. Front Chem 11:1158198. https://doi.org/10.3389/fchem.2023.1158198
Copeland B (2024) Our favorite edible flower recipes that are colorful & delicious. Martha Stewart Blog. https://www.marthastewart.com/edible-flower-recipes-7503243.
Dankhade S, Kale R, Jadhav D, Dhawade MR, Wankhade Y (2024) Nutritional and medicinal knowledge of wild edible flowers amongst rural women. E3S Web of Conferences 477:00080. https://doi.org/10.1051/e3sconf/202447700080
De Lima Franzen F, Rodrigues de Oliveira MS, Lidório HF, Farias Menegaes J, Martins Fries LL (2019) Composición química de pétalos de flores de rosa, girasol y caléndula para su uso en la alimentación humana. Ciencia y Tecnología Agropecuaria 20(1):149–168. https://doi.org/10.21930/rcta.vol20_num1_art:1252
de Pires Jr EO, Di Gioia F, Rouphael Y, Ferreira IC, Caleja C, Barros L, & Petropoulos SA. (2021). The compositional aspects of edible flowers as an emerging horticultural product. In Molecules (26, Issue 22). MDPI. doi:https://doi.org/10.3390/molecules26226940
Rajendra Dikshit A, Srivastava M, Kumar V (2024) The role of edible flowers in enhancing customer experience in luxury hotels. Int J Sci Res Engg Manage (8):9. https://doi.org/10.55041/IJSREM34380
dos Santos IC, Reis SN (2021) Edible flowers: traditional and current use. Ornamental Hortic 27(4):438–445. https://doi.org/10.1590/2447-536X.v27i4.2392
El Motteleb A, Dalia M, El Aleem A, Dalia I (2017) Renoprotective effect of Hypericum perforatum against diabetic nephropathy in rats: insights in the underlying mechanisms. Clin Exp Pharmacol Physiol 44(4):509–521. https://doi.org/10.1111/1440-1681.12729
Elsayed HE, El-Deeb EM, Taha H, Taha HS, Elgindi MR, Moharram FA (2023) Essential oils of Psidium cattleianum Sabine leaves and flowers: anti-inflammatory and cytotoxic activities. Front Chem 11:1120432. https://doi.org/10.3389/FCHEM.2023.1120432/BIBTEX
Fakhri S, Tomas M, Capanoglu E, Hussain Y, Abbaszadeh F, Lu B, Hu X, Wu J, Zou L, Smeriglio A, Simal-Gandara J, Cao H, Xiao J, Khan H (2022) Antioxidant and anticancer potentials of edible flowers: where do we stand? Crit Rev Food Sci Nutr 62(31):8589–8645). Taylor and Francis Ltd. https://doi.org/10.1080/10408398.2021.1931022
Fang Z, Rossato SL, Hang D, Khandpur N, Wang K, Lo CH, Willett WC, Giovannucci EL, Song M (2024) Association of ultra-processed food consumption with all cause and cause specific mortality: population based cohort study. BMJ 385:e078476. https://doi.org/10.1136/bmj-2023-078476
Fernandes L, Ramalhosa E, Baptista P, Pereira JA, Saraiva JA, Casal SIP (2019) Nutritional and nutraceutical composition of pansies (Viola × wittrockiana) during flowering. J Food Sci 84(3):490–498. https://doi.org/10.1111/1750-3841.14482
Food and Nutrition Board Institute of Medicine (2000) Dietary reference intakes for vitamin C, vitamin E, selenium, and carotenoids. National Academies Press, Washington (DC). https://doi.org/10.17226/9810
Forman HJ, Zhang H (2021) Targeting oxidative stress in disease: promise and limitations of antioxidant therapy. Nat Rev Drug Discov 20(9):689–709. https://doi.org/10.1038/s41573-021-00233-1
Gatto MA, Ippolito A, Linsalata V, Cascarano NA, Nigro F, Vanadia S, Venere D, Donato. (2011) Activity of extracts from wild edible herbs against postharvest fungal diseases of fruit and vegetables. Postharvest Biol Technol 61(1):72–82. https://doi.org/10.1016/j.postharvbio.2011.02.005

Guiné RPF, Florença SG, Moya KV, Anjos O (2020) Edible flowers, old tradition or new gastronomic trend: a first look at consumption in Portugal versus Costa Rica. Foods 9(8):977. https://doi.org/10.3390/foods9080977

Guiné RPF, Florença SG, Ferrão AC, Bizjak MČ, Vombergar B, Simoni N, Vieira V (2021) Factors affecting eating habits and knowledge of edible flowers in different countries. Open Agric 6(1):67–81. https://doi.org/10.1515/opag-2020-0208

Hajam YA, Rani R, Ganie SY, Sheikh TA, Javaid D, Qadri SS, Pramodh S, Alsulimani A, Alkhanani MF, Harakeh S, Hussain A, Haque S, Reshi MS (2022) Oxidative stress in human pathology and aging: molecular mechanisms and perspectives. Cells 11(3):552. https://doi.org/10.3390/cells11030552

Hegde AS, Gupta S, Sharma S, Srivatsan V, Kumari P (2022) Edible rose flowers: a doorway to gastronomic and nutraceutical research. Food Res Int 162:111977. https://doi.org/10.1016/j.foodres.2022.111977

Institute of Medicine (US) (2011) Dietary reference intakes for calcium and vitamin D. In: Dietary reference intakes for calcium and vitamin D. National Academies Press. https://doi.org/10.17226/13050

Jabri MA, Wannes D, Hajji N, Sakly M, Marzouki L, Sebai H (2017) Role of laxative and antioxidant properties of Malva sylvestris leaves in constipation treatment. Biomed Pharmacother 89:29–35. https://doi.org/10.1016/j.biopha.2017.02.020

Jadhav HB, Badwaik LS, Annapure U, Casanova F, Alaskar K (2023) A review on the journey of edible flowers from farm to consumer's plate. Appl Food Res 3(2):100312. https://doi.org/10.1016/j.afres.2023.100312

Janarny G, Ranaweera KK, Somathilaka D, Gunathilake KD, Priyantha P (2022) Digestive recovery of polyphenols, antioxidant activity, and anti-inflammatory activity of selected edible flowers from the family Fabaceae. J Food Biochem 46(2):e14052. https://doi.org/10.1111/jfbc.14052

Jena AB, Samal RR, Bhol NK, Duttaroy AK (2023) Cellular Red-Ox system in health and disease: the latest update. Biomed Pharmacother 162:114606. https://doi.org/10.1016/j.biopha.2023.114606

Jürgen D (2022) A pastry chef's guide to working with flowers. Institute of Culinary Education Blog. https://www.ice.edu/blog/edible-flowers-for-cakes#:~:text=Flower%20Infusions,and%20harvest%20all%20that%20flavor

Kandylis P (2022) Phytochemicals and antioxidant properties of edible flowers. Appl Sci 12(19):9937. https://doi.org/10.3390/app12199937

Kang Y, Liu Z, Li W, Li X, Zhang X, Xu Z, Gao Y, Qi Y (2022) Anti-oxidative and anti-inflammatory activities of the ethanol extract of edible flower from Chimonanthus praecox. Front Pharmacol 13:1004520. https://doi.org/10.3389/fphar.2022.1004520

Khomdram SD, Fanai L, Yumkham SD (2019) Local knowledge of edible flowers used in Mizoram. Indian J Tradit Knowl 18, Issue 4:714–723

Król J, Brodziak A, Ślusarczyk L, Matwijczuk A, Chwil M, Matraszek-Gawron R (2024) Yogurt with cornflower (Centaurea cyanus L.) petals as a source of antioxidant compounds and dietary fiber: physicochemical and spectroscopic research during storage. J Dairy Sci 108(3):2243–2263. https://doi.org/10.3168/jds.2024-25628

Kurutas EB (2016) The importance of antioxidants which play the role in cellular response against oxidative/nitrosative stress: current state. Nutr J 15(1):71. https://doi.org/10.1186/s12937-016-0186-5

Lee D, Qi Y, Kim R, Song J, Kim H, Kim HY, Jang DS, Kang KS (2021) Methyl caffeate isolated from the flowers of prunus persica (L.) batsch enhances glucose-stimulated insulin secretion. Biomolecules 11(2):1–13. https://doi.org/10.3390/biom11020279

Li S, Ma J, Wen J, Peng J, Huang P, Zeng L, Chen S, Ji G, Yang X, Wu W (2024) Trends in consumption of ultra-processed foods among adults in southern China: analysis of serial cross-sectional health survey data 2002–2022. Nutrients 16(23):4008. https://doi.org/10.3390/nu16234008

Liu RH (2003) Health benefits of fruit and vegetables are from additive and synergistic combinations of phytochemicals 1-4. Am J Clin Nutr 78(3):517S–520S

Liu J, Wu Y, Ma W, Zhang H, Meng X, Zhang H, Guo M, Ling X, Li L (2023a) Anti-inflammatory activity of panax notoginseng flower saponins quantified using LC/MS/MS. Molecules 28(5):2416. https://doi.org/10.3390/molecules28052416

Liu X, Wang S, Cui L, Zhou H, Liu Y, Meng L, Chen S, Xi X, Zhang Y, Kang W (2023b) Flowers: precious food and medicine resources. Food Sci Hum Well 12(4):1020–1052. https://doi.org/10.1016/j.fshw.2022.10.022

Lobo V, Patil A, Phatak A, Chandra N (2010) Free radicals, antioxidants and functional foods: impact on human health. Pharmacogn Rev 4(8):118–126. https://doi.org/10.4103/0973-7847.70902

Lu B, Li M, Yin R (2016) Phytochemical content, health benefits, and toxicology of common edible flowers: a review (2000–2015). Crit Rev Food Sci Nutr 56:S130–S148. https://doi.org/10.1080/10408398.2015.1078276

Marchioni I, Gabriele M, Carmassi G, Ruffoni B, Pistelli L, Pistelli L, Najar B (2024) Phytochemical, nutritional and mineral content of four edible flowers. Foods 13(6):939. https://doi.org/10.3390/foods13060939

Meysami M, Hashempur MH, Kamalinejad M, Emtiazy M (2021) Efficacy of short term topical Malva sylvestris L. cream in pediatric patients with atopic dermatitis: a randomized double-blind placebo-controlled clinical trial. Endocr Metab Immune Disord Drug Targets 21(9):1673–1678. https://doi.org/10.2174/1871530320666201023125411

Mireya SGE, José AR, del Socorro CCN, Ojeda-Ramírez D, Yadira ZRQ, Juan HÁ, Esther RM (2021) Effect of boiling on nutritional, antioxidant and physicochemical properties of edible plants (Malva parviflora and Myrtillocactus geometrizans). Emir J Food Agric 33(6):510–519. https://doi.org/10.9755/ejfa.2021.v33.i6.2726

Mlcek J, Rop O (2011) Fresh edible flowers of ornamental plants – a new source of nutraceutical foods. Trends Food Sci Technol 22(10):561–569. https://doi.org/10.1016/j.tifs.2011.04.006

Mlcek J, Plaskova A, Jurikova T, Sochor J, Baron M, Ercisli S (2021) Chemical, nutritional and sensory characteristics of six ornamental edible flowers species. Foods 10(9):2053. https://doi.org/10.3390/foods10092053

Moliner C, Cásedas G, Barros L, Finimundy TC, Gómez-Rincón C, López V (2022) Neuroprotective profile of edible flowers of borage (Borago officinalis L.) in two different models: Caenorhabditis elegans and neuro-2a cells. Antioxidants 11(7):1244. https://doi.org/10.3390/antiox11071244

Mortas M (2023) Characterization of a new powdered, milk-based medicinal plant (Alcea rosea) drink product. Sustainability (Switzerland) 15(12):9320. https://doi.org/10.3390/su15129320

Motti R (2022) Wild edible plants: a challenge for future diet and health. Plants 11(3):344. https://doi.org/10.3390/plants11030344

Motti R, Paura B, Cozzolino A, de Falco B (2022) Edible flowers used in some countries of the Mediterranean Basin: an ethnobotanical overview. Plants 11(23):3272. https://doi.org/10.3390/plants11233272

Mulík S, Ozuna C (2020) Mexican edible flowers: cultural background, traditional culinary uses, and potential health benefits. Int J Gastron Food Sci 21:100235. https://doi.org/10.1016/j.ijgfs.2020.100235

Mulík S, Hernández-Carrión M, Pacheco-Pantoja SE, Aguilar-Ruiz N, Ozuna C (2022) Culinary uses of Mexican edible flowers: recipe analysis. Int J Gastron Food Sci 28:100539. https://doi.org/10.1016/j.ijgfs.2022.100539

Mulík S, Hernández-Carrión M, Pacheco-Pantoja SE, Ozuna C (2024) Endemic edible flowers in the Mexican diet: understanding people's knowledge, consumption, and experience. Future Foods 9:100374. https://doi.org/10.1016/j.fufo.2024.100374

Murray M (2022) Petal to plate: cooking with edible flowers|Frank Family Vineyards Blog. https://www.frankfamilyvineyards.com/blog/petal-to-plate-cooking-with-edible-flowers/

Navarro-González I, González-Barrio R, García-Valverde V, Bautista-Ortín AB, Periago MJ (2015) Nutritional composition and antioxidant capacity in edible flowers: characterisation of phenolic compounds by HPLC-DAD-ESI/MSn. Int J Mol Sci 16(1):805–822. https://doi.org/10.3390/ijms16010805

Newman M, Kirker CL (2016) Edible flowers: a global history. Reaktion Books Ltd., London

Nguyen HL, Thuy H, Phuong TT, Nguyen NP, Nguyen P, Linh NH, Nguyen DT, Hien T, Dat V, Truong MN, Nguyen-Hoang TM, Luu TN, Nguyen NP, Nguyen TB, Mai HN (2024) Effect of Camellia flava (Pitard) Sealy flower extract on the degeneration of Islets of Langerhans and insulin resistance in alloxan-induced hyperglycemia model on Swiss albino mice. Pharmacia 71:1–15. https://doi.org/10.3897/pharmacia.71.e126255

Ninama V, Shah H, Kapadia C, Italiya A, Datta R, Singh S, Singh A (2024) Assessment of phytochemicals, nutritional compositions and metabolite profiling using GCMS– from annual edible flowers. Sci Hortic 323:112551. https://doi.org/10.1016/j.scienta.2023.112551

Pagliai G, Dinu M, Madarena MP, Bonaccio M, Iacoviello L, Sofi F (2021) Consumption of ultra-processed foods and health status: a systematic review and meta-analysis. Br J Nutr 125(3):308–318. https://doi.org/10.1017/S0007114520002688

Pahwa R, Goyal A, Jialal I (2023) Chronic inflammation. In: Pathobiology of human disease: a dynamic encyclopedia of disease mechanisms, pp 300–314. https://doi.org/10.1016/B978-0-12-386456-7.01808-6

Pires TCSP, Dias MI, Barros L, Ferreira ICFR (2017) Nutritional and chemical characterization of edible petals and corresponding infusions: valorization as new food ingredients. Food Chem 220:337–343. https://doi.org/10.1016/j.foodchem.2016.10.026

Pires TCSP, Dias MI, Barros L, Barreira JCM, Santos-Buelga C, Ferreira ICFR (2018) Incorporation of natural colorants obtained from edible flowers in yogurts. LWT 97:668–675. https://doi.org/10.1016/j.lwt.2018.08.013

Pizzino G, Irrera N, Cucinotta M, Pallio G, Mannino F, Arcoraci V, Squadrito F, Altavilla D, Bitto A (2017) Oxidative stress: harms and benefits for human health. Oxid Med Cell Longev 2017:8416763. https://doi.org/10.1155/2017/8416763

Polumackanycz M, Kaszuba M, Konopacka A, Marzec-Wróblewska U, Wesolowski M, Waleron K, Buciński A, Viapiana A (2020) Phenolic composition and biological properties of wild and commercial dog rose fruits and leaves. Molecules (Basel, Switzerland) 25(22):5272. https://doi.org/10.3390/MOLECULES25225272

Prabawati NB, Oktavirina V, Palma M, Setyaningsih W (2021) Edible flowers: antioxidant compounds and their functional properties. Horticulturae 7(4):66. https://doi.org/10.3390/horticulturae7040066

Public Health England (2015) McCance and Widdowson's The Composition of Foods Seventh SummaryEdition.https://assets.publishing.service.gov.uk/media/60538b91e90e07527df82ae4/McCance_Widdowsons_Composition_of_Foods_Integrated_Dataset_2021.xlsx

Purohit SR, Rana SS, Idrishi R, Sharma V, Ghosh P (2021) A review on nutritional, bioactive, toxicological properties and preservation of edible flowers. Future Foods 4(100078). https://doi.org/10.1016/j.fufo.2021.100078

Rao V, Poonia A (2024a) Bioactive compounds, nanoparticle synthesis and potential utilization of edible flowers for development of functional dairy products: a review. J Food Sci Technol 61(6):1053–1068. https://doi.org/10.1007/s13197-023-05853-6

Rao V, Poonia A (2024b) Microbial biosynthesis of selenium nanoparticles using probiotic strain and its characterization. J Food Meas Charact 18(7):5477–5488. https://doi.org/10.1007/s11694-024-02581-z

Rivas-García L, Navarro-Hortal MD, Romero-Márquez JM, Forbes-Hernández TY, Varela-López A, Llopis J, Sánchez-González C, Quiles JL (2021) Edible flowers as a health promoter: an evidence-based review. Trends Food Sci Technol 117:46–59. https://doi.org/10.1016/j.tifs.2020.12.007

Rodrigues H, Spence C (2023) Looking to the future, by studying the history of edible flowers. Int J Gastron Food Sci 34:100805. https://doi.org/10.1016/j.ijgfs.2023.100805

Shim JS, Shim SY, Cha HJ, Kim J, Kim HC (2021) Socioeconomic characteristics and trends in the consumption of ultra-processed foods in Korea from 2010 to 2018. Nutrients 13(4):1120. https://doi.org/10.3390/nu13041120

Singh NK, Baranwal J, Pati S, Barse B, Khan RH, Kumar A (2023) Application of plant products in the synthesis and functionalisation of biopolymers. Int J Biol Macromol 237:124174. https://doi.org/10.1016/j.ijbiomac.2023.124174

Skrajda-Brdak M, Dąbrowski G, Konopka I (2020) Edible flowers, a source of valuable phytonutrients and their pro-healthy effects—a review. Trends Food Sci Technol 103:179–199. https://doi.org/10.1016/j.tifs.2020.06.016

Specialty Produce. (n.d.). Fava Flowers Information and Facts. Retrieved 20 Jan 2025 from https://specialtyproduce.com/produce/Fava_Flowers_8169.php

Šťastná K, Sumczynski D, Yalcin E (2021) Nutritional composition, in vitro antioxidant activity and phenolic profile of shortcrust cookies supplemented by edible flowers. Foods 10(11):2531. https://doi.org/10.3390/foods10112531

Stallings VA et al (2019) In: Stallings VA, Harrison M, Oria M (eds) Dietary reference intakes for sodium and potassium. National Academies Press. https://doi.org/10.17226/25353

Stefaniak A, Grzeszczuk ME (2018) Nutritional and biological value of five edible flower species. Notulae Botanicae Horti Agrobotanici Cluj-Napoca 47(1):128–134. https://doi.org/10.15835/nbha47111136

Takahashi JA, Rezende FAGG, Moura MAF, Dominguete LCB, Sande D (2020) Edible flowers: bioactive profile and its potential to be used in food development. Food Res Int 129:108868. https://doi.org/10.1016/j.foodres.2019.108868

Tavallali V, Rahmati S, Bahmanzadegan A, Lasibi MJ, Mahdavi. (2024) Phenolic profile and evaluation of antimicrobial and anticancer activities of Calendula officinalis L. using exogenous polyamines application. Ind Crop Prod 214:118571. https://doi.org/10.1016/j.indcrop.2024.118571

Thilagam E, Chidambaram K, Raviteja C, Vahana T, Vasudevan P (2021) Anti-hyperglycemic and hypolipidemic effects of Saraca asoca (Roxb.) Wild. flowers in alloxan-treated diabetic rats. J Pharm Pharmacogn Res 9(1):58–68. http://jppres.com/jppres

Tiwari S (2021) From medieval monks to Michelin-starred chefs, edible flowers' culinary journey. Food by Gulf News. https://gulfnews.com/food/from-medieval-monks-to-michelin-starred-chefs-edible-flowers-culinary-journey-1.1639300471443

UNESCO (2010) Traditional Mexican cuisine - ancestral, ongoing community culture, the Michoacán paradigm - intangible heritage - Culture Sector - UNESCO. https://ich.unesco.org/en/RL/traditional-mexican-cuisine-ancestral-ongoing-community-culture-the-michoacan-paradigm-00400

Valencia-Cordova MG, Jaguey-Hernández Y, Castañeda-Ovando A, González-Olivares LG, Castañeda-Ovando EP, Añorve-Morga J, de la O-Arciniega, Minarda. (2024) Lesser-explored edible flowers as a choice of phytochemical sources for food applications. Int J Food Sci 2024(1):9265929. https://doi.org/10.1155/2024/9265929

Venkatesh S, Thilagavathi J, Shyam sundar, D. (2008) Anti-diabetic activity of flowers of hibiscus rosasinensis. Fitoterapia 79(2):79–81. https://doi.org/10.1016/j.fitote.2007.06.015

Vitale M, Costabile G, Testa R, D'Abbronzo G, Nettore IC, Macchia PE, Giacco R (2024) Ultra-processed foods and human health: a systematic review and meta-analysis of prospective cohort studies. Adv Nutr 15(1):100121. https://doi.org/10.1016/j.advnut.2023.09.009

Wang Z, Guan Y, Yang R, Li J, Wang J, Jia AQ (2020) Anti-inflammatory activity of 3-cinnamoyltribuloside and its metabolomic analysis in LPS-activated RAW 264.7 cells. BMC Complementary Med Ther 20(1):329. https://doi.org/10.1186/S12906-020-03115-Y

Zagorsky JL, Smith PK (2017) The association between socioeconomic status and adult fast-food consumption in the U.S. Econ Hum Biol 27:12–25. https://doi.org/10.1016/j.ehb.2017.04.004

Zhang XK, Cao GH, Bi Y, Liu XH, Yin HM, Zuo JF, Xu W, Li HD, He S, Hong ZX (2023) Comprehensive analysis of 34 edible flowers by the determination of nutritional composition and antioxidant capacity planted in Yunnan Province China. Molecules *28*(13):5260. https://doi.org/10.3390/molecules28135260

Chapter 4
Mineral Composition of Edible Flowers

Oscar Cruz Álvarez, Marisela Calderón Jurado, Adalberto Benavides Mendoza, Graciela Dolores Ávila Quezada, and Damaris Leopoldina Ojeda Barrios

4.1 Introduction

Flowers are the reproductive structures of plants, that have evolved for sexual reproduction by attracting animal pollinators (mostly inset and birds) through their fragrance, color, and shape (Tuladhar 2021). The flowers of most species are harmless and non-toxic. Since ancient times, they have been seen as offering both nutritional and functional health benefits. Hence, our ancestors have incorporated these flowers into their diets and used them to formulate a range of traditional medicines and dishes in their local cuisines (Figueredo-Urbina et al. 2021; Wen et al. 2025). Similarly, many ancient civilizations, including those of Rome, Greece, China, Egypt, Mesopotamia and the pre-Hispanic cultures of the Americas, supplemented their diets with the tuberous roots and inflorescences of approximately 97 families, 100 genera and 180 species of cultivated and wild plants (Lu et al. 2016; Matyjaszczyk and Śmiechowska 2019; Hernández-Epigmenio et al. 2022).

Over the past two decades, the popularity of edible flowers (EFs) has grown significantly. This has prompted efforts to characterize them, and to reevaluate them as novel culinary ingredients (Pensamiento-Niño et al. 2021). The interest is largely due to their high contents of phytochemical compounds, essential mineral nutrients, and generally low caloric value (Mulík and Ozuna 2020; Pinedo-Espinoza et al. 2020). These nutritional attributes are particularly appealing to health-conscious

O. C. Álvarez · M. C. Jurado · G. D. Á. Quezada · D. L. O. Barrios (✉)
Facultad de Ciencias Agrotecnológicas, Universidad Autónoma de Chihuahua, Chihuahua, Mexico
e-mail: dojeda@uach.mx

A. B. Mendoza
Departamento de Horticultura, Universidad Agraria Antonio Narro, Buenavista, Saltillo, Mexico

A. Poonia et al. (eds.), *Edible Flowers: Source of Phytonutrients, Valorization and Technological Advancements*, https://doi.org/10.1007/978-3-032-00667-7_4

consumer sectors, including vegetarians and vegans. Additionally, edible flowers are increasingly valued for their contributions of color, flavor, sweetness, aroma, and aesthetic appeal to various traditional dishes, including salads, purées, tortillas, and desserts (Castañeda-Rodríguez et al. 2024; Marcos-Gómez et al. 2024). Furthermore, EFs play a pivotal role in the creation of condiments and the development of new gourmet dishes, such as stews, beverages, and other culinary preparations (Pensamiento-Niño et al. 2024). These flowers are an excellent source of antioxidants, vitamins, and minerals. The market for nutraceuticals is driven primarily by increased consumer awareness, improved purchasing power, the popularisation of the beneficial effects of phytochemicals on the human body, consumer readiness to purchase functional foods, and a strong desire to maintain a healthy lifestyle. It is anticipated that dairy-based beverages will propel the market expansion for functional foods (Rao and Poonia 2024a).

The increase in demand and consumption of EFs has fostered the development of a significant economic niche within the horticultural market for fresh, dehydrated, and minimally processed products (Hernández-Epigmenio et al. 2022). Furthermore, the utilization of cultivated plants traditionally valued for their fruits, seeds, or shoots, or grown primarily for the ornamental value of their flowers and inflorescences, has expanded and diversified (Fig. 4.1) (Martínez-Damián et al. 2021; Calderón-Jurado et al. 2023). This trend can also be seen as a strategy for the *in situ* conservation of numerous wild species threatened by habitat alteration

Fig. 4.1 Example of edible flowering plants. (**a**) crab apple cv. Manchurian; (**b**) crab apple cv. John Downie; (**c**) peach cv. Fair Time; (**d**) peach cv. July Flame; (**e**) dahlia; (**f**) marigold; (**g**) French marigold; (**h**) Lemmon's marigold; (**i**) tuberous begonia; (**j**) rose; (**k**) coral tree; and (**l**) garambullo

arising from anthropogenic activities and the increasing adverse effects of climate change (Rivera-Espejel et al. 2019; Pensamiento-Niño et al. 2021).

Plant survival is closely linked to seed and fruit production, which is why not all flowers are edible. Many flowers produce or secrete repellent or toxic substances, including cyanogenic glycosides, tannins, steroids, coumarins, alkaloids, phytates, and saponins. These substances can cause mild to severe physical or mental disorders when consumed fresh (Sotelo et al. 2007; Dusemund et al. 2017; Pinela et al. 2017). However, in certain species, these anti-nutrient compounds in the EFs can be eliminated through cooking (Castañeda-Rodríguez et al. 2024). One of the challenges associated with the production of EFs is the detection of pathogenic bacteria (e.g., *Salmonella* spp.), elevated concentrations of pesticides (e.g., dimethoate and diethyl-meta-toluamide), and certain sulfur derivatives, such as sulfites (Matyjaszczyk and Michalski 2019). These concerns highlight the need for strict agronomic management protocols during cultivation, preservation, transport, and storage. Continuous monitoring and awareness-raising among farmers, distributors, and marketers are essential if consumer confidence and consumer education are to be ensured, along with international trade and compliance (Gupta et al. 2024). Interestingly, there are no official lists of edible and inedible flowers from any health or food safety organization, including the Food and Agriculture Organisation of the United Nations (FAO), the World Health Organisation (WHO), the Food and Drug Administration (FDA), the European Food Safety Authority (EFSA), or the Codex Alimentarius (Fernandes et al. 2020). In regions with intensive industrial activities, such as mining and metallurgy, there is an elevated risk that wild plants will absorb toxic chemicals from the soil, such as cadmium (Cd), mercury (Hg), nickel (Ni), lead (Pb), cobalt (Co), and vanadium (V). An associated risk is from airborne particulate pollutants originating from fossil fuel combustion, forest fires, or volcanic eruption (Mallick et al. 2024; Sik et al. 2024).

One of the key dietary attractions of EFs lies in their low caloric content and valuable nutritional profile, which includes proteins, minerals, vitamins, and a whole range of secondary metabolites, including phenols, flavonoids, carotenoids, and anthocyanins, which influence their color and flavor, while also serving as essential components in pharmaceutical formulations (Takahashi et al. 2020; Mallick et al. 2024). The concentrations of bioactive compounds with antioxidant properties play a pivotal role in the exploration and re-evaluation of flowers from both horticultural and ornamental species (Rodrigues and Spence 2023). The culinary uses of wildflower species are explored in many local cuisines (Sotelo et al. 2007; Hernández-Epigmenio et al. 2022; Castañeda-Rodríguez et al. 2024; Maaloul et al. 2024). However, research on the mineral compositions of EFs and their effect on human health remains limited to a small number of species (Jilani et al. 2012; Kibar and Temel 2016; Grzeszczuk et al. 2018; Drava et al. 2020; Calderón-Jurado et al. 2023; Marchioni et al. 2024). Current research challenges include identifying those germplasms having the highest mineral nutrient contents and evaluating their potential health benefits, considering the diverse climatic and socio-economic conditions of local human populations. Nanotechnology represents an emerging field of scientific exploration focused on the examination and control of matter at the smallest scales, encompassing molecules and atoms (Rao and Poonia 2024b).

This chapter explores the mineral compositions of edible flowers, providing quantitative data and highlighting their potential contributions to a healthy human diet.

4.2 Quantitative Analysis of Minerals in Edible Flowers

Edible flowers have long been complementary ingredients in food preparation, also playing essential roles in traditional medicine, with palaeobotanical records dating back some 4000 years (Rop et al. 2012). Certain species, such as roses, were used to enhance the aesthetic and sensory value of food prepared for the nobility, particularly during receptions, feasts, and banquets (Rodrigues and Spence 2023). However, with population growth and globalization, the diversity of edible plant species by the average person has expanded, including, but to a lesser extent, those with flowers and inflorescences (Sik et al. 2024).

On the other hand, the two major challenges facing the global population are obesity and malnutrition. These are both linked to the excessive consumption of food products of low nutritional value but high caloric content. As a result, the rising interest in flower consumption stems from a desire to improve eating habits and adopt healthier lifestyles, both from a nutritional and an environmental perspective. This is in line with the growing production and dissemination of cookbooks, television programs, culinary magazines, and scientific literature (Demasi et al. 2023; Calderón-Jurado et al. 2023). These include the use of EFs from cultivated plants (ornamentals, and horticultural and medicinal species) and from wild plants. These are incorporated as novel or 'fashionable' ingredients, and often in fresh or minimally processed forms (Pires et al. 2017; Martínez-Damián et al. 2021). However, the commercial production of flowers for human consumption must be carefully managed to minimize health risks associated with pesticides or microbiological contaminants (Matyjaszczyk and Śmiechowska 2019).

While there is a wide variety of flowering plants, not all are suitable for human consumption, and some can be toxic. Hence, accurate identification is crucial to prevent the poisoning of consumers, and allergenic health disorders, including asthma (Takahashi et al. 2020). Furthermore, in some species, certain flower structures (stems, sepals, pistils, or stamens) should be removed before consumption, along with their pollen. For some species, toxicological studies are required to determine the e maximum recommended doses to ensure consumer safety (Mlcek and Rop 2011; Rivas-García et al. 2023).

Functional components, including mineral nutrients, are found in variable concentrations in EFs (Pires Jr et al. 2023; Drava et al. 2020; Marchioni et al. 2024), with some published determinations presented with respect to fresh weight, and others to dry weight (Demasi et al. 2023). This makes comparison between species more complex, as some flowers are usually consumed fresh and others dried. Most published information on plant tissue composition pertains to leaves or roots, with equivalent information for flowers and inflorescences being much more limited and relating to fewer plant species and cultivars (Araújo et al. 2019). This trend continues, despite the growing interest in finding new species with edible flow and which contain minerals and other substances valuable in human nutrition and health (Rop et al. 2012; Kibar and Temel 2016).

Globally, there are multiple issues relating to hunger and inadequate nutrition that contribute to health disorders and premature death, particularly among mothers, infants, and young children. Meanwhile, obesity is increasing, and this is associated with a range of chronic degenerative diseases (Welch 2002; Silva et al. 2019). In this context, EFs consumption can help address mineral nutrient deficiencies, including P, K, Ca, Mg, Fe, Cu, Mn, Mo, Zn, B, and Na (Quintaes and Diez-Garcia 2015). However, the concentration of these minerals in flowers fluctuates depending on a wide range of factors including agronomic practices, soils and climates, on also plant species, flower harvest stage, and postharvest processing. Also, wide variation in mineral composition has been reported even among cultivars of the same species, For example, the two cultivars 'Magic Yellow' and 'Magic Red' of *Mimulus × hybridus* L. contain, respectively, concentrations of K (54.45 and 45.86 g kg^{-1} dw), Fe (368.23 and 683.50 mg kg^{-1} dw), Zn (39.92 and 25.53 mg kg^{-1} dw), and Mn (18.20 and 30.64 mg kg^{-1} dw) (Grzeszczuk et al. 2018). Similarly, wide variations occur between species with the flowers of peach cv. 'July Flame' contains significant concentrations of Mn (42.0 ± 0.02 mg kg^{-1} dw) and Na (20.00 ± 0.04 mg kg^{-1} dw) and the crab apple cv. 'Manchurian' containing Fe (92.10 ± 2.65 mg kg^{-1} dw), Cu (14.50 ± 0.61 mg kg^{-1} dw) and Mn (93.40 ± 1.85 mg kg^{-1} dw) (Calderón-Jurado et al. 2023).

On the other hand, Drava et al. (2020) report notable concentrations of micronutrients (dry weight), including Fe (122.5 ± 38.0 and 130.0 ± 8.2 μg kg^{-1}) and Mn (307.4 ± 169.9 μg kg^{-1} and 83.9 ± 4.7 μg kg^{-1}) in paracress [(*Acmella oleracea* (L.) R.K.Jansen] and garden dahlia (*Dahlia pinnata* Cav.) across multiple species with edible flowers. Additionally, they document Zn concentrations in flowers of courgette (*Cucurbita pepo* L.) (69.6 ± 4.1 μg kg^{-1}), common violet (*Viola odorata* L.) (64.1 ± 1.6 μg kg^{-1}), pumpkin (*Cucurbita moschata* Duchesne) (63.6 ± 0.1 μg kg^{-1}), Tulbagia (*Tulbaghia cominsii* Vosa) (62.3 ± 1.1 μg kg^{-1}), and basil (*Ocimum basilicum* L.) (58.5 ± 1.8 μg kg^{-1}).

Generally, reports on the concentration of mineral nutrients in edible flowers remain scarce, despite their crucial role in brain development and infant growth by promoting cell proliferation, metabolism, and differentiation, and reducing the risk of ailments such as anemia (Welch 2002). An overview of the mineral concentration in edible flowers from multiple plant species is presented in Tables 4.1 and 4.2.

Table 4.1 Overview of the mineral macronutrient concentrations found in edible flowers across a diverse range of plant species

Specie	Units	N	P	K	Ca	Mg	S	References
Peony (*Paeonia officinalis*) "Sarah Bernardt"	g kg^{-1} dw	6.96	6.62	8.15	1.49	1.45		Grzeszczuk et al. (2018)
Peach (*Prunus persica*) "July flame", "Cary Mac", "Fair Time"	g kg^{-1} dw	23.7 23.5 22.6	2.7 3.3 2.8	11.2 11.3 11.2	5.4 5.6 6.2	2.9 4.3 4.5	– – –	Calderón-Jurado et al. (2023)
Peach (*Prunus persica*) "Calanda", 'Catherina', "Babygold5"	% dw	2.7 2.8 2.8	0.45 0.45 0.40	2.1 2.2 2.3	0.70 0.76 0.90	0.21 0.25 0.30	– – –	El-Jendoubi et al. (2013)
Rose (*Rosa hybrid*) "Kardinal", "Anjleeq", "Maria Shever"	mg kg^{-1} dw	– – –	– – –	224.5 203.3 210.0	415.35 458.65 486.60	136.1 138.5 141.7	– – –	Jilani et al. (2012)
Rose (*Rosa hybrid*: Crystalline' × 'Playgirl)	g 100 g^{-1} dw	–	–	1.328	0.15	0.101	–	Mallick et al. (2024)
Rose (*Rosa* spp.)	mg 100 g^{-1} dw	–	–	2948.0	118.8	490.4	–	Araújo et al. (2019)
Rose (*Rosa*) "Gloria Dei"	mg kg^{-1} fw	–	245.1	2033.44	285.58	142.45	–	Mlcek et al. (2021)
Crab apple (*Malus sylvestris*) "John Downie", "Manchurian	g kg^{-1}dw	23.2 22.3	2.9 3.5	22.0 21.5	5.8 5.4	6.5 6.6	– –	Calderón-Jurado et al. (2023)
Pot marigold (*Calendula officinalis*) "Antiqua Orange"	mg kg^{-1} fw	–	268.9	2988.6	294.1	189.6	–	Mlcek et al. (2021)
Pot marigold (*Calendula officinalis*)	mg 100 g^{-1} dw	–	–	7639	188.7	452.7	–	Araújo et al. (2019)
Marigold (*Tagetes erecta*)	mg 100 g^{-1} fw		0.065	0.215	0.110	0.060	0.04	Navarro-González et al. (2015)
Feverfew (*Chrysanthemum parthenium*)	mg kg^{-1} fw	–	501.3	3600.3	341.3	195.2	–	Rop et al. (2012)
French marigold (*Tagetes patula*)	mg kg^{-1} fw	–	478.2	3808.7	346.85	205.2	–	Rop et al. (2012)
French marigold (*T. patula*)	mg 100 g^{-1} dw	–	–	4496.0	524.5	629.1	–	Araújo et al. (2019)
French marigold (*T. patula*)	mg kg^{-1} fw	–	397.1	3623.8	362.95	203.1	–	Mlcek et al. (2021)
Pansy (*V. x wittrockiana*)	mg kg^{-1} fw	–	514.6	3964.8	486.44	190.1	–	Rop et al. (2012)
Pansy (*V. x wittrockiana*)	mg 100 g^{-1}	–	82.4	278.6	123.9	55.9	44.1	González-Barrio et al. (2018)
Pansy (*V. x wittrockiana*) "Series Delta Mix"	g kg^{-1}dw	21.2	9.3	19.4	2.61	1.6	–	de Oliveira et al. (2024)

Pumpkin (*Cucurbita maxima*)	mg 100 g^{-1} dw	–	18.2	4.23	17.6	2.75	1.23	Ghosh and Rana (2021)
Pumpkin (*C. maxima*)	mg g^{-1}dw	–	–	0.39	0.40	0.98	–	Gowthami and Poornima (2017)
Nasturtium (*Tropaeolum majus*)	mg 100 g^{-1} fw	–	0.05	0.225	0.055	0.035	0.040	Navarro-González et al. (2015)
Nasturtium (*T. majus*)	mg kg^{-1}fw	–	452.02	2353.30	317.54	152.89	–	Mlcek et al. (2021)
Nasturtium (T. *majus*)	µg 100 g dw	–	–	6187	225.6	576.6	–	Araújo et al. (2019)
Paracress (*Spilanthes oleracea*)	mg 100 g^{-1} fw	–	0.080	0.355	0.105	0.06	0.060	Navarro-González et al. (2015)
Tuberous begonia (*Begonia x tuberhybrida*)	mg kg^{-1} fw	–	208.59	2008.55	385.75	131.18	–	Mlcek et al. (2021)
Coriander (*Coriandrum sativum*)	mg 100 g^{-1} dw	–	6566		647.2	886.1		Araújo et al. (2019)
Snapdragon (*Antirrhinum majus*)	mg 100 g^{-1}	–	37.6	223	67.9	33.5	22.5	González-Barrio et al. (2018)
Rosemary (*Rosmarinus officinalis*)	mg 100 g^{-1} dw	–	–	4862	166.6	317.4	–	Araújo et al. (2019)
Heartsease (*Viola tricolor*)	mg 100 g^{-1} dw	–	–	7019	185.2	491.1	–	Araújo et al. (2019)
Heartsease (*V. tricolor*) cv Hortensis	mg kg^{-1} dw	27.32	10.02	20.93	3.49	1.81		de Oliveira et al. (2024)
Squash (*Cucurbita moschata*)	mg g^{-1}dw	–	5.95	31.54	47.51	4.98	–	Marchioni et al. (2024)
Chinese pink (*Dianthus chinensis*)	mg g^{-1}dw	–	3.94	30.91	46.63	2.48	–	Marchioni et al. (2024)
Royal fuchsia (*Fuchsia regia*)	mg g^{-1}dw	–	3.55	26.11	90.54	3.75	–	Marchioni et al. (2024)
Horned pansy (*Viola cornuta*)	mg g^{-1}dw	–	4.18	26.49	29.70	4.04	–	Marchioni et al. (2024)
Curry plant (*Helichrysum italicum*)	µg g^{-1} dw µg g^{-1} fw	– –	215.0 130.8	31,000 18,867	32,000 19,475	– –	391.0 238.0	Primitivo et al. (2022)
Tea (*Camellia sinensis*) "QN06"	mg kg^{-1} dw	–	2856.8	7926.7	4305.5	1490.9	2404.5	Jia et al. (2016)
Wild cabbage (*Trachyandra ciliata*)	mg 100 g^{-1}dw	2700	450	3150	2620	350	–	Ngxabi et al. (2025)
Broom (*Cytisus striatus*)	mg kg^{-1}dw	–	2571.7	25,872.5	5535.0	2870.2	–	Caramelo et al. (2024)
Sandkool (*Trachyandra divaricata*)	mg 100 g^{-1}dw	3790	740	5700	610	340	–	Bulawa et al. (2022)
Dahlia (*Dahlia* sp.)	g kg^{-1} dw	20.1	3.51	43.82	2.49	2.00	1.31	Costa et al. (2022)
Dahlia (*Dahlia* sp.)	mg 100 g^{-1}dw	–	4.6	38.1	59.6	7.1	–	Lara-Cortés et al. (2013)
White meranti (*Shorea roxburghii*)	mg 100 g^{-1} dw	–	–	1360	610	–	–	Suksathan et al. (2021)
Guese rose (*Viburnum inopinatum*)	mg 100 g^{-1} dw	–	–	2370	790	–	–	Suksathan et al. (2021)

(continued)

Table 4.1 (continued)

Specie	Units	N	P	K	Ca	Mg	S	References
Beechwood (*Gmelina arborea*)	mg 100 g^{-1} dw	–	–	40	480	–	–	Suksathan et al. (2021)
Licorice (*Glycyrrhiza glabra*)	mg 100 g^{-1} dw	–	–	20	530	–	–	Suksathan et al. (2021)
Common poppy (*Papaver rhoeas* L.)	mg L^{-1}		17.377	261.785	1.60	6.55		Katarzyna et al. (2021)
Century plant (*Agave americana*)	mg 100 g^{-1}dw		240.1	1552.8	361.2	396.5	–	Mulík and Ozuna (2020)
Maguey (*Agave salmiana*)	g 100 g^{-1}dw	1.72	0.32	1.60	0.22	0.12	–	Pinedo-Espinoza et al. (2020)
Aloe (*Aloe vera*)	g 100 g^{-1}dw	2.07	0.22	2.16	0.25	0.15	–	Pinedo-Espinoza et al. (2020)
Coral tree (*Erythrina americana*)	g 100 g^{-1}dw	2.70	0.32	2.26	0.62	0.15	–	Pinedo-Espinoza et al. (2020)
Garambullo (*Myrtillocactus geometrizans*)	g 100 g^{-1}dw	2.0	0.26	2.90	0.21	0.18	–	Pinedo-Espinoza et al. (2020)
Prickly pear (*Opuntia ficus-indica*)	mg g^{-1} dw	–	–	11.99	18.87	–	–	Berrabah et al. (2019)
Milk thistle (*Silybum marianum*)	mg 100 g^{-1} dw	–	–	27.2	166.3	3.9	–	Maaloul et al. (2024)
Ivory thistle (*Silybum eburneum*)	mg 100 g^{-1} dw	–	–	93.6	230.2	13.5	–	Maaloul et al. (2024)
Cardamom (*Amomum maximum*)	mg 100 g^{-1} dw	–	–	278	56	–	–	Rachkeeree et al. (2018)
Galangal (*Alpinia galanga*)	mg 100 g^{-1} dw	–	–	589	41	–	–	Rachkeeree et al. (2018)
Ginger (*Zingiber officinale*)	mg 100 g^{-1} dw	–	–	737	41	–	–	Rachkeeree et al. (2018)
Bitter ginger (*Zingiber ottensii*)	mg 100 g^{-1} dw	–	–	547	140	–	–	Rachkeeree et al. (2018)
Pinecone ginger (*Zingiber zerumbet*)	mg 100 g^{-1} dw	–	–	310	83	–	–	Rachkeeree et al. (2018)
Saffron (*Crocus sativus*)	g 100 g^{-1}dw	–	0.279	1.395	0.139	0.113		Serrano-Díaz et al. (2013)
Moringa (*Moringa oleifera*)	% kg^{-1} dw	–	0.6	3.2	0.5	–	1.05	Monteiro et al. (2022)
Moringa (*M. oleifera*)	mg g^{-1}dw	–	–	0.15	0.52	0.04	–	Gowthami and Poornima (2017)
Tamarind (*Tamarindus indica*)	mg g^{-1}dw	–	–	0.05	0.12	0.07	–	Gowthami and Poornima (2017)
Wax gourd (*Benincasa hispida*)	mg g^{-1}dw	–	–	0.10	1.92	0.38	–	Gowthami and Poornima (2017)
Mahua (*Madhuca longifolia*)	mg 100 g^{-1} dw	–	–	480.0	177.4	95.9	–	Singh et al. (2021)
Elderberry (*Sambacus nigra*)	g kg^{-1}dw	–	–	45.52	4.473	14.61		Barbeş et al. (2020)
St. John's Wort (*Hypericum perforatum*)	g kg^{-1}dw	–	–	0.772	1.262	0.235	–	Barbeş et al. (2020)
Silver linden (*Tilia tomentosa*)	g kg^{-1}dw	–	–	0.840	4.832	0.176	–	Barbeş et al. (2020)

Rose (*Rosa meidrifora*)	g 100 g^{-1} dw	–	–	1.288	0.19	0.102	–	Mallick et al. (2024)
Rose (*Rosa damascene*)	g 100 g^{-1} dw	–	–	1.408	0.15	0.102	–	Mallick et al. (2024)
Rose (*Rosa grandiflora*)	g 100 g^{-1} dw	–	–	1.288	0.19	0.101	–	Mallick et al. (2024)
Rose (*Rosa alba*)	g 100 g^{-1} dw	–	–	0.984	0.19	0.102	–	Mallick et al. (2024)
Rose (*Rosa tradescant*)	g 100 g^{-1} dw	–	–	1.207	0.14	0.101	–	Mallick et al. (2024)
Rose (*Rosa polyantha*)	g 100 g^{-1} dw	–	–	1.247	0.16	0.101	–	Mallick et al. (2024)
Rose (*Rosa floribunda*)	g 100 g^{-1} dw	–	–	1.328	0.18	0.103	–	Mallick et al. (2024)
Rose (Rosa) 'Fragrant Delight'	g 100 g^{-1} dw	–	–	1.288	0.17	0.102	–	Mallick et al. (2024)

– data unavailable

Table 4.2 Overview of the mineral micronutrient concentrations found in edible flowers across a diverse range of plant species

Specie	Units	Fe	Cu	Mn	Mo	Zn	Na	B	References
Peony (*Paeonia officinalis*) "Sarah Bernardt"	mg kg^{-1}	35.66	3.03	10.59	–	12.61	0.04	–	Grzeszczuk et al. (2018)
Rosa (*Rosa hybrida*) "Kardinal", "Anjleeq", Maria shever	mg kg^{-1} dw	69.23 40.7 55.5	26.04 21.5 29.55	44.55 49.65 53.25	– – –	47.14 44.50 24.06	159.4 99.5 133.2	– – –	Jilani et al. (2012)
Peach (*Prunus persica*) "July flame", "Cary Mac", "Fair Time"	mg kg^{-1} dw	182.66 181.50 182.16	11.16 11.56 11.51	42.0 43.0 36.0	– – –	24.50 27.00 26.16	20.0 10.0 10.0	– – –	Calderón-Jurado et al. (2023)
Peach (*Prunus persica*) "Calanda", "Catherina", "Babygold5"	mg kg^{-1} dw	206.2 245.6 278.7	806.6 150.9 175.0	54.4 43.2 50.2	– – –	62.6 52.6 57.4	– – –	– – –	El-Jendoubi et al. (2013)
Crab apple (*Malus sylvestris*) "John Downie", "Manchurian"	mg kg^{-1} dw	83.70 92.10	11.80 14.50	83.70 93.40	– –	43.00 45.40	30.0 30.0	– –	Calderón-Jurado et al. (2023)
Lemmon's marigold (*Tagetes lemmonii*)	µg g^{-1} dw	54.0	9.57	24.3	–	38.0	–	–	Drava et al. (2020)
China pinks (*Dianthus chinensis*)	µg g^{-1} dw	54.6	6.17	33.3	–	54.7	–	–	Drava et al. (2020)
Basil (*Ocimum basilicum*)	µg g^{-1} dw	63.2	16.62	23.3	–	58.5	–	–	Drava et al. (2020)
Pot marigold (*Calendula officinalis*) "Antiqua Orange"	mg kg^{-1} fw	4.62	2.14	7.33	0.59	10.79	86.3	–	Mlcek et al. (2021)
Pot marigold (*Calendula officinalis*)	mg 100 g^{-1} dw	20.0	6.03	1.58	–	6.03	649.2	–	Araújo et al. (2019)
Marigold (*Tagetes erecta*)	mg 100 g^{-1} fw	1.026	0.104	0.303	–	0.568	0.01	–	Navarro-González et al. (2015)
Feverfew (*Chrysanthemum parthenium*)	mg kg^{-1}dw	5.83	2.35	7.33	0.31	5.94	113.3	–	Rop et al. (2012)
French marigold (*Tagetes patula*)	mg kg^{-1}dw	8.72	1.09	7.86	0.37	13.29	114.3	–	Rop et al. (2012)
French marigold (*T. patula*)	mg kg^{-1} fw	8.20	1.24	7.64	0.43	14.91	121.0	–	Mlcek et al. (2021)
French marigold (*T. patula*)	mg 100 g^{-1} dw	24.6	2.94	4.68	–	11.0	73.2	–	Araújo et al. (2019)
Pansy (*Viola x wittrockiana*)	mg kg^{-1}dw	7.29	1.95	7.93	0.84	11.52	132.0	–	Rop et al. (2012)
Pansy (*V. x wittrockiana*)	mg 100 g^{-1} fw	0.84	0.091	1.10	–	1.18	70.9	–	González-Barrio et al. (2018)
Pansy (*V. x wittrockiana*) "Series Delta Mix"	mg kg^{-1}dw	–	3.06	97.58	–	92.17	4.3	–	de Oliveira et al. (2024)
Pumpkin (*Cucurbita maxima*)	mg 100 g^{-1} dw	4.23	0.75	–	–	–	11.5	–	Ghosh and Rana (2021)
Nasturtium (*Tropaeolum majus*)	mg 100 g^{-1} fw	0.55	0.472	0.397	–	0.660	0.01	–	Navarro-González et al. (2015)

Nasturtium (*T. majus*)	mg kg^{-1} fw	6.52	1.22	5.79	0.32	8.89	85.2	–	Mlcek et al. (2021)
Nasturtium (*T. majus*)	mg 100 g^{-1} dw	20.4	1.73	6.85	–	15.5	185.6		Araújo et al. (2019)
Paracress (*Spilanthes oleracea*)	mg 100 g^{-1} fw	1.500	0.165	0.555	–	0.543	0.01	–	Navarro-González et al. (2015)
Tuberous Begonia (*Begonia* ^{x} *tuberhybrida*)	mg kg^{-1} fw	3.12	1.28	5.77	0.76	4.78	70.3	–	Mlcek et al. (2021)
Rose (*Rosa*) "Gloria Dei'	mg kg^{-1} fw	4.02	2.31	3.41	0.70	4.62	79.2	–	Mlcek et al. (2021)
Kapok (*Bombax costatum*)	µg g^{-1}dw	1.5	11.9	22.4	–	67.1	–	–	Maiga et al. (2005)
Snapdragon (*Antirrhinum majus*)	mg 100 g^{-1} fw	0.55	0.032	0.55	–	0.41	5.7	–	González-Barrio et al. (2018)
Snapdragon (*A. majus*)	µg g^{-1} dw	33.9	3.55	32.2	–	22.6	–	–	Drava et al. (2020)
Coriander (*Coriandrum sativum*)	mg 100 g^{-1} dw	28.4	3.57	7.28	–	21.2	395.1	–	Araújo et al. (2019)
Rosemary (*Rosmarinus officinalis*)	mg 100 g^{-1} dw	20.2	2.80	8.97	–	7.08	72.5	–	Araújo et al. (2019)
Heartsease (*Viola tricolor*)	mg 100 g^{-1} dw	38.6	2.11	6.74	–	15.2	145.4	–	Araújo et al. (2019)
Heartsease (*V. tricolor*) "Hortensis"	mg kg^{-1}dw	–	21.63	82.64	–	77.45	3.0	–	de Oliveira et al. (2024)
Roselle (*Hibiscus sabdarifa*)	µg g^{-1} dw	14.0	5.88	8.1	–	25.8	–	–	Drava et al. (2020)
Rose (*Rosa* sp)	mg kg^{-1} dw	21.3	2.25	5.29	–	11.6	79.5	–	Araújo et al. (2019)
Squash (*Cucurbita moschata*)	mg g^{-1}dw	66.84	11.86	48.15	–	67.42	0.15	–	Marchioni et al. (2024)
Royal fuchsia (*Fuchsia regia*)	mg g^{-1}dw	89.24	18.83	40.51	–	72.72	90.5	–	Marchioni et al. (2024)
Chinese pink (*Dianthus chinensis*)	mg g^{-1}dw	73.33	11.23	54.91	–	51.60	0.67	–	Marchioni et al. (2024)
Horned pansies (*Viola cornuta*)	mg g^{-1}dw	61.00	9.24	61.59	–	76.92	0.24	–	Marchioni et al. (2024)
Curry plant (*Helichrysum italicum*)	µg g^{-1} dw µg g^{-1} fw	586.0 356.6	92.0 56.0	216.0 131.5	– –	310.0 188.7	– –	– –	Primitivo et al. (2022)
Tea (*Camellia sinensis*) "QN06"	mg kg^{-1} dw	68.2	5.6	1542.3	–	19.6	–	19.1	Jia et al. (2016)
Wild cabbage (*Trachyandra ciliata*)	mg 100 g^{-1}dw	10.8	0.25	1.95	–	2.3	220.0	–	Ngxabi et al. (2025)
Broom (*Cytisus striatus*)	mg kg^{-1}dw	81.9	20.6	920.2	–	50.9	144.7	–	Caramelo et al. (2024)
Sandkool (*Trachyandra divaricata*)	mg 100 g^{-1}dw	12.3	0.5	4.23	–	10.8	220.0	–	Bulawa et al. (2022)
Dahlia (*Dahlia* spp.)	mg kg^{-1} dw	207.82	6.67	24.39	–	30.58	–	20.6	Costa et al. (2022)
Dahlia (*Dahlia* spp.)	mg 100 g^{-1}dw	3.2	0.2	0.03	–	3.2	15.7	–	Lara-Cortés et al. (2014)
Common poppy (*Papaver rhoeas*)	mg L^{-1}	0.288	0.050	0.037	0.011	0.192	5.2	–	Katarzyna et al. (2021)
White meranti (*Shorea roxburghii*)	mg 100 g^{-1} dw	7.53	–	–	–	–	1.7	–	Suksathan et al. (2021)

(continued)

Table 4.2 (continued)

Specie	Units	Fe	Cu	Mn	Mo	Zn	Na	B	References
Guelder rose (*Viburnum inopinatum*)	mg 100 g^{-1} dw	4.48	–	–	–	–	3.3	–	Suksathan et al. (2021)
Beechwood (*Gmelina arborea*)	mg 100 g^{-1} dw	6.16	–	–	–	–	14.1	–	Suksathan et al. (2021)
Licorice (*Glycyrrhiza glabra*)	mg 100 g^{-1} dw	6.37	–	–	–	–	31.9	–	Suksathan et al. (2021)
Century plant (*Agave americana*)	mg 100 g^{-1}dw	7.7	0.3	1.1	–	4.9	9.5	–	Mulík and Ozuna (2020)
Maguey (*Agave salmiana*)	mg 100 g^{-1}dw	86.60	6.50	29.40	–	46.60	–	55.5	Pinedo-Espinoza et al. (2020)
Aloe (*Aloe vera*)	mg 100 g^{-1}dw	77.29	6.61	37.88	–	50.29	–	60.8	Pinedo-Espinoza et al. (2020)
Coral tree (*Erythrina americana*)	mg 100 g^{-1}dw	64.50	13.20	53.30	–	43.10	–	62.6	Pinedo-Espinoza et al. (2020)
Garambullo (*Myrtillocactus geometrizans*)	mg 100 g^{-1}dw	65.80	6.20	40.90	–	37.70	–	53.1	Pinedo-Espinoza et al. (2020)
Prickly pear (*Opuntia ficus-indica*)	mg g^{-1} dw	–	–	–	–	–	38.6	–	Berrabah et al. (2019)
Milk thistle (*Silybum marianum*)	mg 100 g^{-1} dw	0.3	0.1	0.05	–	–	9.9	–	Maaloul et al. (2024)
Ivory thistle (*Silybum eburneum*)	g 100 g^{-1} dw	0.3	0.1	0.1	–	–	64.9	–	Maaloul et al. (2024)
Ginger (*Zingiber officinale*)	mg 100 g^{-1} dw	0.32	–	–	–	–	–	–	Rachkeeree et al. (2018)
Saffron (*Crocus sativus*)	g 100 g^{-1}dw	0.016	–	–	–	–	0.01	–	Serrano-Díaz et al. (2013)
Moringa (*Moringa oleifera*)	mg kg^{-1} dw	88.9	8.4	30	–	43	–	–	Monteiro et al. (2022)
Mahua (*Madhuca longifolia*)	mg 100 g^{-1} dw	95	–	–	–	–	135.2	–	Singh et al. (2021)
Elderberry (*Sambacus nigra*)	g kg^{-1}dw	–	–	–	–	–	0.10	–	Barbeş et al. (2020)
St. John's Wort (*Hypericum perforatum*)	g kg^{-1}dw	–	–	–	–	–	0.12	–	Barbeş et al. (2020)
Silver linden (*Tilia tomentosa*)	g kg^{-1}dw	–	–	–	–	–	0.14	–	Barbeş et al. (2020)
Rose (*Rosa meidrifora*)	g 100 g^{-1} dw	0.039	–	–	–	–	0.079	–	Mallick et al. (2024)
Rose (*Rosa damascene*)	g 100 g^{-1} dw	0.060	–	–	–	–	0.078	–	Mallick et al. (2024)
Rose (*Rosa grandiflora*)	g 100 g^{-1} dw	0.074	–	–	–	–	0.055	–	Mallick et al. (2024)
Rose (*Rosa alba*)	g 100 g^{-1} dw	0.090	–	–	–	–	0.072	–	Mallick et al. (2024)
Rose (*Rosa hybrid*: Crystalline' × 'Playgirl)	g 100 g^{-1} dw	0.037	–	–	–	–	0.092	–	Mallick et al. (2024)
Rose (*Rosa tradescant*)	g 100 g^{-1} dw	0.020	–	–	–	–	0.065	–	Mallick et al. (2024)
Rose (*Rosa polyantha*)	g 100 g^{-1} dw	0.017	–	–	–	–	0.079		Mallick et al. (2024)
Rose (*Rosa floribunda*)	g 100 g^{-1} dw	0.028	–	–	–	–	0.062		Mallick et al. (2024)
Rose (*Rosa*) 'Fragrant Delight'	g 100 g^{-1} dw	0.017	–	–	–	–	0.085		Mallick et al. (2024)

– data unavailable

4.3 Suggested Flower Consumption Based on Mineral Content

Edible flowers contain mineral elements in varying concentrations and, in some cases, can surpass those found in traditional horticultural crops (Jadhav et al. 2023; Mondal and Saha 2024). However, a key drawback affecting their consumption is the large volume required to meet daily nutritional requirements. Therefore, they should be regarded solely as a complementary source in human nutrition and healthcare.

Multiple reports identify edible flowers as sources of phosphorus (P), potassium (K), calcium (Ca), magnesium (Mg), sulphur (S), manganese (Mn), iron (Fe), copper (Cu), molybdenum (Mo), zinc (Zn), boron (B), and sodium (Na) (Grzeszczuk et al. 2018; Singh et al. 2021; Mallick et al. 2024). It has been observed that the daily requirement for essential nutrients, such as phosphorus, varies between 100 and 1250 mg. In this context, the phosphorus content in flowers of species such as peony (*Paeonia officinalis* cv Sarah Bernardt) (6.62 g kg^{-1} dw) (Grzeszczuk et al. 2018), coriander (*Coriandrum sativum*) (6566 mg 100 g^{-1} dw), and tea (*Camellia sinensis* cv QN06) (2856.8 mg kg^{-1} dw) has been reported (Araújo et al. 2019). Potassium requirements can also be supplemented through the intake of edible flowering species listed in Table 4.1, including peony (*Paeonia officinalis* cv Sarah Bernardt) (8.15 g kg^{-1} dw) (Grzeszczuk et al. 2018), peach (Prunus persica cvs July Flame, Cary Mac, and Fair Time) (11.2–11.3 g kg^{-1} dw) (Calderón-Jurado et al. 2023), crab apple (*Malus sylvestris* cvs. John Downie and Manchurian) (21.5–22.0 g kg^{-1} dw) (Calderón-Jurado et al. 2023), pot marigold (*Calendula officinalis*) (7369 mg 100 g^{-1} dw), tea (*Camellia sinensis* cv QN06) (7926.7 mg kg^{-1} dw), french marigold (*Tagetes patula*) (4496 mg 100 g^{-1} dw) (Araújo et al. 2019), pansy (*Viola* × *wittrockiana*) (3964.8 mg kg^{-1} fw) (Rop et al. 2012), and feverfew (*Chrysanthemum parthenium*) (3600.3 mg kg^{-1} fw) (Rop et al. 2012), among others (Tables 4.1 and 4.3). Similarly, for calcium, the required intake range fluctuates between 200 and 3400 mg, with an interesting option being broom flowers (*Cytisus striatus*) (5535.0 mg kg^{-1} dw) (Caramelo et al. 2024), elderberry (*Sambucus nigra*) (4473 mg kg^{-1} dw) (Barbeş et al. 2020), tea (*Camellia sinensis* cv QN06) (4305.5 mg kg^{-1} dw) (Jia et al. 2016), and silver linden (*Tilia tomentosa*) (4832 mg kg^{-1} dw) (Barbeş et al. 2020). However, the majority of the flowers analysed can be consumed and provide part of the daily calcium requirement. Magnesium is an essential element, with requirements ranging between 30 and 410 mg. The highest needs are associated with adolescents over 14 years of age, women, and pregnant or lactating adolescents (310–410 mg). One of the notable plants rich in magnesium includes broom (*Cytisus striatus*) (2870.2 mg kg^{-1} dw) (Caramelo et al. 2024), tea (*Camellia sinensis* cv QN06) (1490.9 mg kg^{-1} dw) (Jia et al. 2016), coriander (*Coriandrum sativum*) (886.1 mg 100 g^{-1} dw) (Araújo et al. 2019), and french marigold (*Tagetes patula*) (629.1 mg 100 g^{-1} dw) (Araújo et al. 2019), among others.

Among the most important micronutrients with a significant impact on human health are Fe, Cu, Mn, Mo, Zn, Na, and B. It is worth noting that iron requirements

Table 4.3 Essential nutrient requirements by growth stage and sex

Growth stage and sex	P	K	Ca	Mg	Fe	Cu	Mn	Mo	Zn	Na	B
Infants age 6 months	100	400	200	30	0.27	0.2	0.003	0.002	2	120	–
Infants aged 7–12 months	275	860	260	75	11	0.22	0.6	0.003	3	370	–
Children aged 1–3 years	460	2000	700	80	7	0.34	1.2	0.017	3	800	3
Children aged 4–8 years	500	2300	1000	130	10	0.44	1.5	0.022	5	1000	6
Children aged 9–13 years	1250	2500	1300	240	8	0.7	1.9	0.0019	8	1200	11
Girls aged 9–13 years	1250	2300	1300	240	8	0.7	1.6	0.034	8	1200	11
Adolescents aged 14–18 years (minors)	1250	3000	1300	410	11	0.89	2.2	0.043	11	1500	17
Adolescents ageds 14 to18 years (girls)	1250	2300	1300	360	15	0.89	1.6	0.043	9	1500	17
Adults (men) over the age of 19	700	3400	–	–		0.90	2.3	0.045	11	1500	20
Adults (women) over the age of 19	700	2600	–	–		0.90	1.8	0.045	8	1500	20
Pregnant adolescents	1250	2600	1300	400	27	1.0	2.0	0.05	12	1500	17
Pregnant women	700	2900	1000	350–360		1.0	2.0	0.05	11	1500	20
Adolescent lactating mothers	1250	2500	1300	360		1.3	2.6	0.05	13	1500	17
Lactating women	700	2800	1300	310–320		1.3	2.6	0.05	12	1500	20
Adults (19–50 años)	–	–	1000	–	8–18	–	–	–	–	–	–
Male adults (51–70 years)	–	–	1000	–	8	–	–	–	–	–	–
Female adults (51–70 years)	–	–	1200	–	8	–	–	–	–	–	–
Senior adults (≥71 years)	–	–	1200	–	–	–	–	–	–	–	–

The data are presented as the mean daily intake in mg/day, as advised by the National Institutes of Health, an agency of the United States Department of Health and Human Services (https://ods.od.nih.gov/HealthInformation/RecursosEnEspanol.aspx). Regarding sodium, the data are derived from the World Health Organisation (https://www.who.int/publications/i/item/9789241504836)
– data unavailable

are relatively low, fluctuating between 0.27 and 27 mg (Table 4.3). The majority of the flowering species reported in this chapter contain significant levels of Fe that can help meet these micronutrient needs. Examples include peach (*Prunus persica*) flowers (206.2–278.7 and 181.50–182.6 mg kg^{-1} dw) (El-Jendoubi et al. 2013; Calderón-Jurado et al. 2023) and dahlia (*Dahlia* spp.) (207.82 mg kg^{-1} dw) (Costa et al. 2022). In general, all the edible flowers analysed contain concentrations of Cu that exceed the daily requirement for the population (0.2–2.6 mg) and can therefore serve as a natural complementary source. This includes peach inflorescences for all reported cultivars, with the most outstanding being cultivar Calanda (806.6 mg kg^{-1} dw) (El-Jendoubi et al. 2013).

For micronutrients such as manganese (0.003 and 2.6 mg) and molybdenum (0.002–0.05 mg), the daily requirements are relatively small, and all the reported edible flowering plant species can be natural sources of these micronutrients. For Mn, notable examples include the flowers of tea (*Camellia sinensis*) 'QN06' (1542.3 mg kg^{-1} dw) (Jia et al. 2016) and broom (*Cytisus striatus*) (920.2 mg kg^{-1} dw) (Caramelo et al. 2024). With the exception of nasturtium (*Tropaeolum majus*) (0.660 mg 100 g^{-1} fw) (Navarro-González et al. 2015), snapdragon (*Antirrhinum majus*) (0.41 mg 100 g^{-1} fw) (González-Barrio et al. 2018), common poppy (*Papaver rhoeas*) (0.192 mg L^{-1}) (Katarzyna et al. 2021), pansy (*Viola* × *wittrockiana*) (1.18 mg 100 g^{-1} fw) (González-Barrio et al. 2018), and paracress (*Spilanthes oleracea*) (0.543 mg 100 g^{-1} fw) (González-Barrio et al. 2018), the rest of the reported flowers present Zn values within the daily requirement range for this micronutrient. In the case of Na, the values found ranged from 0.01 mg 100 g^{-1} fw (in marigold and paracress) (Navarro-González et al. 2015) to 649.2 mg 100 g^{-1} dw (in pot marigold) (Araújo et al. 2019). However, daily sodium requirements range from 120 to 1500 mg/day. Therefore, flower consumption may only serve as a natural source for this element in infants between 6–12 months and children between 1–3 years of age, with requirements of 120, 370, and 800 mg, respectively. Finally, the body's need for B varies, ranging from 3 mg in children between 1 and 3 years of age, to up to 20 mg in adults over 19 years of age, as well as pregnant and lactating women. Although reports for B are scarce, in general, all species of edible flowers are an excellent option for supplementing the intake of this micronutrient. Notably, some species with edible flowers from arid climates in Mexico, such as maguey, aloe, coral tree, and garambullo, are particularly outstanding, with values of 55.5, 60.8, 62.6, and 53.1 mg 100 g^{-1} dw, respectively (Pinedo-Espinoza et al. 2020).

4.4 Conclusions and Perspectives

Edible flowers from both cultivated and wild plants can serve as complementary natural sources of various phytochemical compounds, such as vitamins, proteins, secondary metabolites, and essential mineral nutrients. However, the mineral composition of most species with edible flowers remains largely unknown. Moreover, the concentration of mineral nutrients shows considerable variability among species and cultivars, influenced by numerous factors, including agronomic management (such as mineral nutrition, irrigation, and others), edaphoclimatic conditions, methods of obtaining, processing, and evaluating samples, as well as how the results are expressed (on a dry or fresh basis), among others.

On the other hand, the daily intake requirements for mineral nutrients, as referenced in this chapter, vary according to age, sex, and the periods of gestation and lactation. Therefore, the consumption of flowers could serve as a complementary source in the daily diet. However, special care must be taken regarding the presence of compounds with anti-nutrient activity, infectious microorganisms, pesticides, heavy metals, and atmospheric pollutants, as these could pose significant health risks. Consequently, the generation and validation of scientific information is crucial to facilitate the development and implementation of standards for the agronomic management, storage, and sale of such food products.

References

Araújo S, Matos C, Correia E, Antunes MC (2019) Evaluation of phytochemicals content, antioxidant activity and mineral composition of selected edible flowers. Qual Assur Saf Crops Foods 11(5):471–478

Barbeş L, Bărbulescu A, Stanciu G (2020) Statistical analysis of mineral elements content in different melliferous plants from the Dobrogea region, Romania. Rom Rep Phys 72:705

Berrabah H, Taïbi K, Ait Abderrahim L, Boussaid M (2019) Phytochemical composition and antioxidant properties of prickly pear (*Opuntia ficus-indica* L.) flowers from the Algerian germplasm. J Food Meas Charact 13(2):1166–1174

Bulawa B, Sogoni A, Jimoh MO, Laubscher CP (2022) Potassium application enhanced plant growth, mineral composition, proximate and phytochemical content in *Trachyandra divaricata Kunth* (Sandkool). Plants 11(22):3183

Calderón-Jurado M, Cruz-Álvarez O, Juárez-López P, Hernández-Rodríguez OA, Orozco-Meléndez LR, Morales-Maldonado ER et al (2023) Bioactive compounds, minerals and antioxidants of edible flowers of peach and apple. Int J Food Prop 26(1):1855–1866

Caramelo D, Pitacas I, Vitória C, Gonçalves J, Gominho J, Gallardo E, Anjos O (2024) Chemical composition and nutritional value of flowers and fruits of *Cytisus striatus* (Hill) Rothm. Plants 13(15):2121

Castañeda-Rodríguez R, Quiles A, Llorca E, Ozuna C (2024) How to cook Yucca spp. flowers? An analysis of their chemical composition, microstructure, and bioactive compound bioaccessibility. Appl Food Res 4(1):100414

Costa PA, Souza DCD, Ossani PC, Mendes MHA, Silva MLDS, Carvalho EEN, Resende LV (2022) Nutritional and functional compounds in dahlia flowers and roots. Braz J Food Technol 25:e2022029

De Oliveira I, Chrysargyris A, Finimundy TC, Carocho M, Santos-Buelga C, Calhelha RC et al (2024) Magnesium and manganese induced changes on chemical, nutritional, antioxidant and antimicrobial properties of the pansy and Viola edible flowers. Food Chem 438:137976

Demasi S, Caser M, Scariot V (2023) Hot and cold drying of edible flowers affect metabolite patterns of extracts and decoctions. Folia Hortic 35(1):193–207

Drava G, Iobbi V, Govaerts R, Minganti V, Copetta A, Ruffoni B, Bisio A (2020) Trace elements in edible flowers from Italy: further insights into health benefits and risks to consumers. Molecules 25(12):2891

Dusemund B, Rietjens IMCM, Cartus A, Schaefer B, Lampen A (2017) Pflanzliche Kontaminanten in Lebensmitteln. Bundesgesundheitsblatt-Gesundheitsforschung-Gesundheitsschutz 60:728–736

El-Jendoubi H, Abadía J, Abadía A (2013) Assessment of nutrient removal in bearing peach trees (*Prunus persica* L. Batsch) based on whole tree analysis. Plant Soil 369:421–437

Fernandes L, Casal S, Pereira JA, Saraiva JA, Ramalhosa E (2020) An overview on the market of edible flowers. Food Rev Intl 36(3):258–275

Figueredo-Urbina CJ, Álvarez-Ríos GD, Cortés Zárraga L (2021) Edible flowers commercialized in local markets of Pachuca de Soto, Hidalgo, Mexico. Bot Sci 100(1):120–138

Ghosh P, Rana SS (2021) Physicochemical, nutritional, bioactive compounds and fatty acid profiling of Pumpkin flower (*Cucurbita maxima*), as a potential functional food. SN Appl Scis 3:1–14

González-Barrio R, Periago MJ, Luna-Recio C, Garcia-Alonso FJ, Navarro-González I (2018) Chemical composition of the edible flowers, pansy (*Viola wittrockiana*) and snapdragon (*Antirrhinum majus*) as new sources of bioactive compounds. Food Chem 252:373–380
Gowthami N, Poornima S (2017) Proximate and mineral analysis of a few edible flowers used in Coimbatore district, Tamil Nadu. Carpathian J Food Sci Technol 9(4):54–60
Grzeszczuk M, Stefaniak A, Meller E, Wysocka G (2018) Mineral composition of some edible flowers. J Elem 23(1):151–162
Gupta A, Sharma S, Rajput D, Patil UK (2024) Medicinal and culinary importance of edible flowers of indian origin: an in-depth review. Discov Food 4(1):148
Hernández-Epigmenio F, García-Mateos MDR, Sosa-Montes E, Mejía-Muñoz JM, Fernández-Pavia YL, Cruz-Álvarez O, Martínez-Damián MT (2022) Phenolic profile and nutritional value of *Dahlia x hortorum* flowers. Revista Chapingo Serie Horticultura 28(3):161–174
Jadhav HB, Badwaik LS, Annapure U, Casonova F, Alaskar K (2023) A review on the journey of edible flowers from farm to consumer's plate. Appl Food Res 3(2):100312
Jia S, Wang Y, Hu J, Ding Z, Liang Q, Zhang Y, Wang H (2016) Mineral and metabolic profiles in tea leaves and flowers during flower development. Plant Physiol Biochem 106:316–326
Jilani MI, Ahmad MI, Hanif R, Nadeem R, Hanif MA, Khan MA et al (2012) Proximate analysis and mineral profile of three elite cultivars of *Rosa hybrida* flowers. Pak J Bot 44(5):1711–1714
Katarzyna J, Karolina J, Patrycja K, Mateusz B, Izabela G (2021) Mineral composition and antioxidant potential in the common poppy (*Papaver rhoeas* L.) petal infusions. Biol Trace Elem Res 199(1):371–381
Kibar B, Temel S (2016) Evaluation of mineral composition of some wild edible plants growing in the Eastern Anatolia Region Grasslands of Turkey and consumed as vegetable. J Food Process Preserv 40(1):56–66
Lara-Cortés E, Osorio-Díaz P, Jiménez-Aparicio A, Bautista-Baños S (2013) Nutritional content, functional properties and conservation of edible flowers. Archivos Latinoamericanos de Nutricion 63(3):197–208
Lara-Cortés E, Martín-Belloso O, Osorio-Díaz P, Barrera-Necha LL, Sánchez-López JA, Bautista-Baños S (2014) Antioxidant capacity, nutritional and functional composition of edible Dahlia flowers. Revista Chapingo. Serie Horticultura 20(1):101–116
Lu B, Li M, Yin R (2016) Phytochemical content, health benefits, and toxicology of common edible flowers: a review (2000-2015). Crit Rev Food Sci Nutr 56(S1):S130–S148
Maaloul S, Mahmoudi M, Mighri H, Ghzaiel I, Bouhamda T, Boughalleb F et al (2024) Tunisian silybum species: important sources of polyphenols, organic acids, minerals, and proteins across various plant organs. Plants 13(7):989
Maiga A, Diallo D, Bye R, Paulsen BS (2005) Determination of some toxic and essential metal ions in medicinal and edible plants from Mali. J Agric Food Chem 53(6):2316–2321
Mallick SR, Hassan J, Hoque MA, Sultana H, Kayesh E, Ahmed M et al (2024) Color, proximate composition, bioactive compounds and antinutrient profiling of rose. Sci Rep 14(1):21690
Marchioni I, Gabriele M, Carmassi G, Ruffoni B, Pistelli L, Pistelli L, Najar B (2024) Phytochemical, nutritional and mineral content of four edible flowers. Foods 13(6):939
Marcos-Gómez R, Vera-Guzmán AM, Pérez-Ochoa ML, Martínez-Martínez L, Hernández-Delgado S, Martínez-Sánchez D, Chávez-Servia JL (2024) Phenolic compounds and antioxidant activity in edible flower species from Oaxaca. Appl Sci 14(8):3136
Martínez-Damián MT, Mejía-Muñoz JM, Colinas-León MT, Hernández-Epigmenio F, Cruz-Alvarez O (2021) Nutritional value, bioactive compounds and capacity antioxidant in edible flowers of dahlia. Acta Scientiarum Polonorum Hortorum Cultus 20(5):63–72
Matyjaszczyk E, Śmiechowska M (2019) Edible flowers. Benefits and risks pertaining to their consumption. Trends Food Sci Technol 91:670–674
Mlček J, Rop O (2011) Fresh edible flowers of ornamental plants – A new source of nutraceutical foods. Trends Food Sci Technol 22(10):561–569
Mlček J, Plaskova A, Jurikova T, Sochor J, Baron M, Ercisli S (2021) Chemical, nutritional and sensory characteristics of six ornamental edible flowers species. Foods 10(9):2053

Mondal P, Saha P (2024) Profiling of mineral content from different edible flowers of West Bengal. J Nat Remed 24(10):2157–2171

Monteiro J, Scotti-Campos P, Pais I, Figueiredo AC, Viegas D, Reboredo F (2022) Elemental composition, total fatty acids, soluble sugar content and essential oils of flowers and leaves of *Moringa oleifera* cultivated in Southern Portugal. Heliyon 8(12):e12647

Mulík S, Ozuna C (2020) Mexican edible flowers: cultural background, traditional culinary uses, and potential health benefits. Int J Gastron Food Sci 21:100235

Navarro-González I, González-Barrio R, García-Valverde V, Bautista-Ortín AB, Periago MJ (2015) Nutritional composition and antioxidant capacity in edible flowers: characterisation of phenolic compounds by HPLC-DAD-ESI/MS^n. Int J Mol Sci 16(1):805–822

Ngxabi S, Jimoh MO, Sogoni A, Laubscher CP, Rautenbach F, Kambizi L (2025) Salinity influenced proximate, minerals, anti-nutrients and phytochemical composition of *Trachyandra ciliata* Kunth (Wild Cabbage): a promising edible halophyte. Food Sci Nutr 13(1):e4755

Pensamiento-Niño CA, Campos-Montiel RG, Añorve-Morga J, Ramírez-Moreno E, Ascacio-Valdés JA, Hernández-Fuentes AD (2021) Nutritional characterization of the functional and antioxidant activity of cactus flowers from Hidalgo, Mexico. Appl Sci 11(13):5965

Pensamiento-Niño CA, Castañeda-Ovando A, Añorve-Morga J, Hernández-Fuentes AD, Aguilar-Arteaga K, Ojeda-Ramírez D (2024) Edible flowers and their relationship with human health: biological activities. Food Rev Intl 40(1):620–639

Pinedo-Espinoza JM, Gutiérrez-Tlahque J, Santiago-Saenz YO, Aguirre-Mancilla CL, Reyes-Fuentes M, López-Palestina CU (2020) Nutritional composition, bioactive compounds and antioxidant activity of wild edible flowers consumed in semiarid regions of Mexico. Plant Foods Hum Nutr 75:413–419

Pinela J, Carvalho AM, Ferreira IC (2017) Wild edible plants: nutritional and toxicological characteristics, retrieval strategies and importance for today's society. Food Chem Toxicol 110:165–188

Pires EDO Jr, Di Gioia F, Rouphael Y, García-Caparrós P, Tzortzakis N, Ferreira IC et al (2023) Edible flowers as an emerging horticultural product: a review on sensorial properties, mineral and aroma profile. Trends Food Sci Technol 137:31–54

Pires TC, Dias MI, Barros L, Ferreira IC (2017) Nutritional and chemical characterization of edible petals and corresponding infusions: valorization as new food ingredients. Food Chem 220:337–343

Primitivo MJ, Neves M, Pires CL, Cruz PF, Brito C, Rodrigues AC et al (2022) Edible flowers of *Helichrysum italicum*: composition, nutritive value, and bioactivities. Food Res Int 157:111399

Quintaes KD, Diez-Garcia RW (2015) The importance of minerals in the human diet. In: Handbook of mineral elements in food. Wiley, West Sussex, pp 1–21

Rachkeeree A, Kantadoung K, Suksathan R, Puangpradab R, Page PA, Sommano SR (2018) Nutritional compositions and phytochemical properties of the edible flowers from selected Zingiberaceae found in Thailand. Front Nutr 5:3

Rao V, Poonia A (2024a) Bioactive compounds, nanoparticle synthesis and potential utilization of edible flowers for development of functional dairy products: a review. J Food Sci Technol 61(6):1053–1068. https://doi.org/10.1007/s13197-023-05853-6

Rao V, Poonia A (2024b) Microbial biosynthesis of selenium nanoparticles using probiotic strain and its characterization. J Food Meas Charact 18(7):5477–5488. https://doi.org/10.1007/s11694-024-02581-z

Rivas-García L, Crespo-Antolín L, Forbes-Hernández TY, Romero-Márquez JM, Navarro-Hortal MD, Arredondo M et al (2023) Bioactive properties of tagetes erecta edible flowers: polyphenol and antioxidant characterization and therapeutic activity against ovarian tumoral cells and caenorhabditis elegans tauopathy. Int J Mol Sci 25(1):280

Rivera-Espejel EA, Cruz-Alvarez O, Mejía-Muñoz JM, García-Mateos MDR, Colinas-León MTB, Martínez-Damián MT (2019) Physicochemical quality, antioxidant capacity and nutritional value of edible flowers of some wild dahlia species. Folia Horticul 31(2):331–342

Rodrigues H, Spence C (2023) Looking to the future, by studying the history of edible flowers. Int J Gastron Food Sci 34:100805

Rop O, Mlcek J, Jurikova T, Neugebauerova J, Vabkova J (2012) Edible flowers -a new promising source of mineral elements in human nutrition. Molecules 17(6):6672–6683

Serrano-Díaz J, Sánchez AM, Martínez-Tomé M, Winterhalter P, Alonso GL (2013) A contribution to nutritional studies on Crocus sativus flowers and their value as food. J Food Compos Anal 31(1):101–108

Sik B, Ajtony Z, Lakatos E, Gál LH, Székelyhidi R (2024) Evaluation of the physical, antioxidant, and organoleptic properties of biscuits fortified with edible flower powders. Food Sci Nutr 12(5):3265–3272

Silva CS, Moutinho C, Ferreira da Vinha A, Matos C (2019) Trace minerals in human health: iron, zinc, copper, manganese and fluorine. Int J Sci Res Methodol 13(3):57–80

Singh V, Singh J, Kushwaha R, Singh M, Kumar S, Rai AK (2021) Assessment of antioxidant activity, minerals and chemical constituents of edible mahua (*Madhuca longifolia*) flower and fruit of using principal component analysis. Nutr Food Sci 51(2):387–411

Sotelo A, López-García S, Basurto-Peña F (2007) Content of nutrient and antinutrient in edible flowers of wild plants in Mexico. Plant Foods Hum Nutr 62:133–138

Suksathan R, Rachkeeree A, Puangpradab R, Kantadoung K, Sommano SR (2021) Phytochemical and nutritional compositions and antioxidants properties of wild edible flowers as sources of new tea formulations. NFS J 24:15–25

Takahashi JA, Rezende FAGG, Moura MAF, Dominguete LCB, Sande D (2020) Edible flowers: bioactive profile and its potential to be used in food development. Food Res Int 129:108868

Tuladhar A (2021) Eating flowers-the consumption and market of edible flowers. 三重大学教養教育院研究紀要 6:25–33

Welch RM (2002) The impact of mineral nutrients in food crops on global human health. Plant Soil 247:83–90

Wen H, Li S, Wei Y, Dong J, Liang Z, Guo L et al (2025) Yunnan edible flowers and their potential in future foods: focus on ethnological applications, chemical and pharmacological research. J Future Foods 5(2):11–133

Chapter 5
Effect of Packaging Materials and Storage Period on Phytonutrient Content of Edible Flowers

Tomás Rivas-García, Felipe Ascencio-Valle Felipe, Ana Gisela Reyes-Alvarado, and Rubí A. Martínez-Camacho

5.1 Introduction

Flowers are inherently short-lived due to their biological programming. Some species, like *Hemerocallis* and *Hibiscus*, complete their entire bloom cycle in just a single day. These short-lived flowers tend to have more active metabolic and respiratory processes compared to other parts of the plant (Serek and Reid 2024). Edible flowers are even more delicate postharvest, as they typically have shorter stems and lack a consistent water source during storage, making them highly susceptible to rapid quality decline (Ikeura et al. 2023). As a result, their handling after harvest must be meticulous to preserve their appearance, texture, and safety for consumption (Marchioni et al. 2020). Without timely and appropriate postharvest treatment, these flowers can quickly degrade due to enzymatic activity, microbial growth, and other biochemical processes (Shantamma et al. 2021).

Interest in edible flowers has risen in recent years, primarily due to their nutritional properties and functional health benefits (Benvenuti and Mazzoncini 2021). In recent years, the antioxidant properties of edible flowers have indeed been extensively documented employing chemical-related antioxidant assays such as DPPH, FRAP, ABTS etc., (Rao and Poonia 2024a). They are rich in health-promoting phytochemicals like phenolic compounds, carotenoids, and tocols. Phenolics, a diverse

T. Rivas-García
SECIHTI-Universidad Autonoma Chapingo, Texcoco, Estado de Mexico, Mexico

F. A.-V. Felipe · A. G. Reyes-Alvarado
SECIHTI-Centro de Investigaciones Biológicas del Noroeste, Instituto Politécnico, La Paz, Baja California Sur, Mexico

R. A. Martínez-Camacho (✉)
Instituto Tecnológico de La Paz, Tecnológico Nacional de México, La Paz, Baja California Sur, Mexico

A. Poonia et al. (eds.), *Edible Flowers: Source of Phytonutrients, Valorization and Technological Advancements*, https://doi.org/10.1007/978-3-032-00667-7_5

group of plant metabolites, include flavonoids, lignans, tannins, phenolic acids, and other subtypes, which are known to support immune function and may offer antimicrobial, antioxidant, anti-inflammatory, and anticancer effects (Balasundram et al. 2006; Benavente-García and Castillo 2008). However, the concentration of these compounds varies across flower types and is influenced by the stage of bloom, often peaking during the early flowering phase (Magri et al. 2020).

Carotenoids are another vital group of phytochemicals found in edible flowers. These pigments, identifiable by their conjugated double-bond systems, influence both color and stability in biological systems (Rodríguez-Bernaldo de Quiros and Costa 2006). Over 600 types have been cataloged in nature, with lutein, zeaxanthin, β-cryptoxanthin, α-carotene, β-carotene, and lycopene among the most prominent (Maiani et al. 2009). These compounds not only contribute to color but are also precursors of vitamin A, support eye health, and protect cells by neutralizing free radicals and singlet oxygen (Rodriguez-Amaya et al. 2008). They are also known to shield skin against UV rays, visible light, and infrared radiation, helping reduce oxidative stress (Zastrow et al. 2009).

In light of their sensitivity and health-promoting potential, enhancing the preservation techniques for edible flowers has become a significant focus. New processing strategies aim to prolong freshness and retain nutritional properties, making them valuable for consumers and producers. This chapter reviews such advancements.

5.2 Packaging Solutions for Edible Flowers

Packaging is critical in protecting edible flowers from dehydration, physical damage, and contamination by environmental agents or microorganisms (Jadhav et al. 2023). Poor handling practices, from harvest through to distribution, can result in major economic losses. Furthermore, packaging influences how consumers perceive the product, affecting their purchasing behavior and overall satisfaction (Versino et al. 2023).

5.2.1 Use of Conventional Plastic Packaging

Edible flowers require packaging solutions that can withstand various physical stresses such as vibration, compression, sudden impacts, and low temperatures during storage and transport, all while preserving their visual and structural integrity. Ideal packaging materials should not only be robust and moisture-resistant but also cost-effective, easy to handle, and preferably reusable. Traditional options frequently include materials like cellophane, cardboard, butter paper, polyethylene, polypropylene, and polyolefin (Versino et al. 2023).

Choosing the right packaging material is critical to controlling the internal atmosphere—specifically, carbon dioxide and oxygen levels-throughout storage. For

Table 5.1 Effectiveness of plastic packaging vs. Edible films and coatings of edible flowers

Effectiveness	Plastic packaging	Edible films
Environmental impact	Traditional plastic packaging is non-renewable, non-biodegradable, and contributes significantly to environmental pollution, requiring landfills for disposal.	These are made from natural, biodegradable materials such as polysaccharides, proteins, and lipids, which can be consumed with food, leaving no waste. They promote sustainability and can be derived from agrifood residues, supporting a circular economy.
Functional properties	Offers strong mechanical properties and effective barriers against moisture and gases, which are crucial for preserving food quality.	While they generally have lower mechanical strength and barrier properties compared to plastics, recent advancements have improved these aspects through the incorporation of nanomaterials and biopolymers. They also provide additional functions such as antioxidant and antimicrobial properties.
Food preservation	Extends shelf life by providing a stable environment for food products.	These can extend the shelf life of perishable foods by maintaining freshness and flavor, and they can be enriched with bioactive compounds to enhance nutritional value. They have been shown to effectively preserve the quality of fruits like strawberries by reducing weight loss and maintaining firmness.

example, *Cymbidium* flowers stored in cellophane have demonstrated a significantly longer vase life, ranging from 28 to 40 days, compared to 18–27 days in unpackaged samples. In another study, flowers wrapped in cellophane maintained freshness for up to 56 days, slightly outperforming those stored in polypropylene and low-density polyethylene (both with 54 days), all exceeding the vase life of unpackaged controls, which lasted around 50 days (De and Singh 2016).

Despite these benefits, conventional plastics may not always meet modern consumer demands regarding food safety, freshness, and sustainability. Research by Van Der Sman et al. (1996) explored how factors like packaging design, ventilation, and box dimensions influence flower quality. Their findings showed that 42% of flowers in specially designed containers and 62% in commercial packaging were impacted by microbial infections. A comparison of conventional plastic packaging with edible films and coatings highlights differences in their ability to protect against such issues, especially concerning moisture and gas barrier properties (Table 5.1).

5.2.2 *Edible Films and Coatings*

In the food sector, two main techniques are employed for preserving perishables: edible films and edible coatings. Edible films are pre-formed thin layers made from biopolymer solutions, which are later applied to the food surface. In contrast, edible

Table 5.2 Comparison of plastic packaging vs. edible films/coatings in terms of barrier properties

Property	Plastic packaging	Edible films/coatings
Oxygen permeability	Low permeability (excellent oxygen barrier).	Higher permeability (it varies based on composition, but is generally more porous than plastics)
Moisture barrier	Excellent moisture barrier (especially polyethylene and polypropylene).	Poor to moderate moisture resistance (hydrophilic nature of many biopolymers makes them susceptible to water absorption).
Ethylene control	Limited ethylene control (some plastics allow ethylene to pass, but specialized films with ethylene absorbers exist).	Moderate to good ethylene control (some coatings contain natural ethylene scavengers like essential oils or polysaccharides).

Source: Elaboration from authors

coatings are prepared as suspensions or emulsions and applied directly onto the food, forming a protective barrier in situ (Wang et al. 2022). Although applied differently, both methods utilize natural components such as polysaccharides, proteins, and lipids. These materials help regulate gas exchange and reduce microbial activity, especially in fresh produce like fruits and vegetables (Martins et al. 2024).

The use of edible coatings (EC) for prolonging the shelf life of edible flowers is gaining traction as a promising preservation method. These coatings serve as semi-permeable barriers that help minimize moisture loss, inhibit gas exchange, and block contaminants, thereby protecting the flowers from physical and environmental damage. Furthermore, they help retain sensory qualities such as color and aroma, making it possible to store edible flowers for extended periods. Despite these benefits, research into edible coatings specifically for flowers like artichokes, broccoli, brassicas, and cauliflower is still in its early stages (Shantamma et al. 2021).

Sodium alginate is one of the most frequently used biopolymers in edible coatings. For instance, applying a 0.5% alginate solution to flowers and then immersing them in a 1% calcium chloride solution for crosslinking was found to maintain freshness for up to 4 days longer than untreated samples (Fernandes et al. 2018a, b). The effectiveness of edible coatings can also be enhanced by integrating them with other non-thermal preservation methods. A notable example involves combining edible coatings with gamma irradiation. In a study on *Brassica oleracea* flowers, an alginate-based coating enriched with natamycin and essential oil, followed by irradiation, significantly extended shelf life when stored at 4 °C (Ben-Fadhel et al. 2017) (Table 5.2).

5.3 Storage Conditions and Their Impact on Phytonutrients

Inadequate practices during harvesting, handling, and storage can result in substantial financial losses for producers. Nonetheless, packaging serves a vital function beyond preservation-it significantly affects how consumers perceive product quality and can impact their buying choices (Versino et al. 2023). For edible flowers, the

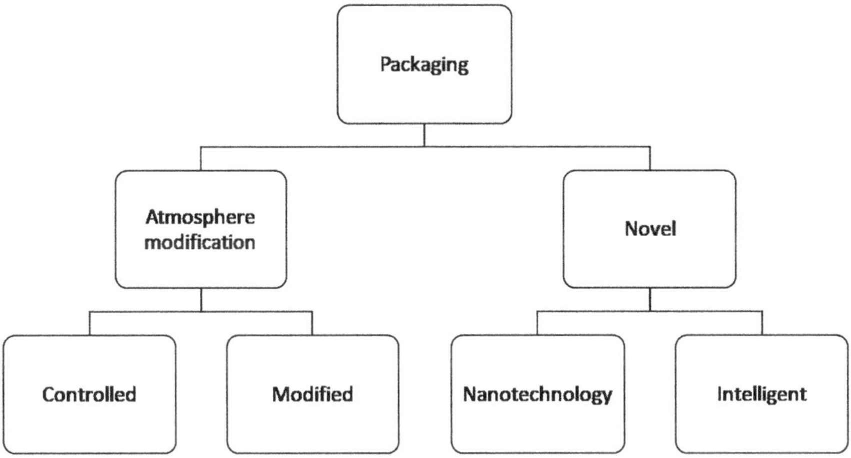

Fig. 5.1 Packaging alternatives for edible flowers preservation

main objectives of packaging are to prevent moisture loss, protect their fragile structure, and serve as a defense against microbial contamination and environmental hazards (Jadhav et al. 2023). An overview of the various packaging methods available for edible flowers is presented in Fig. 5.1, with each option discussed in further detail in the following sections.

5.3.1 Controlled Atmosphere Packaging (CAP)

Controlled Atmosphere Packaging (CAP) involves adjusting and maintaining specific levels of oxygen, carbon dioxide, and nitrogen, alongside carefully managed temperature and humidity, to extend the postharvest life of sensitive products. This technique slows down respiration, reduces microbial activity, and delays enzymatic degradation by creating a stable internal environment.

A study by Fernández-León et al. (2013) examined broccoli (*Brassica oleracea var. italica*) stored at 1–2 °C with 85–90% relative humidity. They evaluated changes in quality parameters such as appearance, chemical composition, and bioactive compounds under different storage conditions: ambient air (control), a controlled atmosphere (10% O_2, 5% CO_2), and 1-MCP treatment (0.6 μL/L). Following storage at room temperature (20 °C) for 2–4 days, broccoli kept under CAP retained its visual and nutritional quality better than those under other conditions, preserving pigments, vitamins, and phenolics.

Likewise, Cefola et al. (2016) tested how various CAP settings influenced quality in fresh zucchini flowers (*Cucurbita pepo* L.). When stored at 5 °C in an atmosphere containing 3% oxygen within nitrogen, the flowers experienced reduced respiration and weight loss. This treatment also helped maintain their color, texture, and nutritional profile. After 9 days, the zucchini flowers were still in marketable condition.

5.3.2 Modified Atmosphere Packaging (MAP)

Modified Atmosphere Packaging (MAP) focuses on modifying the internal gas composition of a package primarily oxygen and carbon dioxide while eliminating unwanted gases. This regulation helps maintain product freshness by slowing metabolic processes, curbing spoilage, and reducing oxidative reactions (Qu et al. 2022; Fang and Wakisaka 2021).

Mohi Alden et al. (2019) demonstrated that cauliflower packed in polyethylene under MAP conditions (92% N_2, 5% CO_2, 3% O_2) remained viable for over 30 days, significantly extending its shelf life compared to conventional air packaging.

In another study, Wang et al. (2020) evaluated the effect of MAP on *Brassica campestris* (Chinese flowering cabbage). They observed that MAP reduced hydrogen peroxide buildup in the stems and enhanced tissue firmness. Moreover, protein expression analysis showed upregulation of 63 proteins, including those involved in antioxidant defense, suggesting MAP's role in reinforcing physiological resilience.

5.3.3 Packaging Enhanced by Nanotechnology

Nanotechnology, which involves engineering materials at the nanometer scale, offers transformative opportunities in food packaging. By embedding nanoparticles into polymers, it is possible to improve packaging durability, barrier efficiency, and sustainability, all while minimizing material usage (Gupta et al. 2024; Mahmud et al. 2022).

Nanomaterials have led to the development of several innovative packaging formats. These include enhanced structural materials (e.g., nanoclays, nanotubes), antimicrobial active packaging using metal nanoparticles (e.g., silver, gold, metal oxides), and smart packaging that incorporates nanosensors for real-time monitoring of freshness or contamination. Bionanocomposites are also being explored for eco-friendly applications, reducing reliance on petrochemical-based plastics (Primožič et al. 2021; Trajkovska Petkoska et al. 2021) (Table 5.3).

5.4 v Degradation During Storage

Edible flowers are appreciated not only for their visual appeal but also for their nutritional properties, as they contain a variety of health-enhancing bioactive compounds (Książkiewicz et al. 2025). Due to their fragile structure, these flowers are particularly vulnerable to rapid deterioration, making careful postharvest handling and storage critical for preserving their nutritional content. This section explores the main categories of phytonutrients found in edible flowers, examines the primary causes of nutrient degradation, and discusses the latest packaging solutions aimed at

Table 5.3 Impact of controlled vs. non-controlled atmospheres on nutrient stability of edible flowers

Food preservation and quality alternatives impact	Impact on food preservation	Mode of action
Controlled Atmospheres (CA)	Nutrient retention	Controlled atmospheres, such as those with specific gas compositions, have been shown to retain higher levels of nutrients like ascorbic acid, polyphenols, and carotenoids in edible flowers. For instance, squash flowers stored under a controlled atmosphere with 5% O_2 and 10% CO_2 retained higher levels of these nutrients compared to those stored in air (Aquino-Bolaños et al. 2013).
	Extended shelf life	CA can significantly prolong the postharvest life of edible flowers by reducing physiological weight loss and maintaining quality attributes such as total sugars and soluble solids (Aquino-Bolaños et al. 2013).
Modified Atmosphere Packaging (MAP)	Quality preservation	Active MAP, which involves altering the gas composition around the flowers, helps maintain the visual and nutritional quality of edible flowers by slowing down decay and preserving phenolic content (Falla et al. 2024).
	Respiration and decay	Active MAP reduces flower respiration rates and sugar consumption, which helps in maintaining the freshness and extending the shelf life of flowers like begonias and violas (Falla et al. 2024).
Non-controlled atmospheres	Rapid quality decline	Flowers stored in non-controlled atmospheres, such as normal air, tend to experience rapid quality degradation, including weight loss and nutrient depletion, leading to a shorter shelf life.
	Increased senescence	Non-controlled conditions can accelerate senescence processes, as seen in Dendrobium orchids, where enzyme activities related to aging were higher compared to those stored under controlled conditions.

reducing nutrient loss during storage. Emphasis is placed on how different storage conditions and durations can impact the stability and retention of these beneficial compounds. Table 5.4. Interaction of packing material and storage period on phytonutrient preservation.

5.4.1 *Major Phytonutrients in Edible Flowers (e.g., Flavonoids, Carotenoids, Anthocyanins)*

Edible flowers are increasingly acknowledged for their nutritional value, particularly due to their content of phytonutrients such as flavonoids, phenolic acids, anthocyanins, carotenoids, and essential vitamins like vitamin C (ascorbic acid) and folate (B9) (Amrouche et al. 2022; Kandylis et al. 2022). Nutrient profiles can vary

Table 5.4 Interaction of packing material and storage period on phytonutrient preservation

Food preservation method	Physico-chemical agent	Impact on phytonutrient preservation
Modified Atmosphere Packaging (MAP)	Active vs. passive MAP	Active MAP, which involves altering the gas composition inside the packaging, has been shown to better maintain the color and slow decay of edible flowers compared to passive MAP, which relies on the natural respiration of the flowers to modify the atmosphere (Falla et al. 2024).
	Impact on nutritional quality	MAP can effectively reduce respiration rates and preserve phenolic content, thereby maintaining the nutritional quality of edible flowers during storage (Falla et al. 2024).
Ethanol and salinity treatments	Ethanol application	While ethanol treatment was tested, it was found to be detrimental, destroying petunia flowers after 14 days of storage.
	Salinity stress	Short-term salinity stress increased certain phytonutrients like polyphenols and antioxidants in petunia flowers, but prolonged storage under these conditions led to deterioration.
Cold storage	Temperature effects	Cold storage at 4 °C is commonly used to extend the shelf life of edible flowers. It helps maintain visual quality and phytonutrient content, such as ascorbic acid and flavonoids, particularly in species like Tagetes lemmonii (Marchioni et al. 2020).
Freezing techniques	Freezing vs. ice cubes	Freezing flowers can maintain or even increase bioactive compounds for up to 1 month, while storing them in ice cubes preserves appearance but not phytonutrient content.
Non-thermal processing	Emerging techniques	Non-thermal methods like irradiation and high hydrostatic pressure are gaining attention for preserving the 'fresh-like' attributes of edible flowers without compromising their nutritional quality.

widely among flower species. For instance, *Viola cornuta* is notable for its high vitamin C and carotenoid levels, *Cucurbita moschata* is abundant in B vitamins and flavonoids, and *Fuchsia regia* provides a rich source of phenolic compounds (Marchioni et al. 2024). Flowers have a high content of natural antioxidants such as anthocyanin, flavonoids and various other phenolic components (Rao and Poonia 2024b). These differences illustrate the broad nutritional spectrum present in various edible flower species.

The vibrant pigments of these flowers, particularly flavonoids and carotenoids, are not only responsible for their striking appearance but also contribute significantly to their antioxidant capacity (Pires Jr et al. 2021). Moreover, edible petals may also offer dietary fiber and vitamin B6, enhancing their potential as functional food ingredients (Kumar et al. 2022). Traditionally, edible flowers have been part of diverse culinary cultures, offering both decorative and nutritional benefits (Samkaria and Kumari 2025).

Flavonoids and phenolics serve as strong antioxidants and have been linked to antimicrobial and anti-inflammatory effects (Breda et al. 2025; Patil et al. 2025; Fakhri et al. 2022). Anthocyanins, known for producing red, purple, and blue hues, are also connected to heart and brain health (Mattioli et al. 2020). Meanwhile, carotenoids contribute to vitamin A synthesis and play a role in eye health. Vitamins like ascorbic acid and folate support immune defense and cellular function (Zhu et al. 2010; Aslam et al. 2017). Despite these benefits, many of these compounds are unstable and highly susceptible to environmental conditions, with temperature shifts and oxygen exposure accelerating nutrient degradation (Enaru et al. 2021; Yin et al. 2022).

5.4.2 Causes of Phytonutrient Degradation

The nutritional quality of edible flowers can decline rapidly after harvest due to their delicate physiology and high respiration rates. Environmental factors such as light, temperature variations, and oxygen exposure play a critical role in the degradation of sensitive bioactives like anthocyanins, carotenoids, and vitamin C.

Anthocyanins, in particular, are unstable in the presence of oxygen and under pH or temperature stress, which can lead to discoloration and a reduction in antioxidant potential (Locatelli et al. 2024). Carotenoids and ascorbic acid are similarly vulnerable to heat and oxidative damage, which compromises both their functionality and the aesthetic value of the flowers (Kandylis 2022). Physical damage during harvesting and processing increases enzymatic activity, hastening the breakdown of compounds like phenolics, especially in cut or minimally processed flowers (Marchioni et al. 2024).

Efforts to reduce degradation often involve storage at low temperatures and the use of modified or controlled atmosphere packaging, which limits oxygen and delays oxidation and enzymatic reactions (Locatelli et al. 2024). Additionally, thermal treatments like steaming have shown promise in preserving lipophilic nutrients better than other cooking methods, such as boiling or frying, which may cause nutrient loss through leaching (Castañeda-Rodríguez et al. 2024). Nevertheless, even with proper postharvest management, the actual absorption of these nutrients during digestion known as bioaccessibility can vary, making it essential to consider both pre- and postharvest conditions to preserve the health benefits of edible flowers (Table 5.5).

5.4.3 Packaging Approaches for Preserving Nutritional Quality

The delicate nature and high water content of edible flowers make them extremely prone to spoilage, necessitating effective packaging to maintain both their visual and nutritional quality. Conventional packaging, such as polyethylene clamshells and rigid plastic containers, offers basic protection against physical damage and

Table 5.5 Comparison of shelf-life extension of edible flowers under different packing and storage conditions

Packaging and storage techniques	*Modified Atmosphere Packaging (MAP)*	Both active and passive MAP have been shown to effectively extend the shelf life of edible flowers by maintaining color and reducing decay. Active MAP, which involves altering the gas composition within the packaging, better preserves petal colors and slows decay compared to passive MAP (Kou et al. 2012; Falla et al. 2024).
	Polymeric film packaging	Using specific polymeric films like D955 and low-density polyethylene (LDPE) can enhance freshness and color retention. For instance, crossandra flowers packaged in D955 film retained freshness for up to 96 h at low temperatures. Similarly, marigold flowers in LDPE bags showed an extended shelf life of 8 days.
	Chemical treatments	Quick dipping in solutions like sodium benzoate and boric acid, followed by low-temperature storage, can significantly extend the shelf life of flowers like tuberose and jasmine.
Effects on quality parameters	*Color and texture*	MAP and specific film packaging help maintain the visual quality of flowers by reducing dehydration and preserving texture (Kou et al. 2012; Falla et al. 2024).
	Nutritional properties	The preservation of nutritional compounds such as phenolics and flavonoids is enhanced under MAP and cold storage conditions. For example, Begonia and Viola flowers maintained higher phenolic content under active MAP (Falla et al. 2024; Marchioni et al. 2020).
	Aroma and sensory attributes	Packaging and storage conditions also influence the aroma profile and sensory attributes of flowers. Cold storage can alter the volatile compounds, affecting the fragrance and taste (Marchioni et al. 2020; Demasi et al. 2021).

moisture loss but is less effective at preventing oxidation and microbial contamination (Wilczyńska et al. 2023b).

To better address these issues, more sophisticated methods like Modified Atmosphere Packaging (MAP) and Controlled Atmosphere Packaging (CAP) have been introduced. These systems help regulate internal oxygen and carbon dioxide levels, thereby reducing metabolic activity and enzymatic reactions that lead to quality deterioration (Pires et al. 2019). MAP, in particular, has been shown to help preserve anthocyanin content and suppress microbial growth. CAP enhances this approach by using components such as oxygen scavengers and CO_2 generators to stabilize internal conditions (Rashed et al. 2024).

In addition, edible coatings based on biopolymers like sodium alginate offer another layer of protection. These coatings limit moisture loss, decrease oxygen transmission, and can also be used to deliver antimicrobial substances (Fernandes et al. 2018a, b). Recent innovations in active packaging such as plant-based

Table 5.6 Packaging strategies for nutrient retention in edible flowers

Packaging strategy	Mechanism of action	Benefits	References
Rigid plastic containers	Basic moisture and structural protection	Widely available; limited microbial protection	Wilczyńska et al. (2023b)
MAP	Alters O_2/CO_2 balance to slow respiration	Reduces enzymatic degradation and microbial growth	Pires et al. (2019)
CAP	Controlled gas exchange using scavengers/emitters	Greater control of internal atmosphere than MAP	Rashed et al. (2024)
Edible coatings	Biopolymer barriers to oxygen and moisture	Extends shelf life, reduces weight loss	Fernandes et al. (2018a, b)
Active packaging	Antimicrobial compounds in packaging matrix	Extends microbial shelf life and preserves sensory quality	Shinde et al. (2023)
Physical field-based	Uses nonthermal energy (e.g., ultrasound, HHP, UV)	Preserves heat-sensitive nutrients and microbial control	Feng et al. (2024)

antimicrobial wrapping materials like Freshness Keeper have further contributed to shelf-life extension by slowing microbial spoilage and maintaining the flowers' sensory attributes over time (Shinde et al. 2023).

Beyond packaging materials themselves, emerging physical preservation techniques are gaining attention. Technologies like pulsed electric fields, high hydrostatic pressure, and electromagnetic treatments can reduce microbial loads and safeguard sensitive phytonutrients by minimizing heat exposure during processing (Feng et al. 2024). Combining these advanced approaches with conventional packaging creates a synergistic effect, allowing for improved nutrient preservation, prolonged freshness, and enhanced consumer appeal of edible flowers (Table 5.6).

The stability of phytonutrients in edible flowers is primarily influenced by storage duration, temperature, humidity, and oxygen exposure. Strategic packaging, particularly using biodegradable and non-toxic materials, can play a significant role in enhancing nutrient retention and extending shelf life. As edible flowers gain commercial popularity, scalable packaging innovations will be crucial to ensuring both consumer safety and product quality throughout the supply chain.

5.5 Safety Profile of Edible Flowers

5.5.1 Potential Risks of Edible Flower Consumption

The use of edible flowers in culinary applications has increased due to their visual appeal, nutritional value, and health-enhancing properties. These flowers are rich in phytochemicals such as flavonoids and phenolic acids, which contribute to antioxidant, anti-inflammatory, hypoglycemic, and neuroprotective effects (Lu et al. 2016; Zheng et al. 2020; Loizzo et al. 2016). Despite these benefits, certain health risks must be considered when incorporating edible flowers into the diet. These include microbiological contamination, allergenic potential, and the presence of anti-nutritional or toxic compounds.

5.5.1.1 Microbiological Risks

Since edible flowers are often consumed raw or minimally processed, they can serve as vectors for pathogenic microorganisms, including *Escherichia coli* and *Staphylococcus aureus*. Although *Salmonella* has not been detected in some investigations, the risk of microbial contamination remains significant, emphasizing the importance of hygienic handling and preparation (Matyjaszczyk and Śmiechowska 2019; Wilczyńska et al. 2021).

5.5.1.2 Allergenic Potential

While allergic reactions to edible flowers are relatively uncommon, some species may provoke responses in sensitive individuals. Although the overall risk is considered low, individuals with known plant allergies or those with occupational exposure should exercise caution (Lucarini et al. 2020).

5.5.1.3 Chemical and Toxicological Concerns

Certain edible flowers may contain anti-nutritional factors such as oxalates and phytates, which can interfere with the body's ability to absorb essential minerals if consumed in excess (Sood et al. 2024). Additionally, toxic compounds like alkaloids may be present in some species, reinforcing the importance of accurate identification and moderate consumption (Nicolau and Gostin 2016; Sood et al. 2024).

5.5.1.4 Trace Elements and Pesticides

Although edible flowers can provide important micronutrients, they may also accumulate trace elements and pesticide residues. Fortunately, studies show that levels of harmful elements like cadmium and lead are typically below dangerous thresholds (Drava et al. 2020). To minimize exposure, it is essential to source flowers from reputable growers operating under controlled conditions (Nicolau and Gostin 2016; Drava et al. 2020).

5.5.2 Microbiological Considerations

While the consumption of edible flowers is generally regarded as safe, their chemical makeup and limited processing can create favorable conditions for microbial growth. Flowers can carry a range of bacteria and fungi, with studies identifying *E. coli* in nasturtium and *S. aureus* in all tested samples. Yeasts and molds were also present in all samples, typically below 6 log CFU/g, though still noteworthy given the fresh nature of these products (Wilczyńska et al. 2021, 2023b).

Other microorganisms found in edible flowers and herbs, such as basil, include *Enterobacter* spp., *Pseudomonas aeruginosa*, and *Bacillus amyloliquefaciens*. Microbial diversity has been shown to vary with cultivation methods, with organic practices generally resulting in greater microbial richness (Wetzel et al. 2010).

Contamination can occur throughout the production chain from field to postharvest due to poor agricultural practices, environmental exposure, or improper handling (Matyjaszczyk and Śmiechowska 2019). Interestingly, studies indicate that packaging and refrigeration do not significantly reduce microbial load, underscoring the need for strict hygiene protocols during cultivation, harvest, and distribution (Wilczyńska et al. 2023b).

5.5.3 Are There Allergens in Edible Flowers?

Edible flowers are generally safe for consumption, but they can contain allergens that may cause reactions in sensitive individuals, particularly among those with occupational exposure to flowers. Although the risk is low, certain species have been identified as potential allergens. For instance, a study on 62 species of edible flowers revealed potential allergens in two species (Lucarini et al. 2020). Some edible flowers, such as those in the Compositae family, including chrysanthemums and marguerites, contain sesquiterpene lactones that can cause allergic contact dermatitis (Andersen et al. 2015). Additionally, pollen from certain ornamental plants, including edible flowers, can trigger allergic reactions, especially in individuals with existing sensitivities. This is particularly true for those in the Asteraceae family, which is known for its allergenic potential (Goldberg et al. 1998). Workers in agriculture or floristry, who have higher exposure to flower pollen, may face a greater risk of developing allergies (Gil et al. 2002). Individuals with hay fever, asthma, or other allergies should exercise caution when consuming edible flowers (Dastidar et al. 2023). While edible flowers pose a low risk of allergenic reactions and offer nutritional benefits, ornamental flowers in urban environments can be significant allergen sources, particularly for those with occupational exposure or sensitivities. Therefore, while edible flowers can generally be enjoyed safely, individuals with known allergies should approach their consumption with care (Lucarini et al. 2020).

5.5.4 Chemical and Toxicological Risks Associated with the Consumption of Edible Flowers

While edible flowers provide nutritional and aesthetic benefits, they also carry potential chemical and toxicological risks. It is essential to understand these risks to ensure safe consumption. Proper identification, preparation, and adherence to safety guidelines are key to minimizing these risks and fully enjoying the benefits. Correctly identifying edible flowers and using appropriate preparation methods are crucial steps to reduce the potential dangers. Traditional methods, such as cooking, can help eliminate or lessen toxic substances (Sotelo et al. 2007; Nicolau and Gostin 2016). However, there is a need for more comprehensive safety regulations and guidelines to guarantee the safe consumption of edible flowers, including more detailed information on their composition and nutritional value (Nicolau and Gostin 2016) (Table 5.7).

5.5.5 Trace Elements and Pesticides Risks Associated with the Consumption of Edible Flowers

The consumption of edible flowers poses potential risks due to trace elements and pesticide residues, as well as impurities from agricultural practices and environmental contamination. While trace elements are typically within safe limits, pesticide residues require careful management to ensure consumer safety. Establishing clear regulations and safety guidelines is crucial to mitigate these risks. Edible flowers can provide essential trace elements such as manganese (Mn), copper (Cu), strontium (Sr), and nickel (Ni), with concentrations varying by species. For instance,

Table 5.7 Allergenic properties of ornamental vs. edible flowers

Ornamental flowers	Edible flowers
Higher allergenicity: Ornamental flowers, particularly those used in urban environments, can be significant sources of allergens. Pollen from ornamental plants, such as those in the Oleaceae family, can cause allergic reactions in sensitive individuals (Goldberg et al. 1998).	**Low allergenicity**: Edible flowers are generally safe for consumption with a low risk of allergic reactions. This is supported by studies that have identified only a few species with potential allergenic properties (Lucarini et al. 2020).
Occupational exposure: Flower growers and individuals frequently exposed to ornamental plants have a higher incidence of allergic reactions, with studies showing a significant percentage of flower growers experiencing occupational asthma and other allergic symptoms (Goldberg et al. 1998).	**Nutritional and health benefits**: Edible flowers are valued for their nutritional content, including antioxidants and bioactive compounds, which contribute to their health benefits.
Cross-reactivity: There is potential for cross-reactivity among different species of ornamental plants, which can exacerbate allergic responses in sensitized individuals.	**Cross-reactivity not observed** specially because the low risk of allergic reactions.

Table 5.8 Chemical and toxicological risks

Nature of the chemical or toxicological agent	Significance of the chemical and toxicological risks
Chemical impurities	Edible flowers can contain chemical impurities from agricultural production and environmental exposure. These include pesticides and other contaminants that may pose health risks if not properly managed (Matyjaszczyk and Śmiechowska 2019; Nicolau and Gostin 2016).
Anti-nutritional compounds	Some edible flowers contain anti-nutritional compounds like oxalates and phytates, which can interfere with mineral absorption and potentially lead to nutritional deficiencies over time. Moderation is advised, especially for individuals with pre-existing health conditions (Sood et al. 2024; Sotelo et al. 2007).
Toxic compounds	Certain flowers may contain toxic alkaloids and other harmful substances. Proper identification and preparation, such as cooking and discarding the broth, can help reduce these risks (Sotelo et al. 2007; Nicolau and Gostin 2016; Pires et al. 2021).
Trace elements	While edible flowers can be a source of essential minerals, they may also contain trace elements like Cd, Co, Ni, and Pb. However, studies have shown that these elements are often present at safe levels, below the maximum permitted levels in foodstuffs (Drava et al. 2020).

Acmella oleracea is rich in manganese, while basil and pumpkin flowers contain higher levels of copper and strontium. However, potentially toxic elements like cadmium (Cd) and lead (Pb) are usually found at very low concentrations, often below detection limits, posing minimal risk from trace elements (Drava et al. 2020). Pesticide residues may transfer from flowers to consumable products, such as tea made from chrysanthemum flowers. While the transfer rates of pesticides like metalaxyl-M and thiamethoxam can be significant, the associated health risks remain low, with risk quotients suggesting minimal concern (Jiang et al. 2020). However, the lack of stringent regulations on pesticide residues in non-edible flowers can lead to overuse in production areas, potentially impacting human health and the environment, including risks such as neurobehavioral issues and reproductive disorders for those working in flower production areas (Pereira et al. 2021) (Table 5.8).

5.6 Conclusions and Perspectives

Edible flowers have emerged as a promising component of the human diet, driven by their health benefits and the growing interest in enhancing people's quality of life, particularly in the post-pandemic era. The industry of molecular and functional foods from natural sources for disease prevention and treatment is rapidly expanding. Edible flowers are rich in proteins, fatty acids, dietary fibers, and a wide range of bioactive compounds, and numerous studies have verified their nutritional potential. These flowers are increasingly being used in high-end kitchens, not only for their ornamental value but also for their health benefits.

However, edible flowers are delicate and perishable, requiring immediate preservation from the moment of harvest until consumption. Techniques like Modified Atmosphere Packaging (MAP) and Controlled Atmosphere Packaging (CAP) have been developed to maintain freshness and preserve their bioactive profile. There is still a need for in-depth research in this area to explore cutting-edge technologies in packaging and preservation methods. Additionally, more extensive studies should focus on the health benefits of underutilized edible flowers, especially their potential in treating rare diseases.

The challenges in developing the edible flower industry include the efficient identification and extraction of bioactive compounds to achieve optimal nutritional profiles. Preservation technologies must also be improved to maintain organoleptic qualities such as color and aroma during processing. Furthermore, a systematic review of safety and nutrient evaluations is necessary. Continued research and the adoption of effective processing and preservation methods will be crucial for the global commercialization of edible flowers.

References

Amrouche TA, Yang X, Capanoglu E, Huang W, Chen Q, Wu L et al (2022) Contribution of edible flowers to the Mediterranean diet: phytonutrients, bioactivity evaluation and applications. Food Front 3(4):592–630

Andersen R, Skov L, Menné T, Johansen J (2015) Clinical consequences of sesquiterpene lactone mix contact allergy in Denmark. Br J Dermatol 172(5):1430–1431. https://doi.org/10.1111/bjd.13584

Aquino-Bolaños T et al (2013) Effect of cold storage on the nutraceutical value of edible flowers. Ital J Food Sci 30:336–344

Aslam F, Muhammad SM, Aslam S, Irfan JA (2017) Vitamins: key role players in boosting up immune response-a mini review. Vitam Miner 6(01):1000153

Balasundram N, Sundram K, Samman S (2006) Phenolic compounds in plants and agri-industrial by-products: antioxidant activity, occurrence, and potential uses. Food Chem 99(1):191–203. https://doi.org/10.1016/j.foodchem.2005.07.042

Benavente-García O, Castillo J (2008) Update on uses and properties of citrus flavonoids: new findings in anticancer, cardiovascular, and anti-inflammatory activity. J Agric Food Chem 56(15):6185–6205. https://doi.org/10.1021/jf8006568

Ben-Fadhel Y, Saltaji S, Khlifi MA, Salmieri S, Dang V, K., & Lacroix, M. (2017) Active edible coating and γ-irradiation as cold combined treatments to assure the safety of broccoli florets (Brassica oleracea L.). Int J Food Microbiol 241:30–38. https://doi.org/10.1016/J.IJFOODMICRO.2016.10.010

Benvenuti S, Mazzoncini M (2021) The biodiversity of edible flowers: discovering new tastes and new health benefits. Front Plant Sci 11:569499. https://doi.org/10.3389/FPLS.2020.569499/PDF

Breda C, Nascimento A, Meghwar P, Lisboa H, Aires A, Rosa E et al (2025) Phenolic composition and antioxidant activity of edible flowers: insights from synergistic effects and multivariate analysis. Antioxidants 14(3):282

Castañeda-Rodríguez R, Quiles A, Llorca E, Ozuna C (2024) How to cook Yucca spp. flowers? An analysis of their chemical composition, microstructure, and bioactive compound bioaccessibility. Appl Food Res 4:100414. https://doi.org/10.1016/j.afres.2024.100414

Cefola M, Amodio ML, Colelli G (2016) Extending postharvest life of ready-to-use zucchini flowers: effects of the atmosphere composition. Acta Hortic 1141:123–130. https://doi.org/10.17660/ActaHortic.2016.1141.13

Dastidar A, Chakraborty S, Saha A, Das S, Zaman Q (2023) Edible flowers: a new source of minerals. IJRASET 11:38–54. https://doi.org/10.22214/ijraset.2023.49312

De LC, Singh DR (2016) Post-harvest management and value addition in orchids. Int J Biol Sci 03(01):14–35. http://www.dnetrw.com

Demasi S et al (2021) Sensory profile, shelf life, and dynamics of bioactive compounds during cold storage of 17 edible flowers. Horticulturae 7(166):1–26

Drava G, Iobbi V, Govaerts R, Minganti V, Copetta A, Ruffoni B, Bisio A (2020) Trace elements in edible flowers from Italy: further insights into health benefits and risks to consumers. Molecules 25(12):2891. https://doi.org/10.3390/molecules25122891

Enaru B, Dreţcanu G, Pop TD, Stănilă A, Diaconeasa Z (2021) Anthocyanins: factors affecting their stability and degradation. Antioxidants 10(12):1967

Fakhri S, Tomas M, Capanoglu E, Hussain Y, Abbaszadeh F, Lu B et al (2022) Antioxidant and anticancer potentials of edible flowers: where do we stand? Crit Rev Food Sci Nutr 62(31):8589–8645

Falla NM et al (2024) Active modified atmosphere packaging helps preserve quality of edible flowers. Agronomy 14:2409. https://doi.org/10.3390/agronomy14102409

Fang Y, Wakisaka M (2021) A review on the modified atmosphere preservation of fruits and vegetables with cutting-edge technologies. Agriculture 11(10):992. https://doi.org/10.3390/AGRICULTURE11100992

Feng M et al (2024) Novel strategies for enhancing quality stability of edible flower during processing using efficient physical fields: a review. Food Chem 448:139077. https://doi.org/10.1016/j.foodchem.2024.139077

Fernandes L et al (2018b) Post-harvest technologies applied to edible flowers: a review. Food Rev Int 35(2):132–154. https://doi.org/10.1080/87559129.2018.1473422

Fernandes L, Pereira JA, Saraiva JA, Casal S, Ramalhosa E (2018a) The effect of different post-harvest treatments on the quality of borage (Borago officinalis) petals. Acta Sci Pol Technol Aliment 17(1):5–10. https://doi.org/10.17306/J.AFS.2018.0533

Fernández-León MF, Fernández-León AM, Lozano M, Ayuso MC, González-Gómez D (2013) Different postharvest strategies to preserve broccoli quality during storage and shelf life: controlled atmosphere and 1-MCP. Food Chem 138(1):564–573. https://doi.org/10.1016/J.FOODCHEM.2012.09.143

Gil M, Hogendijk S, Hauser C (2002) Allergy to eggplant flower pollen. Allergy 57(7):652. https://doi.org/10.1034/j.1398-9995.2002.23749.x

Goldberg A, Confino-Cohen R, Waisel Y (1998) Allergic responses to pollen of ornamental plants: high incidence in the general atopic population and especially among flower growers. J Allergy Clin Immunol 102(2):210–214. https://doi.org/10.1016/S0091-6749(98)70088-0

Gupta RK, Abd El Gawad F, Ali EAE, Karunanithi S, Yugiani P, Srivastav PP (2024) Nanotechnology: current applications and future scope in food packaging systems. Meas: Food 13:100131. https://doi.org/10.1016/J.MEAFOO.2023.100131

Ikeura H, Kobayashi F, Kai T, Tsuchiya Y, Tamaki M (2023) Effects of different storage conditions on the colour, antioxidant activity, and volatile components of edible roses. Sci Hortic 310:111707. https://doi.org/10.1016/J.SCIENTA.2022.111707

Jadhav HB, Badwaik LS, Annapure U, Casanova F, Alaskar K (2023) A review on the journey of edible flowers from farm to consumer's plate. Appl Food Res 3(2):100312. https://doi.org/10.1016/J.AFRES.2023.100312

Jiang M, Zhang W, Zhang T, Liang G, Hu B, Han P, Gong W (2020) Assessing transfer of pesticide residues from chrysanthemum flowers into tea solution and associated health risks. Ecotoxicol Environ Saf 187:109859. https://doi.org/10.1016/j.ecoenv.2019.109859

Kandylis P (2022) Phytochemicals and antioxidant properties of edible flowers. Appl Sci 12(19):9937. https://doi.org/10.3390/app12199937

Kou L et al (2012) Extending the shelf life of edible flowers with controlled release of 1-Methylcyclopropene and modified atmosphere packaging. J Food Sci 77:S188–S193

Książkiewicz M, Karczewska M, Nawrot F, Korybalska K, Studzińska-Sroka E (2025) Traditionally used edible flowers as a source of neuroactive, antioxidant, and anti-inflammatory extracts and bioactive compounds: a narrative review. Molecules 30(3):677

Kumar S, Dalal JS, Punia V, Pandey A (2022) An analytical study on the valuable sources and potential therapeutic applications of selected edible flowers in the new normal. NeuroQuantology 20(22):951

Locatelli S, Zanin G, Sambo P, Nicoletto C (2024) Effects of LED irradiation and non-thermal plasma treatment on horned pansy during flowering: enhancing yield and functional quality of edible flowers. Horticulturae 10(12):1274. https://doi.org/10.3390/horticulturae10121274

Loizzo M, Pugliese A, Bonesi M, Tenuta M, Menichini F, Xiao J, Tundis R (2016) Edible flowers: a rich source of phytochemicals with antioxidant and hypoglycemic properties. J Agric Food Chem 64(12):2467–2474. https://doi.org/10.1021/acs.jafc.5b03092

Lu B, Li M, Yin R (2016) Phytochemical content, health benefits, and toxicology of common edible flowers: a review (2000–2015). Crit Rev Food Sci Nutr 56:S130–S148. https://doi.org/10.1080/10408398.2015.1078276

Lucarini M, Copetta A, Durazzo A, Gabrielli P, Lombardi-Boccia G, Lupotto E, Santini A, Ruffoni B (2020) A snapshot on food allergies: a case study on edible flowers. Sustainability 12(20):8709. https://doi.org/10.3390/su12208709

Magri A, Adiletta G, Petriccione M (2020) Evaluation of antioxidant systems and ascorbate-glutathione cycle in feijoa edible flowers at different flowering stages. Foods 9(1):95. https://doi.org/10.3390/foods9010095

Mahmud J, Sarmast E, Shankar S, Lacroix M (2022) Advantages of nanotechnology developments in active food packaging. Food Res Int 154:111023. https://doi.org/10.1016/J.FOODRES.2022.111023

Maiani G, Periago Caston MJ, Catasta G, Toti E, Cambrodon IG, Bysted A et al (2009) Carotenoids: actual knowledge on food sources, intakes, stability and bioavailability and their protective role in humans. Mol Nutr Food Res 53(S2):S194–S218. https://doi.org/10.1002/mnfr.200800053

Marchioni I, Gabriele M, Carmassi G, Ruffoni B, Pistelli L, Pistelli L, Najar B (2024) Phytochemical, nutritional and mineral content of four edible flowers. Foods 13(6):939. https://doi.org/10.3390/foods13060939

Marchioni I, Pistelli L, Ferri B, Copetta A, Ruffoni B, Pistelli L, Najar B (2020) Phytonutritional content and aroma profile changes during postharvest storage of edible flowers. Front Plant Sci 11:590968. https://doi.org/10.3389/FPLS.2020.590968/BIBTEX

Martins VFR, Pintado ME, Morais RMSC, Morais AMMB (2024) Recent highlights in sustainable bio-based edible films and coatings for fruit and vegetable applications. Foods 13(2):318. https://doi.org/10.3390/FOODS13020318

Mattioli R, Francioso A, Mosca L, Silva P (2020) Anthocyanins: a comprehensive review of their chemical properties and health effects on cardiovascular and neurodegenerative diseases. Molecules 25(17):3809

Matyjaszczyk E, Śmiechowska M (2019) Edible flowers. Benefits and risks pertaining to their consumption. Trends Food Sci Technol 91:670–674. https://doi.org/10.1016/J.TIFS.2019.07.017

Mohi Alden K, Omid M, Rajabipour A, Tajeddin B, Soltani Firouz M (2019) Quality and shelf-life prediction of cauliflower under modified atmosphere packaging by using artificial neural networks and image processing. Comput Electron Agric 163:104861. https://doi.org/10.1016/J.COMPAG.2019.104861

Nicolau A, Gostin A (2016) Safety of edible flowers. In: Regulating safety of traditional and ethnic foods. Academic, Waltham, pp 395–419. https://doi.org/10.1016/B978-0-12-800605-4.00021-9

Patil SA, Pagariya MC, Kulkarni AJ, Jadhav PR, Yadav RS, Kadam GB et al (2025) Phytochemical insights and antioxidant capabilities of rose cultivars: implications for traditional medicine and modern industry. Discov Agric 3(1):23

Pereira P, Parente C, Carvalho G, Torres J, Meire R, Dorneles P, Malm O (2021) A review on pesticides in flower production: a push to reduce human exposure and environmental contamination. Environ Pollut 289:117817. https://doi.org/10.1016/j.envpol.2021.117817
Pires EDO Jr, Di Gioia F, Rouphael Y, Ferreira IC, Caleja C, Barros L, Petropoulos SA (2021) The compositional aspects of edible flowers as an emerging horticultural product. Molecules 26(22):6940
Pires TCSP et al (2019) Edible flowers: emerging components in the diet. Trends Food Sci Technol 93:244–258. https://doi.org/10.1016/j.tifs.2019.09.020
Primožič M, Knez Ž, Leitgeb M (2021) (Bio)nanotechnology in food science—food packaging. Nano 11(2):292. https://doi.org/10.3390/NANO11020292
Qu P, Zhang M, Fan K, Guo Z (2022) Microporous modified atmosphere packaging to extend shelf life of fresh foods: a review. Crit Rev Food Sci Nutr 62(1):51–65. https://doi.org/10.1080/10408398.2020.1811635
Rashed NM et al (2024) An analysis of conventional and modern packaging approaches for cut flowers: a review article. Front Plant Sci 15:1371100. https://doi.org/10.3389/fpls.2024.1371100
Rao V, Poonia A (2024a) Bioactive compounds, nanoparticle synthesis and potential utilization of edible flowers for development of functional dairy products: a review. J Food Sci Technol 61(6):1053–1068. https://doi.org/10.1007/s13197-023-05853-6
Rao V, Poonia A (2024b) Microbial biosynthesis of selenium nanoparticles using probiotic strain and its characterization. J Food Meas Charact 18(7):5477–5488. https://doi.org/10.1007/s11694-024-02581-z
Rodriguez-Amaya DB, Kimura M, Godoy HT, Amaya-Farfan J (2008) Updated Brazilian database on food carotenoids: factors affecting carotenoid composition. J Food Compos Anal 21(6):445–463. https://doi.org/10.1016/j.jfca.2008.04.001
Rodríguez-Bernaldo de Quiros A, Costa HS (2006) Analysis of carotenoids in vegetable and plasma samples: a review. J Food Compos Anal 19(2–3):97–111. https://doi.org/10.1016/j.jfca.2005.04.004
Samkaria S, Kumari P (2025) Wild edible flowers of Indian Himalayan region, their traditional uses and potential health benefits: a way forward for food and nutritional security. Plant Foods Hum Nutr 80(1):60
Serek M, Reid MS (2024) Role of growth regulators in the postharvest life of ornamentals. In: Plant growth regulators in agriculture and horticulture. CRC Press, Boca Raton, pp 147–174. https://doi.org/10.1201/9781003578505-5
Shantamma S, Vasikaran EM, Waghmare R, Nimbkar S, Moses JA, Anandharamakrishnan C (2021) Emerging techniques for the processing and preservation of edible flowers. Future Foods 4:100094. https://doi.org/10.1016/J.FUFO.2021.100094
Shinde SP et al (2023) A way forward for a sustainable active packaging solution for prolonging the freshness and shelf life of Rosa hybrida L. cut flowers. Postharvest Biol Technol 204:112475. https://doi.org/10.1016/j.postharvbio.2023.112475
Sood Y, Lal M, Kalia A, Verma S (2024) Edible flowers: super foods with potential health benefits. Int J Plant Soil Sci 36(3):213–221. https://doi.org/10.9734/ijpss/2024/v36i34417
Sotelo A, López-García S, Basurto-Peña F (2007) Content of nutrient and antinutrient in edible flowers of wild plants in Mexico. Plant Foods Hum Nutr 62:133–138. https://doi.org/10.1007/S11130-007-0053-9
Trajkovska Petkoska A, Daniloski D, D'Cunha NM, Naumovski N, Broach AT (2021) Edible packaging: sustainable solutions and novel trends in food packaging. Food Res Int 140:109981. https://doi.org/10.1016/J.FOODRES.2020.109981
Van Der Sman RGM, Evelo RG, Wilkinson EC, Van Doom WG (1996) Quality loss in packed rose flowers due to Botrytis cinerea infection as related to temperature regimes and packaging design. Postharvest Biol Technol 7(4):341–350. https://doi.org/10.1016/0925-5214(95)00047-X
Versino F, Ortega F, Monroy Y, Rivero S, López OV, García MA (2023) Sustainable and bio-based food packaging: a review on past and current design innovations. Foods 12(5):1057. https://doi.org/10.3390/FOODS12051057

Wang L, Wen M, Chen F, Luo Z, Yin J, Chen Y, Huang H (2020) High oxygen atmospheric packaging (HOAP) reduces H2O2 production by regulating the accumulation of oxidative stress-related proteins in Chinese flowering cabbage. Postharvest Biol Technol 165:111183. https://doi.org/10.1016/J.POSTHARVBIO.2020.111183

Wang Q, Chen W, Zhu W, McClements DJ, Liu X, Liu F (2022) A review of multilayer and composite films and coatings for active biodegradable packaging. Npj Sci Food 6(1):1–16. https://doi.org/10.1038/s41538-022-00132-8

Wetzel K, Lee J, Lee C, Binkley M (2010) Comparison of microbial diversity of edible flowers and basil grown with organic versus conventional methods. Can J Microbiol 56(11):943–951. https://doi.org/10.1139/w10-082

Wilczyńska A et al (2023b) Effect of packaging on microbial quality of edible flowers during refrigerated storage. Pol J Food Nutr Sci 73(1):32–38

Wilczyńska A, Kukułowicz A, Lewandowska A (2021) Preliminary assessment of microbial quality of edible flowers. LWT Food Sci Technol 150:111926. https://doi.org/10.1016/J.LWT.2021.111926

Yin X, Chen K, Cheng H, Chen X, Feng S, Song Y, Liang L (2022) Chemical stability of ascorbic acid integrated into commercial products: a review on bioactivity and delivery technology. Antioxidants 11(1):153

Zastrow L, Groth N, Klein F, Kockott D, Lademann J, Renneberg R et al (2009) The missing link–light-induced (280-1,600 nm) free radical formation in human skin. Skin Pharmacol Physiol 22(1):31–44. https://doi.org/10.1159/000188083

Zheng J, Lu B, Xu B (2020) An update on the health benefits promoted by edible flowers and involved mechanisms. Food Chem 340:127940. https://doi.org/10.1016/j.foodchem.2020.127940

Zhu C, Bai C, Sanahuja G, Yuan D, Farré G, Naqvi S et al (2010) The regulation of carotenoid pigmentation in flowers. Arch Biochem Biophys 504(1):132–141

Chapter 6
Antioxidant Potential of Edible Flowers

Pan Nu Thwe and Wee Sim Choo

6.1 Introduction

Flowers have been appreciated for their beauty throughout history. Edible flowers play a role gastronomy for their ability to enrich foods with unique colors, flavors, and textures. Edible flowers being non-toxic, are considered healthy and safe for human consumption. They have been employed by humans since long time to beautify dishes, supply nutrients, and improve wellness (Rao and Poonia 2024a, b). The use of edible flowers has been documented in traditional Asian, European and Middle Eastern cuisines (Fernandes et al. 2017). Edible flowers also play a role in natural medicine because of their medicinal value and are used in cosmetics (Gokiladevi et al. 2024). The Food and Drug Administration (FDA), Food and Agriculture Organization of the United Nations (FAO), Codex Alimentarius or the European Food Safety Authority (EFSA), and World Health Organization (WHO) do not have specific legislation for edible flowers or publish any official lists of edible and nonedible flowers (Fernandes et al. 2020).

There is renewed or growing interest in edible flowers due to the increasing demand from consumers for natural compounds and healthy foods rich in bioactive compounds. Edible flowers are rich sources of phenolic acids, flavonoids, and other phytochemicals that exhibit significant antioxidant activity. In addition, pigments such as anthocyanins, chlorophylls and carotenoids in edible flowers also exhibit significant antioxidant activity. The chapter discusses the antioxidant properties of edible flowers around the world, along with methods for evaluating their antioxidant capacities.

P. N. Thwe · W. S. Choo (✉)
School of Science, Monash University Malaysia, Selangor, Malaysia
e-mail: choo.wee.sim@monash.edu

A. Poonia et al. (eds.), *Edible Flowers: Source of Phytonutrients, Valorization and Technological Advancements*, https://doi.org/10.1007/978-3-032-00667-7_6

6.2 Phytochemical Content

There are various phytochemicals present in edible flowers. Table 6.1 shows a comprehensive list of flavonoid, phenolic acid, and phenol contents in various edible flowers, whereas Table 6.2 shows a comprehensive list of various types of pigment, their content, and color in various edible flowers. Anthocyanins, carotenoids, or chlorophylls are the major pigments in edible flowers.

6.3 Antioxidant Properties

Antioxidants are substances that prevent oxidation, a chemical reaction that involves oxidizing agents or oxidants, commonly present in the form of free radicals (Kandylis 2022). This oxidation process is undesirable as it damages essential cellular components such as DNA, and lipid membranes, leading to the onset of degenerative diseases (Pisoschi and Pop 2015; Chensom et al. 2019).

An easy and simple way to test the antioxidant potential of edible flowers' extract, fraction or isolated compounds is through *in vitro* assays. These assays are beneficial as they are performed under a controlled environment, allowing the investigation of a substance/compound of interest having its antioxidant effect with minimal confounding variables in play (Madorran et al. 2020). The antioxidants can be classified as primary or secondary antioxidants. Primary antioxidants cause direct reduction of radicals through single electron transfer (SET) or of quenching free radicals through hydrogen atom transfer (HAT) (Kandylis 2022). Meanwhile, secondary antioxidants have an indirect mechanism in preventing free radicals by binding to pro-oxidative metal ions (Boulebd and Spiegel 2023). Thus, with multiple antioxidant mechanisms in play, various assays are essential to be applied.

Common *in vitro* biochemical assays include oxygen radical absorbance capacity (ORAC), ferric reducing antioxidant power (FRAP), 2,2-diphenyl-1-picrylhydrazyl (DPPH), and 2,2′-azino-bis(3-ethylbenzothiazoline-6-sulfonic acid) (ABTS). $DPPH^{\bullet}$ and $ABTS^{\bullet+}$ are radicals that are reduced by a hydrogen atom or an electron transferred from the antioxidants, forming non-radical molecules which are DPPH-H and ABTS, respectively (Apak et al. 2016; Bibi Sadeer et al. 2020). They are considered mixed of SET/HAT where the mechanism that dominates over the other is influenced by the pH, solvent, or other reaction conditions (Apak et al. 2016). In FRAP assay, another SET based assay, instead of radicals, a single electron is transferred to ferric ions (Fe(III)), reducing it to ferrous (Fe(II)) ions (Barros et al. 2020). Collectively, all these assays rely on spectrophotometric or colorimetric technique, where there is change in color of the radical or metal upon its reduction resulting in the end-product that can be measured at a specific wavelength. Meanwhile, the ORAC assay involves antioxidants adopting the HAT mechanism to quench peroxyl radicals that are causing oxidative damage toward fluorescein, a fluorescent probe (Xiao et al. 2020). The degree of fluorescence loss is used as an

Table 6.1 Phytochemical content of various edible flowers

Phyto-chemical	Source	Content		References
Flavonoid				
	Aegle marmelos	TFC	1.50 mg RE/g DW	Janarny et al. (2021)
	Agastache foeniculum	TFC	16.96 mg CAE/g DW	Araújo et al. (2019)
	Agave durangensis	TFC	120.84–1210.48 μg/g dry extract	Barriada-Bernal et al. (2014)
	Agave salmiana	TFC	4.58 mg QE/g DW	Pinedo-Espinoza et al. (2020)
	Aglaia odorata	TFC	3.85 mg CAE/g DW	Zheng et al. (2018)
	Albizia julibrissin	TFC	3.78 mg CAE/g DW	Zheng et al. (2018)
	Allium cepa	TFC	3.60 mg RE/g DW	Janarny et al. (2021)
		TFC	16.92 ppm	Mikołajczak et al. (2020)
	Allium schoenoprasum	TFC	14.92 ppm	Mikołajczak et al. (2020)
		TFC	4.56 mg QE/g DW	Socha et al. (2021)
	Aloe vera	TFC	7.79 mg QE/g DW	Pinedo-Espinoza et al. (2020)
	Amygdalus persica	TFC	16.90 mg CAE/g DW	Zheng et al. (2018)
	Anchusa azurea	TFC	17.30 mg QE/g dry extract	Loizzo et al. (2016)
	Antigonon leptopus	TFC	130.40–307.40 μg/g DW	Kaisoon et al. (2011)
		TFC	32[a]–50[a] mg RE/g DW	Kaisoon et al. (2011)
		TFC	157.26 mg/100 g DW	Kaisoon et al. (2012)
	Aponogeton crispus	TFC	4.60 mg RE/g DW	Janarny et al. (2021)
	Armeniaca mume	TFC	28.50 mg CAE/g DW	Zheng et al. (2018)
	Azadirachta indica	TFC	3.30 mg RE/g DW	Janarny et al. (2021)
	Bauhinia racemosa	TFC	1.8–3.5[a] mg RE/g DW	Kaisoon et al. (2012), Janarny et al. (2021)
	Bauhinia variegata	TFC	6.62 mg CAE/g DW	Zheng et al. (2018)
		TFC	195–281 mg RE/g extract	Suksathan et al. (2021)
	Begonia boliviensis	TFC	1.43–1.47 mg CAE/g FW	Marchioni et al. (2024)
	Begonia tuberhybrida	TFC	243.17–343.70 ppm	Mikołajczak et al. (2020)
		TFC	21.96 mg QE/g DW	Socha et al. (2021)
		TFC	2.32 g RE/kg FW	Mlcek et al. (2021)
	Bellis perennis	TFC	152.22 ppm	Mikołajczak et al. (2020)
	Benincasa hispida	TFC	1.97–28.96 mg QE/g DW	Yan et al. (2022)
	Bombax ceiba	TFC	5.76 mg CAE/g DW	Zheng et al. (2018)
	Borago officinalis	TFC	12.59 mg CAE/g DW	Araújo et al. (2019)
		TFC	0.30[a]–0.79[a] mg QE/g FW	Fernandes et al. (2020)
	Bougainvillea glabra	TFC	7.1–0 mg RE/g DW	Janarny et al. (2021)

(continued)

Table 6.1 (continued)

Phyto-chemical	Source	Content		References
		TFC	173.06 mg/100 g DW	Kaisoon et al. (2012)
	Bougainvillea hybrida	TFC	27.10–140.30 μg/g DW	Kaisoon et al. (2011)
		TFC	10[a]–48[a] mg RE/g DW	Kaisoon et al. (2011)
	Bougainvillea spectabilis	TFC	0.86–1.26 mg QE/g FW	Petrova et al. (2016)
	Calendula officinalis	TFC	9.40 mg CAE/g DW	Araújo et al. (2019)
		TFC	0.70–3.30 mg QE/g FW	Petrova et al. (2016)
		TFC	6.82 mg CAE/g DW	Zheng et al. (2018)
		TFC	86.00 ppm	Mikołajczak et al. (2020)
		TFC	7.81 mg QE/g DW	Socha et al. (2021)
		TFC	2.05 g RE/kg FW	Mlcek et al. (2021)
		TFC	165.49 mg CTH/100 g FW	Dujmović et al. (2022)
		TFC	3.03–19.8 mg RE/g DW	Chen et al. (2015), Janarny et al. (2021)
	Calotropis gigantea	TFC	8.10 mg RE/g DW	Janarny et al. (2021)
	Camellia azalea	TFC	8.76 mg CAE/g DW	Zheng et al. (2018)
	Camellia japonica	TFC	4.36 mg CAE/g DW	Zheng et al. (2018)
	Camellia sinensis	TFC	26.667–60.288 mg CAE/g DW	Zeng et al. (2014)
		TFC	65–130 mg RE/g extract	Suksathan et al. (2021)
	Campsis grandiflora	TFC	9.76 mg CAE/g DW	Zheng et al. (2018)
	Canna edulis	TFC	6.67 mg CAE/g DW	Zheng et al. (2018)
	Capparis spinosa	TFC	19.40 mg QE/g dry extract	Loizzo et al. (2016)
	Carica papaya	TFC	2.10 mg RE/g DW	Janarny et al. (2021)
	Cassia auriculata	TFC	5.60–9.13 mg RE/g DW	Kaisoon et al. (2012), Janarny et al. (2021)
	Cassia fistula	TFC	4.9–8.0[a] mg RE/g DW	Kaisoon et al. (2012), Janarny et al. (2021)
	Cassia siamea	TFC	52.9–133.7 μg/g DW	Kaisoon et al. (2011)
		TFC	9.8[a]–35[a] mg RE/g DW	Kaisoon et al. (2011)
	Centaurea cyanus	TFC	21.82 ppm	Mikołajczak et al. (2020)
		TFC	4.80 mg QE/g DW	Socha et al. (2021)
		TFC	264.90 mg CTH/100 g FW	Dujmović et al. (2022)
	Chamaemelum nobile	TFC	15.97 mg RE/g DW	Chen et al. (2015)
	Chamomilla	TFC	15.66 mg RE/g DW	Chen et al. (2015)
	Chamomilla recutita	TFC	180.83 ppm	Mikołajczak et al. (2020)
	Chimonanthus praecox	TFC	5.96 mg RE/g DW	Chen et al. (2015)
		TFC	6.93 mg CAE/g DW	Zheng et al. (2018)

(continued)

Table 6.1 (continued)

Phyto-chemical	Source	Content		References
	Chrysanthemum	TFC	5.71 mg QE/g DW	Socha et al. (2021)
	Chrysanthemum indicum	TFC	44.63 mg CAE/g DW	Zeng et al. (2014)
		TFC	20.45–25.39 mg CAE/g DW	Zheng et al. (2018)
	Chrysanthemum lavandulifolium	TFC	19.19 mg CAE/g DW	Zheng et al. (2018)
	Chrysanthemum morifolium	TFC	10.84 mg CAE/g DW	Zeng et al. (2014)
		TFC	18.73 mg CAE/g DW	Zheng et al. (2018)
		TFC	1[a]–52[a] mg RE/g DW	Xiong et al. (2014), Chen et al. (2015), and Huang et al. (2017)
	Cichorium intybus	TFC	32.00 mg QE/g dry extract	Loizzo et al. (2016)
	Citrus aurantium	TFC	2.96 mg CAE/g DW	Zheng et al. (2018)
	Clitoria ternatea	TFC	115.5–141.4 μg/g DW	Kaisoon et al. (2011)
		TFC	5[a]–15[a] mg RE/g DW	Kaisoon et al. (2011)
	Cocos nucifera	TFC	0.60 mg RE/g DW	Janarny et al. (2021)
	Coreopsis tinctoria	TFC	29.30 mg CAE/g DW	Zheng et al. (2018)
	Coriandrum sativum	TFC	16.51 mg CAE/g DW	Araújo et al. (2019)
	Cosmos sulphureus	TFC	158.1–502.1 μg/g DW	Kaisoon et al. (2011)
		TFC	23[a]–58[a] mg RE/g DW	Kaisoon et al. (2011)
		TFC	121.71 mg/100 g DW	Kaisoon et al. (2012)
	Crocus sativus	TFC	2.57 mg CAE/g DW	Zheng et al. (2018)
	Crotalaria juncea	TFC	3.9–7[a] mg RE/g DW	Kaisoon et al. (2012), Janarny et al. (2021)
	Cucumis sativus	TFC	0.67 mg CAE/g DW	Zheng et al. (2018)
	Cucurbita pepo	TFC	13.46–23.82 ppm	Mikołajczak et al. (2020)
		TFC	73.59 mg CTH/100 g FW	Dujmović et al. (2022)
		TFC	3.61–14.3 mg QE/g DW	Socha et al. (2021), Yan et al. (2022)
	Cymbidium sinense	TFC	4.06 mg CAE/g DW	Zheng et al. (2018)
	Dahlia campanulata	TFC	0.76–1.47 mg QE/g FW	Espejel et al. (2019)
	Dahlia coccinea	TFC	6.78–11.32 mg QE/g FW	Espejel et al. (2019)
	Dahlia hortorum	TFC	1.81–7.82 mg QE/g FW	Espejel et al. (2019)
	Dahlia merckii	TFC	0.69–1.03 mg QE/g FW	Espejel et al. (2019)
	Dendranthema morifolium	TFC	5.89–13.22 mg RE/g DW	Chen et al. (2015)
	Dianthus	TFC	11.88 mg QE/g DW	Socha et al. (2021)
	Dianthus barbatus	TFC	2.57–7.5 mg/g fresh sample	Gonçalves et al. (2020)
	Dianthus caryophyllus	TFC	1.41 mg RE/g DW	Chen et al. (2015)

(continued)

Table 6.1 (continued)

Phyto-chemical	Source	Content		References
		TFC	53.25 mg CAE/g DW	Zeng et al. (2014)
		TFC	2.45 mg CAE/g DW	Zheng et al. (2018)
		TFC	17.50 mg/g fresh sample	Gonçalves et al. (2020)
		TFC	108.91 ppm	Mikołajczak et al. (2020)
	Diuris brevis	TFC	1.23–1.82 mg QE/g FW	Espejel et al. (2019)
	Durio zibethinus	TFC	0.40 mg RE/g DW	Janarny et al. (2021)
	Epipremnum aureum	TFC	1.62 mg CAE/g DW	Zheng et al. (2018)
	Eriobotrya japonica	TFC	9.60 mg CAE/g DW	Zheng et al. (2018)
	Erythrina americana	TFC	25.34 mg QE/g DW	Pinedo-Espinoza et al. (2020)
	Florists chrysanthemum	TFC	11.25–15.15 mg CAE/g DW	Zheng et al. (2018)
	Flos carthami	TFC	0[a]–8[a] mg RE/g DW	Xiong et al. (2014)
	Flos lonicerae	TFC	1[a]–89[a] mg RE/g DW	Xiong et al. (2014), Huang et al. (2017)
	Flos rosae rugosae	TFC	2[a]–28 mg RE/g DW	Xiong et al. (2014)
	Gentianae radix	TFC	6.87 mg RE/g DW	Chen et al. (2015)
	Geranium macrorrhizum	TFC	0.74–1.52 mg QE/g FW	Petrova et al. (2016)
	Glycyrrhiza glabra	TFC	661–801 mg RE/g extract	Suksathan et al. (2021)
	Gmelina arborea	TFC	857–972 mg RE/g extract	Suksathan et al. (2021)
	Gmelina asiatica	TFC	3.70 mg RE/g DW	Janarny et al. (2021)
	Gomphrena globosa	TFC	0.45 mg RE/g DW	Chen et al. (2015)
		TFC	1.18–2.25 mg CAE/g DW	Zheng et al. (2018)
	Hedychium coronarium	TFC	38.60 mg QE/g dry extract	Loizzo et al. (2016)
		TFC	1.95 mg CAE/g DW	Zheng et al. (2018)
	Helianthus tuberosus	TFC	3.32–8.89 mg QE/g FW	Petrova et al. (2016)
	Helichrysum bracteatum	TFC	28.59 mg CAE/g DW	Zheng et al. (2018)
	Hemerocallis	TFC	3.76 g RE/kg FW	Mlcek et al. (2021)
	Hemerocallis citrina	TFC	0.93 mg CAE/g DW	Zheng et al. (2018)
	Hemerocallis hybrida	TFC	50.09 ppm	Mikołajczak et al. (2020)
		TFC	9.55 mg QE/g DW	Socha et al. (2021)
	Hibiscus rosa-sinensis	TFC	6.20 mg RE/g DW	Janarny et al. (2021)
	Hibiscus sabdariffa	TFC	24.47 mg CAE/g DW	Zeng et al. (2014)
		TFC	2.92 mg CAE/g DW	Zheng et al. (2018)
		TFC	0.5[a]–12[a] mg RE/g DW	Xiong et al. (2014), Chen et al. (2015), and Huang et al. (2017)
	Hylocereus undatus	TFC	0.77 mg CAE/g DW	Zheng et al. (2018)

(continued)

Table 6.1 (continued)

Phyto-chemical	Source	Content		References
	Ixora chinensis	TFC	59.6–251 μg/g DW	Kaisoon et al. (2011)
		TFC	17[a]–38[a] mg RE/g DW	
	Ixora coccinea	TFC	6.90 mg RE/g DW	Janarny et al. (2021)
	Jasminum sambac	TFC	54.36 mg CAE/g DW	Zeng et al. (2014)
		TFC	4.44 mg CAE/g DW	Zheng et al. (2018)
		TFC	1[a]–24[a] mg RE/g DW	Chen et al. (2015), Huang et al. (2017), and Janarny et al. (2021)
	Kalanchoe blossfeldiana	TFC	0.30[a]–0.70[a] mg QE/g FW	Fernandes et al. (2020)
	Lasia spinosa	TFC	2.80 mg RE/g DW	Janarny et al. (2021)
	Lavandula stoechas	TFC	54.88 mg CAE/g DW	Araújo et al. (2019)
	Lavandula angustifolia	TFC	27.43 mg RE/g DW	Chen et al. (2015)
		TFC	12.30 mg CAE/g DW	Zheng et al. (2018)
		TFC	24.01 ppm	Mikołajczak et al. (2020)
		TFC	2.10 mg QE/g DW	Socha et al. (2021)
		TFC	18.63–27.392 mg CAE/g DW	Zeng et al. (2014), Araújo et al. (2019)
	Lavandula pedunculata	TFC	2[a]–70[a] mg RE/g DW	Xiong et al. (2014), Huang et al. (2017)
	Lavandula spica	TFC	3.37 mg/g fresh sample	Gonçalves et al. (2020)
	Leucaena leucocephala	TFC	38.6–93.3 μg/g DW	Kaisoon et al. (2011)
		TFC	5[a]–16^a mg RE/g DW	Kaisoon et al. (2011)
	Lilium brownii	TFC	7.16 mg CAE/g DW	Zheng et al. (2018)
		TFC	0[a]–15[a] mg RE/g DW	Xiong et al. (2014), Chen et al. (2015), and Huang et al. (2017)
	Lilium bulbiferum	TFC	1.63 mg CAE/g DW	Zheng et al. (2018)
	Lilium longiflorum	TFC	22.12 mg CAE/g DW	Zeng et al. (2014)
	Lonicera japonica	TFC	52.51 mg CAE/g DW	Zheng et al. (2018)
		TFC	49.02–34.112 mg CAE/g DW	Zeng et al. (2014), Araújo et al. (2019)
	Luffa cylindrica	TFC	14.27–26.24 mg QE/g DW	Yan et al. (2022)
	Madhuca longifolia	TFC	1.50 mg RE/g DW	Janarny et al. (2021)
	Magnolia denudata	TFC	2.25 mg CAE/g DW	Zheng et al. (2018)
	Magnolia grandiflora	TFC	2.59 mg CAE/g DW	Zheng et al. (2018)
	Malus pumila	TFC	16.89 mg CAE/g DW	Zheng et al. (2018)
	Malus spectabilis	TFC	1.56 mg CAE/g DW	Zheng et al. (2018)
	Malva sylvestris	TFC	12.70 mg QE/g dry extract	Loizzo et al. (2016)

(continued)

Table 6.1 (continued)

Phyto-chemical	Source	Content		References
	Malvaviscus arboreus	TFC	34.8–69.5 μg/g DW	Kaisoon et al. (2011)
		TFC	6[a]–17[a] mg RE/g DW	Kaisoon et al. (2011)
	Manihot esculenta	TFC	4.90 mg RE/g DW	Janarny et al. (2021)
	Matricaria recutita	TFC	17.57 mg RE/g DW	Chen et al. (2015)
		TFC	10.22 mg CAE/g DW	Zheng et al. (2018)
	Matthiola incana	TFC	36.67 mg CAE/g DW	Zeng et al. (2014)
		TFC	7.48 mg CAE/g DW	Zheng et al. (2018)
	Mentha spicata	TFC	92.41 ppm	Mikołajczak et al. (2020)
	Michelia alba	TFC	5.56 mg CAE/g DW	Zheng et al. (2018)
	Musa spp	TFC	0.30 mg RE/g DW	Janarny et al. (2021)
	Myosotis sylvatica	TFC	4.26 mg RE/g DW	Chen et al. (2015)
		TFC	11.06 mg CAE/g DW	Zheng et al. (2018)
	Myrtillocactus geometrizans	TFC	72.40 mg QE/g DW	Pinedo-Espinoza et al. (2020)
	Nelumbo nucifera	TFC	11.7–270.3 μg/g DW	Kaisoon et al. (2011)
		TFC	18[a]–47[a] mg RE/g DW	Kaisoon et al. (2011)
		TFC	7.79 mg CAE/g DW	Zheng et al. (2018)
	Nymphaea stellata	TFC	8.82 mg CAE/g DW	Zheng et al. (2018)
	Ocimum sanctum	TFC	11.60 mg RE/g DW	Janarny et al. (2021)
	Oenothera biennis	TFC	56.86 mg CAE/g DW	Araújo et al. (2019)
	Orchis	TFC	2.2–2.63 mg/g fresh sample	Gonçalves et al. (2020)
	Oroxylum indicum	TFC	3.00 mg CAE/g DW	Zheng et al. (2018)
	Osmanthus fragrans	TFC	20.37 mg CAE/g DW	Zeng et al. (2014)
		TFC	85.31 mg CAE/g DW	Zheng et al. (2018)
		TFC	15[a]–159[a] mg RE/g DW	Huang et al. (2017), Chen et al. (2015)
	Paeonia lactiflora	TFC	13.78 mg RE/g DW	Chen et al. (2015)
		TFC	28.76 mg CAE/g DW	Zeng et al. (2014)
	Paeonia suffruticosa	TFC	38.93 mg CAE/g DW	Zeng et al. (2014)
		TFC	19.20 mg CAE/g DW	Zheng et al. (2018)
		TFC	0.5[a]–48[a] mg RE/g DW	Xiong et al. (2014), Huang et al. (2017)
	Panax ginseng	TFC	16.85 mg CAE/g DW	Zeng et al. (2014)
	Panax notoginseng	TFC	11.12 mg CAE/g DW	Zeng et al. (2014)
	Panax pseudoginseng	TFC	1.44 mg CAE/g DW	Zheng et al. (2018)
	Papaver rhoeas	TFC	47.68 mg CAE/g DW	Zeng et al. (2014)
	Pelargonium grandiflorum	TFC	307.34 ppm	Mikołajczak et al. (2020)
	Plumbago auriculata	TFC	3.63 mg/g fresh sample	Gonçalves et al. (2020)

(continued)

Table 6.1 (continued)

Phyto-chemical	Source	Content		References
	Plumeria obtusa	TFC	46.8–702.5 µg/g DW	Kaisoon et al. (2011)
		TFC	11[a]–46^a mg RE/g DW	Kaisoon et al. (2011)
	Plumeria rubra	TFC	3.83–15.69 mg CAE/g DW	Zheng et al. (2018)
	Prunella vulgaris	TFC	3.58 mg CAE/g DW	Zheng et al. (2018)
	Prunus persica	TFC	60.14 mg CAE/g DW	Zeng et al. (2014)
		TFC	1[a]–28[a] mg RE/g DW	Xiong et al. (2014), Huang et al. (2017)
	Punica granatum	TFC	24.98 mg CAE/g DW	Zheng et al. (2018)
		TFC	3.60 mg RE/g DW	Janarny et al. (2021)
	Pyrus pashia	TFC	51.2–278.8 mg RE/g dry extract	He et al. (2015)
	Red artful plum tea	TFC	0.68 mg RE/g DW	Chen et al. (2015)
	Rhododendron simsii	TFC	20.02 mg CAE/g DW	Zheng et al. (2018)
	Robinia pseudoacacia	TFC	45.30 mg QE/g dry extract	Loizzo et al. (2016)
	Rosa	TFC	6–16.57 mg/g fresh sample	Gonçalves et al. (2020)
		TFC	2.43 g RE/kg FW	Mlcek et al. (2021)
	Rosa canina	TFC	215.90 ppm	Mikołajczak et al. (2020)
	Rosa centifolia	TFC	10.91 mg CAE/g DW	Zheng et al. (2018)
	Rosa chinensis	TFC	1[a]–39 mg RE/g DW	Xiong et al. (2014)
		TFC	24.13 mg CAE/g DW	Zheng et al. (2018)
	Rosa gallica	TFC	13.43 mg CAE/g DW	Zheng et al. (2018)
	Rosa rugosa	TFC	77.31 mg CAE/g DW	Zeng et al. (2014)
		TFC	2.51–23.56 mg CAE/g DW	Zheng et al. (2018)
		TFC	14.36 mg QE/g DW	Socha et al. (2021)
		TFC	3.41–38[a] mg RE/g DW	Chen et al. (2015), Huang et al. (2017)
	Rosa sp.	TFC	2.62 mg CAE/g DW	Araújo et al. (2019)
	Rosmarinus officinalis	TFC	20.30 mg QE/g dry extract	Loizzo et al. (2016)
		TFC	18.83 mg RE/g DW	Chen et al. (2015)
		TFC	27.94 mg CAE/g DW	Araújo et al. (2019)
	Salvia elegans	TFC	24.26 mg CAE/g DW	Araújo et al. (2019)
	Salvia splendens	TFC	13.87 ppm	Mikołajczak et al. (2020)
	Sambucus nigra	TFC	64.20 mg QE/g dry extract	Loizzo et al. (2016)
		TFC	53.59 ppm	Mikołajczak et al. (2020)
		TFC	2.06 mg QE/g DW	Socha et al. (2021)
	Sesbania grandiflora	TFC	0.7–1.21 mg RE/g DW	Kaisoon et al. (2012), Janarny et al. (2021)

(continued)

Table 6.1 (continued)

Phyto-chemical	Source	Content		References
	Shorea roxburghii	TFC	121–197 mg RE/g extract	Suksathan et al. (2021)
	Siraitia grosvenorii	TFC	4.50 mg RE/g DW	Chen et al. (2015)
		TFC	2.85 mg CAE/g DW	Zheng et al. (2018)
	Sophora japonica	TFC	18.18 mg CAE/g DW	Zheng et al. (2018)
	Sophora viciifolia	TFC	53.3–237.2 mg RE/g dry extract	Tai et al. (2011)
	Tagetes	TFC	14.66 mg QE/g DW	Socha et al. (2021)
	Tagetes erecta	TFC	313.1–990.1 μg/g DW	Kaisoon et al. (2011)
		TFC	38[a]–69[a] mg RE/g DW	Kaisoon et al. (2011)
		TFC	0.45–6.15 mg QE/g FW	Petrova et al. (2016)
		TFC	42.45 mg CAE/g DW	Zeng et al. (2014)
		TFC	16.90 mg/g fresh sample	Gonçalves et al. (2020)
		TFC	165.30 mg/100 g DW	Kaisoon et al. (2012)
		TFC	391.09 mg CTH/100 g FW	Dujmović et al. (2022)
		TFC	26.47 mg CAE/g DW	Araújo et al. (2019)
		TFC	307.34 ppm	Mikołajczak et al. (2020)
		TFC	2.02 g RE/kg FW	Mlcek et al. (2021)
	Tamarindus indica	TFC	2–4[a] mg RE/g DW	Kaisoon et al. (2012), Janarny et al. (2021)
	Taraxacum officinale	TFC	0.20[a]–0.72[a] mg QE/g FW	Fernandes et al. (2020)
	Telosma minor	TFC	3.2–126.1 μg/g DW	Kaisoon et al. (2011)
		TFC	4[a]–10[a] mg RE/g DW	Kaisoon et al. (2011)
	Tilia cordata	TFC	205.40 ppm	Mikołajczak et al. (2020)
	Trifolium repens	TFC	91.86 ppm	Mikołajczak et al. (2020)
	Trollius chinensis	TFC	83.80 mg CAE/g DW	Zeng et al. (2014)
	Tropaeolum majus	TFC	24.73 mg CAE/g DW	Zheng et al. (2018)
		TFC	5.13 mg CAE/g DW	Araújo et al. (2019)
		TFC	163.32 ppm	Mikołajczak et al. (2020)
		TFC	0.79 mg QE/g DW	Socha et al. (2021)
		TFC	1.29 g RE/kg FW	Mlcek et al. (2021)
		TFC	198.70 mg CTH/100 g FW	Dujmović et al. (2022)
	Viburnum inopinatum	TFC	161–213 mg RE/g extract	Suksathan et al. (2021)
	Vinca rosea	TFC	11.00 mg RE/g DW	Janarny et al. (2021)
	Viola tricolor	TFC	32.84 mg CAE/g DW	Araújo et al. (2019)
		TFC	1.53 mg QE/g DW	Socha et al. (2021)
		TFC	0.40[a]–1.40[a] mg QE/g FW	Fernandes et al. (2020)
	Viola × wittrockiana	TFC	244.35 ppm	Mikołajczak et al. (2020)
	Wisteria sinensis	TFC	4.35–30.39 mg QE/g DW	Yan et al. (2022)

(continued)

Table 6.1 (continued)

Phyto-chemical	Source	Content		References
Phenolic acid			0.00	
	Acacia	TPAC[b]	35.12 mg/100 g DW	Nowicka and Wojdyło (2019)
	Allium schoenoprasum	TPAC	0.78 mg CFAE/g DW	Socha et al. (2021)
	Anthyllis vulneraria	TPAC[b]	1.47 mg/100 g DW	Nowicka and Wojdyło (2019)
	Antigonon leptopus	TPAC[c]	116.6–183.9 μg/g DW	Kaisoon et al. (2011)
		TPAC[c]	272.89 mg/100 g DW	Kaisoon et al. (2012)
	Arnica	TPAC[b]	214.86 mg/100 g DW	Nowicka and Wojdyło (2019)
	Begonia tuberhybrida	TPAC	8.60 mg CFAE/g DW	Socha et al. (2021)
	Bellis perennis	TPAC[b]	350.40 mg/100 g DW	Nowicka and Wojdyło (2019)
	Bougainvillea glabra	TPAC[c]	228.29 mg/100 g DW	Kaisoon et al. (2012)
	Calendula officinalis	TPAC	1.36 mg CFAE/g DW	Socha et al. (2021)
		TPAC[b]	14.90 mg/100 g DW	Nowicka and Wojdyło (2019)
	Cassia siamea	TPAC[c]	455.5–793.3 μg/g DW	Kaisoon et al. (2011)
	Centaurea cyanus	TPAC	1.66 mg CFAE/g DW	Socha et al. (2021)
		TPAC[b]	47.85 mg/100 g DW	Nowicka and Wojdyło (2019)
	Chrysanthemum	TPAC	2.56 mg CFAE/g DW	Socha et al. (2021)
	Clitoria ternatea	TPAC[c]	243.5–261.5 μg/g DW	Kaisoon et al. (2011)
	Cosmos sulphureus	TPAC[c]	418.5–615.1 μg/g DW	Kaisoon et al. (2011)
		TPAC[c]	210.27 mg/100 g DW	Kaisoon et al. (2012)
	Crataegus	TPAC[b]	245.53 mg/100 g DW	Nowicka and Wojdyło (2019)
	Cucurbita pepo	TPAC	2.01 mg CFAE/g DW	Socha et al. (2021)
	Dianthus	TPAC	1.86 mg CFAE/g DW	Socha et al. (2021)
	Hemerocallis hybrida	TPAC	2.62 mg CFAE/g DW	Socha et al. (2021)
	Ixora chinensis	TPAC[c]	251.1–381.4 μg/g DW	Kaisoon et al. (2011)
	Lamium album	TPAC[b]	151.54 mg/100 g DW	Nowicka and Wojdyło (2019)
	Lavandula	TPAC[b]	201.11 mg/100 g DW	Nowicka and Wojdyło (2019)
	Lavandula angustifolia	TPAC	1.72 mg CFAE/g DW	Socha et al. (2021)
	Leucaena leucocephala	TPAC[c]	134.9–254.5 μg/g DW	Kaisoon et al. (2011)

(continued)

Table 6.1 (continued)

Phyto-chemical	Source	Content		References
	Malva	TPAC[b]	161.53 mg/100 g DW	Nowicka and Wojdyło (2019)
	Malva arborea	TPAC[b]	45.91 mg/100 g DW	Nowicka and Wojdyło (2019)
	Malvaviscus arboreus	TPAC[c]	116.7–324.4 μg/g DW	Kaisoon et al. (2011)
	Matricaria	TPAC[b]	189.82 mg/100 g DW	Nowicka and Wojdyło (2019)
	Nelumbo nucifera	TPAC[c]	147.9–502.9 μg/g DW	Kaisoon et al. (2011)
	Plumeria obtusa	TPAC[b]	281.5–329.9 μg/g DW	Kaisoon et al. (2011)
	Primula	TPAC[b]	9.00 mg/100 g DW	Nowicka and Wojdyło (2019)
	Rosa rugosa	TPAC	6.03 mg CFAE/g DW	Socha et al. (2021)
	Sambucus nigra	TPAC	4.98 mg CFAE/g DW	Socha et al. (2021)
		TPAC[b]	78.00 mg/100 g DW	Nowicka and Wojdyło (2019)
	Tagetes	TPAC	6.63 mg CFAE/g DW	Socha et al. (2021)
	Tagetes erecta	TPAC[c]	549.2–1072.5 μg/g DW	Kaisoon et al. (2011)
		TPAC[c]	942.25 mg/100 g DW	Kaisoon et al. (2012)
	Telosma minor	TPAC[c]	70.4–116.7 μg/g DW	Kaisoon et al. (2011)
	Tilia cordata	TPAC[b]	3.66 mg/100 g DW	Nowicka and Wojdyło (2019)
	Tropaeolum majus	TPAC	3.41 mg CFAE/g DW	Socha et al. (2021)
	Verbascum	TPAC[b]	31.71 mg/100 g DW	Nowicka and Wojdyło (2019)
	Viola tricolor	TPAC	7.41 mg CFAE/g DW	Socha et al. (2021)
Phenols			0.00	
	Aegle marmelos	TPC	14.40 mg GAE/g DW	Janarny et al. (2021)
	Agastache foeniculum	TPC	52.06 mg GAE/g	Araújo et al. (2019)
	Agave salmiana	TPC	4.63 mg GAE/g DW	Pinedo-Espinoza et al. (2020)
	Ageratum conyzoides	TPC	4.63 mg GAE/g FW	Li et al. (2014)
	Aglaia odorata	TPC	55.46 mg GAE/g DW	Zheng et al. (2018)
	Albizia julibrissin	TPC	14.15 mg GAE/g DW	Zheng et al. (2018)
	Allamanda cathartica	TPC	4.16 mg GAE/g FW	Li et al. (2014)
	Allium cepa	TPC	30.00 mg GAE/g DW	Janarny et al. (2021)
		TPC	69.62 ppm	Mikołajczak et al. (2020)
	Allium schoenoprasum	TPC	18.28 mg/g DW	Kucekova et al. (2013)
		TPC	250.06 ppm	Mikołajczak et al. (2020)

(continued)

Table 6.1 (continued)

Phyto-chemical	Source	Content		References
		TPC	7.15 mg GAE/g DW	Socha et al. (2021)
	Allium ursinum	TPC	184.40 mg GAE/100 g FW	Demasi et al. (2021)
	Aloe vera	TPC	4.55 mg GAE/g DW	Pinedo-Espinoza et al. (2020)
	Aloysia citriodora	TPC	222.85 mg GAE/g DW	Coyago-Cruz et al. (2024)
	Amygdalus persica	TPC	51.18 mg GAE/g DW	Zheng et al. (2018)
	Anchusa azurea	TPC	33.60 mg CGAE/g extract	Loizzo et al. (2016)
	Anethum graveolens	TPC	254.66 mg GAE/g DW	Coyago-Cruz et al. (2024)
	Antigonon leptopus	TPC	65[a]–177.2 mg GAE/g DW	Kaisoon et al. (2011), Kaisoon et al. (2012)
	Antirrhinum majus	TPC	0.97 mg GAE/g FW	Kritsi et al. (2022)
		TPC	101.32–200.53 mg GAE/g DW	Coyago-Cruz et al. (2024)
		TPC	3.02 GAE/g FW	Chensom et al. (2019)
	Aponogeton crispus	TPC	36.60 mg GAE/g DW	Janarny et al. (2021)
	Armeniaca mume	TPC	54.65 mg GAE/g DW	Zheng et al. (2018)
	Azadirachta indica	TPC	33.30 mg GAE/g DW	Janarny et al. (2021)
	Bauhinia purpurea	TPC	6.14 mg GAE/g FW	Li et al. (2014)
	Bauhinia racemosa	TPC	25.00 mg GAE/g DW	Janarny et al. (2021)
		TPC	150[a] µmol GAE/g DW	Janarny et al. (2022)
	Bauhinia variegata	TPC	47.98 mg GAE/g DW	Zheng et al. (2018)
		TPC	134–208 mg GAE/g extract	Suksathan et al. (2021)
	Begonia boliviensis	TPC	2.24–3.06 mg GAE/g FW	Marchioni et al. (2024)
	Begonia doblet	TPC	275.17 mg GAE/g DW	Coyago-Cruz et al. (2024)
	Begonia semperflorens	TPC	2.04 GAE/g FW	Chensom et al. (2019)
	Begonia tuberhybrida	TPC	539.57–996.80 ppm	Mikołajczak et al. (2020)
		TPC	33.00 mg GAE/g DW	Socha et al. (2021)
		TPC	4.82 g GAE/kg FW	Mlcek et al. (2021)
	Bellis perennis	TPC	16.20 mg/g DW	Kucekova et al. (2013)
		TPC	553.21 ppm	Mikołajczak et al. (2020)
		TPC	396.30 mg GAE/100 g FW	Demasi et al. (2021)
	Benincasa hispida	TPC	7.59–49.92 mg GAE/g DW	Yan et al. (2022)
		TPC[c]	0.658–9.649 mg/g DW	Yan et al. (2022)
	Bidens pilosa	TPC	8.12 mg GAE/g FW	Li et al. (2014)
	Bombax ceiba	TPC	21.63 mg GAE/g DW	Zheng et al. (2018)

(continued)

Table 6.1 (continued)

Phyto-chemical	Source	Content		References
	Bombax malabaricum	TPC	3.88 mg GAE/g FW	Li et al. (2014)
	Borago officinalis	TPC	16.58 mg GAE/g DW	Araújo et al. (2019)
		TPC	0.4[a]–1.5[a] mg GAE/g FW	Fernandes et al. (2020)
		TPC	163.40 mg GAE/100 g FW	Demasi et al. (2021)
	Bougainvillea hybrida	TPC	45[a]–135[a] mg GAE/g DW	Kaisoon et al. (2011)
	Bougainvillea glabra	TPC	21.4–138.2 mg GAE/g DW	Janarny et al. (2021), Kaisoon et al. (2012)
	Bougainvillea spectabilis	TPC	1.65–6.87 mg GAE/g FW	Petrova et al. (2016), Li et al. (2014)
	Brassica campestris	TPC	3.32–4.37 mg GAE/g FW	Li et al. (2014)
	Brunfelsia acuminata	TPC	4.08 mg GAE/g FW	Li et al. (2014)
	Calendula officinalis	TPC	4104.67 µg/g DW	Chen et al. (2015)
		TPC	0.7–3.30 mg GAE/g FW	Petrova et al. (2016)
		TPC	214.15 ppm	Mikołajczak et al. (2020)
		TPC	189.60 mg GAE/100 g FW	Demasi et al. (2021)
		TPC	3.65 g GAE/kg FW	Mlcek et al. (2021)
		TPC	379.38 mg GAE/100 g FW	Dujmović et al. (2022)
		TPC	13.03–290.74 mg GAE/g DW	Chen et al. (2015), Zheng et al. (2018), Araújo et al. (2019), Janarny et al. (2021), Socha et al. (2021), and Coyago-Cruz et al. (2024)
	Calliandra haematocephala	TPC	14.43 mg GAE/g FW	Li et al. (2014)
	Calotropis gigantea	TPC	52.90 mg GAE/g DW	Janarny et al. (2021)
	Camellia azalea	TPC	61.46 mg GAE/g DW	Zheng et al. (2018)
	Camellia japonica	TPC	21.95 mg GAE/g DW	Zheng et al. (2018)
		TPC	5.14 mg GAE/g FW	Li et al. (2014)
	Camellia sinensis	TPC	187–230 mg GAE/g extract	Suksathan et al. (2021)
	Campsis grandiflora	TPC	29.61 mg GAE/g DW	Zheng et al. (2018)
	Canna edulis	TPC	18.01 mg GAE/g DW	Zheng et al. (2018)
	Canna indica	TPC[c]	532.20 mg/100 g DW	Coyago-Cruz et al. (2024)
		TPC	115.21–261.56 mg GAE/g DW	Coyago-Cruz et al. (2024)
	Capparis spinosa	TPC	33.60 mg CGAE/g extract	Loizzo et al. (2016)

(continued)

Table 6.1 (continued)

Phyto-chemical	Source	Content		References
	Carica papaya	TPC	20.30 mg GAE/g DW	Janarny et al. (2021)
	Cassia auriculata	TPC	39.20 mg GAE/g DW	Janarny et al. (2021)
		TPC	230.60 μmol GAE/g DW	Janarny et al. (2022)
	Cassia fistula	TPC	34.80 mg GAE/g DW	Janarny et al. (2021)
		TPC	215[a] μmol GAE/g DW	Janarny et al. (2022)
	Cassia siamea	TPC	70[a]–90[a] mg GAE/g DW	Kaisoon et al. (2011)
	Catharanthus roseus	TPC	3.51 mg GAE/g FW	Kritsi et al. (2022)
	Centaurea cyanus	TPC	185.97 ppm	Mikołajczak et al. (2020)
		TPC	7.34 mg GAE/g DW	Socha et al. (2021)
		TPC	378.5–647.09 mg GAE/100 g FW	Fernandes et al. (2020), Dujmović et al. (2022)
	Chaenomeles sinensis	TPC	13.93 mg GAE/g FW	Li et al. (2014)
	Chamaemelum nobile	TPC	44594.77 μg/g DW	Chen et al. (2015)
		TPC	28.53 mg GAE/g DW	Chen et al. (2015)
		TPC	16.49 mg GAE/g DW	Coyago-Cruz et al. (2024)
	Chamomilla	TPC	34373.82 μg/g DW	Chen et al. (2015)
		TPC	30.15 mg GAE/g DW	Chen et al. (2015)
	Chamomilla recutita	TPC	653.20 ppm	Mikołajczak et al. (2020)
	Chimonanthus praecox	TPC	12415.63 μg/g DW	Chen et al. (2015)
		TPC	25.77–33.89 mg GAE/g DW	Chen et al. (2015), Zheng et al. (2018)
	Chrysanthemum	TPC	8.71 mg GAE/g DW	Socha et al. (2021)
	Chrysanthemum coronarium	TPC	3.76 mg GAE/g FW	Li et al. (2014)
	Chrysanthemum indicum	TPC	35.98–55.04 mg GAE/g DW	Zheng et al. (2018)
		TPC	1.94–2.13 mg GAE/g FW	Kritsi et al. (2022)
	Chrysanthemum lavandulifolium	TPC	39.44 mg GAE/g DW	Zheng et al. (2018)
	Chrysanthemum morifolium	TPC	26023.68 μg/g DW	Chen et al. (2015)
		TPC[c]	911.2–477.7 μg/g DW	Xiong et al. (2014)
		TPC	24.58–33.59 mg GAE/g DW	Chen et al. (2015), Zheng et al. (2018)
		TPC	4[a]–40[a] mg CGAE/g DW	Xiong et al. (2014), Huang et al. (2017)
	Chrysanthemum × hybrid	TPC[c]	1958.30 mg/100 g DW	Coyago-Cruz et al. (2024)
		TPC	100.89–245.36 mg GAE/g DW	Coyago-Cruz et al. (2024)

(continued)

Table 6.1 (continued)

Phyto-chemical	Source	Content		References
	Cichorium intybus	TPC	43.20 mg CGAE/g extract	Loizzo et al. (2016)
		TPC	24.75 mg/g DW	Kucekova et al. (2013)
		TPC	618.40 mg GAE/100 g FW	Demasi et al. (2021)
	Citrus aurantium	TPC	62.48 mg GAE/g DW	Zheng et al. (2018)
	Citrus × aurantifolia	TPC	65.40 mg GAE/g DW	Coyago-Cruz et al. (2024)
	Clitoria ternatea	TPC	21–60[a] mg GAE/g DW	Kaisoon et al. (2011), Janarny et al. (2021)
		TPC	125[a] µmol GAE/g DW	Janarny et al. (2022)
	Cocos nucifera	TPC	40.80 mg GAE/g DW	Janarny et al. (2021)
	Coreopsis tinctoria	TPC	92.93 mg GAE/g DW	Zheng et al. (2018)
	Coriandrum sativum	TPC	12.17 mg GAE/g DW	Araújo et al. (2019)
	Cosmos sulphureus	TPC	13.08 GAE/g FW	Chensom et al. (2019)
		TPC	85[a]–102.5 mg GAE/g DW	Kaisoon et al. (2011), Kaisoon et al. (2012)
	Crocus sativus	TPC	21.42 mg GAE/g DW	Zheng et al. (2018)
	Crotalaria juncea	TPC	18.90 mg GAE/g DW	Janarny et al. (2021)
		TPC	110[a] µmol GAE/g DW	Janarny et al. (2022)
	Cucumis sativus	TPC	5.48 mg GAE/g DW	Zheng et al. (2018)
	Cucurbita pepo	TPC	267.33–338.68 ppm	Mikołajczak et al. (2020)
		TPC[c]	4.266–12.906 mg/g DW	Yan et al. (2022)
		TPC	110.24 mg GAE/100 g FW	Dujmović et al. (2022)
		TPC	18.8–40.86 mg GAE/g DW	Socha et al. (2021), Yan et al. (2022)
	Cyclamen hederifolium	TPC	5.46–6.39 mg GAE/g FW	Kritsi et al. (2022)
	Cyclamen repandum	TPC	7.35 mg GAE/g FW	Kritsi et al. (2022)
	Cymbidium sinense	TPC	23.29 mg GAE/g DW	Zheng et al. (2018)
	Dahlia pinnata	TPC	280.21–338.6 mg GAE/g DW	Coyago-Cruz et al. (2024)
		TPC	13.36 mg GAE/g FW	Marchioni et al. (2024)
	Dendranthema morifolium	TPC	13449.68–38326.08 µg/g DW	Chen et al. (2015)
		TPC	16.77–24.83 mg GAE/g DW	Chen et al. (2015)
	Dianthus	TPC	15.88 mg GAE/g DW	Socha et al. (2021)
	Dianthus barbatus	TPC	1.87–5.1 mg/g fresh sample	Gonçalves et al. (2020)
	Dianthus carthusianorum	TPC	936.30 mg GAE/100 g FW	Demasi et al. (2021)
	Dianthus caryophyllus	TPC	16022.47 µg/g DW	Chen et al. (2015)

(continued)

Table 6.1 (continued)

Phyto-chemical	Source	Content		References
		TPC	18.17 mg/g fresh sample	Gonçalves et al. (2020)
		TPC	1.30–5.50 mg GAE/g FW	Li et al. (2014), Kritsi et al. (2022)
		TPC	351.87 ppm	Mikołajczak et al. (2020)
		TPC	27.86–30.83 mg GAE/g DW	Chen et al. (2015), Zheng et al. (2018)
	Dianthus chinensis	TPC	5.27 mg GAE/g FW	Li et al. (2014)
		TPC	179.63–248.57 mg GAE/g DW	Coyago-Cruz et al. (2024)
	Dianthus pavonius	TPC	752.80 mg GAE/100 g FW	Demasi et al. (2021)
	Durio zibethinus	TPC	9.00 mg GAE/g DW	Janarny et al. (2021)
	Epipremnum aureum	TPC	12.94 mg GAE/g DW	Zheng et al. (2018)
	Ericaceae rhododendron	TPC	6.33 mg GAE/g FW	Li et al. (2014)
	Eriobotrya japonica	TPC	21.58 mg GAE/g DW	Zheng et al. (2018)
	Erythrina variegata	TPC	3.90 mg GAE/g FW	Li et al. (2014)
	Erythronium dens-canis	TPC	364.30 mg GAE/100 g FW	Demasi et al. (2021)
	Erythrina americana	TPC	7.80 mg GAE/g DW	Pinedo-Espinoza et al. (2020)
	Florists chrysanthemum	TPC	23.2–32.47 mg GAE/g DW	Zheng et al. (2018)
	Flos carthami	TPC[c]	1138.8–29.4 μg/g DW	Xiong et al. (2014)
		TPC	0[a]–25[a] mg CGAE/g DW	Xiong et al. (2014)
	Flos chrysanthemi	TPC	3.75 mg GAE/g FW	Li et al. (2014)
	Flos lonicerae	TPC[c]	17,380–116.9 μg/g DW	Xiong et al. (2014)
		TPC	0[a]–75[a] mg CGAE/g DW	Xiong et al. (2014), Huang et al. (2017)
	Flos rosae rugosae	TPC[c]	7038.1–651.2 μg/g DW	Xiong et al. (2014)
		TPC	25[a]–120[a] mg CGAE/g DW	Xiong et al. (2014)
	Fuchsia magellanica	TPC	174.30 mg GAE/g DW	Coyago-Cruz et al. (2024)
	Gentianae radix	TPC	18020.88 μg/g DW	Chen et al. (2015)
		TPC	25.06 mg GAE/g DW	Chen et al. (2015)
	Geranium macrorrhizum	TPC	9.89–19.79 mg GAE/g FW	Petrova et al. (2016)
	Geranium sylvaticum	TPC	1267.80 mg GAE/100 g FW	Demasi et al. (2021)
	Gerbera jamesonii bolus	TPC	4.89 mg GAE/g FW	Li et al. (2014)
	Gladiolus hybrids	TPC	2.30 mg GAE/g FW	Li et al. (2014)
	Glycyrrhiza glabra	TPC	96–241 mg GAE/g extract	Suksathan et al. (2021)

(continued)

Table 6.1 (continued)

Phyto-chemical	Source	Content		References
	Gmelina arborea	TPC	122–129 mg GAE/g extract	Suksathan et al. (2021)
	Gmelina asiatica	TPC	17.60 mg GAE/g DW	Janarny et al. (2021)
	Gomphrena globosa	TPC	1181.13 μg/g DW	Chen et al. (2015)
	Gomphrena globosa	TPC	8.96–13.16 mg GAE/g DW	Chen et al. (2015), Zheng et al. (2018)
	Hedychium coronarium	TPC	48.40 mg CGAE/g extract	Loizzo et al. (2016)
		TPC	14.19 mg GAE/g DW	Zheng et al. (2018)
	Helianthus annuus	TPC	1.86 mg GAE/g FW	Li et al. (2014)
		TPC	77.42 mg GAE/g DW	Coyago-Cruz et al. (2024)
	Helianthus tuberosus	TPC	3.03–17.36 mg GAE/g FW	Petrova et al. (2016)
	Helichrysum bracteatum	TPC	72.38 mg GAE/g DW	Zheng et al. (2018)
	Hemerocallis	TPC	6.59 g GAE/kg FW	Mlcek et al. (2021)
	Hemerocallis citrina	TPC	8.66 mg GAE/g DW	Zheng et al. (2018)
	Hemerocallis hybrida	TPC	341.41 ppm	Mikołajczak et al. (2020)
		TPC	15.73 mg GAE/g DW	Socha et al. (2021)
	Hibiscus rosa-sinensis	TPC	4.2–6.80 mg GAE/g FW	Li et al. (2014), Kritsi et al. (2022)
		TPC	32.60 mg GAE/g DW	Janarny et al. (2021)
	Hibiscus rosa-sinensis (Malvaceae)	TPC[c]	1706.2–2683.7 mg/100 g DW	Coyago-Cruz et al. (2024)
		TPC	152–353.82 mg GAE/g DW	Coyago-Cruz et al. (2024)
	Hibiscus rosa-sinensis (Rosaceae)	TPC	253.52 mg GAE/g DW	Coyago-Cruz et al. (2024)
	Hibiscus sabdariffa	TPC	6126.18 μg/g DW	Chen et al. (2015)
		TPC[c]	8444.2–22.8 μg/g DW	Xiong et al. (2014)
		TPC	2[a]–30.0[a] mg CGAE/g DW	Xiong et al. (2014), Huang et al. (2017)
		TPC	16.41–21.09 mg GAE/g DW	Chen et al. (2015), Zheng et al. (2018)
	Hylocereus undatus	TPC	9.66 mg GAE/g DW	Zheng et al. (2018)
	Impatiens walleriana	TPC	7.62 mg GAE/g FW	Li et al. (2014)
	Ipomoea cairica	TPC	1.77 mg GAE/g FW	Li et al. (2014)
	Iris japonica	TPC	0.63 mg GAE/g FW	Li et al. (2014)
	Ixora chinensis	TPC	85[a]–95[a] mg GAE/g DW	Kaisoon et al. (2011)
	Ixora coccinea	TPC	43.40 mg GAE/g DW	Janarny et al. (2021)
	Jasminum nudiflorum	TPC	3.08 mg GAE/g FW	Li et al. (2014)

(continued)

Table 6.1 (continued)

Phyto-chemical	Source	Content		References
	Jasminum officinale	TPC	4.59 mg GAE/g FW	Kritsi et al. (2022)
	Jasminum sambac	TPC	1761.98 μg/g DW	Chen et al. (2015)
		TPC	10[a]–44.0[a] mg CGAE/g DW	Huang et al. (2017)
		TPC	12.66–43.6 mg GAE/g DW	Chen et al. (2015), Zheng et al. (2018), and Janarny et al. (2021)
	Jatropha integerrima	TPC	17.22 mg GAE/g FW	Li et al. (2014)
	Kalanchoe blossfeldiana	TPC	0.6[a]–1.8[a] mg GAE/g FW	Fernandes et al. (2020)
	Lantana camara	TPC	1.4–3.5 mg GAE/g FW	Li et al. (2014), Kritsi et al. (2022)
		TPC	182.40 mg GAE/g DW	Coyago-Cruz et al. (2024)
	Lantana viburnoides	TPC	65.98–259.28 mg GAE/g DW	Coyago-Cruz et al. (2024)
	Lasia spinosa	TPC	35.10 mg GAE/g DW	Janarny et al. (2021)
	Lathyrus odoratus	TPC	3.02–3.43 GAE/g FW	Chensom et al. (2019)
	Lavandula angustifolia	TPC	14068.04 μg/g DW	Chen et al. (2015)
		TPC	154.13 ppm	Mikołajczak et al. (2020)
		TPC	396.00 mg GAE/100 g FW	Demasi et al. (2021)
		TPC	7.41–50.8 mg GAE/g DW	Chen et al. (2015), Zheng et al. (2018), Araújo et al. (2019), and Socha et al. (2021)
	Lavandula pedunculata	TPC[c]	3287.6–90.6 μg/g DW	Xiong et al. (2014)
		TPC	11[a]–46.0[a] mg CGAE/g DW	Xiong et al. (2014), Huang et al. (2017)
	Lavandula spica	TPC	7.93 mg/g fresh sample	Gonçalves et al. (2020)
	Lavandula stoechas	TPC	46.26 mg GAE/g DW	Araújo et al. (2019)
	Leucaena leucocephala	TPC	60[a] mg GAE/g DW	Kaisoon et al. (2011)
	Leucanthemum vulgare	TPC	448.80 mg GAE/100 g FW	Demasi et al. (2021)
	Ligustrum sinense	TPC	6.22 mg GAE/g FW	Li et al. (2014)
	Lilium brownii	TPC	5055.55 μg/g DW	Chen et al. (2015)
		TPC[c]	1083.4–2.0 μg/g DW	Xiong et al. (2014)
		TPC	1.27 mg GAE/g FW	Li et al. (2014)
		TPC	18.44–34.75 mg GAE/g DW	Chen et al. (2015), Zheng et al. (2018)
		TPC	2[a]–25[a] mg CGAE/g DW	Xiong et al. (2014), Huang et al. (2017)

(continued)

Table 6.1 (continued)

Phyto-chemical	Source	Content		References
	Lilium bulbiferum	TPC	12.14 mg GAE/g DW	Zheng et al. (2018)
	Lilium candidum	TPC	0.92 mg GAE/g FW	Kritsi et al. (2022)
	Limonium sinuatum	TPC	34.17 mg GAE/g FW	Li et al. (2014)
	Lonicera japonica	TPC	41.2–87.48 mg GAE/g DW	Zheng et al. (2018), Araújo et al. (2019)
	Loropetalum chinense	TPC	11.46 mg GAE/g FW	Li et al. (2014)
	Luffa cylindrica	TPC[c]	3.407–8.529 mg/g DW	Yan et al. (2022)
		TPC	26.05–44.52 mg GAE/g DW	Yan et al. (2022)
	Madhuca longifolia	TPC	18.20 mg GAE/g DW	Janarny et al. (2021)
	Magnolia denudata	TPC	19.60 mg GAE/g DW	Zheng et al. (2018)
	Magnolia grandiflora	TPC	10.35 mg GAE/g DW	Zheng et al. (2018)
	Magnolia soulangeana	TPC	5.30 mg GAE/g FW	Li et al. (2014)
	Malus pumila	TPC	60.58 mg GAE/g DW	Zheng et al. (2018)
	Malus spectabilis	TPC	8.23 mg GAE/g DW	Zheng et al. (2018)
	Malva sylvestris	TPC	30.40 mg CGAE/g extract	Loizzo et al. (2016)
	Malvaviscus arboreus	TPC	2.25–3.12 mg GAE/g FW	Li et al. (2014), Kritsi et al. (2022)
		TPC	42[a]–178.51 mg GAE/g DW	Kaisoon et al. (2011), Coyago-Cruz et al. (2024)
	Manihot esculenta	TPC	30.70 mg GAE/g DW	Janarny et al. (2021)
	Matricaria recutita	TPC	39883.82 μg/g DW	Chen et al. (2015)
		TPC	26.41–26.53 mg GAE/g DW	Chen et al. (2015), Zheng et al. (2018)
	Matthiola incana	TPC	43.28 mg GAE/g DW	Zheng et al. (2018)
		TPC	1.70 mg GAE/g FW	Li et al. (2014)
	Mentha aquatica	TPC	1061.70 mg GAE/100 g FW	Demasi et al. (2021)
	Mentha spicata	TPC	553.66 ppm	Mikołajczak et al. (2020)
	Michelia alba	TPC	18.90 mg GAE/g DW	Zheng et al. (2018)
	Mirabilis jalapa	TPC	119.36 mg GAE/g DW	Coyago-Cruz et al. (2024)
	Musa spp	TPC	6.70 mg GAE/g DW	Janarny et al. (2021)
	Myosotis sylvatica	TPC	8688.17 μg/g DW	Chen et al. (2015)
		TPC	25.93–64.02 mg GAE/g DW	Chen et al. (2015), Zheng et al. (2018)
	Myrtillocactus geometrizans	TPC	28.69 mg GAE/g DW	Pinedo-Espinoza et al. (2020)
	Nelumbo nucifera	TPC	63[a]–158[a] mg GAE/g DW	Kaisoon et al. (2011)
	Nymphaea stellata	TPC	69.57 mg GAE/g DW	Zheng et al. (2018)
	Ocimum sanctum	TPC	44.50 mg GAE/g DW	Janarny et al. (2021)

(continued)

Table 6.1 (continued)

Phyto-chemical	Source	Content		References
	Oenothera biennis	TPC	34.96 mg GAE/g DW	Araújo et al. (2019)
	Oncidium varicosum	TPC	4.46 mg GAE/g FW	Li et al. (2014)
	Orchis	TPC	3.57–4.87 mg/g fresh sample	Gonçalves et al. (2020)
	Orostachys fimbriata	TPC	12.36 mg GAE/g FW	Li et al. (2014)
	Oroxylum indicum	TPC	46.00 mg GAE/g DW	Zheng et al. (2018)
	Osmanthus fragrans	TPC	80621.33 μg/g DW	Chen et al. (2015)
		TPC	68.19–134.05 mg GAE/g DW	Chen et al. (2015)
		TPC	30[a]–110[a] mg CGAE/g DW	Huang et al. (2017)
		TPC	16.00 mg GAE/g FW	Li et al. (2014)
	Oxalis corymbosa	TPC	2.20 mg GAE/g FW	Li et al. (2014)
	Paeonia lactiflora	TPC	222.01 mg GAE/g DW	Chen et al. (2015)
		TPC	89784.78 μg/g DW	Chen et al. (2015)
	Paeonia officinalis	TPC	1930.50 mg GAE/100 g FW	Demasi et al. (2021)
	Paeonia suffruticosa	TPC[c]	37952.5–820 μg/g DW	Xiong et al. (2014)
		TPC	240.03 mg GAE/g DW	Zheng et al. (2018)
		TPC	12[a]–230[a] mg CGAE/g DW	Xiong et al. (2014), Huang et al. (2017)
	Panax pseudoginseng	TPC	7.66 mg GAE/g DW	Zheng et al. (2018)
	Pelargonium grandiflorum	TPC	1028.16 ppm	Mikołajczak et al. (2020)
	Pelargonium hortorum	TPC	25.68 mg GAE/g FW	Li et al. (2014)
	Pelargonium hortorum (Caryophyllaceae)	TPC[c]	992.40 mg/100 g DW	Coyago-Cruz et al. (2024)
		TPC	110.74–525.39 mg GAE/g DW	Coyago-Cruz et al. (2024)
	Pelargonium hortorum (Geraniaceae)	TPC	155.99–316.99 mg GAE/g DW	Coyago-Cruz et al. (2024)
	Pentas lanceolate	TPC	6.52 GAE/g FW	Chensom et al. (2019)
	Petunia × hybrida	TPC	1.47 GAE/g FW	Chensom et al. (2019)
	Phaseolus vulgaris	TPC	1.86 mg GAE/g FW	Li et al. (2014)
	Platycodon grandiflorus	TPC	4.57 mg GAE/g FW	Li et al. (2014)
	Plumbago auriculata	TPC	18.27 mg/g fresh sample	Gonçalves et al. (2020)
		TPC	6.25 mg GAE/g FW	Kritsi et al. (2022)
	Plumeria obtusa	TPC	40[a]–90[a] mg GAE/g DW	Kaisoon et al. (2011)
	Plumeria rubra	TPC	20.4–56.02 mg GAE/g DW	Zheng et al. (2018)

(continued)

Table 6.1 (continued)

Phyto-chemical	Source	Content		References
	Primula × polyantha	TPC	7.32 GAE/g FW	Chensom et al. (2019)
	Primula veris	TPC	1044.90 mg GAE/100 g FW	Demasi et al. (2021)
	Primula vulgaris	TPC	602.90 mg GAE/100 g FW	Demasi et al. (2021)
		TPC	9.53 mg GAE/g DW	Zheng et al. (2018)
	Prunus mume	TPC	149.90 mg GAE/g DW	Shi et al. (2009)
	Prunus persica	TPC[c]	11135.4–93.9 μg/g DW	Xiong et al. (2014)
		TPC	10[a]–42[a] mg CGAE/g DW	Xiong et al. (2014), Huang et al. (2017)
	Punica granatum	TPC	39.2–245.84 mg GAE/g DW	Zheng et al. (2018), Janarny et al. (2021)
	Pyrus pashia	TPC	28.1–168.9 mg GAE/g extract	He et al. (2015)
	Raphanus raphanistrum	TPC	278.38 mg GAE/g DW	Coyago-Cruz et al. (2024)
	Red artful plum tea	TPC	1391.30 μg/g DW	Chen et al. (2015)
		TPC	8.59 mg GAE/g DW	Chen et al. (2015)
	Rhapniolepis indica.	TPC	7.97 mg GAE/g FW	Li et al. (2014)
	Rhododendron simsii	TPC	249.82 mg GAE/g DW	Zheng et al. (2018)
		TPC	1.76–6.75 mg GAE/g FW	Li et al. (2014), Kritsi et al. (2022)
	Rhoeo discolor	TPC	2.56 mg GAE/g FW	Li et al. (2014)
	Robinia pseudoacacia	TPC	75.20 mg CGAE/g extract	Loizzo et al. (2016)
		TPC	203.80 mg GAE/100 g FW	Demasi et al. (2021)
	Rosa	TPC	17.6–27.53 mg/g fresh sample	Gonçalves et al. (2020)
		TPC	4.45 g GAE/kg FW	Mlcek et al. (2021)
	Rosa 'Purple Fragrance'	TPC	12.36 GAE/g FW	Chensom et al. (2019)
	Rosa 'Yves Piaget'	TPC	9.99 GAE/g FW	Chensom et al. (2019)
	Rosa banksiae	TPC	199.68 mg GAE/g DW	Coyago-Cruz et al. (2024)
	Rosa canina	TPC	857.72 ppm	Mikołajczak et al. (2020)
		TPC	194.49 mg GAE/g DW	Coyago-Cruz et al. (2024)
		TPC	1396.60 mg GAE/100 g FW	Demasi et al. (2021)
	Rosa centifolia	TPC	108.94 mg GAE/g DW	Zheng et al. (2018)
	Rosa chinensis	TPC[c]	11923.5–738.0 μg/g DW	Xiong et al. (2014)
		TPC	20[a]–209[a] mg CGAE/g DW	Xiong et al. (2014)
		TPC	284.80 mg GAE/g DW	Zheng et al. (2018)

(continued)

Table 6.1 (continued)

Phyto-chemical	Source	Content		References
	Rosa chinensis	TPC	18.55 mg GAE/g FW	Kritsi et al. (2022)
	Rosa damascene	TPC	107.02 mg GAE/g DW	Coyago-Cruz et al. (2024)
	Rosa gallica	TPC	111.34 mg GAE/g DW	Zheng et al. (2018)
	Rosa hybrida	TPC	35.84 mg GAE/g FW	Li et al. (2014)
	Rosa pendulina	TPC	1773.70 mg GAE/100 g FW	Demasi et al. (2021)
	Rosa rugosa	TPC	9039.18–16106.83 µg/g DW	Chen et al. (2015)
		TPC	28[a]–108[a] mg CGAE/g DW	Huang et al. (2017)
		TPC	25.79–312.21 mg GAE/g DW	Chen et al. (2015), Zheng et al. (2018), and Socha et al. (2021)
	Rosa sp.	TPC	9.89 mg GAE/g DW	Araújo et al. (2019)
	Rosa × hybrid	TPC[c]	1644.70 mg/100 g DW	Coyago-Cruz et al. (2024)
		TPC	69.23–586.37 mg GAE/g DW	Coyago-Cruz et al. (2024)
	Rosa × hybrid roseta	TPC	96.51–323.12 mg GAE/g DW	Coyago-Cruz et al. (2024)
	Rosmarinus officinalis	TPC	11522.43 µg/g DW	Chen et al. (2015)
		TPC	38.80 mg CGAE/g extract	Loizzo et al. (2016)
		TPC	23.02–24.94 mg GAE/g DW	Chen et al. (2015), Araújo et al. (2019)
	Ruda chalepensis	TPC	217.59 mg GAE/g DW	Coyago-Cruz et al. (2024)
	Ruellia simpex	TPC	4.30 mg GAE/g FW	Kritsi et al. (2022)
	Rumex acetosa	TPC	42.74 mg/g DW	Kucekova et al. (2013)
	Salvia 'Purple Queen	TPC	3.65 mg GAE/g FW	Marchioni et al. (2024)
	Salvia elegans	TPC	40.86 mg GAE/g DW	Araújo et al. (2019)
	Salvia farinacea	TPC	3.20 mg GAE/g FW	Marchioni et al. (2024)
	Salvia jamensis	TPC	2.58 mg GAE/g FW	Marchioni et al. (2024)
	Salvia microphylla	TPC	293.39 mg GAE/g DW	Coyago-Cruz et al. (2024)
	Salvia pratensis	TPC	16.97 mg/g DW	Kucekova et al. (2013)
		TPC	314.70 mg GAE/100 g FW	Demasi et al. (2021)
	Salvia splendens	TPC	115.98 ppm	Mikołajczak et al. (2020)
		TPC	2.57 mg GAE/g FW	Li et al. (2014)
	Sambucus nigra	TPC	228.50 mg CGAE/g extract	Loizzo et al. (2016)
		TPC	12.88 mg/g DW	Kucekova et al. (2013)
		TPC	250.51 ppm	Mikołajczak et al. (2020)

(continued)

Table 6.1 (continued)

Phyto-chemical	Source	Content		References
		TPC	14.36 mg GAE/g DW	Socha et al. (2021)
		TPC	508.70 mg GAE/100 g FW	Demasi et al. (2021)
	Sesbania grandiflora	TPC	1.10 mg GAE/g DW	Janarny et al. (2021)
		TPC	6.40 μmol GAE/g DW	Janarny et al. (2022)
	Shorea roxburghii	TPC	128–138 mg GAE/g extract	Suksathan et al. (2021)
	Siraitia grosvenorii	TPC	9257.05 μg/g DW	Chen et al. (2015)
		TPC	12.85–22.18 mg GAE/g DW	Chen et al. (2015), Zheng et al. (2018)
	Sophora japonica	TPC	84.23 mg GAE/g DW	Zheng et al. (2018)
	Sophora viciifolia	TPC	29.3–143.8 mg GAE/g dry extract	Tai et al. (2011)
	Spilanthes oleracea	TPC	6.64 mg GAE/g FW	Navarro-González et al. (2014)
	Strelitzia reginae	TPC	9.40 mg GAE/g FW	Li et al. (2014)
	Tagetes	TPC	89.22 mg GAE/g DW	Socha et al. (2021)
	Tagetes erecta	TPC	17.47 mg/g fresh sample	Gonçalves et al. (2020)
		TPC	898.19 mg GAE/100 g FW	Dujmović et al. (2022)
		TPC	0.35–26.63 mg GAE/g FW	Navarro-González et al. (2014), Petrova et al. (2016), and Kritsi et al. (2022)
		TPC	65[a]–212.9 mg GAE/g DW	Kaisoon et al. (2011, 2012)
	Tagetes patula	TPC	79.78–303.61 mg GAE/g DW	Araújo et al. (2019), Coyago-Cruz et al. (2024)
		TPC	717.74 ppm	Mikołajczak et al. (2020)
		TPC[c]	2827.40 mg/100 g DW	Coyago-Cruz et al. (2024)
		TPC	470.80 mg GAE/100 g FW	Demasi et al. (2021)
		TPC	4.78 g GAE/kg FW	Mlcek et al. (2021)
	Tamarindus indica	TPC	19.40 mg GAE/g DW	Janarny et al. (2021)
		TPC	115[a] μmol GAE/g DW	Janarny et al. (2022)
	Taraxacum officinale	TPC	11.97 mg/g DW	Kucekova et al. (2013)
		TPC	0.8[a]–1.8[a] mg GAE/g FW	Fernandes et al. (2020)
		TPC	159.40 mg GAE/100 g FW	Demasi et al. (2021)
	Tecomaria capensis	TPC	2.49 mg GAE/g FW	Kritsi et al. (2022)
	Telosma minor	TPC	30[a]–70[a] mg GAE/g DW	Kaisoon et al. (2011)
	Thunbergia alata	TPC	2.02 mg GAE/g FW	Kritsi et al. (2022)
	Tifolium alpinum	TPC	464.60 mg GAE/100 g FW	Demasi et al. (2021)

(continued)

Table 6.1 (continued)

Phyto-chemical	Source	Content		References
	Tilia cordata	TPC	1003.62 ppm	Mikołajczak et al. (2020)
	Torenia fournieri	TPC	2.37–4.85 GAE/g FW	Chensom et al. (2019)
	Tragopogon pratensis	TPC	11.72 mg/g DW	Kucekova et al. (2013)
	Trifolium repens	TPC	30.98 mg/g DW	Kucekova et al. (2013)
		TPC	573.66 ppm	Mikołajczak et al. (2020)
		TPC	167.76 mg GAE/g DW	Coyago-Cruz et al. (2024)
	Tropaeolum majus	TPC	444.58 ppm	Mikołajczak et al. (2020)
		TPC	12.95 mg GAE/g FW	Navarro-González et al. (2014)
		TPC	336.96–355.80 mg GAE/100 g FW	Demasi et al. (2021), Dujmović et al. (2022)
		TPC	3.23 g GAE/kg FW	Mlcek et al. (2021)
		TPC	11.78–62.67 mg GAE/g DW	Zheng et al. (2018), Araújo et al. (2019), and Socha et al. (2021)
		TPC	2.39 GAE/g FW	Chensom et al. (2019)
	Tulbaghia simmleri	TPC	1.89 mg GAE/g FW	Marchioni et al. (2024)
	Tulbaghia violacea	TPC	2.09 mg GAE/g FW	Marchioni et al. (2024)
	Veronica arvensis	TPC	24.29 mg/g DW	Kucekova et al. (2013)
	Viburnum inopinatum	TPC	120–176 mg GAE/g extract	Suksathan et al. (2021)
	Vinca rosea	TPC	57.40 mg GAE/g DW	Janarny et al. (2021)
	Viola odorata	TPC	428.40 mg GAE/100 g FW	Demasi et al. (2021)
	Viola tricolor	TPC	46.78–63.43 mg GAE/g DW	Araújo et al. (2019)
		TPC	1.8[a]–7.15 mg GAE/g FW	Fernandes et al. (2020), Kritsi et al. (2022)
	Viola × wittrockiana	TPC	809.10 ppm	Mikołajczak et al. (2020)
	Wedelia trilobata	TPC	3.85 mg GAE/g FW	Li et al. (2014)
	Wisteria sinensis	TPC[c]	1.355–22.921 mg/g DW	Yan et al. (2022)
		TPC	19.77–37.83 mg GAE/g DW	Yan et al. (2022)
	Youngia japonica	TPC	1.11 mg GAE/g FW	Li et al. (2014)
	Zantedeschia aethiopica	TPC	3.07 mg GAE/g FW	Li et al. (2014)

TFC total flavonoid content, *TPC* total phenolic content *TPAC* total phenolic acid content *DW* dry weight, *FW* fresh weight, *RE* rutin equivalents, *QE* quercetin equivalents, *CFAE* caffeic acid equivalents, *GAE* gallic acid equivalents, *CAE* catechin equivalents, *CTH* catechol equivalents, *CGAE* chlorogenic acid equivalents

[a]Approximated value from graph

[b]Data from ultra performance liquid chromatography (UPLC)

[c]Data from high performance liquid chromatography (HPLC)

Table 6.2 Pigment content of various edible flowers

Pigment	Source	Color	Content		References
Anthocyanin					
	Aegle marmelos	Green-white	Total monomeric anthocyanin	270.4 μg CGE/g DW	Janarny et al. (2021)
	Ageratum houstonianum	Blue	Total anthocyanin	2.99 mg CGE/100 g FW	Benvenuti et al. (2016)
	Allium cepa	White, purple	Total monomeric anthocyanin	334.1 μg CGE/g DW	Janarny et al. (2021)
	Anchusa azurea	Blue	Total anthocyanin	3.7 mg/g of extract	Li et al. (2014)
	Antirrhinum majus	Red, rose, white, orange-pink	Total anthocyanin	0.70–12.17 mg CGE/100 g FW	Benvenuti et al. (2016), Stefaniak and Grzeszczuk (2019)
	Aponogeton crispus	White, pink, light purple	Total monomeric anthocyanin	50.1 μg CGE/g DW	Janarny et al. (2021)
	Azadirachta indica	White, pale yellow	Total monomeric anthocyanin	30.1 μg CGE/g DW	Janarny et al. (2021)
	Bauhinia racemosa	White, pink	Total monomeric anthocyanin	68.1 μg CGE/g DW	Janarny et al. (2021)
		White, cream-yellow	Total anthocyanin	180[a] nmol CGE/g DW	Janarny et al. (2022)
	Begonia boliviensis	Pink, white	Total anthocyanin	0.01–0.22 mg ME/g FW	Marchioni et al. (2024)
	Begonia semperflorens	Red	Total anthocyanin	5.09 mg CGE/100 g FW	Benvenuti et al. (2016)
	Borago officinalis	Blue	Total anthocyanin	1.43 mg CGE/100 g FW	Benvenuti et al. (2016)
	Bougainvillea glabra	Purple, pink, red, orange, yellow, white	Total monomeric anthocyanin	37.1 μg CGE/g DW	Janarny et al. (2021)
	Calendula officinalis	Orange	Total anthocyanin	0.47 mg CGE/100 g FW	Benvenuti et al. (2016)
		Yellow, orange	Total monomeric anthocyanin	21.1 μg CGE/g DW	Janarny et al. (2021)

(continued)

Table 6.2 (continued)

Pigment	Source	Color	Content		References
	Calotropis gigantea	White, lavender	Total monomeric anthocyanin	162.3 μg CGE/g DW	Janarny et al. (2021)
	Capparis spinosa	White, pink-white	Total anthocyanin	5.4 mg/g of extract	Li et al. (2014)
	Carica papaya	Pale yellow, white	Total monomeric anthocyanin	1405.1 μg CGE/g DW	Janarny et al. (2021)
	Cassia auriculata	Yellow	Total monomeric anthocyanin	22 μg CGE/g DW	Janarny et al. (2021)
		Yellow	Total anthocyanin	49 nmol CGE/g DW	Janarny et al. (2022)
	Cassia fistula	Yellow	Total monomeric anthocyanin	99.1 μg CGE/g DW	Janarny et al. (2021)
		Yellow	Total anthocyanin	210[a] nmol CGE/g DW	Janarny et al. (2022)
	Centaurea cyanus	Blue, purple, pink, maroon, violet	Total anthocyanin	1012.09 mg/kg	Dujmović et al. (2022)
		White with a little violet	Total anthocyanin	1.68 mg/100 g dm	Nowicka and Wojdyło (2019)
	Cichorium intybus	Blue-purple	Total anthocyanin	1.2 mg/g of extract	Li et al. (2014)
	Clitoria ternatea	Blue, shades of purple	Total anthocyanin	0.62–9.61 mg CGE/g	Jeyaraj et al. (2021), Vidana Gamage (2023)
	Cocos nucifera	Green-yellow, light yellow	Total monomeric anthocyanin	48.2 μg CGE/g DW	Janarny et al. (2021)
	Crotalaria juncea	Yellow	Total monomeric anthocyanin	68.1 μg CGE/g DW	Janarny et al. (2021)
		Yellow	Total anthocyanin	185[a] nmol CGE/g DW	Janarny et al. (2022)
	Dahlia campanulata	White with light-purple hue	Total anthocyanin	0.24–0.43 mg CGE/100 g FW	Espejel et al. (2019)

(continued)

Table 6.2 (continued)

Pigment	Source	Color	Content		References
	Dahlia coccinea	Orange	Total anthocyanin	1.17–1.52 mg CGE/100 g FW	Espejel et al. (2019)
	Dahlia hortorum	Cherry, yellow, white, purple, red	Total anthocyanin	0.06–14.74 mg CGE/100 g FW	Espejel et al. (2019)
	Dahlia merckii	Violet	Total anthocyanin	1.53–1.78 mg CGE/100 g FW	Espejel et al. (2019)
	Dahlia pinnata	Orange-yellow	Total anthocyanin	0.19 mg ME/g FW	Marchioni et al. (2024)
	Dianthus barbatus	Red-white, red	Total anthocyanin	0.3–0.33 mg/g Mv3GlcE	Gonçalves et al. (2020)
	Dianthus caryophyllus	Red	Total anthocyanin	4.47 mg/g Mv3GlcE	Gonçalves et al. (2020)
	Dianthus chinensis	Dark maroon/ almost black, white edged	Total anthocyanin	443.47 mg CGE/100 g FW	Stefaniak and Grzeszczuk (2019)
	Dianthus × *barbatus*	Red, rose, white	Total anthocyanin	0.73–13.35 mg CGE/100 g FW	Benvenuti et al. (2016)
	Diuris brevis	Violet	Total anthocyanin	2.11–2.24 mg CGE/100 g FW	Espejel et al. (2019)
	Durio zibethinus	White, yellow-white	Total monomeric anthocyanin	20.1 μg CGE/g DW	Janarny et al. (2021)
	Fuchsia hybrida	Red	Total anthocyanin	7.58 mg CGE/100 g FW	Benvenuti et al. (2016)
	Gmelina asiatica	Yellow	Total monomeric anthocyanin	86.2 μg CGE/g DW	Janarny et al. (2021)
	Hedychium coronarium	White	Total anthocyanin	0.7 mg/g of extract	Li et al. (2014)
	Hemerocallis × *hybrida*	Orange-yellow	Total anthocyanin	2.77 mg CGE/100 g FW	Stefaniak and Grzeszczuk (2019)
	Hibiscus rosa-sinensis	Red, white, pink, peach, orange, yellow, blue, purple	Total monomeric anthocyanin	2003.5 μg CGE/g DW	Janarny et al. (2021)

(continued)

Table 6.2 (continued)

Pigment	Source	Color	Content		References
	Ixora coccinea	Red, orange, yellow, pink, white	Total monomeric anthocyanin	1573.1 μg CGE/g DW	Janarny et al. (2021)
	Jasminum sambac	White	Total monomeric anthocyanin	279.2 μg CGE/g DW	Janarny et al. (2021)
	Lamium album	White	Total anthocyanin	247.56 mg/100 g dm	Nowicka and Wojdyło (2019)
	Lasia spinosa	Yellow, brown	Total monomeric anthocyanin	248 μg CGE/g DW	Janarny et al. (2021)
	Lavandula	Pale purple	Total anthocyanin	8.37 mg/100 g dm	Nowicka and Wojdyło (2019)
	Lavandula spica	Lilac	Total anthocyanin	0.4 mg/g Mv3GlcE	Gonçalves et al. (2020)
	Madhuca longifolia	Cream-white	Total monomeric anthocyanin	6 μg CGE/g DW	Janarny et al. (2021)
	Malva	Rose-purple	Total anthocyanin	13.38 mg/100 g dm	Nowicka and Wojdyło (2019)
	Malva arborea	Pink-purple with dark purple veins	Total anthocyanin	286.14 mg/100 g dm	Nowicka and Wojdyło (2019)
	Malva sylvestris	Rose-purple	Total anthocyanin	0.3 mg/g of extract	Li et al. (2014)
	Manihot esculenta	Purple, pink, yellow, white	Total monomeric anthocyanin	110.1 μg CGE/g DW	Janarny et al. (2021)
	Mimulus × hybridus	Yellow, red	Total anthocyanin	15.71–30.33 mg CGE/100 g FW	Stefaniak and Grzeszczuk (2019)
	Monarda didyma	Red	Total anthocyanin	204.42 mg CGE/100 g FW	Stefaniak and Grzeszczuk (2019)
	Musa spp	White, yellow, red, purple	Total monomeric anthocyanin	8.2 μg CGE/g DW	Janarny et al. (2021)

(continued)

Table 6.2 (continued)

Pigment	Source	Color	Content		References
	Ocimum sanctum	White, pale pink, lavender	Total monomeric anthocyanin	234.1 μg CGE/g DW	Janarny et al. (2021)
	Orchis	Pink, yellow	Total anthocyanin	0.3 mg/g Mv3GlcE	Gonçalves et al. (2020)
	Pelargonium peltatum	Red	Total anthocyanin	12.52 mg CGE/100 g FW	Benvenuti et al. (2016)
	Petunia × hybrida	Red, rose, white	Total anthocyanin	2.5–14.44 mg CGE/100 g FW	Benvenuti et al. (2016)
	Plumbago auriculata	Blue	Total anthocyanin	0.6 mg/g Mv3GlcE	Gonçalves et al. (2020)
	Punica granatum	Red, orange	Total monomeric anthocyanin	1181.1 μg CGE/g DW	Janarny et al. (2021)
	Robinia pseudoacacia	Cream-white	Total anthocyanin	3.6 mg/g of extract	Li et al. (2014)
	Rosa	Pink, red, orange	Total anthocyanin	0.4–3.07 mg/g Mv3GlcE	Gonçalves et al. (2020)
	Rosa 'Amadeus'	Light pink with yellow hue	Total anthocyanin	0.01[a] mg CGE/g FW	Friedman et al. (2010)
	Rosa 'brandy'	Orange-yellow	Total anthocyanin	0.02[a] mg CGE/g FW	Friedman et al. (2010)
	Rosa 'English sachet'	Pink	Total anthocyanin	0.02[a] mg CGE/g FW	Friedman et al. (2010)
	Rosa 'Eterna'	Pink-white	Total anthocyanin	0.032[a] mg CGE/g FW	Friedman et al. (2010)
	Rosa 'Maxim'	White with pink hue	Total anthocyanin	0.04[a] mg CGE/g FW	Friedman et al. (2010)
	Rosa 'San Francisco'	Red	Total anthocyanin	2.5[a] mg CGE/g FW	Friedman et al. (2010)
	Rosa 'sweet surrender'	Light pink	Total anthocyanin	0.035[a] mg CGE/g FW	Friedman et al. (2010)
	Rosa 'Trier 2000'	Pink	Total anthocyanin	0.07[a] mg CGE/g FW	Friedman et al. (2010)
	Rosmarinus officinalis	Blue-violet	Total anthocyanin	1.1 mg/g of extract	Li et al. (2014)
	Salvia 'Purple Queen'	Purple	Total anthocyanin	0.69 mg ME/g FW	Marchioni et al. (2024)
	Salvia farinacea	Violet-blue	Total anthocyanin	0.56 mg ME/g FW	Marchioni et al. (2024)
	Salvia jamensis J. Compton	Magenta	Total anthocyanin	0.4 mg ME/g FW	Marchioni et al. (2024)

(continued)

Table 6.2 (continued)

Pigment	Source	Color	Content		References
	Sambucus nigra	White	Total anthocyanin	0.9 mg/g of extract	Li et al. (2014)
	Sesbania grandiflora	White, pink, red	Total monomeric anthocyanin	32 μg CGE/g DW	Janarny et al. (2021)
		White, pink, red	Total anthocyanin	80[a] nmol CGE/g DW	Janarny et al. (2022)
	Tagetes erecta	Orange	Total anthocyanin	0.75 mg CGE/100 g FW	Benvenuti et al. (2016)
		Orange	Total anthocyanin	0.9 mg/g Mv3GlcE	Gonçalves et al. (2020)
	Tamarindus indica	Yellow with orange or red streaks	Total monomeric anthocyanin	38 μg CGE/g DW	Janarny et al. (2021)
		Yellow with orange or red streaks	Total anthocyanin	90[a] nmol CGE/g DW	Janarny et al. (2022)
	Tropaeolum majus	Orange	Total anthocyanin	8.27 mg CGE/100 g FW	Benvenuti et al. (2016)
	Tulbaghia simmleri 'Alba'	White	Total anthocyanin	0.08 mg ME/g FW	Marchioni et al. (2024)
	Tulbaghia violacea 'Alba'	White	Total anthocyanin	0.04 mg ME/g FW	Marchioni et al. (2024)
	Verbascum	White with a little violet	Total anthocyanin	0.74 mg/100 g dm	Nowicka and Wojdyło (2019)
	Vinca rosea	Pink, white, red, pink-white	Total monomeric anthocyanin	715.2 μg CGE/g DW	Janarny et al. (2021)
	Viola × wittrockiana	Red, blue, yellow, white	Total anthocyanin	0.35–13.6 mg CGE/100 g FW	Benvenuti et al. (2016)
Carotenoid					
	Acacia	Yellow	Total carotenoids	31.33 mg/100 g dm	Nowicka and Wojdyło (2019)
	Aegle marmelos	Green-white	Total carotene content	4.6 mg β-carotene/g DW	Janarny et al. (2021)

(continued)

Table 6.2 (continued)

Pigment	Source	Color	Content		References
	Agave salmiana	Yellow	Carotenoids (red)	0.36 mg/g DW	Pinedo-Espinoza et al. (2020)
		Yellow	Carotenoids (yellow)	0.04 mg/g DW	Pinedo-Espinoza et al. (2020)
	Allium cepa	White, purple	Total carotene content	19.8 mg β-carotene/g DW	Janarny et al. (2021)
		White, purple	Total carotenoids	0.16 ppm	Mikołajczak et al. (2020)
	Allium schoenoprasum	Pink, pale purple	Total carotenoids	0.03 ppm	Mikołajczak et al. (2020)
	Aloe vera	Yellow	Carotenoids (red)	0.32 mg/g DW	Pinedo-Espinoza et al. (2020)
		Yellow	Carotenoids (yellow)	0.27 mg/g DW	Pinedo-Espinoza et al. (2020)
	Aloysia citriodora	Fuchsia	Total carotenoids	173.28 mg β-carotene/100 g DW	Coyago-Cruz et al. (2024)
	Amaranthus hypochondriacus	Purple	Total carotenoids	0%	Barros et al. (2020)
	Anchusa azurea	Blue	Total carotenoids	3 mg/g of extract	Li et al. (2014)
	Anethum graveolens	Yellow	Total carotenoids	467.62 mg β-carotene/100 g DW	Coyago-Cruz et al. (2024)
	Anthyllis vulneraria	Yellow	Total carotenoids	74.43 mg/100 g DW	Nowicka and Wojdyło (2019)
	Antirrhinum majus	Red, yellow	Total carotenoids	230.88–399.89 mg β-carotene/100 g DW	Coyago-Cruz et al. (2024)
		Orange-pink	Total carotenoids	64.03 μg/g FW	Stefaniak and Grzeszczuk (2019)
	Aponogeton crispus	White, pink, light purple	Total carotene content	71.4 mg β-carotene/g DW	Janarny et al. (2021)
	Arnica	Yellow	Total carotenoids	558.51 mg/100 g DW	Nowicka and Wojdyło (2019)

(continued)

Table 6.2 (continued)

Pigment	Source	Color	Content		References
	Azadirachta indica	White, pale yellow	Total carotene content	5.1 mg β-carotene/g DW	Janarny et al. (2021)
	Bauhinia racemosa	White, pink	Total carotene content	45.3 mg β-carotene/g DW	Janarny et al. (2021)
	Begonia boliviensis	Pink, white	Total carotenoids	3.58–20.83 μg/g FW	Marchioni et al. (2024)
	Begonia doblet	Pink	Total carotenoids	194.69 mg β-carotene/100 g DW	Coyago-Cruz et al. (2024)
	Begonia semperflore	Pink	β-Carotene	1 μg/g FW	Chensom et al. (2019)
	Begonia tuberhybrida	White, red	Total carotenoids	4.42–12.34 ppm	Mikołajczak et al. (2020)
	Bellis perennis	White, pink	Total carotenoids	0.25 ppm	Mikołajczak et al. (2020)
		Pink, red	Total carotenoids	174.22 mg/100 g DW	Nowicka and Wojdyło (2019)
	Bougainvillea glabra	Purple, pink, red, orange, yellow, white	Total carotene content	64.5 mg β-carotene/g DW	Janarny et al. (2021)
	Calendula officinalis	Yellow	Total carotenoids	5745.28 mg β-carotene/100 g DW	Coyago-Cruz et al. (2024)
		Yellow, orange	Total carotenoids	0.42 mg/g	Dujmović et al. (2022)
		Yellow, orange	Total carotene content	243.9 mg β-carotene/g DW	Janarny et al. (2021)
		Yellow, orange	Total carotenoids	0.87 ppm	Mikołajczak et al. (2020)
		Yellow, orange	Total carotenoids	721.49 mg/100 g DW	Nowicka and Wojdyło (2019)
		Yellow, orange	Total carotenoids	57.2 μg/g FW	Petrova et al. (2016)
	Calotropis gigantea	White, lavender	Total carotene content	27.6 mg β-carotene/g DW	Janarny et al. (2021)
	Canna indica	Red	Total carotenoids	78.86–1223.51 mg β-carotene/100 g DW	Coyago-Cruz et al. (2024)
		Red	Total carotenoids	2043.3 mg/100 g DW	Coyago-Cruz et al. (2024)

(continued)

Table 6.2 (continued)

Pigment	Source	Color	Content		References
	Capparis spinosa	White, pink-white	Total carotenoids	3.4 mg/g of extract	Li et al. (2014)
	Carica papaya	Pale yellow, white	Total carotene content	30.8 mg β-carotene/g DW	Janarny et al. (2021)
	Cassia auriculata	Yellow	Total carotene content	282.3 mg β-carotene/g DW	Janarny et al. (2021)
	Cassia fistula	Yellow	Total carotene content	98.5 mg β-carotene/g DW	Janarny et al. (2021)
	Centaurea cyanus	Blue, lavender	Total carotenoids	0.21 ppm	Mikołajczak et al. (2020)
		White with a little violet	Total carotenoids	30 mg/100 g DW	Nowicka and Wojdyło (2019)
	Chamaemelun nobile	White	Total carotenoids	63.89 mg β-carotene/100 g DW	Coyago-Cruz et al. (2024)
	Chamomilla recutita	White with yellow centers	Total carotenoids	0.83 ppm	Mikołajczak et al. (2020)
	Chrysanthemum × *hybrid*	Pink, orange, yellow	Total carotenoids	57.65–1103.46 mg β-carotene/100 g DW	Coyago-Cruz et al. (2024)
	Cichorium intybus	Blue-purple	Total carotenoids	2.5 mg/g of extract	Li et al. (2014)
	Citrus × *aurantifolia*	White	Total carotenoids	96.61 mg β-carotene/100 g DW	Coyago-Cruz et al. (2024)
	Clitoria ternatea	Blue, shades of purple	Total carotene content	82.1 mg β-carotene/g DW	Janarny et al. (2021)
	Cocos nucifera	Green-yellow, light yellow	Total carotene content	5.2 mg β-carotene/g DW	Janarny et al. (2021)
	Cosmos sulphureus	Yellow	β-Carotene	10 μg/g FW	Chensom et al. (2019)
		Yellow	β-Cryptoxanthin	28 μg/g FW	Chensom et al. (2019)
	Crataegus	White	Total carotenoids	472.6 mg/100 g DW	Nowicka and Wojdyło (2019)
	Crotalaria juncea	Yellow	Total carotene content	243.8 mg β-carotene/g DW	Janarny et al. (2021)

(continued)

Table 6.2 (continued)

Pigment	Source	Color	Content		References
	Cucurbita pepo	Orange-yellow	Total carotenoids	0.28 mg/g	Dujmović et al. (2022)
		Yellow, orange	Total carotenoids	0.58–0.74 ppm	Mikołajczak et al. (2020)
	Dahlia pinnata	Fuchsia, orange, red	Total carotenoids	78.64–931.92 mg β-carotene/100 g DW	Coyago-Cruz et al. (2024)
		Orange-yellow	Total carotenoids	96.01 μg/g FW	Marchioni et al. (2024)
	Dianthus caryophyllus	Light pink	Total carotenoids	0.26 ppm	Mikołajczak et al. (2020)
	Dianthus chinensis	Pink, red	Total carotenoids	49.05–75.9 mg β-carotene/100 g DW	Coyago-Cruz et al. (2024)
		Dark maroon/almost black, white edged	Total carotenoids	261.59 μg/g FW	Stefaniak and Grzeszczuk (2019)
	Durio zibethinus	White, yellow-white	Total carotene content	9.2 mg β-carotene/g DW	Janarny et al. (2021)
	Erythrina americana	Red	Carotenoids (red)	0.8 mg/g DW	Pinedo-Espinoza et al. (2020)
	Fuchsia magellanica	Pink	Total carotenoids	89.3 mg β-carotene/100 g DW	Coyago-Cruz et al. (2024)
	Gmelina asiatica	Yellow	Total carotene content	244.6 mg β-carotene/g DW	Janarny et al. (2021)
	Hedychium coronarium	White	Total carotenoids	2.7 mg/g of extract	Li et al. (2014)
	Helianthus annuus	Yellow	Total carotenoids	1187.7 mg/100 g DW	Coyago-Cruz et al. (2024)
		Yellow	Total carotenoids	5154.64 mg β-carotene/100 g DW	Coyago-Cruz et al. (2024)
	Helianthus tuberosus	Yellow	Total carotenoids	15.6 μg/g FW	Petrova et al. (2016)
	Hemerocallis × *hybrida*	Orange-yellow	Total carotenoids	227.11 μg/g FW	Stefaniak and Grzeszczuk (2019)
		Orange-yellow	Total carotenoids	2.22 ppm	Mikołajczak et al. (2020)

(continued)

Table 6.2 (continued)

Pigment	Source	Color	Content		References
	Hibiscus rosa-sinensis	Red, white, pink, peach, orange, yellow, blue, purple	Total carotene content	13 mg β-carotene/g DW	Janarny et al. (2021)
	Hibiscus rosa-sinensis (Malvaceae)	Orange, pink	Total carotenoids	116.47–2468.5 mg β-carotene/100 g DW	Coyago-Cruz et al. (2024)
		Orange, pink	Total carotenoids	2.5–108.3 mg/100 g DW	Coyago-Cruz et al. (2024)
	Hibiscus rosa-sinensis (Rosaceae)	Orange, pink	Total carotenoids	116.47 mg β-carotene/100 g DW	Coyago-Cruz et al. (2024)
	Ixora coccinea	Red, orange, yellow, pink, white	Total carotene content	97.6 mg β-carotene/g DW	Janarny et al. (2021)
	Jasminum sambac	White	Total carotene content	21.9 mg β-carotene/g DW	Janarny et al. (2021)
	Lamium album	White	Total carotenoids	48.15 mg/100 g DW	Nowicka and Wojdyło (2019)
	Lantana camara	Red, yellow, pink, white	Total carotenoids	2947.72 mg β-carotene/100 g DW	Coyago-Cruz et al. (2024)
	Lantana viburnoides	Red, red-orange, yellow, white	Total carotenoids	42.81–658.02 mg β-carotene/100 g DW	Coyago-Cruz et al. (2024)
	Lasia spinosa	Yellow, brown	Total carotene content	192.4 mg β-carotene/g DW	Janarny et al. (2021)
	Lathyrus odoratus	Pink	β-Carotene	8.0–9.0 μg/g FW	Chensom et al. (2019)
	Lavandula	Pale purple	Total carotenoids	380.61 mg/100 g DW	Nowicka and Wojdyło (2019)
	Lavandula angustifolia	Violet-blue	Total carotenoids	0.41 ppm	Mikołajczak et al. (2020)
	Madhuca longifolia	Cream-white	Total carotene content	29.7 mg β-carotene/g DW	Janarny et al. (2021)

(continued)

Table 6.2 (continued)

Pigment	Source	Color	Content		References
	Malva	Rose-purple	Total carotenoids	203.91 mg/100 g DW	Nowicka and Wojdyło (2019)
	Malva arborea	Pink-purple with dark purple veins	Total carotenoids	40.29 mg/100 g DW	Nowicka and Wojdyło (2019)
	Malva sylvestris	Rose-purrple	Total carotenoids	1.3 mg/g of extract	Li et al. (2014)
	Malvaviscus arboreus	Red	Total carotenoids	88.76 mg β-carotene/100 g DW	Coyago-Cruz et al. (2024)
	Manihot esculenta	Purple, pink, yellow, white	Total carotene content	147.4 mg β-carotene/g DW	Janarny et al. (2021)
	Matricaria	White with yellow centers	Total carotenoids	51.49 mg/100 g DW	Nowicka and Wojdyło (2019)
	Mentha spicata	Pink-white	Total carotenoids	0.63 ppm	Mikołajczak et al. (2020)
	Mimulus × *hybridus*	Yellow, red	Total carotenoids	473.42–529.68 μg/g FW	Stefaniak and Grzeszczuk (2019)
	Mirabilis jalapa	Fuchsia	Total carotenoids	277.88 mg β-carotene/100 g DW	Coyago-Cruz et al. (2024)
	Monarda didyma	Red	Total carotenoids	167.2 μg/g FW	Stefaniak and Grzeszczuk (2019)
	Musa spp	White, yellow, red, purple	Total carotene content	4.9 mg β-carotene/g DW	Janarny et al. (2021)
	Myrtillocactus geometrizans	Green-white	Carotenoids (red)	0.42 mg/g DW	Pinedo-Espinoza et al. (2020)
		Green-white	Carotenoids (yellow)	0.02 mg/g DW	Pinedo-Espinoza et al. (2020)
	Ocimum sanctum	White, pale pink, lavender	Total carotene content	201.2 mg β-carotene/g DW	Janarny et al. (2021)

(continued)

Table 6.2 (continued)

Pigment	Source	Color	Content		References
	Pelargonium grandiflorum	Red	Total carotenoids	1.37 ppm	Mikołajczak et al. (2020)
	Pelargonium hortorum (Caryophyllaceae)	Fuchsia, orange, pink, pink-fuchsia, pink-white, red, white	Total carotenoids	32.38–202.73 mg β-carotene/100 g DW	Coyago-Cruz et al. (2024)
	Pelargonium hortorum (Geraniaceae)	Fuchsia, pink, red, white, red-white	Total carotenoids	58.98–131.47 mg β-carotene/100 g DW	Coyago-Cruz et al. (2024)
	Pentas lanceolate	Red	β-Carotene	9 μg/g FW	Chensom et al. (2019)
	Petunia × *hybrida*	Pink	β-Carotene	3 μg/g FW	Chensom et al. (2019)
	Primula	White	Total carotenoids	34.1 mg/100 g DW	Nowicka and Wojdyło (2019)
	Primula × *polyantha*	Yellow	β-Carotene	23 μg/g FW	Chensom et al. (2019)
		Yellow	β-cryptoxanthin	20 μg/g FW	Chensom et al. (2019)
	Punica granatum	Red, orange	Total carotene content	9.8 mg β-carotene/g DW	Janarny et al. (2021)
	Raphanus raphanistrum	Pink	Total carotenoids	270.83 mg β-carotene/100 g DW	Coyago-Cruz et al. (2024)
	Robinia pseudoacacia	Cream-white	Total carotenoids	2.8 mg/g of extract	Li et al. (2014)
	Rosa 'Purple Fragrance'	Pink	β-carotene	1 μg/g FW	Chensom et al. (2019)
	Rosa banksiae	White	Total carotenoids	150.51 mg β-carotene/100 g DW	Coyago-Cruz et al. (2024)
	Rosa canina	Pink	Total carotenoids	93.39 mg β-carotene/100 g DW	Coyago-Cruz et al. (2024)
		Pale pink	Total carotenoids	0.63 ppm	Mikołajczak et al. (2020)
	Rosa damascene	Pink	Total carotenoids	132.55 mg β-carotene/100 g DW	Coyago-Cruz et al. (2024)

(continued)

Table 6.2 (continued)

Pigment	Source	Color	Content		References
	Rosa × hybrid	Pink, red, white, yellow, orange-yellow, orange, purple	Total carotenoids	52.16–660.13 mg β-carotene/100 g DW	Coyago-Cruz et al. (2024)
	Rosa × hybrid roseta	Orange-yellow, orange, purple, pink, red, white, pink-white, yellow	Total carotenoids	56.49–361.8 mg β-carotene/100 g DW	Coyago-Cruz et al. (2024)
	Rosmarinus officinalis	Blue-violet	Total carotenoids	1.5 mg/g of extract	Li et al. (2014)
	Ruda chalepensis	Yellow	Total carotenoids	923.26 mg β-carotene/100 g DW	Coyago-Cruz et al. (2024)
	Salvia 'Purple Queen'	Purple	Total carotenoids	6.12 μg/g FW	Marchioni et al. (2024)
	Salvia farinacea	Violet-blue	Total carotenoids	7.37 μg/g FW	Marchioni et al. (2024)
	Salvia jamensis J. Compton	Magenta	Total carotenoids	5.7 μg/g FW	Marchioni et al. (2024)
	Salvia microphylla	Blue	Total carotenoids	134.76 mg β-carotene/100 g DW	Coyago-Cruz et al. (2024)
	Salvia splendens	Purple	Total carotenoids	0.31 ppm	Mikołajczak et al. (2020)
	Sambucus nigra	White	Total carotenoids	1.7 mg/g of extract	Li et al. (2014)
		Yellow-white, cream	Total carotenoids	0.63 ppm	Mikołajczak et al. (2020)
		Pink-white	Total carotenoids	33.33 mg/100 g DW	Nowicka and Wojdyło (2019)
	Sesbania grandiflora	White, pink, red	Total carotene content	18.2 mg β-carotene/g DW	Janarny et al. (2021)
	Tagetes erecta	Orange, pale yellow	Total carotenoids	0.58 mg/g	Dujmović et al. (2022)
		Yellow	Total carotenoids	6.3 μg/g FW	Petrova et al. (2016)

(continued)

Table 6.2 (continued)

Pigment	Source	Color	Content		References
	Tagetes patula	Orange, yellow	Total carotenoids	703.1 mg/100 g DW	Coyago-Cruz et al. (2024)
		Orange, yellow	Total carotenoids	500.62–2057.79 mg β-carotene/100 g DW	Coyago-Cruz et al. (2024)
		Orange, yellow	Total carotenoids	2.07 ppm	Mikołajczak et al. (2020)
	Tamarindus indica	Yellow with orange or red streaks	Total carotene content	19.9 mg β-carotene/g DW	Janarny et al. (2021)
	Tilia cordata	Yellow-white	Total carotenoids	0.95 ppm	Mikołajczak et al. (2020)
		Yellow-white	Total carotenoids	130.7 mg/100 g DW	Nowicka and Wojdyło (2019)
	Torenia fournieri	Blue, violet	β-carotene	1 μg/g FW	Chensom et al. (2019)
		Blue, violet	β-cryptoxanthin	1 μg/g FW	Chensom et al. (2019)
	Trifolium repens	White	Total carotenoids	260.37 mg β-carotene/100 g DW	Coyago-Cruz et al. (2024)
		White, pink-white	Total carotenoids	0.06 ppm	Mikołajczak et al. (2020)
	Tropaeolum majus	Orange	Total carotenoids	53.23–73.88%	Barros et al. (2020)
		Orange	β-carotene	6 μg/g FW	Chensom et al. (2019)
		Orange	β-cryptoxanthin	5 μg/g FW	Chensom et al. (2019)
		Yellow, orange	Total carotenoids	0.28 mg/g	Dujmović et al. (2022)
		Orange	Total carotenoids	2.12 ppm	Mikołajczak et al. (2020)
	Tulbaghia simmleri 'Alba'	White	Total carotenoids	5.33 μg/g FW	Marchioni et al. (2024)
	Tulbaghia violacea 'Alba'	White	Total carotenoids	10.28 μg/g FW	Marchioni et al. (2024)
	Verbascum	White with a little violet	Total carotenoids	115.27 mg/100 g DW	Nowicka and Wojdyło (2019)

(continued)

Table 6.2 (continued)

Pigment	Source	Color	Content		References
	Vinca rosea	Pink, white, red, pink-white	Total carotene content	34.8 mg β-carotene/g DW	Janarny et al. (2021)
	Viola × *wittrockiana*	Purple-yellow	Total carotenoids	0.61 ppm	Mikołajczak et al. (2020)
Chlorophyll					
	Acacia	Yellow	Total chlorophyll	5.48 mg/100 g DW	Nowicka and Wojdyło (2019)
	Allium cepa	White, purple	Total chlorophyll	1.03 ppm	Mikołajczak et al. (2020)
	Allium schoenoprasum	Pink, pale purple	Total chlorophyll	2.15 ppm	Mikołajczak et al. (2020)
	Anthyllis vulneraria	Yellow	Total chlorophyll	9.23 mg/100 g DW	Nowicka and Wojdyło (2019)
	Antirrhinum majus	Orange-pink	Total chlorophyll	79.99 μg/g FW	Stefaniak and Grzeszczuk (2019)
	Arnica	Yellow	Total chlorophyll	38.42 mg/100 g DW	Nowicka and Wojdyło (2019)
	Begonia tuberhybrida	White, red	Total chlorophyll	5.76–9.01 ppm	Mikołajczak et al. (2020)
	Bellis perennis	White, pink	Total chlorophyll	4.15 ppm	Mikołajczak et al. (2020)
		Pink, red	Total chlorophyll	14.21 mg/100 g DW	Nowicka and Wojdyło (2019)
	Calendula officinalis	Yellow, orange	Total chlorophyll	0.75 mg/g	Dujmović et al. (2022)
		Yellow, orange	Total chlorophyll	1.27 ppm	Mikołajczak et al. (2020)
		Yellow, orange	Total Chla+b	22.5 μg/g FW	Petrova et al. (2016)
	Centaurea cyanus	Blue, lavender	Total chlorophyll	1.13 ppm	Mikołajczak et al. (2020)
	Chamomilla recutita	White with yellow centers	Total chlorophyll	1.39 ppm	Mikołajczak et al. (2020)
	Crataegus	White	Total chlorophyll	17.07 mg/100 g DW	Nowicka and Wojdyło (2019)

(continued)

Table 6.2 (continued)

Pigment	Source	Color	Content		References
	Cucurbita pepo	Orange-yellow	Total chlorophyll	0.04 mg/g	Dujmović et al. (2022)
		Yellow, orange	Total chlorophyll	0.97–1.08 ppm	Mikołajczak et al. (2020)
	Dianthus caryophyllus	Light pink	Total chlorophyll	1.1 ppm	Mikołajczak et al. (2020)
	Dianthus chinensis	Dark maroon/almost black, white edged	Total chlorophyll	369.78 μg/g FW	Stefaniak and Grzeszczuk (2019)
	Geranium macrorrhizum	Pink	Total Chla+b	41.5 μg/g FW	Petrova et al. (2016)
	Helianthus tuberosus	Yellow	Total Chla+b	0.5 μg/g FW	Petrova et al. (2016)
	Hemerocallis × *hybrida*	Orange-yellow	Total chlorophyll	215.19 μg/g FW	Stefaniak and Grzeszczuk (2019)
		Orange-yellow	Total chlorophyll	1.31 ppm	Mikołajczak et al. (2020)
	Lavandula	Pale purple	Total chlorophyll	59.45 mg/100 g DW	Nowicka and Wojdyło (2019)
	Lavandula angustifolia	Violet-blue	Total chlorophyll	2.53 ppm	Mikołajczak et al. (2020)
	Malva	Rose-purple	Total chlorophyll	28.84 mg/100 g DW	Nowicka and Wojdyło (2019)
	Matricaria	White with yellow centers	Total chlorophyll	7.25 mg/100 g DW	Nowicka and Wojdyło (2019)
	Mentha spicata	Pink-white	Total chlorophyll	6.8 ppm	Mikołajczak et al. (2020)
	Mimulus × *hybridus*	Yellow, red	Total chlorophyll	29.08–34.32 μg/g FW	Stefaniak and Grzeszczuk (2019)
	Monarda didyma	Red	Total chlorophyll	461.67 μg/g FW	Stefaniak and Grzeszczuk (2019)
	Pelargonium grandiflorum	Red	Total chlorophyll	2.52 ppm	Mikołajczak et al. (2020)

(continued)

Table 6.2 (continued)

Pigment	Source	Color	Content		References
	Rosa canina	Pale pink	Total chlorophyll	1.24 ppm	Mikołajczak et al. (2020)
	Salvia splendens	Purple	Total chlorophyll	1.64 ppm	Mikołajczak et al. (2020)
	Sambucus nigra	Yellow-white, cream	Total chlorophyll	1.52 ppm	Mikołajczak et al. (2020)
		Pink-white	Total chlorophyll	2.14 mg/100 g DW	Nowicka and Wojdyło (2019)
	Tagetes erecta	Orange, pale yellow	Total chlorophyll	0.32 mg/g	Dujmović et al. (2022)
		Yellow	Total Chla+b	23.6 μg/g FW	Petrova et al. (2016)
	Tagetes patula	Orange, yellow	Total chlorophyll	1.95 ppm	Mikołajczak et al. (2020)
	Tilia cordata	Yellow-white	Total chlorophyll	1.92 ppm	Mikołajczak et al. (2020)
		Yellow-white	Total chlorophyll	19.7 mg/100 g DW	Nowicka and Wojdyło (2019)
	Trifolium repens	White, pink-white	Total chlorophyll	6.34 ppm	Mikołajczak et al. (2020)
	Tropaeolum majus	Yellow, orange	Total chlorophyll	0.06 mg/g	Dujmović et al. (2022)
		Orange	Total chlorophyll	1.23 ppm	Mikołajczak et al. (2020)
	Viola × *wittrockiana*	Purple-yellow	Total chlorophyll	3.16 ppm	Mikołajczak et al. (2020)

DW dry weight, *FW* fresh weight, *CGE* cyanidin-3-glucoside equivalents, *ME* malvin chloride equivalents, *Mv3GlcE* malvidin-3-glucoside equivalents

[a]Approximated value from graph

indicator to determine the extent of oxidative damage, with reduced fluorescence indicating poor antioxidant capacity (Xiao et al. 2020).

Among all antioxidant assays performed for different edible flowers in previous studies, those from Rosaceae are one of the flower families standing to have high antioxidant activity over Caryophyllaceae and Gentianaceae across different in vitro assays, including FRAP and DPPH (Zhang et al. 2023). Similarly, the highest antioxidant activity was reported for pink rose (*Rosa* L.) from Rosaceae family at 15.77 mg/g Trolox equivalent (TE) and 9.27 mg/g TE determined using DPPH and ABTS assays, respectively (Gonçalves et al. 2020). This high antioxidant capacity was correlated to its high total phenols and anthocyanin content. Likewise, Red China roses (*Rosa chinensis* Jacq.) exhibiting the highest total phenolic content reported the greatest antioxidant activity of 130.96 mg Fe(II)/g for FRAP and 43.88 mg gallic acid equivalent (GAE)/g for ABTS, comparatively out of 30 flowers (Kritsi et al. 2022). Antioxidant activities are attributed to plants' composites of polyphenols which not only includes phenolic acids but also flavonoids (Kaisoon et al. 2011), anthocyanins and carotenoids (Benvenuti et al. 2016). However, there are exceptions, flowers like Sweet William (*Dianthus barbatus* L.) with its red color indicative of rich anthocyanin content had low antioxidant value of 1.53 mg/g GAE and 1.33 mg/g GAE, determined using DPPH and ABTS assays, respectively (Gonçalves et al. 2020). Meanwhile, *Myrtillocactus geometrizans*, which is white in color and indicative of poor carotenoid content had a strong antioxidant capacity of 675.06 μmol TE/g and 819.90 μmol TE/g, determined using DPPH and FRAP assays, respectively (Pinedo-Espinoza et al. 2020). This high antioxidant capacity was comparable to other strong candidates including *Paeonia lactiflora* Pall. (836.65 μmol TE/g) and *Osmanthus fragrans* (Thunb.) Lour. (913.58 μmol TE/g), determined using FRAP assay (Chen et al. 2015). These findings reveal that the antioxidant potential of edible flowers is influenced by the composition of phytochemicals, where the increase or decrease in a single presence of polyphenols may not predict its true antioxidant potential. There may be a synergistic or antagonistic effect that ultimately defines its level of antioxidant activity. Further, a single type of assay does not define its true antioxidant potential, as different assays investigate different antioxidant mechanisms. For instance, in the ORAC assay, red capuzin (*Tropaeolum majus*) had the highest antioxidant capacity, followed by orange capuzin (*Tropaeolum majus*) and purple amaranth (*Amaranthus hypondriacus*), with the lowest being jambu (*Spilanthes oleracea* L.) (Barros et al. 2020). In contrast, the order does not remain the same in DPPH and FRAP assays where the lowest antioxidant capacity were red *Tropaeolum majus* and orange *Tropaeolum majus*, respectively (Barros et al. 2020).

Nonetheless, antioxidant results from biochemical assays cannot be solely reliant on as they may not translate toward its effect in biological settings. As an intermediate to using expensive *in vivo* assays, cell-based assays are used (Nwachukwu et al. 2021). For measuring the inhibition of intracellular radical production such as reactive oxygen species (ROS), the fluorogenic probe 2′-7′-dichlorodihydrofluorescein diacetate (DCFH-DA) is widely utilized (Nwachukwu et al. 2021). This assay involves intracellular conversion of a non-fluorescent DCFH-DA to form a fluorescent DCF upon oxidation by ROS (Yu et al. 2021). The resulting fluorescence intensity correlates to the amount of intracellular ROS present. For instance, the method utilized resulted in *Tagetes erecta* flower extract exhibiting the highest cellular antioxidant activity in human liver cells (Hep G2 cells) with an EC_{50} value of 413 μM quercetin equivalent (QE)/g DW, which is a lower value than *Cosmos sulphureus* (966 μM QE/g DW), *Antigonon leptopus* (967 μM QE/g DW), and *Bougainvillea glabra* (859 μM QE/g DW) (Kaisoon et al. 2012). This indicates the lower the EC_{50} value, the lower amount of the extract, required to inhibit half of ROS production. Using the same cell line, the highest cellular antioxidant activity was found for *Paeonia suffruticosa* (32.27 μmol QE/100 g) and *Rosa rugosa* flowers (31.74 μmol QE/100 g) and the lowest for *Chrysanthemum morifolium* (7.60 μmol QE/100 g)

(Huang et al. 2017). Alternative to assessing cellular antioxidant activity, the extent of oxidative damage towards cells such as lipid peroxidation, can be assessed to understand the effectiveness of the antioxidants. The biomarkers measured are the lipid peroxidation byproducts such as malondialdehyde (MDA) and thiobarbituric acid reactive substances (TBARS). In the liver and kidney tissue taken from mice that underwent fasting to induce lipid peroxidation, the 20 μg/mL ethanolic extract dose of *Pyrus pashia* and *Sophora viciifolia* led to a marked decrease in TBARS (He et al. 2015) and increased inhibition of MDA formation (78.9% for liver and approximately 65% for kidney), respectively (Tai et al. 2011). These findings indicate that significant inhibition of lipid peroxidation has taken place. Another parameter that can be measured in cell-based assays is the endogenous antioxidant system which includes the enzymes glutathione peroxidase (GPx), catalase (CAT), and superoxide dismutase (SOD) and the peptide, glutathione (GSH). These elements are essential in mitigating oxidative stress in biological systems through conversion of radicals toward less reactive forms (Nwachukwu et al. 2021). In THP–1, a macrophage cell line, the *Trifolium repens* red flower extract increased the SOD (44.54 U/mL) and CAT activity (328.81 nmol/min/mg) levels higher than untreated sodium fluoride-induced oxidative stress cells (8.69 U/mL and 46.32 mmol/min/mg) (Jakubczyk et al. 2021). Other edible flowers with a considerable elevation of SOD enzyme included *Carthamus tinctorius*, *Chrysanthemum morifolium*, *Gardenia jasminoides*, *Jasminum sambac*, *Lonicera japonica* and *Rosa rugosa* Thunb. in the range of 19.66–34.32 U/mL investigated in rat cardiac microvascular endothelial cells (Wang et al. 2017). These findings indicate that edible flowers have the ability to restore the endogenous antioxidant defense system.

Regardless of the promising results of *in vitro* biochemical and cell-based assays, *in vivo* assays are essential to translate edible flowers' antioxidant effect in tackling various oxidative stress-led diseases. The assays involve administering antioxidants of a certain dose to diseased animal models, such as rodents or *Caenorhabditis elegans*. After a period, they are sacrificed of certain body parts for the assay to be conducted (Alam et al. 2013), measuring similar parameters as a cell-based method. In a rodent model with ethanol-induced gastric injury, pretreatment of *Dendrobium officinale* flower extract markedly increased the CAT and SOD activity levels, and decreased MDA level, compared to without pretreatment (Fan et al. 2024). This was supported by evident gastric tissue recovery based on histopathological analysis, indicating the extract's protective nature in attenuating oxidative damage of the tissue. A boost in antioxidant enzymes was observed in pancreatic tissue from *ex vivo* oxidative stress-induced rats treated with *Hibiscus sabdoriffa* ethanolic flower extract (Mohamed et al. 2022) and in the serum of hyperlipidemia-induced oxidative stress rats treated with aqueous flower extract of either *Carthamus tinctorius* L., *Chrysanthemum morifolium* Ramat., *Eugenia caryophyllata* Thunb., *Lonicera japonica* Thunb., *Magnolia officinalis* Rehd. et Wils., or *Rosmarinus officinalis* L. (Wang et al. 2017). Further, treatment with 800 mg/kg BW of *Clitoria ternatea* flower extract reversed the markedly reduced SOD level (untreated control: 5.08 U/mL, treated: 11.39 U/mL) in the liver of rat model induced with diabetes mellitus and dyslipidemia to a level similar to that of a healthy model (approximately 12.5 U/

mL), with a comparatively higher CAT level (Widowati et al. 2024). Likewise, in aluminium chloride-induced Alzheimer's disease in Wister rats, *Calendula officinalis* L. flower extract restored the SOD, CAT and GSH activity levels in the hippocampus and cortex region, indicating repairment of the antioxidant defense system (Jasoria et al. 2024). The treatment also reduced levels of nitrite and MDA, with visible cognitive improvement of the rat model. Alternative to the rodent model, in Moliner et al. (2018), *Tagetes erecta* L. ethanolic extract had a significant increase in survival rate of oxidative stress induced *Caenorhabditis elegans* at higher doses of 125 and 250 μg/mL. The survival rate increased from 4% to 28% and 2.8% to 21% for the orange and yellow cultivars, respectively. Further, the increase in the survival rate of *C. elegans* by approximately 10% was observed using *Cirsium japonicum* methanolic flower extract at 10 mg/mL, where higher doses did not provide a beneficial effect (Yen et al. 2023). Broadly, these *in vivo* studies reflect various diseases or oxidative damaged injuries that are alleviated through the use of edible flower extracts to counter oxidative stress.

Overall, there are plenty of edible flowers with promising antioxidant potential displayed in *in vitro* and *in vivo* assays. Future studies, however, remain on isolating individual compound(s) to investigate the main antioxidant players and their mechanism of action in synergy and in isolation.

6.4 Conclusion

There are many edible flowers from around the world that have significant antioxidant activities and, therefore, good potential for various applications in the food, nutraceutical, cosmetic, and pharmaceutical industries. The extensive biological and culinary potential of many edible flowers has yet to be discovered; therefore, interest from researchers in edible flowers will continue to grow in the coming years.

References

Alam MN, Bristi NJ, Rafiquzzaman M (2013) Review on *in vivo* and *in vitro* methods evaluation of antioxidant activity. Saudi Pharm J 21(2):143–152

Apak R, Özyürek M, Güçlü K, Çapanoğlu E (2016) Antioxidant activity/capacity measurement. 1. Classification, physicochemical principles, mechanisms, and electron transfer (ET)-based assays. J Agric Food Chem 64(5):997–1027

Araújo S, Matos C, Correia E, Antunes M (2019) Evaluation of phytochemicals content, antioxidant activity and mineral composition of selected edible flowers. Qual Assur Saf Crops Foods 11(5):471–478

Barriada-Bernal LG, Almaraz-Abarca N, Delgado-Alvarado EA, Gallardo-Velázquez T, Ávila-Reyes JA, Torres-Morán MI, González-Elizondo MS, Herrera-Arrieta Y (2014) Flavonoid composition and antioxidant capacity of the edible flowers of *Agave durangensis* (Agavaceae). CYTA J Food 12(2):105–114

Barros RGC, Andrade JKS, Pereira UC, de Oliveira CS, Rezende YRRS, Silva TOM, Nogueira JP, Gualberto NC, Araujo HCS, Narain N (2020) Phytochemicals screening, antioxidant capacity and chemometric characterization of four edible flowers from Brazil. Food Res Int 130:108899

Benvenuti S, Bortolotti E, Maggini R (2016) Antioxidant power, anthocyanin content and organoleptic performance of edible flowers. Sci Hortic 199:170–177

Bibi Sadeer N, Montesano D, Albrizio S, Zengin G, Mahomoodally MF (2020) The versatility of antioxidant assays in food science and safety – chemistry, applications, strengths, and limitations. Antioxidants 9(8):709

Boulebd H, Spiegel M (2023) Computational assessment of the primary and secondary antioxidant potential of alkylresorcinols in physiological media. RSC Adv 13(42):29463–29476

Chen GL, Chen SG, Xie YQ, Chen F, Zhao YY, Luo CX, Gao YQ (2015) Total phenolic, flavonoid and antioxidant activity of 23 edible flowers subjected to *in vitro* digestion. J Funct Foods 17:243–259

Chensom S, Okumura H, Mishima T (2019) Primary screening of antioxidant activity, total polyphenol content, carotenoid content, and nutritional composition of 13 edible flowers from Japan. Prev Nutr Food Sci 24(2):171

Coyago-Cruz E, Alarcón A, Guachamin A, Méndez G, Osorio E, Heredia-Moya J, Zuñiga-Miranda J, Beltrán-Sinchiguano E, Vera E (2024) Functional, antioxidant, antibacterial, and antifungal activity of edible flowers. Antioxidants 13(11):1297

Demasi S, Caser M, Donno D, Enri SR, Lonati M, Scariot V (2021) Exploring wild edible flowers as a source of bioactive compounds: new perspectives in horticulture. Folia Horticulturae 33(1):27–48

Dujmović M, Radman S, Opačić N, Fabek Uher S, Mikuličin V, Voća S, Šic Žlabur J (2022) Edible flower species as a promising source of specialized metabolites. Plants 11(19):2529

Espejel EAR, Alvarez OC, Muñoz JMM, Mateos MRG, León MTBC, Damián MTM (2019) Physicochemical quality, antioxidant capacity and nutritional value of edible flowers of some wild dahlia species. Folia Horticulturae 31(2):331–342

Fan P, Xie S, Zhang Z, Yuan Q, He J, Zhang J, Liu X, Liu X, Xu L (2024) *Dendrobium officinale* flos water extract ameliorates ethanol-induced acute gastric mucosal injury via inhibiting oxidative stress and inflammation. J Sci Food Agric 104(14):8593–8603

Fernandes L, Casal S, Pereira JA, Saraiva JA, Ramalhosa E (2017) Edible flowers: a review of the nutritional, antioxidant, antimicrobial properties and effects on human health. J Food Compos Anal 60:38–50

Fernandes L, Casal S, Pereira JA, Saraiva JA, Ramalhosa E (2020) An overview on the market of edible flowers. Food Rev Int 36(3):258–275

Friedman H, Agami O, Vinokur Y, Droby S, Cohen L, Refaeli G, Resnick N, Umiel N (2010) Characterization of yield, sensitivity to *Botrytis cinerea* and antioxidant content of several rose species suitable for edible flowers. Sci Hortic 123(3):395–401

Gokiladevi R, Varsini K, Shafrin K, Nithya R (2024) Edible flowers: blooming delicacies for health and wealth. Futur Trends Agric Eng Food Sci 3:77–88

Gonçalves F, Gonçalves JC, Ferrão AC, Correia P, Guiné RP (2020) Evaluation of phenolic compounds and antioxidant activity in some edible flowers. Open Agric J 5(1):857–870

He J, Yin T, Chen Y, Cai L, Tai Z, Li Z, Liu C, Wang Y, Ding Z (2015) Phenolic compounds and antioxidant activities of edible flowers of *Pyrus pashia*. J Funct Foods 17:371–379

Huang W, Mao S, Zhang L, Lu B, Zheng L, Zhou F, Zhao Y, Li M (2017) Phenolic compounds, antioxidant potential and antiproliferative potential of 10 common edible flowers from China assessed using a simulated *in vitro* digestion–dialysis process combined with cellular assays. J Sci Food Agric 97(14):4760–4769

Jakubczyk K, Łukomska A, Gutowska I, Kochman J, Janił J, Janda K (2021) Edible flowers extracts as a source of bioactive compounds with antioxidant properties – *in vitro* studies. Appl Sci 11(5):2120

Janarny G, Ranaweera K, Gunathilake K (2021) Antioxidant activities of hydro-methanolic extracts of Sri Lankan edible flowers. Biocatal Agric Biotechnol 35:102081

Janarny G, Ranaweera KKDS, Gunathilake KDPP (2022) Digestive recovery of polyphenols, antioxidant activity, and anti-inflammatory activity of selected edible flowers from the family Fabaceae. J Food Biochem 46(2):e14052

Jasoria Y, Agrawal M, Kumar S, Chaudhary H, Sahu KK, Singhal M, Arora S, Chandolia P, Saha S, Singh K, Mahour S, Akram W, Jain D (2024) *In-vivo* evaluation of neuroprotective effect of Chinese plant *Calendula officinalis* Linn. flower extract against aluminium chloride-induced Alzheimer's in Wistar rats. Pharmacol Res Mod Chin Med 12:100458

Jeyaraj EJ, Lim YY, Choo WS (2021) Effect of organic solvents and water extraction on the phytochemical profile and antioxidant activity of *Clitoria ternatea* flowers. ACS Food Sci Technol 1:1567–1577

Kaisoon O, Siriamornpun S, Weerapreeyakul N, Meeso N (2011) Phenolic compounds and antioxidant activities of edible flowers from Thailand. J Funct Foods 3(2):88–99

Kaisoon O, Konczak I, Siriamornpun S (2012) Potential health enhancing properties of edible flowers from Thailand. Food Res Int 46(2):563–571

Kandylis P (2022) Phytochemicals and antioxidant properties of edible flowers. Appl Sci 12(19):9937

Kritsi E, Tsiaka T, Ioannou A-G, Mantanika V, Strati IF, Panderi I, Zoumpoulakis P, Sinanoglou VJ (2022) *In vitro* and in silico studies to assess edible flowers' antioxidant activities. Appl Sci 12(14):7331

Kucekova Z, Mlcek J, Humpolicek P, Rop O (2013) Edible flowers – antioxidant activity and impact on cell viability. Open Life Sci 8(10):1023–1031

Li AN, Li S, Li HB, Xu DP, Xu XR, Chen F (2014) Total phenolic contents and antioxidant capacities of 51 edible and wild flowers. J Funct Foods 6:319–330

Loizzo MR, Pugliese A, Bonesi M, Tenuta MC, Menichini F, Xiao J, Tundis R (2016) Edible flowers: a rich source of phytochemicals with antioxidant and hypoglycemic properties. J Agric Food Chem 64(12):2467–2474

Madorran E, Stožer A, Bevc S, Maver U (2020) *In vitro* toxicity model: upgrades to bridge the gap between preclinical and clinical research. Bosnian J Basic Med Sci 20(2):157–168

Marchioni I, Rodolfi M, Najar B, Ruffoni B, Machado J, Pistelli L (2024) Phytonutritional compounds and antioxidant activity of eight new edible flowers. Nat Prod Res 38(21):3728–3734

Mikołajczak N, Sobiechowska DA, Tańska M (2020) Edible flowers as a new source of natural antioxidants for oxidative protection of cold-pressed oils rich in omega-3 fatty acids. Food Res Int 134:109216

Mlcek J, Plaskova A, Jurikova T, Sochor J, Baron M, Ercisli S (2021) Chemical, nutritional and sensory characteristics of six ornamental edible flowers species. Foods 10(9):2053

Mohamed AI, Salau VF, Erukainure OL, Islam MS (2022) *Hibiscus sabdariffa* L. polyphenolic-rich extract promotes muscle glucose uptake and inhibits intestinal glucose absorption with concomitant amelioration of Fe^{2+}-induced hepatic oxidative injury. J Food Biochem 46(12):e14399

Moliner C, Barros L, Dias MI, López V, Langa E, Ferreira ICFR, Gómez-Rincón C (2018) Edible flowers of *Tagetes erecta* L. as functional ingredients: phenolic composition, antioxidant and protective effects on *Caenorhabditis elegans*. Nutrients 10(12):2002

Navarro-González I, González-Barrio R, García-Valverde V, Bautista-Ortín AB, Periago MJ (2014) Nutritional composition and antioxidant capacity in edible flowers: characterisation of phenolic compounds by HPLC-DAD-ESI/MS^n. Int J Mol Sci 16(1):805–822

Nowicka P, Wojdyło A (2019) Anti-hyperglycemic and anticholinergic effects of natural antioxidant contents in edible flowers. Antioxidants 8(8):308

Nwachukwu ID, Sarteshnizi RA, Udenigwe CC, Aluko RE (2021) A concise review of current *in vitro* chemical and cell-based antioxidant assay methods. Molecules 26(16):4865

Petrova I, Petkova N, Ivanov I (2016) Five edible flowers–valuable source of antioxidants in human nutrition. Int J Pharmacogn Phytochem Res 8(4):604–610

Pinedo-Espinoza JM, Gutiérrez-Tlahque J, Santiago-Saenz YO, Aguirre-Mancilla CL, Reyes-Fuentes M, López-Palestina CU (2020) Nutritional composition, bioactive compounds and antioxidant activity of wild edible flowers consumed in semiarid regions of Mexico. Plant Foods Hum Nutr 75:413–419

Pisoschi AM, Pop A (2015) The role of antioxidants in the chemistry of oxidative stress: a review. Eur J Med Chem 97:55–74

Rao V, Poonia A (2024a) Bioactive Compounds, nanoparticle synthesis and potential utilization of edible flowers for development of functional dairy products: a review. J Food Sci Technol 61(6):1053–1068. https://doi.org/10.1007/s13197-023-05853-6

Rao V, Poonia A (2024b) Microbial biosynthesis of selenium nanoparticles using probiotic strain and its characterization. J Food Meas Charact *18*(7):5477–5488. https://doi.org/10.1007/s11694-024-02581-z

Shi J, Gong J, Wu X, Zhang Y (2009) Antioxidant capacity of extract from edible flowers of *Prunus mume* in China and its active components. LWT-Food Sci Technol 42(2):477–482

Socha R, Kałwik J, Juszczak L (2021) Phenolic profile and antioxidant activity of the selected edible flowers grown in Poland. Acta Univ Cibiniensis Ser E: Food Technol 25(2):185–200

Stefaniak A, Grzeszczuk ME (2019) Nutritional and biological value of five edible flower species. Not Bot Horti Agrobot Cluj-Napoca 47(1):128–134

Suksathan R, Rachkeeree A, Puangpradab R, Kantadoung K, Sommano SR (2021) Phytochemical and nutritional compositions and antioxidants properties of wild edible flowers as sources of new tea formulations. NFS J 24:15–25

Tai Z, Cai L, Dai L, Dong L, Wang M, Yang Y, Cao Q, Ding Z (2011) Antioxidant activity and chemical constituents of edible flower of *Sophora viciifolia*. Food Chem 126(4):1648–1654

Vidana Gamage GC, Choo WS (2023) Hot water extraction, ultrasound, microwave and pectinase-assisted extraction of anthocyanins from blue pea flower. Food Chem Adv 2:100209

Wang F, Miao M, Xia H, Yang LG, Wang SK, Sun GJ (2017) Antioxidant activities of aqueous extracts from 12 Chinese edible flowers *in vitro* and *in vivo*. Food Nutr Res 61(1):1265324

Widowati W, Darsono L, Utomo HS, Sabrina AHN, Natariza MR, Tarigan ACV, Waluyo NW, Gleyriena AM, Siahaan BH, Oktaviani R (2024) Antidiabetic and hepatoprotection effect of butterfly pea flower (*Clitoria ternatea* L.) through antioxidant, anti-inflammatory, lower LDH, ACP, AST, and ALT on diabetes mellitus and dyslipidemia rat. Heliyon 10(8):e29812

Xiao F, Xu T, Lu B, Liu R (2020) Guidelines for antioxidant assays for food components. Food Front 1(1):60–69

Xiong L, Yang J, Jiang Y, Lu B, Hu Y, Zhou F, Mao S, Shen C (2014) Phenolic compounds and antioxidant capacities of 10 common edible flowers from China. J Food Sci 79(4):C517–C525

Yan G, Zhou Y, Zhao L, Wang W (2022) Phenolic contents and antioxidant activities of solvent extracts from four edible flowers. Int Food Res J 29(5):1155–1167

Yen PL, Lin TA, Chuah WL, Chang CY, Tseng YH, Huang CY, Yang JC, Hsu FL, Liao VHC (2023) Methanol extracts from *Cirsium japonicum* DC. var. *australe* Kitam. and their active components reduce intracellular oxidative stress in *Caenorhabditis elegans*. Molecules 28(19):6923

Yu D, Zha Y, Zhong Z, Ruan Y, Li Z, Sun L, Hou S (2021) Improved detection of reactive oxygen species by DCFH-DA: new insight into self-amplification of fluorescence signal by light irradiation. Sens Actuators B Chem 339:129878

Zeng Y, Deng M, Lv Z, Peng Y (2014) Evaluation of antioxidant activities of extracts from 19 Chinese edible flowers. Springerplus 3:315

Zhang XK, Cao GH, Bi Y, Liu XH, Yin HM, Zuo JF, Xu W, Li HD, He S, Zhou XH (2023) Comprehensive analysis of 34 edible flowers by the determination of nutritional composition and antioxidant capacity planted in Yunnan province China. Molecules 28(13):5260

Zheng J, Yu X, Maninder M, Xu B (2018) Total phenolics and antioxidants profiles of commonly consumed edible flowers in China. Int J Food Prop 21(1):1524–1540

Chapter 7
Bioactive Compounds and Therapeutic Potential of Edible Flowers

Muhammad Afzaal, Muhammad Awais Saleem, Aftab Ahmed, Farhan Saeed, Muhammad Aqib Saeed, Haseeba Khalil Janjua, and Tuba Mustafa

7.1 Introduction

The usage of flowers in cooking was utilized back in ancient cultures. Since they were thought to be therapeutic plant foods with positive health effects, they have been a part of the human diet in a variety of ways for millennia (Navarro-González et al. 2015). According to earlier research, flowers were employed in cooking by people from Rome, China, the Middle East, and India. During Queen Victoria's reign, edible flowers obtained fame in both North America and Europe (Chen and Wei 2016). Chinese people have been using flowers as components in a wide range of dishes since 3000 BCE. In ancient China, edible flowers were used in herbal remedies as well as culinary ingredients (Fernandes et al. 2019a, b). According to early accounts, the edible roses and violet flowers were utilized in omelets and purees, while lavender was used in sauces in ancient Rome. Calendula (Calendula officinalis) flowers were utilized as a salad element in medieval France. The inflorescences of fried elder (Sambucus nigra) were frequently eaten in Central Europe (Vinokur et al. 2006; Mlcek and Rop 2011).

The world is learning the worth of flowers as food if consumers are looking for new natural sources of bioactive chemicals, even if edible flowers have historically

M. Afzaal (✉) · F. Saeed · T. Mustafa
Department of Food Science, Government College University Faisalabad, Faisalabad, Pakistan
e-mail: muhammadafzaal@gcuf.edu.pk

M. A. Saleem · M. A. Saeed · H. K. Janjua
Department of Human Nutrition and Dietetics, Mirpur University of Science and Technology Mirpur, Mirpur, Pakistan

A. Ahmed
Department of Nutritional Sciences, GC University Faisalabad, Faisalabad, Pakistan

A. Poonia et al. (eds.), *Edible Flowers: Source of Phytonutrients, Valorization and Technological Advancements*, https://doi.org/10.1007/978-3-032-00667-7_7

been consumed mostly for their scent and aesthetic appeal (Nowicka and Wojdyło 2019). The growing interest in edible flowers has led to a deeper evaluation of certain species, including hibiscus (*Hibiscus rosa-sinensis* L.) (Antarkar et al. 2019), chrysanthemum (*Chrysanthemum morifolium Ramat.*) (Zhang et al. 2019), pot marigold (Calendula officinalis L.) (Chitrakar et al. 2019), rose (Rosa spp.) (Qiu et al. 2019), and centaurea (Centaurea cyanus L.) (Lockowandt et al. 2019). Furthermore, edible flowers present appealing opportunities for the development of functional foods due to the growing understanding of their chemical composition and biological characteristics, their potential for toxicity (Katisart and Konsue 2019), and the availability of cutting-edge analytical tools like HPLC to identify their chemical components (Piovesana et al. 2019). Acute and chronic illnesses can be prevented, controlled, or treated by consuming functional foods that contain naturally occurring bioactive chemicals in the right amounts.

Particularly when eaten in their natural state or with little processing, edible flowers preserve their vitamins, proteins, essential oils, and antioxidants (Fernandes et al. 2019a, b). In line with this pattern, edible flowers have also been linked to several biological qualities recently, including the reduction of ulcerative colitis, anti-hyperglycemic and anticholinergic activity (Nowicka and Wojdyło 2019), defense against oxidative effects in erythrocytes (Yang et al. 2019), and even anticancer activity (Nguyen et al. 2019). Secondary metabolites, which are bioactive compounds found in edible flowers, can also be found in other plant parts and provide adaptive benefits, such as preventing plant pathogen infection and mitigating the effects of environmental stressors like heat, dehydration, and excessive UV radiation.

Petals and other flower parts are abundant in vitamins, minerals, and substances that are physiologically active. Edible flowers are an excellent source of phenolic compounds, according to studies on phytochemicals and their possible health benefits (Fernandes et al. 2017). By altering the genes involved in the crucial regulation process, these compounds have been linked to the prevention of various forms of cancer as well as chronic health conditions like heart disease, Type 2 diabetes, inflammatory disease, and Alzheimer's disease. This is because they control both the onset and progression of these illnesses (Gutiérrez-Grijalva et al. 2016).

7.2 History, Diversity, and Availability

As mentioned in several ancient texts, flowers were utilized in cooking in the past. Different cultures use edible flowers in different ways. To support the claim that edible flowers have been a component of cuisines throughout for thousands of years, Mlcek and Rop (Mlcek and Rop 2011) examined over 50 studies (Table 7.1). Flowers have been utilized in Chinese cuisine since 3000 BC (Stradley 2015). In Rome, lavender was used to make sauces, while violets and roses were utilized as garnishes. Flowers have been used by humans in the Middle East and India from ancient times, much like in China and Rome (Deka and Nath 2014). The squash

Table 7.1 Typical edible flowers, consumable components, and applications throughout continents

Name of the continent	Name of the flower	Figure	Edible parts	Flavors	Uses	References
Asia	Chrysanthemums		Except stamens whole	Slightly sweet	Medicinal + cooking	Tai and Chen (2000)
	Daylilies		Whole	Citrus	Garnishing + cooking	Fernandes et al. (2019a, b)
	Moringa		Whole	Little bitter	Cooking + medicinal	
	Banana		White parts inside the flower	Slightly sweet	Cooking + medicinal	Nadumane and Timsina (2014)
	Pumpkin		Whole	Sweet	Cooking	
Africa	Wildmalva		Petals	Sweet	Medicinal + garnishing	
	Peppermint-scented pelargonium		Whole	Peppermint	Garnishing + cooking	Demarne and van der Walt (1990)
	Wild sorrel	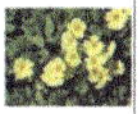	Petals	Sharp taste	Cooking + garnishing	
Australia	Nasturtium		Whole	Slightly sour	Medicinal + garnishing	
	French marigold		Petal	Spice	Cooking	Kanwar et al. (2016)
Europe	Cornflower		Petals	Sweet to spicy	Medicinal + garnishing	Jauron et al. (2013)
South America	Snapdragon		Petals	Sweet	Medicinal + garnishing	Al-Snafi (2015)

plant and pumpkin blooms were quite popular in North America. The Zuni people are well-known for their ardent devotion to squash blossoms (Murphy 2015). The use of edible flowers was common in North America, England, and other parts of Europe throughout the Victorian era (Fernandes et al. 2017).

Despite being inexpensive and a good source of nutrients, EFs are still underutilised because of several socioeconomic issues. Tribal belts are more likely than urban markets to consume EF. Additionally, EF's availability varies seasonally (Fernandes et al. 2019a, b). There is no statistical information available on the worldwide production of EF. However, there is little information on the production of certain EF (flowers that aren't classified as flowers, such saffron, broccoli, and cauliflower) (Fernandes et al. 2019a, b). Between 2000 and 2014, Asia accounted for 76.6% of the world's production of broccoli and cauliflower, whereas Europe only made up 11.9%. Portugal generated 38,968 tonnes of these two items in 2014 (FAOSTAT 2019). Asia-Pacific's excellent position in the export of floricultural products is mostly due to favorable agroclimatic conditions and low labor costs. Furthermore, there is a discernible need for EFs across Europe because of the growing popularity of cut flowers in regions of France, Germany, and the Netherlands. The Netherlands is the top supplier of floral arrangements and bulbs in Europe. In 2013, Colombia's exports of carnations, roses, and chrysanthemums made it the second-biggest exporter of cut flowers (Fernandes et al. 2019a, b).

7.3 Edible Flower Nutritional Composition

Edible flowers are becoming more and more popular as a new, healthful food trend (Benvenuti and Bortolotti 2016; Mlcek and Rop 2011). Native Americans' diets depend heavily on these edible blooms from wild plants, at least for a brief period after they bloom. In any event, it is critical to understand the nutritional makeup and other characteristics of those deemed acceptable in terms of human nourishment. According to their proximate organization, Water makes well over 80% of edible flowers and their fat and protein content are low. Depending on the kind of bloom, they also include varying levels of minerals, dietary fiber, and total sugars (Mlcek and Rop 2011; Sotelo et al. 2007).

Numerous studies have looked at the nutritional qualities of edible flowers. The dietary components of several edible flowers, including *Euphorbia radians benth, Arbutus xalapensis, Agave salmiana, Yucca filifera, Erythrina americana, Erythrina caribaea, Cucurbita pepo,* and *Aloe vera,* were investigated by Sotelo et al. According to the research on the nutritional composition and antioxidant potential of edible flowers, the dietary components of edible blooms were comparable to those of other plants, with a lower concentration of protein and total fat and a larger quantity of water and total fiber from the diet. The flowers' moisture content varied between 860 and 932 g kg^{-1}, while their crude protein and crude fiber contents ranged between 113 and 276 g/kg of dried matter (DM) and 103 and 176 g/kg of DM, respectively (Navarro-González et al. 2015). The nutritional makeup of maize flower, calendula, rose and dahlia petals and infusions was investigated by (Pires et al. 2017). They said that the most common macronutrient was carbohydrate, which was followed by ash and proteins. The highest organic acid content was found in calendula and rose petal infusions, mostly as a result of the existence of quinic and malic acids, respectively. Juárez-Trujillo et al. (2018) found that the total dietary fiber and unsaturated fatty acids, such as oleic, palmitic, and linoleic acid, were abundant in the carpels and petals of azote (*Yucca elephantipes*) flowers.

According to (Fernandes et al. 2019a, b), who examined the nutritionary makeup of pansy (Viola × Wittrockiana) at various flowering phases, completely open flowers contained the highest levels of protein and carbs, with 8.0 g/100 g and 2.00 g/100 g fresh weight, respectively. Stefaniak and Grzeszczuk (2019) examined the nutritional and biological properties of five edible flowers: Monarda didyma, Hemerocallis × hybrida, Dianthus chinensis, Antirrhinum majus, and Mimulus × hybridus. M. didyma flowers were the highest source of dry matter (18.86 ± 0.20% Fresh Weight), total ash (1.563 ± 0.11% FW), crude fiber (1.928 ± 0.20% FW), and total protein (7.816 ± 0.68% FW). Grzeszczuk et al. (Grzeszczuk et al. 2017) also looked at the mineral composition of dried flowers from the following plants: Monarda didyma, Hemerocallis × hybrida, Dianthus chinensis, Mimulus × hybridus, and Antirrhinum majus. Their research showed that the blooms of Mimulus and Monarda contained a significant quantity of both macro and micro components. The most prevalent macroelement detected in all flowers was potassium (30.10 g kg^{-1} dry matter), whereas iron (155.89 mg kg − 1 dry matter) was the most prevalent microelement.

7.4 Phytochemical Makeup of Edible flowers

Phytochemicals are naturally occurring, physiologically non-nutritive, and active chemical substances present in plants and have therapeutic effects on human health. Edible Flowers have a high content of natural antioxidants such as anthocyanin, flavonoids and various other phenolic components (Rao and Poonia 2024a, b). They shield plants from illness and environmental harm while enhancing their color, flavor, and scent. Carotenoids, phenolic acids, and flavonoids found in edible flowers are the most typical physiologically active substances (Mlcek and Rop 2011; Navarro-González et al. 2015). Carotenoids and flavonoids are responsible for the color of flowers and provide them with antioxidant properties (Chen et al. 2017). It has been demonstrated that the phytochemicals in marigolds, particularly lutein and flavonoids, have anti-inflammatory, cytotoxic, and free radical-scavenging effects on a range of cancerous cells.

7.4.1 Phenolic Acids

Phenolic acids, which fall among the groups of hydroxycinnamic acid (C6-C3) and hydroxybenzoic acids (C6-C1), are distinguished by their inclusion of phenolic functionality in the layout of the structure. Derivatives of hydroxybenzoic acid have been found in palatable blossoms of many kinds. However, blossoms of *Calendula officinalis L.*, *Sesbania grandiflora L.*, and *Butea monosperma* have been shown to contain hydroxycinnamic acid. Orange-hue samples of the Mexi pan marigold (Tagetes erecta L.) were likewise found to contain 15 phenolic acids. The most prevalent hydroxybenzoic acids in floral extraction (methanol and water (1:4)) were gallic acid and pyrogallol. Phenolic acid derivatives, specifically trans-ferulic acids, p-Coumaric, and chlorogenic were the main substances found in the 75% ethanol extraction of Hibiscus roseus Thore blossoms.

Lu et al. (2020) found five phenolic acid variants while examining the impact of methods for drying on the phenolic amount of chrysanthemum flower. Following the dehydration of the flowers, phenolic acids were shown to have greater resilience than flavonoids in the same research. Ferulic acids and P-coumaric were first identified in all the species examined when examining the extract of ethanol (12.1%) from four different flowers that are consumed in Brazil (Tropaeolum majus L. (red), *Amaranthus hypochondriacus* L., *Spilanthes oleracea* L. and *Tropaeolum majus* L. (orange)).

7.4.2 Flavonoids

The greatest class of secondary metabolic products discovered in plants is the flavonoids family, which is classified into the families of chalcones, anthocyanins, flavonols, flavones, and flavanones. Flavonoids are often found in flowers. They are identified as non-nitrogenous plant pigments and are regarded as low molecular weight substances. Because of the locations and quantities of hydroxyl and methoxyl groups found in each unique molecule, the molecular structure of flavonoids directly affects their antioxidant capacity. With the highest dispersion and wide range of chemicals found in the environment, they are thought to be the most important naturally occurring phenolics. *Ocimum sanctum* L., *Calendula officinalis* L., *Hibiscus rosa-sinensis L.*, *Vinca rosea* L., *Hibiscus sabdariffa* L., *Erythrina americana Mill.*, *Opuntia ficus-indica (*L.*) Mill, Plumeria rubra* L., *Tagetes erecta* L., *Agave durangensis Gentry*, *Artemisia spp.* and Dahlia spp. were among the flowers that were palatable and were found to contain flavonoids. The proportion of flavonoids in total in an ethanol extract (60.0%, v/v) of dried *Agave duragenensis L.* florals was assessed by Barriada-Bernal et al. They found that the whole flowers contributed more flavonoids (1210.4 g/g of dry extraction) than draws from the tepals, altars, and pollen. Furthermore, the chemicals hesperidin, quercitrin, luteolin, quercetin, hesperetin, kaempferol, and apigenin were identified as the primary flavonoid constituents in a study including 70 Chinese edible flower types.

7.4.3 Chalcones

Chalcones are substances from the flavonoids group that are mostly found in species in the The legumes family. They also appear in vegetables, fruits, and flowers belonging to the Compositae and Moraceae families. They may give some medicinal plants their yellow color, particularly in their petals, and they also help draw pollinators like insects and birds Often referred to as open-chain flavonoids, chalcones are distinguished by the lack of a "C-ring" in their molecular organization and are primarily represented by variants of phloretin, arbutin, chalconaringenin and phloridzin. Chalcones were found in the yellow hue of blooms of Coreopsis spp. and are thought to be the compounds that give flowers of Dahlia variabilis (Willd.) Desf. (butein and isoliquiritigenin) and Dianthus caryophyllus L. (chalcone 2-O-glucoside) their striking yellow hue. According to research by Li et al. (2021), the primary dihydrochalcone constituents discovered in the edible petals of Malus pumila Mill. are phloretin and phlorizin.

7.4.4 Flavanones

Flavanones, often referred to as 2-phenyl-chroman-4-ones, are polyphenolic substances with a fundamental framework of 2,3-dihydroflavone that are symmetric at the C2 position and missing the dual bond linking C2 and C3. Examples of these compounds include naringenin, hesperidin, heridictol and isosacuratenin. The way flavonoids communicate with physiological receptors and hence impact their bioactive qualities is greatly influenced by this change in their molecular arrangement. These substances, which are primarily linked to the concentration of hesperetin, neohesperidin, hesperidin,and naringin derivatives, are linked to a variety of health advantages in palatable flowers, including anti-aging properties. According to earlier research, Hylocereus undatus (Haw.) Britton & Rose (2162.3 g/g), Chrysanthemum indicum L. (2653.7 g/g), Chrysanthemum morifolium Ramat. (748.78 g/g), Gomphrena globosa L. (143.42 g/g) and Prunus persica (L.) Batsch (850.54 g/g), all contain hesperetin and its related compounds. Additionally, Dahlia spp. was shown to have two variants of naringenin molecules, naringenin hexoside-acetylhexoside and naringenin-3-O-glucoside, at 0.81 and 2.91 mg/g dw, respectively [1]. The detection of naringenin for lilac (20.11 g/g dw), orange (6.90 g/g dw), and violet (1.83 g/g dw) flowers was noted by Lara Cortés et al. (2016) in their study of a similar species (Dahlia sp.).

7.4.5 Flavones

Flavones are distinguished by the bonding of the B ring to C-2 and the presence of two bonds connecting C-2 and C-3 within their exterior organization. Their primary constituents are luteolin, acacetin, apigenin, chrysoeriol, and their glucosides, and they constitute the second biggest family of flavonoids in edible flowers. They are extensively distributed in nature. For instance, they have been found in the flowers of Florists' chrysanthemum (4.52 g/g), Dendran thema lavandulifolium (Fischer ex Trautv.) Kitam. (2.13 g/g), Matthiola incana (L.) R.Br. (10.45 g/g), Rosa rugosa

Thunb. (1.364 g/g), Tropaeolum majus L. (53.55 g/g), and Tropaeolum majus L. (53.45 g/g). The most prevalent culinary flowers featuring flavone chemicals are dandelion, rose, and peony.

7.4.6 Flavonols

Because of the hydroxyl pair that is located at the third position of C ring, flavonols have been identified as by-products of alcohol of particular flavones. Because of their framework resemblance to flavones, these substances commonly occur in veggies and fruits, drinks, and herbal products. They are also known as 3-hydroxy-2 phenylchromen-4-one. Flavonoids are a group of compounds most prominently embodied by quercitin and kaempferol, which are commonly found in culinary flowers, in spite of their resemblance to flavones and the fact that they are really proanthocyanin elements. Numerous attractive plants have been found to contain flavonol derivatives, such as kaempferol, which was found in Rhododendron indicum var. simsii (Planch.). Some delicious blossoming plants also contain significant amounts of quercitin and rutin which are two of the primary flavonols in the blooms of Citrus aurantium L. (quercetin, 185.371 0.112 g/g DW; rutin, 362.7 0.022 g/g DW) and Malus pumila Mill. Lastly, Pires et al. (2018) discovered flavonoid compounds such quercitin and kaempferol in four different flowers: *Rosa gallica* L., *Calendula officinalis* L., *Cyanus segetum Hill and Dahlia migno*n L.

7.4.7 Flavanols

Because of the absence of an extra bond connecting C2 and C3 and the group of carbonyls in ring C, flavanols also referred to as catechins-have dual aromatic carbons (C2 and C3). Although these are the primary component of *Camellia sinensis* and *C. assumica*, they may be found in many plant components. Furthermore, the floral isolates of Camelia nitidissima Chi are an excellent origin of catechin and its related compounds, according to Yang et al. (2018). Among plant components, buckwheat (Fagopyrum Esculentum Moench) flowers have the greatest levels of catechin, while the blooms of highbush blueberries (Vaccinium a gustifolium L.) and Cucurbita pepo L. are also rich sources of (+)-catechin and (−) epicatechin. While Li et al. (2018) observed that epicatechin and catechin constituted amongst the greatest number of phenolic substances amony culinary blossoms of various kinds, Liang et al. (2018) proposed that (−) epicatechin was the most prevalent phenolic component in the smaller flowers and the florals of Lilium pumilum.

7.5 Flowers as Cultural and Regional Foods

Several ancient writings mention the cultural eating of flowers. Dandelion is mentioned as a herb bitter in taste in the bible that is used in unleavened bread and salads. Assyrian tablets from 668–626 BCE, Egyptian papyri from 1550 BCE, the Solomon song, and a Sumerian Herbarium from 2500 BCE each mention saffron (*Crocus sativus L.*). It was also discovered in 1500 BCE artifacts discovered in Greece.

The Greeks grew a variety of flowers, including lotus (*Nelumbo nucifera Gaertn.*), carnation (*Dianthus caryophyllus* L.), poppy (*Papaver rhoeas L.*), and artichoke (*C. cardunculus*). Accounts of aniseed (*Pimpinella anisum L.*), nasturtium (*Tropaeolum majus*), saffron (*C. sativus*), and calendula (*C. officinalis*) are in ancient Rome. On bits of silk and bamboo, ancient recipes utilizing lotus petals (N. nucifera) were discovered in China. Calendula flowers (*C. officinalis*) used to be consumed in salads in medieval France, and there is proof that the Aztecs in the Americas utilized squash blossoms (*Curcubita pepo L.*) and manioc flowers (*Manihot esculenta Crantz*) (Kirker and Newman 2016; Mlcek and Rop 2011).

Flowers are used in common foods in many cultures as part of festivals and rituals Sotelo et al. (2007) state that before Spanish colonisation, indigenous people in Mexico ate certain flowers, such as yucca (*Yucca filifera Engelm*), sun spurge (*Euphorbia radians Benth.*), coral tree (Erythrina caribaea and Erythrina americana Mill), and squash (*C. pepo*). Information on nutritional content is currently limited, even though some locals have continued this practice.

There have also been reports of wild plant ingestion in developed North American and European nations. Dourson and Dourson (2019) published a guide on recognizing edible plants, including Eastern Kentuky edible flowers. Sheep milk and rennet derived from cardoon flowers (C. cardunculus) are used to make traditional cheese in Portugal and Spain (Gomes et al. 2019).

Unfortunately, information about the historic and cultural use of edible flowers tends to fade because of urbanisation and the young population's acceptance of fast food. Researching the edible plants that indigenous people in rural or forest settings eat can help maintain local biodiversity and traditional knowledge while also examining the nutritional diversity and economic opportunity offered by wild species (Harmayani et al. 2019).

Sanskrit texts from 3000 BCE mention flowers and gardens in India, indicating the nation's long-standing custom of using flowers primarily for ornamentation and decoration. Even though flowers have been a part of Indian culture for a very long time, the country's floriculture has only recently gained commercial status because of the increasing need for loose and trimmed flowers, which is encouraged by the increased impact of Western culture (Mother's Day, Valentine's Day, Friendship Day, etc.). That market is predicted to grow over the next several years (Imarcgroup 2019). In addition to its decorative use, several scholars have shown that indigenous communities in India also consume flowers.

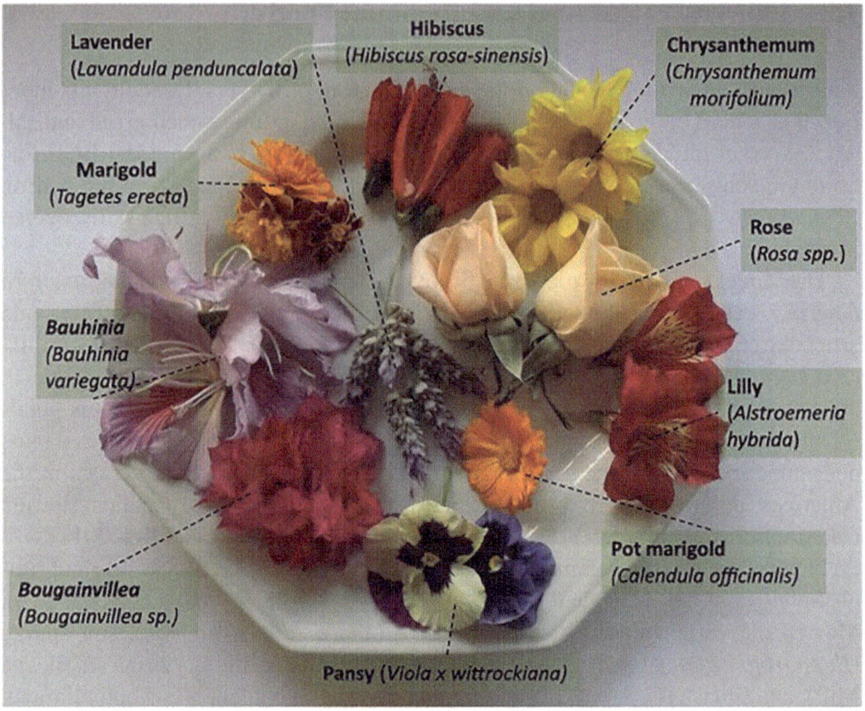

Fig. 7.1 List a few edible flowers together with their common and scientific names. (Takahashi et al. 2020)

Pinakin et al. (2020) evaluated edible flowers in India (2020). According to the writers, tree blooms have a significant historical significance in the nation and are connected to immortality. Because of this, the tribal group uses various tree sections for diverse purposes. Tribal people use *Rhododendron arboretum, B. variegata, Delonix,* and *Moringa oleifera regia* as food and medicine because of their many health advantages (Fig. 7.1).

7.6 Positive Aspects of Health of Edible Flowers

Numerous intriguing health-promoting qualities, such as anticancer, antimicrobial, neuroprotective, anti-diabetic, hepatoprotective, anti-obesity, and hypoglycemia, are present in edible flowers (Kaisoon and Konczak 2012, Skrajda-Brdak et al. 2020, Petrova and Petkova 2016). These effects are mainly occurring sure to the presence of numerous phytochemicals found in these flowers (Fig. 7.2).

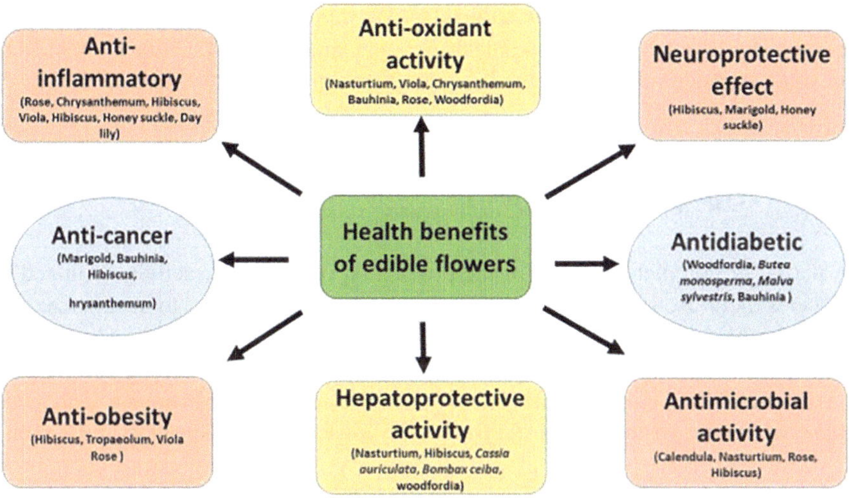

Fig. 7.2 Health benefits of edible flowers

7.6.1 *Cardioprotective*

Globally, cardiac conditions are said to be a leading contributor to a higher death rate. According to Wang et al. (2014), the hazard of cardiac collapse starts with coronary artery stenosis and progresses to myocardial necrosis brought on by acute chronic ischemia and hypoxia. Angina and myocardial infarction (MI), also known as ischemic heart disease, is caused by a deprivation of blood flow to the heart's cells, which leads to cell death from coronary artery blockage. According to Parcha et al. (2017), this typically occurs because of thrombus development over an atherosclerotic plaque that has dissected, ulcerated, or hemorrhage. According to Micucci et al. (2015), hypertension is another long-term medical condition that increases the risk of peripheral artery disease, cardiac arrest, ischemic stroke, and ruptured aorta. Research shows that controlling hypertension and rehabilitation of the ischemic myocardium can effectively treat ischemic heart disease. Through various methods, phytochemicals derived from edible flowers have helped reduce illness and modify cardiac risk factors.

The increased blood marker enzymes of myocardial infarction in rats treated with isoproterenol were considerably decreased by ethanolic extracts of Tecoma stans flowers. Additionally, these extracts have stopped the decline of catalase, and superoxide dismutase, and reduced glutathione levels, all of which serve as cardioprotectants (Ittagi et al. 2014). According to a study by Wang et al. (2014), aqueous extracts of lavender have been shown to significantly improve the resume of ischemia/reperfusion (I/R) altered cardiac function in rats by affecting the maximum rate of left ventricular pressure, the reduction of pressure developed in the left ventricle, and the coronary flow reduced by I/R injury. Additionally, the superoxide anion dismutase activity, malondialdehyde level, and cardioprotection content were all significantly modulated by the lavender extract treatment (Wang et al. 2014).

7.6.2 Hypoglycaemic Effects

A partial or complete absence of the production of insulin or activity is linked to diabetes mellitus, a metabolic disorder. Due to the body's failure to metabolize glucose, the blood sugar level rises. Extracts from edible flowers can reduce blood glucose levels that are increased when compared to glibenclamide and insulin. (Arya et al. 2015; Kumar et al. 2015). When Streptozotocin-nicotinamide-induced diabetic rats are given *Woodfordia fructicosa* flower extract, their blood insulin and C-peptide levels rise (Arya et al. 2015). Malva sylvestris herbal extract suppressed both α-amylase and α-glucosidase, with an IC50 of 7.8 and 11.3 μg/mL, respectively. (Loizzo et al. 2016). When compared to the common medication glibenclamide, ethanolic extracts of Catharanthus roseus leaves and flowers showed a dose-dependent reduction in blood sugar levels. The increased hepatic glucose utilization was the cause of the hypoglycemic effect (Mishra and Verma 2017). Since Butea monosperma flower water-soluble extract greatly improves Saccharomyces cerevisiae's ability to absorb glucose, it holds promise for the development of antidiabetic drugs (Khan and Gupta 2017). Additionally, an Ethanolic extract of Bauhinia variegata flowers was shown to have antidiabetic benefits by Tripathi et al. (2019). The Extract of rose plant stimulated the protein kinase B (AKT)/ phosphatidylinositol 3-kinase (PI3K) signaling route and showed an increased antioxidant reaction in addition to glucose tolerance and high insulin resistance, according to studies conducted on rats having diabetes by Liu et al. (2017).

Liu et al. (2017) displayed that the flower bud extract of Lonicera japonica exhibited more inhibiting properties against α-glucosidase and α-amylase in vitro. In rats with diabetes caused by streptozotocin STZ, there was a reduction in blood oxidative stress and lipid metabolism, in addition to a decrease in glucose levels resistance of insulin. It has been proven that the rich rutin flowers of Cichorium intybus and Robinia pseudo-acacia block α-amylase and α-glucosidase in a dose-dependent way (Loizzo et al. 2016).

7.6.3 Anti-cancerous Effects

Edible flowers showed potent effects against a lot of cancer types. Recent studies have shown the therapeutic value of *Bauhinia tomentosa* blossoms and extracts of Tagetes erecta petal in opposition to liver lines of cancer cells, namely MCF-7 and HePG2, using the MTT (3-(4,5-dimethylthiazol-2-yl)-2,5-diphenyltetrazolium bromide) assay (Solomon et al. 2016). Nowadays, everyone is concerned about skin conditions including photoaging and skin cancer. In light of this, Sun et al. (2016) conducted research utilizing extract from wild chrysanthemums and demonstrated its potential to prevent photoaging and cell death. According to the authors, skin disease may be prevented by the bioactive chemicals found in wild chrysanthemum extract. The anticancer potential of Hibiscus sabdariffa, a plant rich in flavonoids, has been thoroughly investigated. Its potential for toxicity on T47D cells of breast cancer was observed by researchers (Kaulika and Febriansah 2019). As mentioned previously, a small number of research also demonstrated that edible rose blooms had anticancer properties (Nanda 2019). All this research showed that edible flowers could be a more effective way to get anticancer drugs.

7.6.4 Antioxidant Activity

The physiological processes that take place inside the human body result in the production of free radicals. Because they are extremely reactive compounds, they cause lipids, proteins, and ultimately DNA to be destroyed oxidatively. It is known that edible flowers contain a variety of beneficial compounds, such as well-known antioxidants such as carotenoids, phenolics, and vitamins (Chensom et al. 2019). Because of their antioxidant activity, which positively impacts persistent diseases including obesity, heart disease, hyperglycemia, and neurodegenerative diseases, natural food sources that contain phenolics are extremely important (Dantas et al. 2019). *Rosa rugosa, Paeonia suffruticosa*, and *Paeonia lactiflora* extract showed greater DPPH (2,2-diphenyl-1-picrylhydrazyl) free radicals shielding ability (94.221 ± 0.102; 93.739 ± 0.424, and 94.244 ± 0.163%, respectively), according to Zeng et al. (2014). Additionally, they discovered a favorable association between these flowers' strong antioxidant activity and polyphenolic content. The antioxidant qualities of Nasturtium officinale lower cellular lipid peroxidation, and superoxide anion as well as shielding against free radicals. Additionally, it has been demonstrated that frequent consumption of it lowers DNA damage and raises the level of antioxidants in the blood of individuals who are healthy (Chaudhary et al. 2018). Delonix regia's phytochemicals are the best for treating antioxidants, and the plant's ethyl acetate extract has strong antioxidant properties (Ramakrishnan 2018). With a 99.53% removal of radicals' activity, the aqueous extract of Rosa Macdub shows strong antioxidant ability. The abundance of flavonoid and phenol compounds is the cause of this. Nonetheless, *Clitoria ternatea* floral ethanol extract showed 90.29% antioxidant activity (Nanda 2019). The dark-colored Ashwini variation has an antioxidant act level of 512.71 μmol Trolox/g, according to studies carried out on 50 Indian rose varieties (Kumari et al. 2017). Additionally, the scientists discovered a positive relationship between the antioxidant capacity of the rose extract and its anthocyanin content. Because the bristly locust blossom (Robinia hispida) contains anthocyanins and flavonoid elements, it has been shown to have potent antioxidant qualities. (Hallmann 2020). Similarly, myrcetin, quercetin, and gallic acid were found in *Madhuca longifolia* flowers, demonstrating potent antioxidant qualities (Singh et al. 2020).

7.6.5 Anti-inflammatory

The immune system is activated by the biochemical mechanism of inflammation (Lasselin and Capuron 2014). It is a vital physiological response that shields the host against invasive diseases and toxins. First, come acute inflammatory signs, then chronic ones, and finally the host's wellness is irreparably damaged. Honeysuckle (Liu et al. 2020), chrysanthemum (Ryu et al. 2019), day lily, roselle (El Bayani et al. 2018), and Portulaca oleracea (Miao et al. 2019) are among the edible flowers that have been found to contain potent protection against inflammation. A cell modeling system was used to assess this feature of the edible floral extracts. The Hibiscus sabdariffa having methanolic extract has been found to have possible anti-inflammatory properties when administered at a dose of 500 milligrams per kilogram of body weight. Rats' plasma and hippocampus IL-1β/IL-1ra ratios were maintained at the same levels. Therefore, defects in memory for spatial consolidation may be evaluated by giving 500 mg per kilogram of body mass of Hibiscus sabdariffa extraction (El Bayani et al. 2018). A significant drop in IL-4 levels was noted following a 200 mg/kg body weight dose of pansy extract. Compared to dexamethasone, asthma symptoms, eosinophil infiltration, and peribronchial inflammation are all successfully decreased (Harati et al. 2018).

7.6.6 Anti-obesity

The accumulation of excess fat, lipid synthesis, and energy imbalance are the causes of obesity, which is associated with adipogenesis and a rise in health-related problems. Obesity can be overcomed using bioactive compounds and elements. According to studies, Hibiscus sabdariffa's bioactive compounds decrease adipogenesis, prevent fat accumulation, and appear to lower body weight, making them useful in the fight against obesity (Ojulari et al. 2019). Tropaeolum majus extract, at a dose of 300 μg/mL, was able to effectively inhibit the synthesis of molecules that are used for the regulation of adipogenesis and lipogenesis to assess buildup of lipids and triglycerides content in 3 T3-L1 adipocytes (Kim et al. 2017). It was shown that Portulaca oleracea methanolic extract has anti-obesity properties because it reduces fat buildup and glucose intake (Lee et al. 2019a, b).

7.6.7 Anti-Alzheimer's Effect

Neurological disorders, such as cerebrovascular and neurodegenerative illnesses, are among the prevalent health concerns that have lately received attention. Among these, Alzheimer's disease is frequently found in older people. According to Castillo-Bautista et al. (2019), oxidative stress, senile plaque deposition, cholinergic dysfunction, neurofibrillary tangle formation, cerebral oxidative stress, neuro-inflammation, and cholinergic insufficiency are some of the variables that contribute to this kind of dementia. Research indicates that the primary cause of Alzheimer's disease is the deposition of amyloid-beta protein and the loss of cholinergic synapses in the neocortex and hippocampal regions. Acetylcholinesterase is also implicated in the progression of Alzheimer's disease. According to research, bioactive substances derived from plant extracts may help alleviate the symptoms of Alzheimer's disease, and there is confirmation that edible flowers include a range of bioactives that may help treat the condition.

According to Jung et al. (2010), chemicals isolated from *N. nucifera* stamens inhibited acetylcholinesterase (AchE) and butylcholinesterase (BchE) to provide a main anti-Alzheimer's action. Vanilloloside, p-hydroxybenzoic acid, and cycloartenol are among the substances found in these stamens. Likewise, *P. granatum* extracts have demonstrated an anti-Alzheimer effect by inhibiting AchE and BchE (Bekir et al. 2016). Rosa damascene extracts have been shown to enhance spatial learning and memory characteristics, increase the expression of neurotrophic genes, induce adult neurogenesis in the adult hippocampus, and support synaptic plasticity and remodeling (Esfandiary et al. 2014).

7.7 Flowers with Other Biological Activities

A wide variety of Edible flowers have long been utilized for medical purposes. For instance, roses were once used to treat circulatory issues, reduce menstrual pain, and, more recently, prevent cancer cells from forming (Yang and Shin 2017). According to other studies, edible flowers can help treat diabetes (Morittu et al. 2019), Alzheimer's disease, bacteria, and fungus. The antioxidant qualities of several of their constituents, such as proteins, lipids, carbohydrates, and DNA, are intimately linked to these biological consequences. Consuming foods high in antioxidants has been linked to a reduction in a lot of illnesses, such as diabetes, CVDs, lung conditions, and neurological disorders.

Certain edible flower species have indeed shown antioxidant activity in vitro using a variety of methods, including a test called the "Ferric Reducing Antioxidant Power" (FRAP) (Bao et al. 2018; Zhang et al. 2019; Wang et al. 2014). Whether or whether they are mixed, polar solvents including methanol, ethanol and water are the most often mentioned solvents for removing antioxidant substances from blossoms. The powerful polarities of specific phenolic chemicals, which are often molecules with a large number of hydroxy or even glycosyl groups within their molecular makeup, are linked to the solvent selection of polar substances. Rao and Poonia (Rao and Poonia 2024a, b) studied the microbial biosynthesis of selenium nanoparticles using probiotic strain and reported the major functional groups identified by FTIR analysis that includes conjugated ketones (C=O stretching) at 1648.82 cm^{-1} to 1688.54 cm^{-1}, halo compounds (C–Br stretching) at 448.85 cm^{-1} to879.54 cm^{-1}, and fluoro compounds (C–F stretching) at 668.10 cm^{-1} to 1065.89 cm^{-1}. The various findings reported for various plants might be related to the production process of the extracts or their concentrations of bioactive chemicals. Since many chemicals are thermosensitive, the typical extraction temperature is 25 °C, while in certain instances, higher temperatures were effectively employed (Lovrić et al. 2017). Ohmic accelerated steam distillation was one of the new methods that allowed for faster extraction without compromising bioactivity (Gavahian and Chu 2018).

Since many infectious illnesses have a high societal burden worldwide, eatable flowers have previously been studied as antibacterial agents to create new medicines that will benefit society. Accordingly, *S. nigra* flower extracts tested against *Salmonella poona*, Bacillus cereus, Pseudomonas aeruginosa, and Staphylococcus sp. shown in culture antibiotic act. While dahlia and rose, flowers decreased tumor cell growth in the cervix and hepatocellular carcinoma, respectively, B. variegata flowers demonstrated toxicity against carcinoma cells of the cervical (Villavicencio et al. 2017; Pires et al. 2017).

7.8 Toxic and Poisonous Compounds

Even while edible flowers have been used for generations in cooking, they may contain harmful substances that, when digested, might hurt human health (Purohit et al. 2021). However, these chemicals are often eliminated, or their amount is reduced below the safety criteria by standard home cooking methods based on traditional knowledge. Additionally, using edible flowers as a garnish in a variety of cuisines reduces the possibility of excessive consumption and harmful side effects. Recent developments in analytical chemistry have made it possible to identify novel chemicals and assess their harmful effects on people. Among these substances are oxalic acid, cyanogenic glucosides, hemagglutinins, alkaloids, and trypsin inhibitors. Glykosides, alkaloids, terpenoids, terpenes, and inhibitory enzymes are among the toxic, poisonous, or antinutritional components found in a variety of ornamental plants, according to a recent analysis (Piovesana et al. 2019). Consequently, it is recommended that a thorough examination of the chemical composition be carried out to detect any possible harmful substances or food allergies, given the large number of species that are frequently used for their edible flowers and the genotypic variations among the different cultivars within the same species (Matyjaszczyk et al. 2019). Furthermore, the source of edible flowers is crucial because the excessive use of agrochemicals in commercial cultivation systems may raise the risk of high concentrations of hazardous compounds, and environmental contamination (such as heavy metals in contaminated soils) or external impurities should also be considered for safety and legal purposes. Commercial production of species to harvest flowers as the end product is necessary to meet the demand of a niche market formed by the growing interest in culinary items that contain edible flowers (Matyjaszczyk et al. 2019).

Considering the local or regional passion for many species, another problem with the safe use of edible flowers is that the information that is currently available about the precautions and any processing that may be required before consumption is only available in languages other than English. As a result, scientists should compile and update this dispersed knowledge and provide it in widely recognized languages (Kalemba-Drożdż 2018). Additionally, certain species-particularly those found in the wild-may have many common names, and other species with distinct chemical profiles may share a common name, thus leading to misidentification (Colombo et al. 2010). Therefore, even for known species, it is crucial to examine the chemical makeup of novel genotypes before recommending them for use in food.

7.9 Family Farming and Local Food Trends

Regional cuisine is popular worldwide because it lowers greenhouse gas emissions and logistical costs while also making food producer transparency, traceability, and quality control easier. Edible flowers may benefit greatly from this industry while boosting the regional economy, enhancing traveling, increasing taxation income, and creating work possibilities (Natcvetova 2019).

Because of the unique qualities of each species, they are also regarded as specialty items and offer a chance to diversify family agricultural revenue (Garrison 2019; Givens and Dunning 2018). Because of their increased spoilage, flowers require a short chain of food, due to which tiny and middle-sized producers are still struggling to maintain quality requirements because of the high costs of shipping and distributing edible flowers. (turgidity, scent, and color). North Carolina State University (USA) launched an intriguing project on this topic that encouraged collaboration between regional farmers and food services to create a farm-to-restaurant chain of value with lower intermediary distribution expense (Givens and Dunning 2018).

The increasing quantity of studies on the dietary and bioactive substances included in edible blossoms, which is connected to the rise in the quantity of culinary flower dishes, publications, and blogs that the public sees, has made the market more substantial and promising. International magazines like The New York Times "Cooking" section include dishes that use edible flowers like lavender (a drink called "Lavender Hill Cob"), a cocktail called "Brooklyn Sunrise" that combines botanical flavors, and a dessert called "Berry tiramisu" (Times 2019). Additionally, palatable florals are the subject of undergraduate culinary school finalized projects, for example, Ebla and Estefanía (2019) devised a variety of recipes to promote the use of edible flowers and fruits.

According to Howell (2019), the establishment of an outdoor space at the Manzo Elementary School to promote sentiments of belonging and well-being was another intriguing flower-growing event. Considering that mental health issues that include depression, anxiety, and eating disorders are on the rise today, flower gardens might be implemented in a variety of settings, including workplaces, colleges, and schools, to foster well-being. Similar to this, agroforestry-inspired edible forest gardens with a variety of plant species and sizes help to create peaceful, pleasant spaces in both urban and rural areas while also supplying fresh products for eating all during the growing season, including leaves, seeds, blossoms, fruit, produce, and nuts (Björklund et al. 2018).

References

Al-Snafi A (2015) The pharmacological importance of Bellis perennis – a review. Int J Phytother 5:63–69

Antarkar S, Sharma A, Bhargava A, Gupta H, Tomar R, Srivastava S (2019) Physico-chemical and nutritional evaluation of cookies with different levels of rosehip and hibiscus powder substitution. Arch Curr Res Int:1–10

Arya A et al (2015) Extract of Woodfordia Fruticosa flowers ameliorates hyperglycemia, oxidative stress and improves B-cell function in Streptozotocin–nicotinamide induced diabetic rats. J Ethnopharmacol 175:229–240

Bao Y, Qu Y, Li J, Li Y, Ren X, Maffucci K, Li R, Wang Z, Zeng R (2018) In vitro and in vivo antioxidant activities of the flowers and leaves from Paeonia rockii and identification of their antioxidant constituents by UHPLC-ESI-HRMSn via pre-column DPPH reaction. Molecules 23(2):392

Bayani GFE, Marpaung NLE, Simorangkir DAS, Sianipar IR, Ibrahim N, Kartinah NT, Mansur IG, Purba JS, Ilyas EII (2018) Anti-inflammatory effects of hibiscus sabdariffa Linn. On the IL-1β/IL-1ra ratio in plasma and hippocampus of Overtrained rats and correlation with spatial memory. Kobe J Med Sci 64(2):E73–E83

Bekir J, Cazaux S, Mars M, Bouajila J (2016) In vitro anticholinesterase and anti-hyperglycemic activities of flowers extracts from seven pomegranate varieties. Ind Crop Prod 81:176–179

Benvenuti E, Bortolotti R (2016) Maggini antioxidant power, anthocyanin content and organoleptic performance of edible flowers. Sci Hortic 199:170–177

Björklund J, Eksvärd K, Schaffer C (2018) Exploring the potential of edible forest gardens: experiences from a participatory action research project in Sweden. Agrofor Syst 93(3):1107–1118

Castillo-Bautista CM, Torres-Tapia LW, Rangel-Méndez JA, Peraza-Sánchez SR, Cortés D, Velasco I, Moo-Puc RE (2019) Neuroprotective effect of Mayan medicinal plant extracts against glutamate-induced toxicity. J Nat Med 73:672–678

Chaudhary S, Hisham H, Mohamed D (2018) A review on phytochemical and pharmacological potential of watercress plant. Asian J Pharm Clin Res 11(12):102–107

Chen N, Wei S (2016) Factors influencing consumers' attitudes towards the consumption of edible flowers. Food Qual Prefer 56:93–100

Chen G, Chen S, Xiao Y, Fu N (2017) Antioxidant capacities and total phenolic contents of 30 flowers. Ind Crop Prod 111:430–445

Chensom S, Okumura H, Mishima T (2019) Primary screening of antioxidant activity, Total polyphenol content, carotenoid content, and nutritional composition of 13 Edible flowers from Japan. Prevent Nutrit Food Sci 24(2):171–178

Chitrakar B, Zhang M, Bhandari B (2019) Edible flowers with the common name "marigold": their therapeutic values and processing. Trends Food Sci Technol 89:76–87

Colombo ML, Assisi F, Puppa TD, Moro P, Sesana FM, Bissoli M, Borghini R, Perego S, Galasso G, Banfi E et al (2010) Most commonly plant exposures and intoxications from outdoor toxic plants. J Pharm Sci Res 2:417–425

Dantas AM, Mafaldo IM, Oliveira PML, Lima MDS, Magnani M, Borges GDSC (2019) Bioaccessibility of phenolic compounds in native and exotic frozen pulps explored in Brazil using a digestion model coupled with a simulated intestinal barrier. Food Chem 274:202–214

Deka K, Nath N (2014) Documentation of edible flowers of Western Assam. Am J Phytomed Clin Ther 2(10):1124–1140

Demarne F-E, van der Walt JJA (1990) Pelargonium tomentosum: a potential source of peppermint-scented essential oil. S Afr J Sci 7(1):36–39. https://doi.org/10.1080/02571862.1990.1063453

Dourson D, Dourson J (2019) Wild yet tasty: a guide to edible plants of eastern Kentucky. University Press of Kentucky

Ebla S, Estefanía C (2019) Creación de un recetario con el uso de flores comestibles pertenecientes a la clasificación de tubérculos y árboles frutales. Bachelor's thesis, Quito: Universidad de las Américas

Esfandiary E, Karimipour M, Esmaeili A (2014) Novel effects of Rosa damascena extract on memory and neurogenesis in a rat model of Alzheimer's disease. J Neurosci Res 92:517–530

FAOSTAT Food and Agriculture Organization of the United Nations Statistics Division (FAOSTAT) (2019)

Fernandes L, Casal S, Pereira JA, Saraiva JA, Ramalhosa E (2017) Edible f lowers: a review of the nutritional, antioxidant, antimicrobial properties and effects on human health. J Food Compos Anal 60:38–50

Fernandes L, Casal S, Pereira JA, Malheiro R, Rodrigues N, Saraiva JA, Ramalhosa E (2019a) Borage, calendula, cosmos, Johnny jump up, and pansy f lowers: volatiles, bioactive compounds, and sensory perception. Eur Food Res Technol 245(3):593–606

Fernandes L, Saraiva JA, Pereira JA, Casal S, Ramalhosa E (2019b) Post-harvest technologies applied to edible flowers: a review: Edible flowers preservation. Food Rev Intl 35(2):132–154

Garrison H (2019) Food studies reimagined: putting experience on the page. Gastronom J Food Cult 19(2):65–70. https://doi.org/10.1525/gfc.2019.19.2.65

Gavahian M, Chu Y (2018) Ohmic accelerated steam distillation of essential oil from lavender in comparison with conventional steam distillation. Innovative Food Sci Emerg Technol 50:34–41

Givens G, Dunning R (2018) Distributor intermediation in the farm to food service value chain. Renew Agric Food Syst 34(03):268–270

Gomes S, Belo AT, Alvarenga N, Dias J, Lage P, Pinheiro C et al (2019) Characterization of *Cynara cardunculus* L. flower from Alentejo as a coagulant agent for cheesemaking. Int Dairy J 91:178–184

Grzeszczuk M, Grzeszczuk M, Meller E, Stefaniak A, Wysocka G (2017) Mineral composition of some edible flowers. J Elem

Gutiérrez-Grijalva EP, Ambriz-Pére DL, Heredia JB (2016) Review: dietary phenolic compounds, health benefits and bioaccessibility. Archivos Latinoamericanos De Nutricion 66:87–100

Hallmann. (2020) Quantitative and qualitative identification of bioactive compounds in edible flowers of black and bristly locust and their antioxidant activity. Biomolecules 10(12):1603

Harati E et al (2018) Effects of Viola tricolor flower hydroethanolic extract on lung inflammation in a mouse model of chronic asthma. Iran J Allergy Asthma Immunol:409–417

Harmayani E, Anal AK, Wichienchot S, Bhat R, Gardjito M, Santoso U, Siripongvutikorn S, Puripaatanavong J, Payyappallimana U (2019) Healthy food traditions of Asia: exploratory case studies from Indonesia, Thailand, Malaysia, and Nepal. J Ethnic Foods 6

Howell JA (2019) Creating sustainable spaces: a school garden case study. Doctoral dissertation, The University of Arizona

Imarcgroup (2019) Indian floriculture market: industry trends, share, size, growth, opportunity and forecast 2019–2024

Ittagi S, Merugumolu VK, Siddamsetty RS (2014) Cardioprotective effect of the hydroalcoholic extract of Tecoma stans flowers against isoproterenol-induced myocardial infarction in rats. Asian Pacific J Trop Dis 4

Juárez-Trujillo N, Monribot-Villanueva JL, Jiménez-Fernández VM et al (2018) Phytochemical characterization of Izote (Yucca elephantipes) flowers. J Appl Bot Food Qual 91:202–210

Jung HA, Jung YJ, Hyun SK, Min B-S, Kim D-W, Jung JH, Choi JS (2010) Selective cholinesterase inhibitory activities of a new monoterpene diglycoside and other constituents from Nelumbo nucifera stamens. Biol Pharm Bull 33:267–272

Kaisoon I, Konczak S (2012) Siriamornpun. Potential health enhancing properties of edible flowers from Thailand. Food Res Int 46(2):563–571

Kalemba-Drożdż M (2018) Comment on article "are wild and cultivated flowers served in restaurants or sold by local producers in Denmark safe for the consumer?". Food Chem Toxicol 120:129–142. https://doi.org/10.1016/j.fct.2018.07.007

Kanwar P, Gupta YC, Dhiman SR, Dogra RK, Sharma Madhu, Bansal M (2016) Heterobeltiosis in French Marigold (Tagetes patula L.). J Agric Sci 8(8):2016. https://doi.org/10.5539/jas.v8n8p182

Katisart T, Konsue A (2019) Acute toxicity of flower extracts from Dolichandrone serrulata in mice. Pharm Res 11(3):230

Kaulika N, Febriansah R (2019) Chemopreventive activity of roselle's hexane fraction against breast cancer by in-vitro and in-silico study. In: Third international conference on sustainable innovation 2019–health science and nursing (IcoSIHSN 2019). Atlantis Press. https://doi.org/10.2991/icosihsn-19.2019.16

Khan S, Gupta SA (2017) Toxicology of the aqueous extract from the flowers of Butea monosperma lam. and it's metabolomics in yeast cells. Food Chem Toxicol 108:486–497

Kim G-C et al (2017) Anti-adipogenic effects of Tropaeolum Majus (nasturtium) ethanol extract on 3T3-L1 cells. Food Nutr Res 61(1):1339555

Kirker L, Newman M (2016) Edible flowers: a global history, 1st edn. Reaktion Books, London

Kumar S, Janadri S, Kumar S (2015) Evaluation of antidiabetic activity of alcoholic extract of flower *Sesbania grandiflora* in alloxan induced diabetic rats. Asian J Pharm Pharmacol 1(1):21–26

Kumari P, Raju DVS, Prasad KV et al (2017) Quantification and correlation of anthocyanin pigments and their antioxidant activities in rose (*Rosa hybrida*) varieties. Indian J Agric Sci 87(10):1340–1346

Lara-Cortés E, Troncoso-Rojas R, Hernández-López M, Bautista-Baños S (2016) Evaluation of the antimicrobial activity of cinnamaldehyde in the preservation of edible dahlia flowers, under different storage conditions. Revista Chapingo. Serie Hortic 22(3):177–189

Lasselin J, Capuron L (2014) Chronic low-grade inflammation in metabolic disorders: relevance for behavioral symptoms. Neuroimmunomodulation 21(2–3):95–101

Lee J, Park G, Chang YH (2019a) Nutraceuticals and antioxidant properties of Lonicera japonica Thunb. as affected by heating time. Int J Food Prop 22(1):630–645

Lee JI et al (2019b) Evaluation of anti-adipogenic active homoisoflavonoids from portulaca oleracea. Zeitschrift Für Naturforschung C 74(9–10):265–273

Liang Z-X, Zhang J-Z, Sun M-Y, et al (2018) Variation of phenolic compounds and antioxidant capacities in different organs of lilium pumilum. Nat Prod Commun 13(6). https://doi.org/10.1177/1934578X1801300616

Li Y, Li L, Lund MN, Li B, Hu Y, Zhang X (2018) Reduction of Nε-(carboxymethyl) lysine by (−)-epicatechin and (−)-epigallocatechin gallate: The involvement of a possible trapping mechanism by catechin quinones. Food Chem 266:427–434

Li Y, Sun H, Li J, Qin S, Niu Z, Qiao X, Yang B (2021) Influence of genetic background, growth latitude and bagging treatment on phenolic compounds in fruits of commercial cultivars and wild types of apples (Malus sp.). Eur Food Res Technol 247(5):1149–1165

Liu L, Tang D, Zhao H, Xin X, Aisa HA (2017) Hypoglycemic effect of the polyphenols rich extract from Rose rugosa Thunb on high fat diet and STZ induced diabetic rats. J Ethnopharmacol 200:174–181

Liu M, Yu Q, Yi Y et al (2020) Antiviral activities of Lonicera japonica Thunb. components against grouper iridovirus in vitro and in vivo. Aquaculture 519:734882. https://doi.org/10.1016/j.aquaculture.2019.734882

Lockowandt L, Pinela J, Roriz CL, Pereira C, Abreu RM, Calhelha RC et al (2019) Chemical features and bioactivities of cornflower (Centaurea cyanus L.) capitula: the blue flowers and the unexplored non-edible part. Ind Crop Prod 128:496–503

Loizzo MR, Pugliese A, Bonesi M, Tenuta MC, Menichini F, Xiao J, Tundis R (2016) Edible flowers: a rich source of phytochemicals with antioxidant and hypoglycemic properties. J Agric Food Chem 64(12):2467–2474

Lovrić V, Putnik P, Kovačević DB, Jukić M, Dragović-Uzelac V (2017) The effect of microwave-assisted extraction on the phenolic compounds and antioxidant capacity of blackthorn flowers. Food Technol Biotechnol 55(2)

Lu Q, Wang S, Xue S, Yang D, Li L (2020) Effects of drying methods on phenolic components in different parts of Chrysanthemum morifolium flower. J Food Process Preserv 44(12):e14982

Matyjaszczyk E, Śmiechowska M, Edible flowers. (2019) Benefits and risks pertaining to their consumption. Trends Food Sci Technol 91:670–674

Miao L et al (2019) The anti-inflammatory potential of Portulaca oleracea L. (purslane) extract by partial suppression on NF-κB and MAPK activation. Food Chem 290:239–245

Micucci M, Malaguti M, Gallina Toschi T, Di Lecce G, Aldini R, Angeletti A, Chiarini A, Budriesi R, Hrelia S (2015) Cardiac and vascular synergic protective effect of Olea europea L. leaves and Hibiscus sabdariffa L. flower extracts. Oxidative Med Cell Longev 2015(1):318125

Mishra JN, Verma NK (2017) A brief study on Catharanthus roseus: a review. Int J Pharm Sci Res 2(2):20–23

Mlcek J, Rop O (2011) Fresh edible flowers of ornamental plants-a new source of nutraceutical foods. Trends Food Sci Technol 22:561–569. https://doi.org/10.1016/j.tifs.2011.04.006

Morittu VM, Musco N, Mastellone V, Bonesi M, Britti D, Infascelli F, Loizzo MR, Tundis R, Sicari V, Tudisco R, Lombardi P (2019) In vitro and in vivo studies of Cucurbita pepo L. flowers: chemical profile and bioactivity. Nat Prod Res 35(17):2905–2909. https://doi.org/10.1080/14786419.2019.1672067

Murphy H (2015) Foods indigenous to the Western hemispher. American Indian Health Diet Project

Nanda BL (2019) Antioxidant and anticancer activity of Edible flowers. J Drug Delivery Ther 9(3-s):290–295

Nadumane VK, Timsina B (2014) Anti-cancer potential of banana flower extract: An in vitro study. Bangl. J Pharmacol 9(4):628–635

Natcvetova A (2019). Impacts of local food on nutrient cycling in Finland. Bachelor's thesis, Tampere University of Applied Sciences, Tampere

Navarro-González I, González-Barrio R, García-Valverde V, Bautista-Ortín A, Periago M (2015) Nutritional composition and antioxidant capacity in edible flowers: characterisation of phenolic compounds by HPLC-DAD-ESI/MSn. Int J Mol Sci 16:805–822

Nguyen C, Baskaran K, Pupulin A, Ruvinov I, Zaitoon O, Grewal S et al (2019) Hibiscus flower extract selectively induces apoptosis in breast cancer cells and positively interacts with common chemotherapeutics. BMC Complement Altern Med 19(1):98

Nowicka P, Wojdyło A (2019) Anti-hyperglycemic and anticholinergic effects of natural antioxidant contents in edible flowers. Antioxidants 8(8):308

Ojulari OV et al (2019) Beneficial effects of natural bioactive compounds from hibiscus sabdariffa L. on obesity. Molecules 24(1):210

Parcha V, Yadav N, Sati A, Dobhal Y, Sethi N (2017) Cardioprotective effect of various extract of rhododendron arboriumsm flower on albino rats. J Pharma Phytochem 6(4):1703–1707

Petrova N, Petkova I (2016) Ivanov five edible flowers–valuable source of antioxidants in human nutrition. Int J Pharma Phytochem Res 8(4):604–610

Pinakin DJ, Kumar V, Kumar S, Kaur S, Prasad R, Sharma BR (2020) Influence of pre-drying treatments on physico-chemical and phytochemical potential of dried mahua flowers. Plant Foods Hum Nutr 75(4):576–582

Piovesana A, Rodrigues E, Noreña CPZ (2019) Composition analysis of carotenoids and phenolic compounds and antioxidant activity from hibiscus calyces (Hibiscus sabdariffa L.) by HPLC/DAD/MS/MS. Phytochem Anal 30(2):208–217

Pires TC, Dias MI, Barros L, Ferreira IC (2017) Nutritional and chemical characterization of edible petals and corresponding infusions: valorization as new food ingredients. Food Chem 220:337–343

Pires TC, Dias MI, Barros L, Calhelha RC, Alves MJ, Oliveira MBP, Ferreira IC (2018) Edible flowers as sources of phenolic compounds with bioactive potential. Food Res Int 105:580–588

Purohit SR, Rana SS, Idrishi R, Sharma V, Ghosh P (2021) A review on nutritional, bioactive, toxicological properties and preservation of edible flowers. Futur Foods 4:100078

Qiu L, Zhang M, Ju R, Wang Y, Chitrakar B, Wang B (2019) Effect of different drying methods on the quality of restructured rose flower (Rosa rugosa) chips. Dry Technol:1–12

Ramakrishnan RB (2018) Evaluation of antioxidant and phytochemical activity in solvent extracts from Delonix regia flowers. Int J Green Pharm 12(03)

Rao V, Poonia A (2024a) Bioactive compounds, nanoparticle synthesis and potential utilization of edible flowers for development of functional dairy products: a review. J Food Sci Technol 61(6):1053–1068. https://doi.org/10.1007/s13197-023-05853-6

Rao V, Poonia A (2024b) Microbial biosynthesis of selenium nanoparticles using probiotic strain and its characterization. J Food Meas Charact 18(7):5477–5488. https://doi.org/10.1007/s11694-024-02581-z

Ryu J, Nam B, Kim BR, Kim SH, Jo YD, Ahn JW, Kim JB, Jin CH, Han AR (2019) Comparative analysis of phytochemical composition of gamma-irradiated mutant cultivars of *Chrysanthemum morifolium*. Molecules (Basel, Switzerland) 24(16):3003

Singh J, Singh R, Kushwaha M, Singh S, Kumar AKR (2020) Assessment of antioxidant activity, minerals and chemical constituents of edible mahua (*Madhuca longifolia*) flower and fruit of using principal component analysis. Nutrit Food Sci

Skrajda-Brdak M, Dąbrowski G, Konopka I (2020) Edible flowers, a source of valuable phytonutrients and their pro-healthy effects–a review. Trends Food Sci Technol 103:179–199

Solomon S, Muruganantham N, Senthamilselvi MM (2016) Anti-cancer activity of Bauhinia tomentosa (flowers) against human liver cancer. J Pharma Phytochem 5:287–290

Sotelo A, López-garcía S, Basurto-Peña F (2007) Content of nutrient and antinutrient in edible flowers of wild plants in Mexico. Plant Foods Hum Nutr 62:133–138

Stefaniak A, Grzeszczuk ME (2019) Nutritional and biological value of five edible flower species. Notulae Botanicae Horti Agrobotanici Cluj-Napoca 47(1):128–134

Stradley L (2015) Edible flowers are the new rage in haute cuisine, What's Cooking America website

Sun S, Jiang P, Su W, Xiang Y, Li J, Zeng L, Yang S (2016) Wild chrysanthemum extract prevents UVB radiation-induced acute cell death and photoaging. Cytotechnology 68(2):229–240

Tai CY, Chen BH (2000) Analysis and stability of carotenoids in the flowers of daylily (hemerocallis d isticha) as affected by various treatments. J Agr Food Chem 48(12):5962–5968

Takahashi JA, Rezende FAGG, Moura MAF, Dominguete LCB, Sande D (2020) Edible flowers: bioactive profile and its potential to be used in food development. Food Res Int 129:108868

The New York Times (2019) Cooking. https://cooking.nytimes.com/search?q=edible+flower. Accessed 13 June 2019

Tripathi AK, Gupta PS, Singh SK (2019) Antidiabetic, anti-hyperlipidemic and antioxidant activities of Bauhinia variegata flower extract. Biocatal Agric Biotechnol 19:101142

Villavicencio AL, Heleno SA, Calhelha RC, Santos-Buelga C, Barros L, Ferreira IC (2017) The influence of electron beam radiation in the nutritional value, chemical composition and bioactivities of edible flowers of Bauhinia variegata L. var. candida alba Buch.-Ham from Brazil. Food Chem 241:163–170

Vinokur Y, Rodov V, Reznick N, Goldman G, Horev B, Umiel N, Friedman H (2006) Rose petal tea as an antioxidant-rich beverage: cultivar effects. J Food Sci 71(1):42–47

Wang D, Guo X, Sun X (2014) Cardioprotective effect of the aqueous extract of the lavender flower against myocardial ischemia/reperfusion injury. J Chem. https://doi.org/10.1155/2014/368376

Yang H, Shin Y (2017) Antioxidant compounds and activities of edible roses (Rosa hybrida spp.) from different cultivars grown in Korea. Appl Biol Chem 60(2):129–136

Yang R, Guan Y, Wang W, Chen H, He Z, Jia AQ (2018) Antioxidant capacity of phenolics in Camellia nitidissima Chi flowers and their identification by HPLC Triple TOF MS/MS. PLoS One 13(4):e0195508. https://doi.org/10.1371/journal.pone.0195508

Yang PF, Yang YN, Feng ZM, Jiang JS, Zhang PC (2019) Six new compounds from the flowers of Chrysanthemum morifolium and their biological activities. Bioorg Chem 82:139–144

Zeng Y, Deng M, Lv Z, Peng Y (2014) Evaluation of antioxidant activities of extracts from 19 Chinese edible flowers. Springerplus 3(1):315

Zhang N, He Z, He S, Jing P (2019) Insights into the importance of dietary chrysanthemum flower (Chrysanthemum morifolium cv. Hangju)-wolfberry (Lycium barbarum fruit) combination in antioxidant and anti-inflammatory properties. Food Res Int 116:810–818

Chapter 8
Toxicological Effects of Edible Flowers

Anwesha Mukherjee, V. S. Anjana, Saranga Rajesh, M. Karthik, and Hannah R. Vasanthi

8.1 Introduction

For ages, edible flowers have been considered an important component of the human diet around the globe (Pires et al. 2018). Previously, they were known and eaten by the people of ancient Rome and Greece as an alternative medicine or medicinal food product in addition to the traditional eating of fruits, seeds, leaves, and roots of vegetables (Takahashi et al. 2020). Flowers are also used to enhance the attraction and appeal of dishes by contributing to their aesthetic aspects (Deepika et al. 2014). Edible flowers are consumed mainly because of their scent and appearance. However, in the modern world, it has received attention as people are looking for bioactive compounds from natural sources (Benvenuti and Mazzoncini 2021), owing to their health benefits for a healthy lifestyle (Pires et al. 2019).

The study on phytocompounds has shown that flowers contain many phytochemicals beneficial to human health (Gayathri et al. 2013). Their potential medicinal and health-promoting properties are conferred by their rich phytochemical profile, which includes several bioactive compounds such as polyphenols, flavonoids, carotenoids, anthocyanins, and antioxidants (Takahashi et al. 2020). The low-fat content and diverse array of bioactive principles in flowers are exploited in several pharmacological applications for managing diseases like cancer, cough, jaundice, cold, diabetes, cardiovascular diseases, pox, and microbial infections, mediated by its anti-inflammatory and antioxidant action. They also contain other bioactive compounds that act on various problems associated with the liver, prostate, bladder, colon, and breast (Purohit et al. 2021). The primary and secondary metabolites in the pollen, nectar, and corolla of the flowers help prevent problems related to nutritional deficiencies and aid in functional activities in humans (Benvenuti and

A. Mukherjee · V. S. Anjana · S. Rajesh · M. Karthik · H. R. Vasanthi (✉)
Department of Biotechnology, Pondicherry University, Puducherry, India

A. Poonia et al. (eds.), *Edible Flowers: Source of Phytonutrients, Valorization and Technological Advancements*, https://doi.org/10.1007/978-3-032-00667-7_8

Mazzoncini 2021). Across the world, there are numerous edible flowers, but only a few have been studied for their beneficial attributes. Some of the commonly used edible flowers around the globe and in India are tabulated in Tables 8.1 and 8.2, respectively.

Although edible flowers exhibit several beneficial effects, few flowers may not be suitable for consumption, which may have a high possibility of causing disorders such as hay fever, asthma, and allergies and can finally lead to death when taken by humans (Purohit et al. 2021). Plant toxicity can occur because of both endogenous and exogenous factors. The present article outlines the different causes for the toxicity of these edible flowers, the different types of toxicity affecting the various human organs, regulatory guidelines to check the toxicity of edible flowers, and the technologies to overcome the toxicities of edible flowers.

Table 8.1 Commonly consumed edible flowers in India

Scientific name	Common name	Biological potential	References
Musa paradisiaca Linn	Banana	Antioxidant, anti-cancer properties	Chowdary et al. (2022)
Dendranthema grandiflora	Chrysanthemums	Antidiabetic, anticancer properties	Patil et al. (2024)
Cucurbita maxima	Pumpkin	Anti-inflammatory, hypotensive, Hepato- protective, anti-diabetic, antioxidant, anti- microbial properties	Muruganantham et al. (2016)
Tamarindus indica	Tamarind	Antimicrobial activity, antioxidant activity	Kishore et al. (2022)
Hibiscus rosasinensis	Hibiscus	Anti-aging, Cardioprotective, anti-inflammatory and hypertensive, anti-Dyslipidemia, wound healing and insecticidal activity	Upadhyay (2017)
Jasminum officinale	Jasmine	Anti-microbial activity, Thrombolytic activity, analgesic activity	Thakur et al. (2024)
Rosa spp.	Rose	Antioxidant activity	Singh et al. (2023)
Crocus sativus	Saffron crocus	Antiparasitic, anti-bactericidal, antioxidant, hypotensive, hypolipidemic, anxiolytic, antidepressant, anticonvulsant, antinociceptive, anti-inflammatory, diuretic, cytotoxic	Sharma and Kumar (2022)
Nelumbo nucifera	Indian lotus	Anticancer, hepatoprotective, antioxidant, antiviral, hypolipidemic, anti-obesity, antipyretic, hypoglycaemic, antifungal, anti-inflammatory and antibacterial activities.	Sharma and Kumar (2022)
Moringa pterygosperma	Drumstick flower	Antioxidants, antifungal, antibacterial, anti-abortifacient, anti-inflammatory, antiulcer, diuretics activities	Kumar Bargah (2015)

(continued)

Table 8.1 (continued)

Scientific name	Common name	Biological potential	References
Tagetes erecta	Marigold	Anti-nociceptive, antiinflammatory, antioxidant, insecticidal, larvicidal, hepatoprotective, antipyretic, wound healing, antibacterial, antimicrobial, antiepileptic and antifungal	Singh et al. (2020)
Clitoria ternatea	Butterfly pea	Antimicrobial, antipyretic, anti-inflammatory, analgesic, diuretic, local anesthetic, antidiabetic, insecticidal, blood platelet aggregation-inhibiting	Mukherjee et al. (2008)
Raphanus sativus	Radish flower	Anticancer, anti-diabetics; antioxidants, hepatoprotective	Manivannan et al. (2019)
Butea monosperma	Palash	Hepatoprotective, antifertility, antifilarial,anti-diabetic, antiviral,anthelmintic, anticonvulsant, antifungal, antimicrobial, antiestrogenic, anticancer, antiinflammatory, antioxidant, antiulcer, wound healing, anti-diarrhoeal, anti-implantation, antidopaminergic, antimycobacterial, osteogenic and osteoprotective	Tiwari et al. (2019)
Syzygium aromaticum	Clove	Antibacterial, antifungal, herbicidal, nematicidal, antitumor and anti-inflammatory	Kaur and Kaushal (2019)
Abelmoschus esculentus	Okra flower	Antioxidant, antispasmodic; demulcent; diaphoretic; diuretic	Jain et al. (2012)

Table 8.2 Commonly consumed edible flowers around the globe

Scientific name	Common name	Origin	Biological potential	References
Camellia sinensis L.	Tea flower	Southeast Asia	Anti-cancer (anti- proliferative)	Esghaei et al. (2018)
B. perennis	Common daisy	Europe and Western Asia	Anti-inflammatory, rheumatism, anti-hyperglycemic, anti-aging, anti-bacterial, Anti-cancer	Albien and Stark (2023)
Calendula officinalis	Marigold	Mediterranean region and Central Europe	Antioxidant, Gastroprotective, Hepatoprotective, anti-inflammatory, sedative, antispasmodic, bactericidal. Treatment of, injuries, ulcers and skin's afflictions. Measles, smallpox	AshwlayanVD and Verma (2018), Albien and Stark (2023)

(continued)

Table 8.2 (continued)

Scientific name	Common name	Origin	Biological potential	References
Centaurea cyanus	Cornflower	Europe	Antioxidant, antimicrobial, antidiabetic, anti-obesity, anti-proliferative, Hepatoprotective, Cardioprotective and renal-protective Activity	Pires et al. (2018)
Antirrhinum majus	Snapdragon	South America	Antioxidant, antimicrobial, anticancer	Kumar (2022)
Lavendula angustifolia	Lavender	Mediterranean	Antioxidant and antimicrobial properties	Leong et al. (2021), Leong et al. (2021)
Tulipa spp.	Tulip	Central Asia	Antioxidant activity, antiradical capacity	Shoji et al. (2005)
Viola odorata	Violets	Europe, Asia, North America	Antioxidant, antimicrobial, thrombolytic, alpha-amylase inhibition and lipase inhibition potential	Ibraheem et al. (2018)
T. majus	Nasturtium, monk cress	South America (Peru)	Antimicrobial, antiseptic, purgative, depurative, diuretic emmenagogue, expectorant, laxative, stimulant, antitumor, antithrombotic, hypertensive, anti-viral, sore throat	Jurca et al. (2018)

8.2 Causes for Toxicity in Edible Flowers

The mild flavor and rich aroma of the edible flowers, added to their health benefits, have made them one of the preferred ingredients in different food items. However, thorough biochemical investigations have revealed potential toxicity associated with these floral additions to our diet. Numerous factors lead to the toxicity of edible flowers, including the source and origin of the flowers. Further, the presence of phytochemicals, pesticides, pollen allergens, natural toxins, pathogens, and the exposure period of the edible flowers may lead to toxicity. These factors are vital to ensure the safety of the utilization of edible flowers in culinary practices.

8.3 Phytochemicals in Edible Flowers

Phytochemicals are naturally occurring compounds in plants, including edible flowers, that contribute to their color, flavor, taste, and health benefits. These flowers are an excellent source of antioxidants, vitamins, and minerals. The market for nutraceuticals is driven primarily by increased consumer awareness, improved purchasing power, the popularisation of the beneficial effects of phytochemicals on the human body, consumer readiness to purchase functional foods, and a strong desire to maintain a healthy lifestyle (Rao and Poonia 2024a, b).

Out of the vast array of phytochemicals reported in edible flowers, such as flavonoids, carotenoids, terpenes, and phenolic acids, some of the other phytochemicals reported in lesser concentrations include alkaloids, nitrogen-containing compounds, saponins, tannins, inhibitors of amylase, and phytic acid which prove to be toxic making the flowers undesirable for consumption (Coyago-Cruz et al. 2023). These toxic elements in edible flowers often lead to detrimental outcomes upon consumption. Some of the toxic elements responsible for the toxicity of edible flowers are schematically shown in Fig. 8.1.

8.3.1 Organic Phytocompounds in Edible Flowers

Most phytochemicals found in plants pose as natural poisons based on dose; thus, identifying the precise flower and its usage for safe consumption is paramount. The natural toxins in edible flowers can harm one's health if eaten excessively or improperly prepared. The bioaccumulation of the toxic compounds, including alkaloids, cyanogenic glycosides, oxalic acid, trypsin inhibitors, hemagglutinins, and some flavonoids, could be dangerous to humans for instance, when plant material is chewed or digested, cyanogenic glycosides can emit hydrogen cyanide, a potent poison (Purohit et al. 2021). Likewise, acute toxicity from alkaloids and oxalic acid can result in vomiting and convulsions, as well as more severe health problems like seizures and even death. Similarly, pyrrolyzine alkaloids are obtained from the floral parts of *Heliotropium amplexicaule* and *Echium vulgare* and are poisonous, leading to toxic effects (Benvenuti and Mazzoncini 2021). Alkenylbenzenes such as methyl eugenol, 1,8 cineole (eucalyptol), safrole, and thujone are some of the widely reported phytotoxins in edible flowers (Madurangi and Maduwanthi 2024). Plant species such as *Sassafras*, *Ocotea pretiosa*, *Piper auritum*, *Cinammomum parthenoxylon* of *Lauraceae*, *Piperaceae*, and *Aristolochiaceae* families possess the alkaloid safrole, which is a potent phytotoxin in the flowers (Van Quan et al. 2020). Dose-dependent toxicity has been observed for these organic compounds. Many of these phytocompounds can be safely taken in minimal doses, can be medicinal, and may not cause serious harm. However, they can build up to levels detrimental to human health in more significant amounts or when the blossoms are not digested properly. For instance, improperly cooked flowers containing cyanogenic glycosides can cause poisoning due to cyanide, which can be fatal (Deepika et al. 2014). Also, in some of the flowers, such as marigold and rose, the bitter base part of the petal is advised to be removed before consumption (Nicolau and Gostin 2016). Therefore, to reduce the risks connected with natural toxins in edible flowers, following the proper preparation, cooking, and other processing methods is crucial to lessen these toxins' levels.

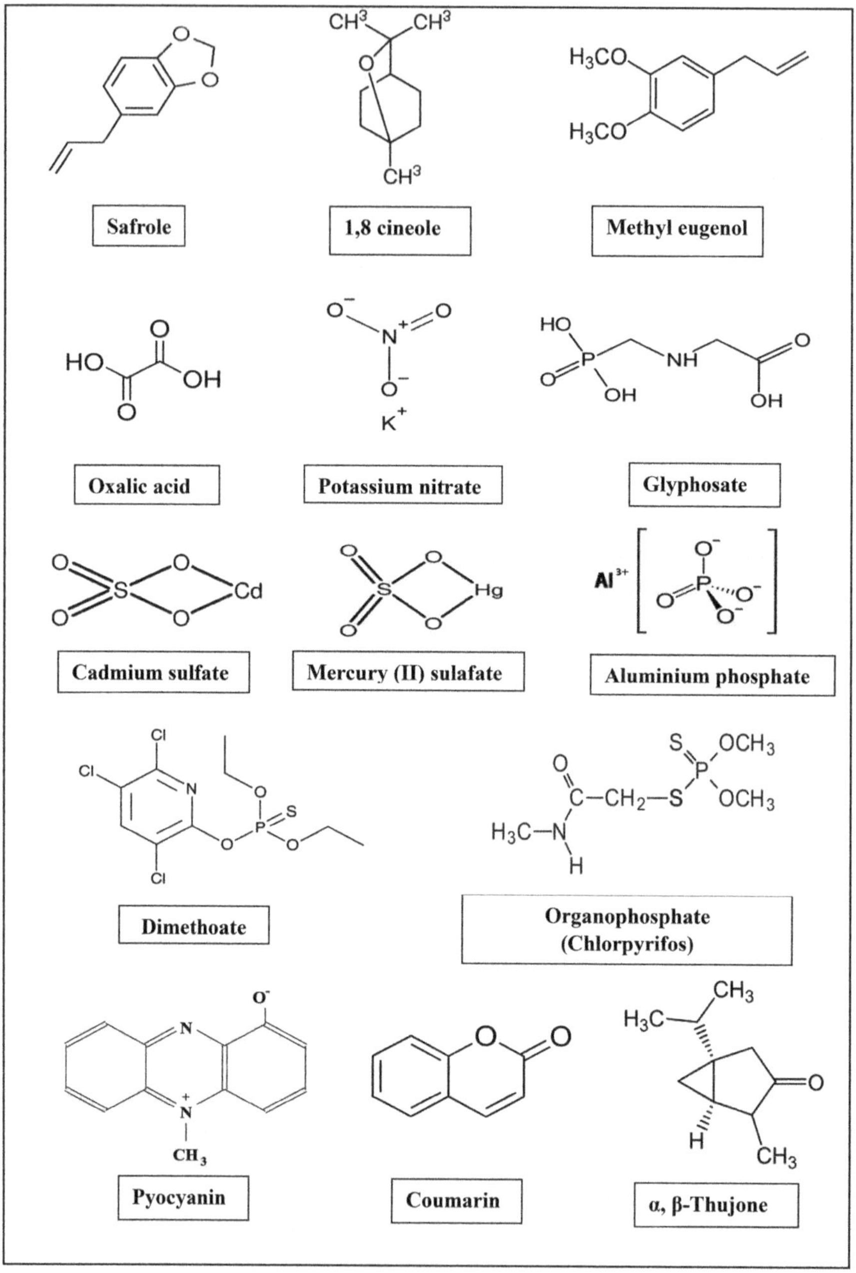

Fig. 8.1 Major toxic elements in edible flowers

8.3.2 *Inorganic Phytocompounds in Edible Flowers*

Chemical impurities traced from agricultural production and the environment are essential toxic agents in plants. Particularly when they acquire heavy metals and other inorganic compounds from fertilizers and environmental pollution, the inorganic character of phytochemicals in edible flowers might increase their toxicity. Cultivation-related fertilizers that contain lead, cadmium, arsenic, methylated sulfates, and organophosphates, can transfer these harmful elements into the flowers, offering serious health dangers such as increased cancer risk and neurological impairment (Fishbein 1974). Heavy metals such as Cd, Cr, Cu, Ni, Pb, and Zn were found to be concentrated in different parts of the plants, such as roots, leaves, stalks, and flowers, and they gradually build up over time due to bioaccumulation. Inorganic compounds from sources such as potassium and phosphate fertilizers can also facilitate the buildup of toxic chemicals in edible flowers. Pumpkin flowers have the highest potassium content, to the tune of 14.412 mg/g. Similarly, night jasmine, drumstick, and Nongmangkha flowers showed potassium content of 11.956, 12.912, and 12.401 mg/g, respectively (Chetia et al. 2024). Uptake of high amounts of heavy metals from the soil, such as cadmium, also poses a serious threat when it passes on to the latter levels in the food chain (Jackson and Alloway 2017). So, these inorganic compounds can function as poisons when ingested, resulting in several harmful health consequences.

8.4 Man-Made Chemical Products in Edible Flowers

Artificial chemicals like pesticides, herbicides, and fertilizers raise serious concerns regarding the toxicity of edible flowers. The recent trend of incorporating edible flowers for daily consumption based on the popular notion that anything natural is safe has urged to increase their shelf-life through various processes of freeze-drying, irradiation technology, with protective coatings over the edible flower part, or by applying hydrostatic pressure (Fernandes et al. 2019). Although these processes of storage and preservation have helped in retaining its bioactive ingredients, they have been identified with several chemical compounds in amounts not suitable for food products, such as insecticide (dimethoate), sulfites, pesticide residues, and insect repellent (diethyl-methazolamide) (Jadhav et al. 2023).

Fertilizers that contain heavy metals like cadmium, lead, and arsenic can accumulate within the flowers and pose health risks upon consumption (Dodangeh et al. 2018). These toxic heavy metals are associated with various health problems, including potential neurological harm and higher risks of cancer. For example, *Epimedii folium* has significant copper, arsenic, lead, and cadmium concentrations. On the other hand, *Cordyceps sinensis* has increased concentrations of these elements. Metals such as mercury and copper are concentrated in distillates, although eating raw plant matter seems to increase the danger of metal buildup more than herbal distillates. Well-known hazardous metals like mercury, lead, and cadmium have been found to build up in concentration and may provide the worst risks in edible flowers (Charen and Harbord 2020).

The pesticides and herbicides applied in growing edible flowers can leave harmful residues if ingested. Although these chemicals are intended to eliminate pests and weeds, they can also harm humans. Residues from pesticides like organophosphates and herbicides such as glyphosate may persist on the flowers even after they have been washed. Long-term exposure to these residues has been linked to multiple health issues, such as endocrine disruption, developmental complications, and elevated cancer risk. Reducing these risks requires the use of organic fertilizers and a reduction in synthetic pesticides and herbicides. After properly cleaning and preparing, edible flowers may also leave fewer chemical residues. To ensure the safety of edible flowers, it is also essential to adhere to the most recent regulations and recommendations for the use of pesticides in agriculture.

8.5 Allergens in Edible Flowers

Many variables contribute to the toxicity of edible flowers associated with allergies. First, certain people may experience allergic reactions due to proteins in the pollen that edible flowers produce. When inhaled or consumed, these proteins can cause respiratory problems, itching, and sneezing. Sometimes, the oral mucosa is affected due to the allergic reaction, resulting in oral allergic syndrome (OAS) (Lis and Bartuzi 2023). For instance, the daisy flowers cause allergies in people who have asthma. The potential allergens in edible flowers could be in any form, such as house dust mites, mold, or pollen, or with parts of the flower, such as its stamen, which could develop an allergic response when touched, consumed, or inhaled. There is also a tolerable dosage limit for ingesting edible flowers with respect to the compounds present in them, beyond which they could be potentially hazardous and evoke allergic responses in a person. For example, the daily consumption limit of *Achillea millefolium* flowers is 18 g for not exceeding thujone content in them, while the advisable limit of *Galium odoratum* consumption is about 7 g not to exceed the coumarin limit (Matyjaszczyk and Śmiechowska 2019).

Additionally, cross-reactivity is possible, wherein individuals allergic to specific plants might react to edible flowers from similar species. An example includes chamomile, potentially causing an allergic response in individuals allergic to ragweed due to their botanical relationship (Takahashi et al. 2020). Roses and violets are two examples of edible flowers that contain specific proteins that might trigger allergic responses. The severity of these allergic reactions might vary from minor symptoms like itching and hives to more serious ones like anaphylaxis. Therefore, it is necessary to correctly identify edible flowers and get medical advice from doctors for people with known allergies to reduce the possibility of allergic reactions from eating these flowers.

8.6 Pathogenic Organisms in Edible Flowers

The toxicity of edible flowers due to pathogenic agents can arise from microbial contamination during cultivation, harvesting, processing, and storage, which can introduce harmful bacteria, such as *Salmonella,* causing foodborne illness. The Rapid Alert System for Food and Feed (RASFF) has recorded salmonella outbreaks associated with edible flowers, underscoring the significance of strict safety protocols (Purohit et al. 2021). RASFF monitors and reports such incidents to reduce risks and protect public health. Inadequate handling and packaging may not be able to eradicate infections or shield the flowers from outside pollutants. Hazardous pathogens can be introduced to the flowers during handling and preparation via biofertilizers, biopesticides, heavy metal buildup, and cross-contamination from polluted water, utensils, or surfaces. For instance, in multiple studies, *P. aeruginosa* and *P. putida* utilised for bio-fertilizer and bioremediation contains plasmid-mediated antibiotic-resistant determinants. A pathogenic factor, pyocyanin, is a toxin produced by *P. aeruginosa,* used in the preparation of biofertilizers and is harmful for consumption since it damages the alveolar and myocytes causing severe disease manifestations (Tariq et al. 2022). Another study carried out to evaluate the microbial contamination in selected edible flowers such as *Bellis*, *Tropaeolum* L., *Calendula officialis* L., *Dianthus caryophyllus* L., and *Hemerocallis* L. reported the presence of *Staphylococcus aureus*, a major causative pathogen for foodborne infections, some yeasts, and molds, but the microbial load varied from flower to flower (Guo 2021). Microbial contamination could happen at any stage, from cultivation to harvesting or processing, which urges a better understanding of their safety before consumption. Thus, proper handling, storing, and treating of edible flowers is essential to reduce these hazards and guarantee their safe ingestion.

8.7 Dose/Period of Exposure

The toxic phytochemicals in edible flowers may cause toxicological outcomes. Therefore, the dosage or period of exposure to edible flowers significantly affects their toxicity. Many edible flowers are considered non-toxic and safe for consumption, but their safety is dose-dependent. For instance, certain edible flowers can be highly toxic and even fatal in higher doses, causing severe symptoms such as convulsions, sedation, photosensitivity, hypotension and hypoglycemia, hormonal imbalance, electrolyte imbalance, hepatotoxicity, and even death upon consumption (Skrajda-Brdak et al. 2020). Some edible flowers interact with other drugs like anti-diabetic and blood pressure medications and show these symptoms (Nowicka and Wojdyło 2019). Conversely, in smaller amounts, these flowers possess beneficial medicinal effects. This dose-dependent toxicity is also observed with other edible flowers that contain bioactive compounds. While these compounds can offer health benefits such as antioxidant, anti-inflammatory, cardioprotective, and anticancer properties in appropriate doses, excessive consumption can lead to various side effects. Joint FAO/WHO Expert Committee on Food Additives (JECFA) guidelines determine the daily ingestible amounts of several edible flowers for humans. The daily recommended dosage of erucic acid, thujone, and coumarins is reported to be 39.5 g, 18 g, and 7 g, respectively (Lucarini et al. 2020). Specific studies on rats showed that edible flowers like *Hibiscus rosa sinensis* at a dose of 1600 mg/kg showed 20% mortality (Meena et al. 2014). Therefore, it is crucial to consume edible flowers in moderation and be aware of their potential toxicity at higher doses.

8.8 Toxicological Effects of Edible Flowers

Edible flowers like *Rosa rugosa, Carpobrotus edulis, Rhododendron simsii, Agave salmiana, Limonium sinuatum, Iris japonica, Siraitia grosvenorii, Osmanthus fragrans, Hibiscus syriacus* L., *Tropaeolum majus,* and *Robinia pseudoacacia* contain numerous phytochemicals that have potential health advantages. Research suggests that although edible flowers typically have positive health effects, some may also be toxic if taken in excess or due to some antinutritional compounds present in them (Lu et al. 2016). Some of the severe toxicological manifestations include hepatotoxicity, hypersensitivity, nephrotoxicity, cardiotoxicity, and other toxic effects. Only a few studies and research on the toxicity of edible flowers exist. The type, the cultivation, and the presence of toxic compounds in the floral tissues of edible flowers determine their toxicity (Lucarini et al. 2020).

8.8.1 Hepatotoxicity

The term "hepatotoxicity"describes the harmful effects that some drugs or therapies have on the liver. Although the medicinal properties of edible flowers have increased their demand as emerging food components, there is an immediate need and caution to be exercised when it comes to certain plant toxins that are naturally found in some edible flowers and that, if taken in large quantities, might have harmful effects, including hepatotoxicity (Johnson 2019). Studies found that some edible flower species, like *Hibiscus sabdariffa* and *Hemerocallis* spp., have hepatotoxic substances, even though their health advantages are widely known (Madurangi and Maduwanthi 2024).

A study conducted by (Akindahunsi and Olaleye 2003) in Wistar rats identified that prolonged exposure to the aqueous extract of *Hibiscus sabdariffa* resulted in liver damage, hepatic system poisoning, chronic hepatitis-like symptoms and ultimately led to muscular dystrophy. The daylily flower (*Hemerocallis* spp.), an edible flower popularly utilized in garnishing meals and salads, was studied for its potential hepatotoxicity in cats, wherein it was found that the ingestion of daylilies in cats caused gastrointestinal distress and acute renal failure with hepatotoxicity (Hadley et al. 2003).

8.8.2 Hypersensitivity

Hypersensitivity reactions are unfavorable immunological reactions that lead to minor allergic symptoms and severe anaphylactic events. Allergic reactions can occur by ingesting certain edible flowers which contain specific substances that pose as allergens. Many bioactive substances found in edible flowers have the potential to cause hypersensitivity reactions in certain people (Prabawati et al. 2021). Some individuals may also experience hypersensitive reactions if they consume infusions or meals colored with specific edible flowers, including *Clitoria ternatea* (Lis and Bartuzi 2023). Although eating edible flowers is generally thought to carry a low risk of allergic reactions, care should still be taken, particularly when consuming large amounts of them (Lucarini et al. 2020). For example, flowers in the daisy (*Bellis perennis*) family may provoke several allergic reactions, such as hay fever or asthma. Therefore, it is advised to be avoided by people with underlying allergic conditions (Deepika et al. 2014). Likewise, extracts from butterfly pea (*Clitoria ternatea*) flowers have been shown to trigger hypersensitivity reactions in specific individuals (Jamil et al. 2018). Even people who have never been exposed to flowers from the *Asteraceae* (daisy) family, like *Echinacea*, may experience unanticipated adverse reactions, such as delayed hypersensitivity (Denisow-Pietrzyk et al. 2019), and some edible flowers, such as *Chrysanthemum* have been linked to delayed hypersensitivity reactions, in individuals who have an impaired IgE-mediated mechanism (de Jong et al. 1998).

8.8.3 Nephrotoxicity

Several herbal therapies and herbal medicines have already been reported to cause adverse renal effects and potential toxicity to the kidneys due to several contaminants, such as heavy metals present within the extracts. Toxicity of the kidneys could happen through several mechanisms, such as compartmental damage of the nephrons, toxin accumulation and poor infiltration, urinary flow obstruction, or disruption of blood flow through the kidney. Likewise, consumption of certain edible flowers in more than the advisable limit could cause severe kidney problems due to the accumulation of compounds. In a study conducted to study the probable toxic effects of *Rosa damascene*, the infusion of the flowers was administered to experimental dogs orally for 10 days, and minimal nephrotoxic effects were reported. The study reports that ingesting a higher dose could be toxic, implying that exceptionally high infusion dosages could lead to nephrotoxic and hepatotoxic effects and could not be readily absorbed by the kidney (Akbari et al. 2013).

8.8.4 Cardiotoxicity

Cardiotoxicity refers to the detrimental effects on the heart caused by certain medications or substances that could lead to dysfunction of the heart. The presence of certain toxins or compounds in edible flowers could negatively impact the functioning of the heart. Some studies reveal that certain compounds in edible flowers could potentially lead to cardiotoxic effects. Although most flavonoids exhibit cardioprotective effects, substances like rutin and some flavonoids in elderberry flowers (*Sambucus nigra*) are the leading cause of cardiotoxic effects (El Mahdy et al. 2017) and might probably be due to the hormesis effect exhibited by many phytochemicals and is dose-dependent. *Hibiscus sabdariffa* exhibits cardioprotective effect and anti-obesity effects, but when co-administered with certain drugs due to prolonged periods of use, it leads to cardiotoxic effects in animal models (Ojulari et al. 2019) (de Arruda et al. 2016). Specific undesirable saponins in *Yucca filifera* exhibit hemolytic effects leading to anemia. Likewise, flowers of *Agave salmiana* show hemagglutinating activity (Sotelo et al. 2007).

Alkaloids and other secondary metabolites found in elderberry flowers may play a significant role in their toxic effects. Notably, the literature does not always precisely identify the elderberry blossom regions with the highest quantities of these harmful chemicals. Since not all portions of the elderberry bloom are acceptable to eat, it is typically advised to use caution while eating (Colombo et al. 2010).

8.8.5 Other Toxic Effects of Edible Flowers

Besides the above-mentioned effects, excessive consumption of edible flowers could also have other toxic effects on our health. For example, *Hibiscus sabdariffa* calyx is frequently used as an aphrodisiac in West African traditional medicine. Researchers from Nigeria have lately investigated the effects of sub-chronic administration of aqueous plant extracts on the testes. It was found that there was a steady decrease in body weight after the treatment period, but no notable changes were observed in the relative or absolute weights of the testes. However, the higher dosages resulted in a notable reduction in epididymal sperm, tubule histological distortion, typical testicular epithelial organization disturbance, and sperm cell disintegration (Ali et al. 2005). Similarly, the flowers of *Erythrina americana* and *Erythrina caribaea* possess trypsin inhibition enzymes that cause indigestion (Barriada-Bernal et al. 2014). Some edible flowers may contain toxic molecules that help plants prevent damage, and those compounds may become toxic to humans. These secondary metabolites are used as functional agents in nutraceutical, pharmaceutical, and cosmetic products (Lucarini et al. 2020). Some parts of the flowers, such as pistils, stamen, and sepals, should be avoided for consumption because of their unappealing flavors and potential health concerns such as gastrointestinal problems (Jadhav et al. 2023). The *Bougainvillea glabra*, a paper flower, is eaten as a refreshment. However, its slightly bitter taste was due to its toxic effects on zebrafish embryos. The results revealed that the extracts of the pink, purple, and dark pink flowers of *B. glabra* were mildly toxic and produced mild alterations in the phenotypic characteristics of the embryo along with edema and hypopigmentation in the yolk sac of the embryos (Teh et al. 2019).

8.9 Regulatory Guidelines for Evaluation of Toxicity

It is intriguing that very few studies are available on the toxicity of edible flowers despite their biological activities and culinary use. However, the majority of the studies include Ames mutagenicity assay and some specific preclinical studies on experimental animals. It is satisfying to note that the flower extracts of *Nasturtium officinale*, *Hibiscus rosa-sinensis*, marigold flower, *Butea monosperma, Bombax ceiba* are safe for consumption and exhibit non-toxic effects (Kandylis 2022). There are several guidelines followed in various countries for assessing the safety of several nutraceutical factors, which include the OECD (Organization for Economic Co-operation and Development), WHO (World Health Organization) guidelines, and the FDA (Food and Drug Administration) guidelines for food ingredients which elaborates on the safety requirements and regulatory guidelines intended for human use. These guidelines focus on testing compounds on animals before humans and then correlating their effects to human conditions.

8.10 OECD Guidelines

The Organization for Economic Co-operation and Development (OECD) formulated a set of regulatory test guidelines to estimate the negative impact of chemicals on human health and the environment. It is accepted internationally as a standard method for testing chemicals so that they can be utilized safely. They were first published in 1981 and subsequently modified. Section 8.4 of the guidelines discusses the health effects criteria to be fulfilled through acute toxicity and repeated dose toxicity studies. Sections 401–405 enlist the acute toxicity criteria for oral, dermal, inhalation, dermal irritation/corrosion, and eye irritation/corrosion test, respectively. Likewise, sections 407 and 408 emphasize repeated dose studies on rodent oral toxicity for 28 and 90 days, respectively. Several other sub-sections of the 400 series set the guidelines to be followed for sub-chronic toxicity, chronic toxicity, carcinogenicity, toxicokinetics, and delayed neurotoxicity. The potential effects of the edible flowers at their different dosages could be tested and classified according to their possible effects noted under these guidelines.

As per regulatory toxicology, animal models are generally employed for acute and short-term toxicity studies involving studies for a shorter period (24, 48, 72, or 96 hrs), wherein the animals are dosed in a step-wise procedure at pre-specified fixed doses of 5, 50, 300, or 2000 mg/kg. Three animals of female sex per step are used (Royapuram Parthasarathy et al. 2022). Single dose used in acute toxicity assessment refers to each test animal receiving a test dose from the tested product. The sample volume administered on experimental animals is restricted to exceed the threshold limit of 1 ml/100 g of body weight. At the same time, in the case of aqueous solutions, it could be raised to 2 ml/100 g of body weight. Timely monitoring of the animals during the study period is vital, and the changes in skin, fur, eyes, mucus membranes, the occurrence of secretions, and autonomic activities such as lacrimation, piloerection, pupil size, and respiratory patterns need to be timely monitored and noted. The changes in gait, posture, and response to handling or any bizarre movements will also be recorded. Toxicity symptoms and animal death are documented periodically throughout the examination of 14 days to establish the Maximally Tolerated Dose (MTD) and the Lethal Dose (LD_{50}). Finally, after the 14th day, gross pathology of the excised organs is noted to identify any toxic changes.

Sub-acute toxicity studies are performed to evaluate the effect of toxicants on various organ systems for evaluating test samples that are intended for short-term consumption. An oral dose of test substance administered at one dose level per group for 28 days is studied following OECD 407 guidelines as per regulatory toxicity (Gayathri et al. 2013). At least 10 animals (5 males and 5 females) are closely monitored and noted for any signs of toxicity. Chronic or long-term toxicity studies are repeated dose studies performed to identify target organs for toxicity, to determine the inter-relationship between aging and toxicity, and investigate treatment effects using a range of doses. This involves long-term studies in rodents ranging from 18 months to 2 years. The safety level of the test sample (in this case, edible flowers) can be confirmed by histopathological analysis of the various vital organs (Sangeetha et al. 2013). In the sub-acute and chronic toxicity assessment, the animal is given repeated exposure to the tested products daily after a particular duration (28 days-OECD 407, 90 days-OECD 408, 12 months-OECD 452). Hazard characterizations include No Observed Adverse Effect Level (NOAEL) and No Observed Effect Level (NOEL), Lowest Observed Adverse Effect Level (LOAEL), and Lowest Observed Effect Level (LOEL) (Kumar Meena et al. 2014).

8.11 WHO Guidelines

The safety and quality of herbal medicines, such as herbal medicinal plants, including edible flowers, herbal materials, herbal preparations, and herbal dosage forms, have become a significant concern to healthcare authorities and the public. Hence, a set of guidelines to ensure the proper development of a methodology to monitor or ensure the safety, quality, and efficacy of herbal medicines have been framed by the WHO and published as the WHO guidelines for evaluating the safety and efficacy of herbal medicines (1993).

The guidelines to assess the quality of herbal medicines concerning contaminants and residues have been newly introduced into the existing guidelines to ensure their quality from batch to batch, which could be due to intrinsic or extrinsic factors. They majorly include acute and chronic toxicity tests for which rodents and non-rodents of male and female sex are used. In rodents, an adequate number of dose levels should be tested to arrive at a rough estimate of LD50. Later, signs, appearance, development, and, when reversible, recovery of these signs should relate to dose and time. At least three doses are administered to the animal groups exposed to long-term toxicity wherein the mid-dose is the optimal therapeutic dose, and the higher dose is 5 times and 10 times higher. Within this range, the addition of at least one more dose may increase the chances of detecting the dose-toxicity relationship for the toxic signs. According to the WHO guidelines, 20 animals of either sex are used for each dose.

8.12 Technologies to Overcome Toxicities of Edible Flowers

Flowers are the most perishable organ of plants because they have higher transpiration rate, metabolism rate, and respiration rate (da Costa et al. 2021). Hence, it is challenging to preserve them for a longer time. This can result in organoleptic changes such as petal drop, browning, flower wilting, dehydrated tissue, and senescence, which contributes to the source of toxicity in addition to phytochemicals (Benvenuti and Mazzoncini 2021). Therefore, the primary strategy for preserving floral tissue is to place them below 5 °C, which can control respiration and ethylene formation (Scariot et al. 2022). Traditionally, edible flowers like *Erythrina* species are cooked or boiled, and the water is drained to reduce the alkaloid content, thereby diminishing the toxic effects (García-Mateos et al. 1996). Nanotechnology represents an emerging field of scientific exploration focused on the examination and control of matter at the smallest scales, encompassing molecules and atoms (Rao and Poonia 2024a, b) in preserving and processing the edible flowers for consumer market.

The preservation and processing of edible flowers can be performed through three main procedures: pre-processing, processing, and packaging (Purohit et al. 2021). Pre-treatment includes nitric oxide fumigation and chlorine dioxide treatment. Nitric oxide can be used to improve the quality of flowers as well as to extend their shelf life. Fumigation of nitric oxide at a high concentration of 2% at 2 °C for 3 h showed high control of western flower thrips (*Frankliniella occidentalis*) and lettuce aphid (*Liriomyza langei*) (Liu 2016). Chlorine dioxide, which is strongly oxidated, can kill microorganisms. It can be used alone or in combination with other preservative chemicals in aqueous form or gaseous form to develop a longer shelf-life of edible flowers. Some processing techniques, like high-pressure processing (HPP), edible coating using nanotechnology, ultrasound and drying, and irradiation, have been tested for preservation. The flowers' phenolic contents and antioxidant activity were reported to be highest after irradiation. Fresh edible flowers of *D.*

chineses L. irradiated by an electron beam with 0.8 and 1.0 kGy exhibited effective capacity to inhibit β-carotene bleaching. Treating with maximum doses of 10.0 kGy showed increased phenolic content in the petals of *Tropaeolum majus, Cammellia sinensis, Illx paraguariensis* (Koike 2019). Gamma irradiation at high doses may negatively impact flowers and beneficial phytochemicals; thus, it should be performed with other treatments, such as NaDCC treatment, to attain low damage to edible flowers and also to store without degrading the active ingredients (Chu et al. 2015). High-pressure processed and frozen species showed improved textural properties, reduced drip loss, and minimum cellular structural changes.

The primary function of the edible coating is to control moisture by acting as a barrier, regulating the exchange of gases for color and flavor retention and enzymatic oxidation (Shantamma et al. 2021). The most effective method is drying, which has been used for preservation for ages. Due to modern innovation, many contemporary drying techniques have emerged, such as freeze, microwave, hybrid, vacuum, and hot/dehumidified air drying (Marchioni et al. 2021). This method favors microbial growth inhibition without influencing the enzyme activity. Packaging, which is post-processing storage, is the most crucial factor that can be used for processing methods and longer shelf life. Conditions inside the packaging are changed to reduce the respiration rate, which helps regulate the heat produced metabolically by flowers. Active Modified Atmosphere Packaging (MAP) along with cold storage showed successful results in preserving the colouration of the petals and slow down the decay of *Begonia grandiflora, Tropaeolum majus L.*, and *Viola cornuta L.* flowers by controlling the respiration rate (Kandylis 2022).

8.13 Conclusions

The beneficial properties of edible flowers are higher and more important than the toxic effects. As mentioned above, the risks associated with their consumption can be effectively managed and overcome with the aid of modern innovative techniques. The vast number of edible flower species grown across the globe could have immense potential as a food product or ingredient due to their immense nutritional potential if they are rightly classified under different categories based on their toxicity. The authorities and guidelines discussed above also ensure the regulation of toxic effects and their safe usage. However, there are only a few studies regarding the safety of these edible flowers, and more studies are warranted with a specific focus on anti-nutritional compounds in these flowers. This will help explore and manipulate more natural molecules from flowers or plants that can be used as functional foods for nutritive and medicinal applications, enhancing health and enriching culinary practices.

References

Akbari M, Kazerani HR, Kamrani A, Mohri M (2013) A preliminary study on some potential toxic effects of Rosa damascena Mill. Iran J Vet Res 14(3):232–236

Akindahunsi AA, Olaleye MT (2003) Toxicological investigation of aqueous-methanolic extract of the calyces of Hibiscus sabdariffa L. J Ethnopharmacol 89(1):161–164

Albien A-L, Stark TD (2023) (Bio) active compounds in daisy flower (Bellis perennis). Molecules 28(23):7716

Ali BH, Al Wabel N, Blunden G (2005) Phytochemical, pharmacological and toxicological aspects of Hibiscus sabdariffa L.: a review. Phytother Res 19(5):369–375

AshwlayanVD KA, Verma M (2018) Therapeutic potential of Calendula officinalis. Pharm Pharmacol Int J 6(2):149–155

Barriada-Bernal LG, Almaraz-Abarca N, Delgado-Alvarado EA, Gallardo-Velázquez T, Ávila-Reyes JA, Torres-Morán MI et al (2014) Flavonoid composition and antioxidant capacity of the edible flowers of Agave durangensis (Agavaceae). CyTA-Jo Food 12(2):105–114

Benvenuti S, Mazzoncini M (2021) The biodiversity of edible flowers: discovering new tastes and new health benefits. Front Plant Sci 11:569499

Charen E, Harbord N (2020) Toxicity of herbs, vitamins, and supplements. Adv Chronic Kidney Dis 27(1):67–71

Chetia I, Das AJ, Badwaik LS (2024) Assessment of nutritional and bioactive properties of selected edible flowers: characterisation of phenolic compounds by reversed-phase high performance liquid chromatography. J Chromatogr Open 6:100167

Chowdary MY, Rana SS, Ghosh P (2022) Banana inflorescence and their potential health benefits as future food. Qual Assur Saf Crops Foods 14(2):131–136

Chu EH, Shin EJ, Park HJ, Jeong RD (2015) Effect of gamma irradiation and its convergent treatment for control of postharvest Botrytis cinerea of cut roses. Radiat Phys Chem 115:22–29

Colombo ML, Assisi F, Della Puppa T, Moro P, Sesana FM, Bissoli M, Borghini R, Perego S, Galasso G, Banfi E (2010) Most commonly plant exposures and intoxications from outdoor toxic plants. J Pharm Sci Res 2(7):417

Coyago-Cruz E, Moya M, Méndez G, Villacís M, Rojas-Silva P, Corell M, Mapelli-Brahm P, Vicario IM, Meléndez-Martínez AJ (2023) Exploring plants with flowers: from therapeutic nutritional benefits to innovative sustainable uses. Foods 12(22):4066

da Costa LC, de Araujo FF, Ribeiro WS, de Sousa Santos MN, Finger FL (2021) Postharvest physiology of cut flowers. Ornam Hortic 27(03):374–385

de Arruda A, Cardoso CAL, Vieira M et al (2016) Safety assessment of Hibiscus sabdariffa after maternal exposure on male reproductive parameters in rats. Drug Chem Toxicol 39(1):22–27

de Jong NW, Vermeulen AM, Van Wijk RG, de Groot H (1998) Occupational allergy caused by flowers. Allergy 53(2):204–209

Deepika SD, Lakshmi SG, Sowmya LK, Sulakshana M (2014) Edible flowers–a review article. Int J Adv Res Sci Technol 3(1):51–57

Denisow-Pietrzyk M, Pietrzyk Ł, Denisow B (2019) Asteraceae species as potential environmental factors of allergy. Environ Sci Pollut Res 26:6290–6300

Dodangeh H, Rahimi G, Fallah M, Ebrahimi E (2018) Investigation of heavy metal uptake by three types of ornamental plants as affected by application of organic and chemical fertilizers in contaminated soils. Environ Earth Sci 77:1–15

El Mahdy C, Popescu S, Borda C (2017) Plants that can be poisonous for cows. A review. Bull Univ Agric Sci Vet Med Cluj-Napoca Anim Sci Biotechnol 74(2)

Esghaei M, Ghaffari H, Esboei BR, Tapeh ZE, Salim FB, Motevalian M (2018) Evaluation of anticancer activity of Camellia sinensis in the Caco-2 colorectal cancer cell line. APJCP 19(6):1697

Fernandes L, Saraiva JA, Pereira JA, Casal S, Ramalhosa E (2019) Post-harvest technologies applied to edible flowers: a review: edible flowers preservation. Food Rev Intl 35(2):132–154

Fishbein L (1974) Mutagens and potential mutagens in the biosphere I. DDT and its metabolites, polychlorinated biphenyls, chlorodioxins, polycyclic aromatic hydrocarbons, haloethers. Sci Total Environ 2(4):305–340

García-Mateos R, Lucas B, Zendejas M, Soto-Hernández M, Martínez M, Sotelo A (1996) Variation of total nitrogen, non-protein nitrogen content, and types of alkaloids at different stages of development in Erythrina americana seeds. J Agric Food Chem 44(10):2987–2991

Gayathri V, Ananthi S, Vasanthi HR (2013) Antihyperlipidemic potential of polyphenol and glycoside rich Nerium oleander flower against triton WR-1339-induced hyperlipidemia in experimental Sprague Dawley rats. J Chem 2013(1):825290

Guo Y (2021) Microbial contamination of edible flowers. Nature Food 2(7):455

Hadley RM, Richardson JA, Gwaltney-Brant SM (2003) A retrospective study of daylily toxicosis in cats. Vet Hum Toxicol 45(1):38–39

Ibraheem RM, Mhawesh AA, Abood KW (2018) Estimation of the whole flavonoid, antioxidant, anti-bacterial challenge concerning Viola odorata (Banafsha) methanolic extract. Iraqi J Agric Sci 49(4)

Jackson AP, Alloway BJ (2017) The transfer of cadmium from agricultural soils to the human food chain. Biogeochem Trace Metals:121–170

Jadhav HB, Badwaik LS, Annapure U, Casanova F, Alaskar K (2023) A review on the journey of edible flowers from farm to consumer's plate. Appl Food Res 3(2):100312

Jain N, Jain R, Jain V, Jain S (2012) A review on: Abelmoschus esculentus. Pharmacia 1(3):84–89

Jamil N, Zairi MNM, Nasim NAM, Pa'ee F (2018) Influences of environmental conditions to phytoconstituents in Clitoria ternatea (butterfly pea flower) – a review. J Sci Technol 10(2)

Johnson JC (2019) Efficacy and potential toxicity of herbal extracts (Hemerocallis, Spp.) elegant explosion daylily flower. Alabama Agricultural and Mechanical University

Jurca T, Baldea I, Filip GA, Olteanu D, Clichici S, Pallag A, Vicas L, Marian E, Micle O, Muresan M (2018) The effect of Tropaeolum majus L. on bacterial infections and in vitro efficacy on apoptosis and DNA lesions in hyperosmotic stress. J Physiol Pharmacol 69(3)

Kandylis P (2022) Phytochemicals and antioxidant properties of edible flowers. Appl Sci 12(19):9937

Kaur K, Kaushal S (2019) Phytochemistry and pharmacological aspects of Syzygium aromaticum: a review. J Pharma Phytochem 8(1):398–406

Kishore K, Kanupriya C, Samant D, Acharya GC, Singh HS, Sahu A (2022) Phenological description and thermal time requirement of tamarind (Tamarindus indica) in tropical conditions. Ann Appl Biol 181(3):309–320

Koike ACR (2019) Antioxidant activity of Dianthus chinensis flowers processed by ionizing radiation. Braz J Radiat Sci 7(2A Suppl)

Kumar G (2022) A review of the chemical constituents and pharmacological activities of Antirrhinum majus (snapdragon). IP Int J Compr Adv Pharmacol 7(2):72–76

Kumar Bargah R (2015) Preliminary test of phytochemical screening of crude ethanolic and aqueous extract of Moringa pterygosperma Gaertn. J Pharma Phytochem 4(1):7–9

Leong W-H, Lai K-S, Lim S-HE (2021) Combination therapy involving Lavandula angustifolia and its derivatives in exhibiting antimicrobial properties and combatting antimicrobial resistance: current challenges and future prospects. PRO 9(4):609

Lis K, Bartuzi Z (2023) Plant food dyes with antioxidant properties and allergies – friend or enemy? Antioxidants 12(7):1357

Liu Y-B (2016) Nitric oxide fumigation for control of western flower thrips and its safety to postharvest quality of fresh fruit and vegetables. J Asia Pac Entomol 19(4):1191–1195

Lu B, Li M, Yin R (2016) Phytochemical content, health benefits, and toxicology of common edible flowers: a review (2000–2015). Crit Rev Food Sci Nutr 56(sup1):S130–S148

Lucarini M, Copetta A, Durazzo A, Gabrielli P, Lombardi-Boccia G, Lupotto E, Santini A, Ruffoni B (2020) A snapshot on food allergies: a case study on edible flowers. Sustainability 12(20):8709

Madurangi A, Maduwanthi S (2024) Edible flowers in herbal beverage development. Asian Food Sci J 23:1–21

Manivannan A, Kim J-H, Kim D-S, Lee E-S, Lee H-E (2019) Deciphering the nutraceutical potential of Raphanus sativus – a comprehensive overview. Nutrients 11(2):402

Marchioni I, Dimita R, Gioè G, Pistelli L, Ruffoni B, Pistelli L, Najar B (2021) The effects of post-harvest treatments on the quality of Agastache aurantiaca edible flowers. Horticulturae 7(4):83

Matyjaszczyk E, Śmiechowska M (2019) Edible flowers. Benefits and risks pertaining to their consumption. Trends Food Sci Technol 91:670–674

Meena AK, Jain A, Pandey K, Singh RK (2014) Acute toxicity and genotoxic activity of Hibiscus rosa-sinensis flower extract. Am J Phytomed Clin Ther 2(4):524–529

Mukherjee PK, Kumar V, Kumar NS, Heinrich M (2008) The Ayurvedic medicine Clitoria ternatea – from traditional use to scientific assessment. J Ethnopharmacol 120(3):291–301

Muruganantham N, Solomon S, Senthamilselvi MM (2016) Antimicrobial activity of Cucurbita maxima flowers (pumpkin). J Pharm Phytochem 5(1):15–18

Nicolau AI, Gostin AI (2016) Safety of edible flowers. In: Regulating safety of traditional and ethnic foods. Elsevier, pp 395–419

Nowicka P, Wojdyło A (2019) Anti-hyperglycemic and anticholinergic effects of natural antioxidant contents in edible flowers. Antioxidants 8(8):308

OECD/OCDE 420 OECD (2001) Guideline for testing of chemicals acute oral toxicity-fixed dose procedure introduction

Ojulari OV, Lee SG, Nam J-O (2019) Beneficial effects of natural bioactive compounds from Hibiscus sabdariffa L. on obesity. Molecules 24(1):210

Patil SA, Kulkarni AJ, Jadhav PR, Yadav RS, Pagariya MC, Saha TN, Shabeer TPA, Sarode DK, Sharma AK, Prasad KV (2024) Potential assessment of chrysanthemum flowers from various cultivars as sources of natural antioxidants and bioactive compounds. Genet Resour Crop Evol:1–19

Pires TCSP, Dias MI, Barros L, Calhelha RC, Alves MJ, Oliveira MBPP, Santos-Buelga C, Ferreira ICFR (2018) Edible flowers as sources of phenolic compounds with bioactive potential. Food Res Int 105:580–588

Pires TC, Barros L, Santos-Buelga C, Ferreira IC (2019) Edible flowers: emerging components in the diet. Trends Food Sci Technol 93:244–258

Prabawati NB, Oktavirina V, Palma M, Setyaningsih W (2021) Edible flowers: antioxidant compounds and their functional properties. Horticulturae 7(4):66

Purohit SR, Rana SS, Idrishi R, Sharma V, Ghosh P (2021) A review on nutritional, bioactive, toxicological properties and preservation of edible flowers. Future Foods 4:100078

Rao V, Poonia A (2024a) Bioactive compounds, nanoparticle synthesis and potential utilization of edible flowers for development of functional dairy products: a review. J Food Sci Technol 61(6):1053–1068. https://doi.org/10.1007/s13197-023-05853-6

Rao V, Poonia A (2024b) Microbial biosynthesis of selenium nanoparticles using probiotic strain and its characterization. J Food Meas Charact 18(7):5477–5488. https://doi.org/10.1007/s11694-024-02581-z

Royapuram Parthasarathy P, Manikandamathavan VM, Chandronitha C, Vasanthi HR, Mohan VK, Vijayakumar V et al (2022) Synthesis, characterization, and in vivo toxicological evaluation of copper (II) oxide containing herbometallic siddha nanocomplex "Thamira Parpam". Front Bioeng Biotechnol 10:849441

Sangeetha MK, Priya CM, Vasanthi HR (2013) Anti-diabetic property of Tinospora cordifolia and its active compound is mediated through the expression of Glut-4 in L6 myotubes. Phytomedicine 20(3–4):246–248

Scariot V, Ferrante A, Romano D (2022) Edible flowers: understanding the effect of genotype, preharvest, and postharvest on quality, safety, and consumption. In: Frontiers in plant science, vol 13. Frontiers Media SA, p 1025196

Shantamma S, Vasikaran EM, Waghmare R, Nimbkar S, Moses JA, Anandharamakrishnan C (2021) Emerging techniques for the processing and preservation of edible flowers. Future Foods 4:100094

Sharma S, Kumar D (2022) Chemical composition and biological uses of Crocus sativus L.(saffron). In: Edible plants in health and diseases: volume II: phytochemical and pharmacological properties. Springer, pp 249–277

Shoji K, Ubukata M, Momonoi K, Tsuji T, Morimatsu T (2005) Anther-specific production of antimicrobial tuliposide B in tulips. J Jpn Soc Hort Sci 74(6):469–475

Singh Y, Gupta A, Kannojia P (2020) Tagetes erecta (Marigold) – a review on its phytochemical and medicinal properties. Curr Med Drug Res 4(1):1–6

Singh K, Sharma YP, Gairola S (2023) Distribution status and ethnomedicinal importance of genus Rosa L.(Rosaceae) in India. Ethnobot Res Appl 25:1–22

Skrajda-Brdak M, Dąbrowski G, Konopka I (2020) Edible flowers, a source of valuable phytonutrients and their pro-healthy effects – a review. Trends Food Sci Technol 103:179–199

Sotelo A, López-García S, Basurto-Peña F (2007) Content of nutrient and antinutrient in edible flowers of wild plants in Mexico. Plant Foods Hum Nutr 62:133–138

Takahashi JA, Rezende FAGG, Moura MAF, Dominguete LCB, Sande D (2020) Edible flowers: bioactive profile and its potential to be used in food development. Food Res Int 129:108868

Tariq M, Jameel F, Ijaz U, Abdullah M, Rashid K (2022) Biofertilizer microorganisms accompanying pathogenic attributes: a potential threat. Physiol Mol Biol Plants 28(1):77–90

Teh LE, Kue CS, Ng CH, Lau BF (2019) Toxicity effect of Bougainvillea glabra (Paper Flower) water extracts on Zebrafish Embryo

Thakur M, Verma R, Kumar D, Manickam S, Ullah R, Ibrahim MA, Bari A, Lalhenmawia H, Kumar D (2024) Hypoglycemic and antioxidant activities of Jasminum officinale L. with identification and characterization of phytocompounds. Heliyon 10(21)

Tiwari P, Jena S, Sahu PK (2019) Butea monosperma: phytochemistry and pharmacology. Acta Sciet Pharma Sci 3(4):19–26

Upadhyay RK (2017) Nutritional, therapeutic, and pharmaceutical potential of plant gums: a review. Int J Green Pharmacy 11(01)

Van Quan N, Dang Xuan T, Teschke R (2020) Potential hepatotoxins found in herbal medicinal products: a systematic review. Int J Mol Sci 21(14):5011

Chapter 9
Nanoparticle Synthesis from Edible Flowers and Their Food Applications

Arati Yadav and Amrita Poonia

9.1 Introduction

Edible flowers being non-toxic, are considered healthy and safe for human consumption. They have been employed by humans since long time to beautify dishes, supply nutrients, and improve wellness (Rao and Poonia 2024a, b). Nanotechnology involves the manipulation, fabrication, and development of experimental particles within the nanoscale range of 1–100 nm. While the purposeful manipulation of nanomaterials began in the 1980s, subsequent advancements in the field have led to diverse branches of research, resulting in innovative researches in the twenty-first century. The creation of metallic nanoparticles has gained significant momentum in the recent times rendering their exclusivity of exhibiting a peculiar minute size and increased surface to volume ratio thus stimulating better chemical and biochemical action. Conventional nanoparticle synthesis through physical and chemical methods has enabled extensive applications; however, these approaches are often associated with limitations such as sustainability concerns, impurities, radiation exposure, and toxicity. Consequently, there has been a growing inclination towards alternative green synthesis routes, stimulating the fabrication of nanoparticles through eco-friendly, non-toxic, and sustainable methods by utilization of biological resources i.e. bacteria, algae, fungi, and plants (Duan et al. 2015). Biological entities for synthesis have emerged as a particularly promising approach due to its reliability, sustainability, and efficiency. Plant components including leaves, roots, fruits, seeds, stems, and flowers contain abundant bioactive compounds that facilitate nanoparticle synthesis, further reinforcing the potential of green nanotechnology. Utilized in

A. Yadav · A. Poonia (✉)
Department of Dairy Science and Food Technology, Institute of Agricultural Sciences, Banaras Hindu University, Varanasi, Uttar Pradesh, India
e-mail: amrita12@bhu.ac.in

A. Poonia et al. (eds.), *Edible Flowers: Source of Phytonutrients, Valorization and Technological Advancements*, https://doi.org/10.1007/978-3-032-00667-7_9

traditional delicacies, culinary applications and formulated for medicinal purpose, the consumption of flowers signifies the lineage of cultural diversity. Nearly 180 species, 100 geneses and 97 families of edible flowers are present in diverse regions of the world (Lu et al. 2016), which are majorly classified into fruit flowers and non-fruit (ornamental and medicinal) flowers. Edible flowers, apart from signifying the traditional usage patterns, cater to consumers seeking natural actives in their diet. Therefore, the populous demand of edible flowers today is not merely limited to the conventional usage patterns, rather they enjoy diverse applications rendering their high bio-active component profile, physico-chemical benefits and varied utilization patterns.

The active compounds of edible flowers such as jasmine, rose pansy, hibiscus, begonia, marigold etc. impart antimicrobial, nephroprotective, anti-inflammatory, hepatoprotective, antidiabetic, antihypertensive effects along with other biological benefits. The pertinence of Edible flowers in nanotechnology rely in the high nutritional and phytochemical profile, which facilitate formulation of nanomaterials, imbibed with potential health advantages. The benefits provided by active components of the edible flowers paves the way for formulation of edible flowers-based nanoparticles. Moreover, nanoparticulation may enhance the release of volatile compounds, intensifying the flavor and aroma of edible flowers (Mayedwa et al. 2021). The phytochemical analysis of these edible flowers determines the presence of active constituents, such as phenolics, indole alkaloids, flavonoids, reducing sugars, tannins, saponins etc. Edible flower mediated green-synthesized nanoparticles possess significant biocompatibility and the peculiar capacity for controlled substance release, thus broadening the application range across various fields.

9.2 Significance of Nanoparticles in Food Science

Edible-flower-mediated nanoparticles offer significant potential for diverse applications across various industries, particularly within the food sector. The concept of integrating nanotechnology into food industries was first highlighted by the US Department of Agriculture's roadmap in September 2003 (Rashidi and Khosravi-Darani 2011). With the dynamicity of food processing industry and simultaneous evolution, the integration of nanotechnology at every stage from production and processing to packaging and marketing focuses on creating safe, fresh, stable and authentic food products with extended shelf life and superior quality. These flower-mediated nanoparticles bring distinct and innovative benefits, enhancing smart packaging, enabling time-controlled delivery of functional components through nano-encapsulation, acting as powerful anti-caking and gelling agents, enhancing flavor, improving bioavailability, development of novel products, fortifying agents, regulation of nutrient delivery and further ensure food safety among other applications. Apart from these characteristic utilization patterns, the edible flower mediated nanoparticles also exhibit potent antimicrobial, antibacterial and antioxidant activity thus imparting anti-diabetic, anti-cancerous, anti-inflammatory, cardio-protective and other related health benefits when utilized in packaging's (particularly carbon dots, Ag, TiO2, ZnO etc.). Additionally, EFNPs contribute to sustainability by reducing reliance on artificial additives/input agents, by promoting eco-friendly food production.

9.3 Phytochemicals in Edible Flowers

9.3.1 Phytochemical Composition of Edible Flowers for Nanoparticle Synthesis

Phytochemicals are non-nutrient chemical compounds derived from diverse sources including plants which impart positive health and stimulate well-being. Based on the functionality in plants, phytochemicals are bifurcated into primary and secondary metabolites. While the primary metabolites (carbohydrates, proteins, lipids and nucleic acids) hold significance in plant's life, secondary metabolites are obtained from the primary metabolic pathways (Hussein and El-Anssary 2019). Apart from being ingredients of delicacies, cuisines and food garnishes, edible flowers are valued for their distinct phytochemical profile (Fig. 9.1) and nutritional composition. Edible flowers exhibit distinct phytochemical and nutritional profile in accordance to different flower parts with moisture being the major component followed by other

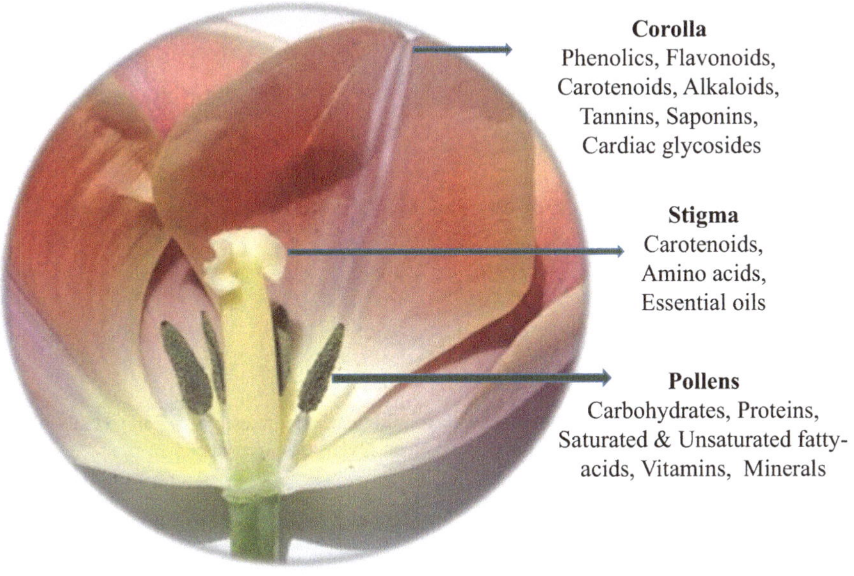

Fig. 9.1 Major bio-active components in different components of flower (Petal, Stigma and Pollens)

macronutrients such as fibers and proteins. The lipid content of the flowers is relatively low, however presence of essential polyunsaturated fatty acids such as linoleic acid and α-linolenic acid is observed. The petals are found to be rich in antioxidants, vitamins and minerals, while the pollen part is rich in protein, carbohydrates, saturated and unsaturated fatty acids (de Pascual-Teresa and Sanchez-Ballesta 2008). Edible flowers contain diverse bioactive profile including phenolics, xanthones, terpenoids, flavonoids (anthocyanins and catechins), dietary fibers, which can be utilized as precursor for NPs synthesis (Bachheti et al. 2020). Moreover, other substances, such as alkaloids, nitrogen-containing substances, and organosulfur compounds, may also be found in trace amounts (Pires Jr et al. 2021). The flower extract imbued with diverse phytochemicals, apart from acting as natural stabilizing agents, also act as reducing agents during the formation of NPs (Hano and Abbasi 2021).

9.3.2 *Bio-active Components Present in Edible Flowers (Table 9.1)*

Table 9.1 Bioactive components present in common edible flowers used for synthesis of nanoparticles

Edible flowers	Active agents constitution	References
Hummingbird tree (*Sesbania grandiflora* L.) 	Phenolic acids, flavonoids, aldehydes	Baessa et al. (2019)
Daisy (*B. perennis*) 	Phenolic acids, carotenoids, triterpenoids,	Nowicka and Wojdylo (2019)
Nasturtium (*Tropaeolum majus*) 	Phenolic compounds, flavonoids, carbohydrates, glucosinolates, reducing sugar, glycosides, steroids, saponins	Faizy et al. (2021)

(continued)

Table 9.1 (continued)

Edible flowers	Active agents constitution	References
Chrysanthemum (*Dendranthema grandiflora*) 	Carotenoids, phenolic acids, anthocyanins, glucosides, flavonoids, triterpenes, alkaloids	Hotta et al. (2021), Nishina et al. (2019)
Calenndula (*Calendula officinalis*) 	Phenolic acids, terpenes, sterols, saponins, flavonoids including quercetin, isorhamnetin, isoquercetin, quinones, glycoside, narcissin, calendoflaside	Hotta et al. (2021)
Lotus (*Nelumbo Nucifera*) 	Phenolic acids, flavonoids, alkaloids, steroids, isoquercetin, luteolin glucoside, kaempferol	Paudel and Panth (2015)
Camellia sinensis L. Kuntze 	Triterpenoid, saponins	Shen et al. (2017), Wang et al. (2017)
Rhodendron (*Rhodendron arboreum*) 	Anthocyanins, flavonoids including Delphinidin, cyanidin malvidin, sorhamnetin, kaempferol, quercetin, and their glycosides	Du et al. (2018)
Hibiscus (*Hibiscus rosa sinensis*) 	Phenolic compounds, flavonoids, carbohydrates, alkaloids, glycosides, phytosterols, proteins, terpenoids, lignins, saponins	Bhaskar and Nithya (2012)

(continued)

Table 9.1 (continued)

Edible flowers	Active agents constitution	References
Marigold (*Tagetes erectus*) 	Terpenoids, quercetin, phenolics, syringic acid, gallic acid, lutein, kaempferol	Chaudhari et al. (2024)
Kachnar (*Bauhinia variegata*) 	Alkaloids, flavonoids (quercetin, myricetin and kaempferol), terpenoids (-sitosterol, amyrin and lupeol)	Sharma et al. (2021)

Other compounds such as antinutritional factors including trypsin inhibitors, hemagglutinins alkaloids, cyanogenic glucosides, and oxalic acid may also be present indicating the necessity of cautious and precautionary utilization patterns

9.4 Role of Phytochemicals as Reducing and Stabilizing Agents

The composition, concentration of phytochemicals/active components, and the extraction technique play a vital role in determining the characteristics of synthesized nanoparticles, including their stability, bioactivity, morphology, and other physicochemical properties. A diverse range of nanoparticles, such as gold, silver, copper, iron, selenium, and magnesium, have been synthesized, each exhibiting unique biological activities with applications across various fields. Edible flowers serve as a rich source of bioactive compounds, including phenolic acids, flavonoids, proteins, organic acids etc. which function as reducing agents, capping agents, and natural stabilizers during nanoparticle biosynthesis. These active components are instrumental in preventing agglomeration, regulating nanoparticle size, and controlling morphology. Specifically, phytochemicals facilitate the reduction of mono-, di-, and tri-valent metal ions into metal nanoparticles, and during the capping stage, they assemble on the nanoparticle surface, enhancing stability and defining the final structural attributes. Furthermore, Biocompatible capping agents exhibit the capacity to alter the surface features of nanoparticles to stimulate interaction with the target cells for various therapeutic applications. This highlights the potential of plant-derived phytochemicals in the sustainable and controlled synthesis of nanoparticles for biomedical and pharmaceutical advancements. Moreover, the active components play a significant role in prevention of agglomeration, controlling the size as well as the morphology of nanoparticles. The phytochemicals cause the reduction of mono/di/tri-valent ions into metal nanoparticles and further during the capping stage assemble on the nanoparticle surface thus stabilizing the surface and giving a final shape to the formulated nanoparticle. The enveloped phytochemicals provide electrostatic repulsion and spatial resistance. The presence of polyphenolic groups in the nanoparticles is evident in the FTIR studies. Considering flavonoids (contains phenolic groups) for instance, during silver nanoparticle synthesis, these active components cause reduction of Ag^+ to AgNPs by keto-enol tautomerism (by de-bonding of the -O-H bond of enol form and provision of reactive hydrogen to the Ag^+ ion causing formation of AgNPs). The causative formation of AgNPs is probably due to BDE (O-H group of the catechol moiety of flavonoid < another O-H group of aromatics). With the reduction of the metal ion into nanoparticles, the phenolic moiety converts into keto group which gets attached on the surface of nanoparticles causing prevention of aggregation of nanoparticles and further stabilized nanoparticles are obtained (Nayem et al. 2020). PS Priyadharshini et al. (2024) concluded through FT-IR, for proteins being responsible for stabilization of AgNPs synthesized by N. nucifera flower extract. Functional groups including carboxyl and amine bound on silver nanoparticles could have acted as substrate, reducing and capping agents, thus confirming the participation of bio-active components in fabrication of nanoparticles. Palithya et al. (2022) in FTIR analysis of AgNPs, formulated from Aerva Lanata flower confirmed the participation of proteins, carboxylic acids and polyphenols in the reduction of silver ions (Ag+ to AgNPs) due to presence of bio-compounds including flavonoids, saponins, phenols, alkaloids and alcohols, aiding in the stabilization (inhibiting agglomeration) of AgNPs by acting as capping agents. Building on the similar concept Ibrahim et al. (2019) confirmed the role of flavonoids in reduction step of safflower mediated AgNPs, ZnNPs, CuNPs, and FeNPs synthesis (Fig. 9.2).

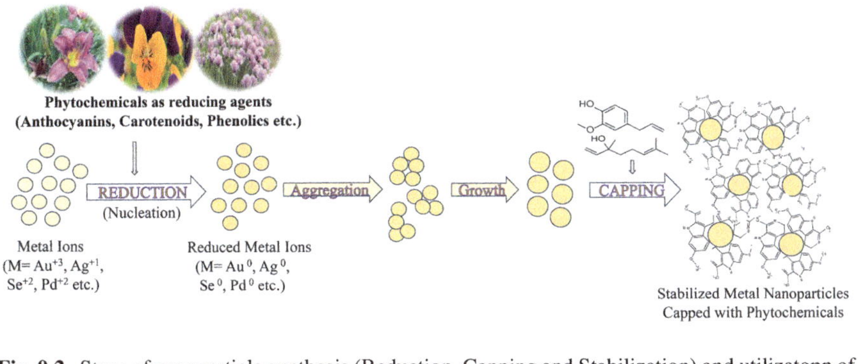

Fig. 9.2 Steps of nanoparticle synthesis (Reduction, Capping and Stabilization) and utilizatopn of Phytochemicals in the mediated steps

9.5 Methods of Nanoparticle Synthesis

9.5.1 Common Techniques for Nanoparticle Synthesis

Apart from anthropological formation and occurrence of nanoparticles in volcanic dust, mineral ores, smoke, soil water tides etc., they can also be synthesized through various other methods with varying dimensions and composition. Nanoparticulation causes creation of multi-component system from already existing simpler components along with imbibing the essence of resources in final systems. The preparation of nanomaterials is a multifaceted approach however the broad synthesis procedures (of inorganic nanomaterials derived from edible flowers) includes chemical, physical/mechanical and green synthesis with the utilization of "top-down" and "bottom-up" approaches (Fig. 9.3).

The physical method utilizes mechanical forces and evaporation in controlled experimental conditions to synthesize nanomaterial (Singh et al. 2023a, b) by using "top-down" approach. The initiation occurs from bulk materials which then undergoes size reduction/disintegration via physical/mechanical processes or decomposed into a smaller molecule to produce nanoparticles (Shah et al. 2015). They encompass lithographic techniques, such as crushing, splitting, and milling in methods like ball milling, laser ablation, inert gas condensation, engraving etc. In chemical methods nanoparticulation is caused by elementary unit entities, i.e. atoms or smaller particles, assembling together to formulate nanoparticles, more appropriately termed as "bottomup" methods. The methods under this category are chemical reduction, sono-chemical methods, electrochemical methods, solvo-thermal methods, micelles, nano-emulsions, etching, thermolysis methods (e.g., pyrolysis and

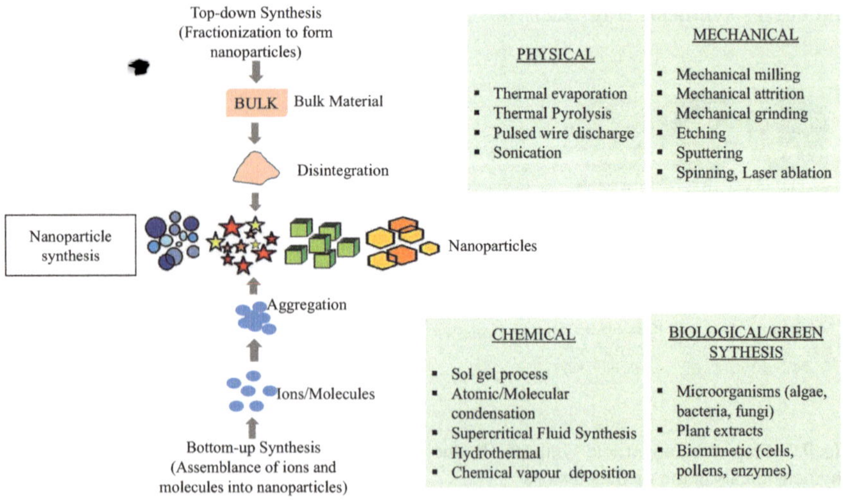

Fig. 9.3 Top down and bottom-up approaches of nanoparticle synthesis

spray pyrolysis), co-precipitation (mainly for semiconductors and oxides), solvated metal atom dispersion, and hybrid methods that methodically combine several of the aforementioned methods to form complex structures on the nanoscale. Green synthesis from bacteria, fungi and plants also utilizes the bottom-up approach for the formulation of nanoparticles. Chemical and green nanoparticulations are observed through color change on reduction of metal ion precursors to their elemental states.

Nanomaterials formulated from Bottom-up approaches allows spatial control of the properties and composition of the individual blocks along with yielding nanomaterials with less defects. On the contrary, top-down methods have a slow production rate thus hindering the commercial production of nanomaterials.

9.5.2 Biological-Synthesis of Nanoparticles

Biological Methods compensate for the limitations including toxicity (arising from utilization of harmful solvents and compositional and morphological characteristics), higher production cost etc. exhibited by physical and chemical methods by utilizing bottom-up approach for synthesis of nanomaterials. The bio-resources utilized for green synthesis include:

- ***Microorganisms***: nanoparticulation occurs when unicellular/multicellular microorganisms (algae, yeasts, bacteria, fungi, viruses) interact, extract, capture and transform the metal ions in their vicinity environment into native metals utilizing extracellular/intracellular enzymes (byproducts of inherent biochemical processes).
- ***Plants***: Almost all plant components including roots, stems, leaves, flowers and fruits, exhibit bioactive agents which exhibit the potential for green formulation of nanoparticles. The functional groups including polyphenols, polysaccharides, flavonoids, proteins, organic acid, vitamins and various secondary metabolites aid in reducing, stabilization and formation of full-fledged nanoparticles with peculiar and exclusive characteristics. Concentration and composition of active components define the inherent characteristics of the created nanoparticles.

9.5.3 Green Synthesis of Nanoparticles and Benefits

Green chemistry arises with the concept of sustainable process development with an inclined approach towards minimizing environmental and synthesis risks associated with devising of the nanoscale materials. In contrast to the chemical procedures which generally utilize expensive, toxic and hazardous raw materials posing environmental risks and unsustainability green synthesis approach employs a safe, sustainable, biocompatible, scalable and eco-friendly method for synthesizing bioinspired materials. Microorganisms such as bacteria, fungi, algae along with plants, act as precursors for the green synthesis. Diverse plant components including leaves, fruits, roots, stems, seeds and flowers are utilized for the synthesis of particles in nanoscale region. Flowers as raw materials for formulation of nanoparticles offers a clean and environmentally acceptable 'green chemistry' procedure due to their biocompatibility and sustainability. The method for nanoparticles synthesis can be performed via intracellular or extracellular methods. While intracellular synthesis is carried out in plant's cells, utilizing its growth conditions and further extraction of nanoparticles by breaking down the plant structure, extracellular method involves the generation of nanoparticles in controlled outside environment (Saim et al. 2021). Phytochemicals mediated synthesis is based on extracellular method which not only aids in reduction of metal ions but also stabilization of nanoparticles, thus imparting final shape to the full-fledged nanoparticles.

Green synthesis of nanoparticles through flowers occurs in following steps (Figs. 9.2 and 9.4):

(i) ***Initial phase***: Obtaining the reaction medium i.e., mixture of alcoholic/aqueous extract of edible flower species and the precursor salt (source of metal ions).
(ii) ***Activation phase***: Reduction of mono/di/tri-valent metal ions to zero-valent metal atom (elemental state) (facilitated by the phytochemicals present in the flower) and generation of nucleation centers.

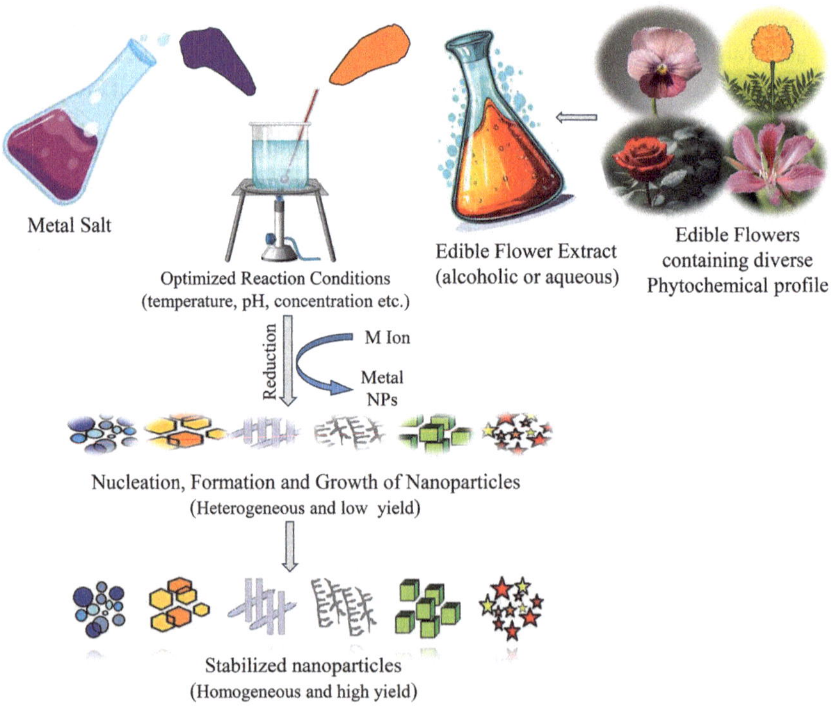

Fig. 9.4 Synthesis of edible flower mediated uniform and stabilized metal nanoparticles

(iii) ***Growth phase***: Small nanoparticles generated after nucleation, rapidly aggregate and clump together into large oligomeric particles causing Ostwald ripening, as per the controlled conditions (temperature, type of active components, concentration, pH reaction time etc.).
(iv) ***Capping and Stabilization***: The final favorable morphology and shape of the nanoparticles is determined after capping of nanoparticles with phytochemicals leading to stabilization, and thus marking the termination of synthesis procedure.

9.5.4 Benefits of Green Synthesis

- Few or no toxic wastes are generated.
- High synthesis yields, thus supporting the maximal utilization of raw materials.
- Synthesis is generally carried out at minimal or room temperatures thus chances of phytochemical/active component degradations are minimal.
- No additional sub/undesirable/side-products are created which may hinder the synthesis procedure.
- One step and quick synthesis occurs due to no requirement of supplementary reducing/capping/stabilization agents (phytochemicals itself act as reducing agents).
- Significant control over the synthesis reactions which can be feasibly terminated during the duration of experimental process, if required.
- Risks of mis-happenings is minimized since the reactions are performed at ambient conditions (especially temperature).
- Highly energy efficient procedure

9.6 Parameters Affecting Nanoparticles Synthesis

Diverse factors are associated with restoration of metals ions to form nanoparticles utilizing different flower extracts, which have a direct and indirect influence on the formulation of nanoparticle rendering them with peculiar features. The preparation of nanoparticles initiates with nucleation and thereafter formation followed by further growth of stable nanoparticles. Several factors thus prominent during the nucleation and growth stage significantly influence the growth of nanoparticles such as:

- ***Technique or synthesis method***: Nanoparticles are conventionally formulated using physical and chemical processes however green synthesis utilizing bioresources, is sustainably advantageous over conventional method of synthesis.
- ***Content, combination, and concentration of biomolecules***: Polyphenols, Proteins and carbohydrates are the major elements influencing nanoparticle synthesis (reduction, growth, capping and stabilization). Different edible flowers

vary in the type, amount and combination of bio-active components (phenols, aldehydes, alkaloids, aromatics, flavonoids) they consist, consequently variation in characteristics of synthesized nanoparticle occurs. Moreover, the plant extract's concentration utilized in the fabrication of metallic NPs significantly enhances the size and shape along with production (Maniam et al. 2020) and stability.

- ***pH of plant extract***: as per various researches, pH significantly effects the binding of metal ions to the phytochemicals, directly influencing the nucleation centers and stability. Moreover, size and texture of nanoparticle also get influenced with the pH of the solution (Harish et al. 2022). According to Singh et al. (2023a, b) pH 3–6, yields NPs with low stability due to probable aggregation while pH <3 and pH >7 yields stable nanoparticles. Muthu and Priya (2017) observed increased size of AgNPs was observed on decreasing the pH during synthesis of NPs from *Cassia auriculata* flower extract. Furthermore, the specificity of alkaline and acidic environments significantly influences the formation of nanoparticles for instance, alkaline pH favours formation of magnetic nanoparticles while metal oxide nanoparticles are usually synthesized in acidic or neutral Ph (Handayani et al. 2020). Chinnappan et al. (2018) also observed efficient synthesis of Ag nanoparticles by complete reduction of silver at pH 7.0 and the decreased reduction rate in acidic and alkaline pH which could be attributed to denaturation of bio-active components in the flower on exposure to altered pH.
- ***Temperature***: in contrast to conventional synthesis methods i.e. physical and chemical methods which require higher synthesis temperature (>350 °C and <350 °C) (Harish et al. 2022), green synthesis is generally limited to temperatures below 100 °C or room temperatures. Room temperature operations although preserve the stability for long term, higher temperatures along with potentially catalyzing the synthesis process, stimulate the interaction of bio-active agents with the surface of nanoparticles, thus preventing further clumping and agglomeration. The reaction temperature determines the nature (size, shape, synthesis) of synthesized nanomaterials (Chauhan et al. 2012). Higher temperatures may stimulate the nucleation process, which could distress the secondary recovery procedure, causing the metal to condense on the surface of developing nanoparticles. This explains formation of different shaped gold nanoparticles from alcoholic Nyctanthes arbor-tristis (night jasmine) flower extract at room temperature, while formation of spherical nanoparticles at 80 °C (Das et al. 2011).
- ***Reaction time***: duration of the synthesis reaction significantly influences nucleation, formation, growth and thus the stability of NPs. Moreover, the constructed nanoparticles exhibit the tendency to enlarge (De Oliveira et al. 2020) or shrink if stored for a prolonged time, influencing their functioning/activity.
- ***Metal Salt Concentration***: metal salts control the crystal growth and direct the synthesis in a shape- and size-controlled manner by acting as surfactants (Aljabali et al. 2022).
- ***Pore Size***: Molecules of different biological origin tend to adhere to different nanomaterials having varying pore sizes, which hold significant significance in applications involving drug delivery.

- ***Ion electrochemical potential***: Ions with a large positive electrochemical potential (eg, Ag+) may exhibit higher capacity to regenerate metal ions. Therefore, a plant extract's capacity to efficiently regenerate metal ions can be significantly greater for ions with a high positive electrochemical potential (like Ag+) than for ions with a low electrochemical potential like [Ag(S2O3)2]3-)] (Haverkamp and Marshall 2009).

Therefore, careful monitoring and controlling these reaction conditions significantly influence the properties of green synthesized nanoparticles.

9.7 Edible Flowers Mediated Nanoparticles and Their Characterization

9.7.1 Diverse Inorganic Nanoparticles

The unique functional properties exhibited by the diverse edible flower varieties offer the usefulness for preparation of nanoparticles. The flower-mediated synthesized NPs include various M and MO nanoparticles such as iron, silver, gold, copper, platinum, palladium, iron oxide, copper oxide etc. with exclusive features. Employing flowers as biological agents potentially delivers the unique properties of flowers into the formulated nanomaterials. Silver and Gold nanoparticles are majorly nanoparticulated materials due to large surface area, diverse biochemical reactivity and exclusive targeted delivery potential.

The synthesis of AgNPs occurs by reduction of monovalent Ag^+ ions to Ag with color changing to brownish. Alcoholic and aqueous extracts from flowers are utilized for the synthesis of AuNPs involving Au^{1+}/Au^{3+} reduction to Au. The green synthesis of gold nanoparticles occurs with appearance of ruby red colour from pale pink (influenced by concentration of precursor gold metal salt). The nanoparticles range from 4 to 100 nm in size and are spherical shaped. The gold nanoparticles (AuNPs) exhibit good optical characteristics, and biocompatibility (Elia et al. 2014) thus stimulating their wider usages. The green synthesis of TiO_2 is confirmed by the reduction of Ti ions with the general colour change to light-green Srinivasan et al. (2019). The ZnNPs exhibit sizes smaller than 60 nm and are mostly spherical. Commonly utilized Zn precursors are zinc nitrate and zinc acetate. The successful formation of zinc nanoparticles is observed by a white or yellowish paste (Pillai et al. 2020). The formation of CuNPs involves reduction of Cu^{+2} ions to Cu through a color change from yellowish to brownish or blue to brown. Commonly utilized precursors include cupric nitrate trihydrate, Cu^{+2} sulfate pentahydrate, copper chloride, and Cu^{+2} acetate. NiNPs synthesis involves the colour changes to dark brown from green (Abdullah et al. 2022) with the commonly utilized precursors nickel nitrate, nickel sulfate and nickel chloride. The green synthesis of MgONPs proceeds with the reduction Mg^{+2} ions to Mg with the color change from brownish to dark brownish-red or pale green to brown. Precursors involved in the synthesis of nanoparticles include magnesium chloride and magnesium nitrate. The formation of SeNPs proceeds with reduction of Se ions identifiable by a color change to light pink or light green to brick red from initially being colorless. FeONPs. The various precursors included in the synthesis of nanoparticles are Ferric chloride hexahydrate, ferric nitrate and ferrous sulfate. The reduction of ion nanoparticles is marked by appearance of black precipitate during the synthesis.

9.8 Techniques for Nanoparticle Characterization

The diversity of edible flowers mediated nanoparticles make the characterization an indispensable parameter to determine their properties which in turn determine the long term applicability of the nanoparticles in the market. Various analytical methods are utilized for determination of properties of nanoparticles for their shape, size, orientation, homogenous-ity, disparity, crystallinity etc. Commonly utilized methods for the characterization of edible flower derived nanoparticles include (Tables 9.2 and 9.3):

Table 9.2 Characterization techniques of nanoparticles

Characterization techniques	Parameters under study
SEM, TEM, HR-TEM, UHR-TEM	NPs morphology (size and shape), allography, arrangement of microstructures and particle size distribution
UV-VIS spectroscopy	Wavelengths utilized for nanoparticle characterization: UV range- 190 to 380 nm and visible range- 380 to 800 nm. eg. AgNPs usually are detected at peak ranges between 400 and 450 nm
Dynamic light scattering (DLS)	Quantification of surface charge and size distribution analysis
Zeta potential	Charge study
FTIR analysis	Surface study to detect phytochemicals and active -agents present in terms of functional groups
EDS/EDX mapping	Elemental composition of nanoparticles
XRD	Crystalline or amorphous structural characteristics with size and orientation patterns
Raman spectroscopy	Detection of vibrational molecules
Superconducting-quantum-interference device magnetometry. VSM (vibrating sample magnetometry)	Magnetic properties

Table 9.3 Characteristics of diverse nanoparticles derived from edible flowers

Metal NPs from edible flower	Color change	SEM/TEM/HRTEM/UHRTEM		FTIR	XRD	TGA/DTA	UV visible spectroscopy	DLS	Raman spectroscopy	EDX	References
		Size (nm)	Shape								
Selenium (*Cassia javanica*)	Yellow → light orange (sodium selenite →SeNPs)	35—100 nm	Spherical and irregular	3232 cm^{-1} → O-H stretching vibrations, 1588 cm^{-1} → tertiary amide, 1423 cm^{-1} → COO- group, 725 cm^{-1} → C-H stretching vibrations, 447 cm^{-1} → selenium in the nano-solution	Crystalline	–	–	–	–	–	Soliman et al. (2024)
Cadmium (*Tagetas* and (*Roseous*)	Marigold → straw yellow color Rose → reddish brown solution	–	Absolutes spherical/ roughly spherical with smooth edges	3300 cm^{-1} → alcoholic/ phenolic –OH, 2930 cm^{-1} → amide, 1643 cm^{-1} → C–C and 1100 cm^{-1} → –OCH3 groups Other metabolites such as flavonoids, tannins, alkaloids, and carotenoids (cause variation in peaks) also take part in nanoparticulation of CdNPs.	–	–	Distinct peaks at →387.33 nm for both flowers mediated green synthesis	–	–	–	Hajra et al. (2016)
Copper (*Coccinia grandalis*)	Green → brown	Average particle size: 18–20 nm	Polycrystalline in nature	3435 cm^{-1} → alcoholic–OH and amines N-H group, 2360 cm^{-1} and 1640 cm^{-1} → → –OCH3 groups and >C=O stretching	Crystalline (FCC structure) Average crystallite size → 22.8 nm	–	Strong Plasmon resonance at ~550–560 nm	–	–	–	Devi and Ahmaruzzaman (2017)
Gold (*Polianthes tuberosa*)	Deep purple-red color formation	Average size → 38.76 nm	Spheres, triangles, pentagons, hexagons, and rods.	3434 cm^{-1} → –OH groups,1582 cm^{-1} → amide group, 1403 cm^{-1} → C=C aromatic rings, which might (presence of phenolic compounds). 1113 cm^{-1} → C-O hydroxyl ester group, 695 cm^{-1} → was confined to the alkyl halides group. (nanoparticle synthesis→ due to the presence of proteins and bioactive metabolites).	Crystalline (FCC structure)	–	Strong Plasmon resonance at → 543 nm	–	–	Characteristic absorption band of gold was observed. (weak bands for chlorine, potassium, and chloride)	Alghuthaymi et al. (2023)

(continued)

Table 9.3 (continued)

Metal NPs from edible flower	Color change	SEM/TEM/HRTEM/UHRTEM		FTIR	XRD	TGA/DTA	UV visible spectroscopy	DLS	Raman spectroscopy	EDX	References
		Size (nm)	Shape								
Silver (*Nyctanthes arbortristis*)	Appearance of brown color	5–20 nm (without any secondary nucleation or agglomeration)	Spherical and oval	3430 cm^{-1} → OH, 2917 cm^{-1} & 2848 cm^{-1} → alkanes, 1696 cm^{-1} → carbonyl, 1435 cm^{-1} → geminal methyl, 1044 cm^{-1} → aliphatic amines. The FTIR results confirmed the capping of phytochemicals on the surface of the AgNPs.	Crystalline (FCC structure)	70% of organic material → degraded at higher temperature.	SPR peaks → broad (at EFE conc. 1–2% (v/v)) SPR peaks → sharp and narrower (at slightly increased EFE conc), SPR peaks → broadened (at very higher conc. of EFE (8–10%). Thus, at very high and low EFE conc., nanoparticle formation is anisotropic.	–	–	–	Gogoi et al. (2015)
Silver (*Bougainvillea glabra*)	Light yellow → brown color	10–50 nm	Predominantly sphericaloval	3490–3500 cm^{-1} → stretching of O-H and H- bonds (alcoholic and phenolic groups), 1450 and 1500 cm^{-1} → stretching vibrations of C=O and C=C,1400 and 1550 cm^{-1} → N-H stretching	Crystalline (FCC structure)	–	Strong Plasmon resonance (sharp peak) at ~418–422 nm	Particle size → 20–100 nm Charge → −15.2 ± 0.5 Polydispersity index → 0.21.	Visible D and G bands → Confirming biosynthesis process of AgNPs formation.	–	Oves et al. (2023)

Silver (*Aerva lanata*)	Colorless → dark brown	2–10 nm Average size → 7 ± 3 nm	Spherical	3841 cm^{-1} → N–H (secondary amines), 3735 cm^{-1} and 3613 cm^{-1} → O–H (alcohols), 3200 cm^{-1} → N–H (amides group), 2434 cm^{-1} → C ≡ C (carbonyl), 2352 cm^{-1} → C ≡ C (carbonyl), 1640 cm^{-1} → C=C (aromatic groups), 1530 cm^{-1}, 1395 cm^{-1} → CH– and C–O stretching vibrations of amide linkage of proteins, 1242 cm^{-1} and 1032 cm^{-1} → C–OH and C–C stretching vibrations of carboxylic acids. 668 $cm^{-1}$and 534 cm^{-1} → C–Cl and C–Br stretching vibrations of halides groups.	Crystalline (FCC structure)	–	Strong Plasmon resonance at → 443 nm	Particle size → 5–15 nm (average size → 7.6 nm) Polydispersity index → 0.419 Zeta potential → −18.7 mV	–	Elemental Silver with weight percentage → 63.44% Weak bands for sodium, oxygen, chlorine etc. were traced.	Palithya et al. (2022)
Silver (*Bauhinia purpurea*)	Brown color obtained	5–50 nm Average diameter →20 nm	Spherical	3443.45 cm − 1 → alcohols and phenolic O–H stretching, 2924.05 cm^{-1} → aldehydic C–H stretching, 1739.81 cm^{-1} → C=O stretching of the carbonyl group	Crystalline (FCC structure)	–	Strong Plasmon resonance band → 450 nm.	–	–	Silver element → 100% without any contaminants	Chinnappan et al. (2018)
Silver (*T. erecta*)	Yellow → dark brown	10–90 nm (average size → 46.11 nm)	Spherical, hexagonal and irregular in shape with moderate variation in size	3740.10 cm^{-1} → N-H amide stretching, 727.19 cm^{-1} → C-H stretching strong aromatic mono-substituted benzene. 793.73 cm^{-1} → C-Cl stretching alkyl halides, 693.43 cm^{-1} → C-H Stretching strong vinyl di-substituted alkenes.	Crystalline (FCC structure)	–	Strong Plasmon resonance at → 430 nm	Zeta Potential → −27.63 mV	–	Presence of elemental silver	Padalia et al. (2015)

(continued)

Table 9.3 (continued)

Metal NPs from edible flower	Color change	SEM/TEM/HRTEM/UHRTEM		FTIR	XRD	TGA/DTA	UV visible spectroscopy	DLS	Raman spectroscopy	EDX	References
		Size (nm)	Shape								
Silver (*Malva sylvestris*)	Pink → dark Brown	Measurement by: SEM → 32–54 nm and average diameter → 30 nm Measurement by: TEM → 20–30 nm.	Spherical (average diameter → 30 nm)	3413 cm^{-1} → OH-stretching in alcohols and phenols, 2920 cm^{-1} → C-H stretching vibration, 2851 cm^{-1} → aldehyde group, 1634 cm^{-1} → stretching and bending vibration of carbonyl amide, 1318 cm^{-1} → stretching vibration of C-N and bending vibration of OH bond (presence of phenol amide, amino, carbonyl and flavonoid led to formation of AgNPs)	Crystalline with cubic structure Average size → 20.19 nm	–	Strong Plasmon resonance for AgNPs at →445 nm	–	–	–	Mehdizadeh et al. (2021)
Silver (*Achillea biebersteinii*)	Color change to dark brown	5 to 35 nm (mean size → 12.58 nm).	Different shapes such as hexagonal, pentagonal and spherical structures	–	–	–	Strong Plasmon resonance at → 460 nm	–	Zeta potential at pH 7.2 → −20 to −40 mV.	Sharp optical absorption peak at~3 KeV and strong signal in the silver region	Baharara et al. (2014)
Silver (*Nelumbo nucifera*)	Colorless → Reddish brown	~200 nm Diameter → 63.47–70.72 nm	Spherical and cubic	2081.97 cm^{-1} → C–H stretching vibration of methyl, methylene, and methoxy groups, 1688.19 cm^{-1} → C=O stretching of amide I bands of peptide linkage. 1394.43 cm^{-1} → stretching vibration of N–H groups and the bending vibration of C–N groups, amide II and III bands, in the proteins, 1112.49 cm^{-1} → aliphatic amides (proteins are responsible for stabilizing the nanoparticles).	Crystalline (FCC structure)	–	Strong Plasmon resonance at ~440.90 nm	–	–	Presence of elemental Silver with weight percentage → 64.42%.	Priyadharshini et al. (2024)

Silver (*Taraxacum officinale*)	Yellow → Dark brown rust	45–55 nm	Spherical with smooth surfaces	–	Crystalline with cubic structure	–	Strong Plasmon resonance at ~452 nm	–	–	Strong absorption peak of Ag in the range of 3 keV → 97.9% silver Weak bands of nitrogen, sodium, oxygen.	Yousefzadeh Valendeh et al. (2023)
Metal oxide nanoparticles											
Copper oxide (organic veranera or bougainvillea)	Appearance of brown color	36 to 54 nm	Predominantly spherical-oval	–	Crystalline with mean crystallite size of ca. → 51.36 nm	18% weight loss (25 °C and 190 °C) → due to evaporation ~31% weight loss (300–700 °C) → due to soil organic breakdown	Strong Plasmon resonance (sharp peak) at ~270–390 nm	30 to 80 nm Majority of NPs → 36–54 nm Zeta potential → −37.8 mV	–	–	Natrayan et al. (2023)

(continued)

Table 9.3 (continued)

Metal NPs from edible flower	Color change	SEM/TEM/HRTEM/UHRTEM		FTIR	XRD	TGA/DTA	UV visible spectroscopy	DLS	Raman spectroscopy	EDX	References
		Size (nm)	Shape								
Iron oxide (*Hibiscus rosasinensis*)	Red → black	Average particle size → 65 nm	Spinel shape (cubic crystal system)	3434.77 cm^{-1} → OH stretch, 2939.45 cm^{-1} → C–H methylene asymmetrical stretch, 2978.04 cm^{-1} → C–H methyl asymmetrical stretch, 1670.46 cm^{-1} → imino group. 1422.01 cm^{-1} → sulphate moiety. 1374.91 cm^{-1} → –OH bend of phenol or tertiary alcohol, 1298.41 cm^{-1} → OH group of primary or secondary alcohol, 1117.9 cm^{-1} → C–O–CN stretch, 818.02 cm^{-1} → -C–O– of epoxy groups, 617.83 cm^{-1} → S–S stretch of disulphides, 478.73 cm^{-1} → S–S stretch of polysulfides. 567.94 cm^{-1} → Fe–O stretching vibration for Fe_3O_4 nanoparticle in the sample, 540 to 575 cm^{-1} → band of spinel ferrite.	Crystalline with cubic structure Crystallite size → 6.19 nm	–	Strong Plasmon resonance at → 229 nm	–	–	–	Razack et al. (2020)

Zinc oxide (*Nyctanthes arbortristis*)	White precipitate	12–32 nm	–	3340.6 cm^{-1} and 3258.2 cm^{-1} → H bonded OH stretch and N-H stretch 2127.8 cm^{-1} → stretching vibrations of C ≡ C stretch, 1641.1 cm^{-1} → stretching bands of C=O, 1456.7 cm^{-1} → amine -NH vibration stretch in protein amide linkages, 1362.5 cm^{-1} → aromatic amines, 1040 cm^{-1} and 1026.8 cm^{-1} → C-N stretch of aliphatic amines. The 746.25 cm^{-1} and 620.65 cm^{-1} → alkanes and C-H bend in alkynes	Crystalline (hexagonal phase of ZnO) Crystallite size → 16.58 nm	–	–	Hydrodynamic diameter → 74.36 nm Polydispersity index value → 0.488	–	–	Jamdagini et al. (2018)

9.9 Nanoparticles in Food Science

The inclined approach of processing, preservation, packaging, transportation, distribution, and storage, rely on food safety and quality (detection and elimination of harmful microorganisms or toxic metabolites), food enrichment/fortification (minerals, vitamins, antioxidants, and essential oils) and shelf-life extension. From raw materials to final products, food processing includes a whole hierarchy of unit operations. The diverse applications of inorganic nanoparticles are rendered to their elevated surface area, precise action, slow & targeted release on active sites, improved sensory performances and the imbibed exclusivity from the bio-resources. Dual patterns of nanoparticle utilization are observed, where, in first, nanomaterials can be used as nutritional supplements (nano-insiders) to alter the food quality directly by direct incorporation in the food matrix (coloring agents, preservatives, aromas, antioxidants, vitamins etc.) and second utilization is observed as food packaging's (nano-outsiders) (Ranjan et al. 2014) to aid in sustenance of food quality for longer durations. Physical and molecular functional forms of various antioxidants, antimicrobials, preservatives, flavouring agents, coloring agents are utilized in the food industry. Additives and food ingredients encapsulation signify the efficiency of applications and decide the utilization patterns. Deterioration of food quality due to water permeability, exposure to air, microbial infection etc. acts as major compromisal factor on ineffective packaging conditions thus the utilization of polymer-cum-nanoparticle based packaging serves as an innovative, efficient and feasible alternative. Subcellular-sized nanoparticles enhance bioavailability, offering a promising approach for subcellular nano-encapsulation methods. Silver nanoparticles are the most commonly utilized as antimicrobial agent in the fabrication of active food packaging while ZnO NPs are utilized as food forticants, in dietary supplements and can also be utilized to improve packaging barrier. TiO_2 NPs (chemically stable & effective photocatalyst) are also significantly incorporated in food packagings due to their inherent ability to block UV radiations and enhancing the barrier potential of the formulated films (Mohr et al. 2019) against various microbes particularly against *Escherichia coli*. Moreover, TiO_2 NPs also find their applications as coloring agent in processed food due to their brightness, color stability and high refraction index. SiO2NPs are often utilized as an anticaking agent in powdered products (e.g., dried milk, dry mixes, icing sugar etc.). Nanostructured materials also find their applications in gelling agents processing to improve food texture (Bajpai et al. 2018). Rao and Poonia (2024a, b) studied the microbial biosynthesis of selenium nanoparticles using probiotic strain and reported the major functional groups identified by FTIR analysis that includes conjugated ketones (C=O stretching) at 1648.82–1688.54 cm^{-1}, halo compounds (C–Br stretching) at 448.85–879.54 cm^{-1}, and fluoro compounds (C–F stretching) at 668.10–1065.89 cm^{-1}. Based on strain-specific measurements using SEM and TEM, the size of the nanoparticles ranged from 16 to 68.96 nm in diameter. Both extracellular and intracellular synthesis of the particles were identified. Significant amount of selenium was found using EDX analysis, with *Lactobacillus rhamnosus* and mix culture having the highest and the lowest concentrations at 97.50% and 90.58% by weight, respectively.

9.10 Applications of Edible Flower-Derived Nanoparticles

9.10.1 Fortifying Agents

Razack et al. (2020) utilized iron oxide nanoparticles, formulated from aqueous extract of *Hibiscus rosa-sinensis* in wheat biscuits as fortificants. As evident from the results, on incorporation of nanoparticles, physical properties remained unaltered as confirmed from the spread ratios (determine flour's rising capacity during baking and quality of flour for making biscuits) which remained almost same after incorporation of nanoparticles. However, the fortified biscuits were somewhat more brittle probably due to poor binding efficiency of FeONPs with the wheat flour. Moreover, the study demonstrated no significant changes in proximate characteristics on fortification, however increment in chemical content was observed along with which it was devised that fortification also did not affect storage and the chewability of the biscuits. Elevated iron content was observed in iron oxide nanoparticles fortified biscuits i.e. 1.5 mg of iron/g in contrast to the control biscuits which exhibited lower iron content i.e. 0.039 mg of iron/g. The bacteriological study reported delayed contamination of fortified biscuits which might be probably due to Iron oxide nanoparticles acting as antimicrobial agent and rendering the biscuits with long shelf life, stability and sound conditions.

9.10.2 In Packaging to Increase Shelf Life of Food Products

Singh (2022) prepared nanocomposites hybrid films by incorporation of bimetallic (Cu/Zn) nanoparticles derived from Hibiscus rosa sinensis flowers into the hydrocolloids i.e. chitosan (1.5 wt%) and pectin (0.5 wt%). The films prepared had high antimicrobial efficacy as evaluated via disc diffusion approach and therefore find applications for utilization in packaging concerning fruit safety as confirmed by the author. Singh et al. (2023a, b) confirmed the enhanced storage stability of white chocolate utilizing locust protein-based films incorporated with varying concentrations (0%, 1.0%, 1.5%, and 2.0%) of *E. purpurea* flower mediated nanoparticles in 90 days trial. The EFPNs apart from imparting bioactive properties also increased the antioxidant and antimicrobial profile as observed through various parameters including total phenolic and flavonoid contents, ABTS, DPPH, FRAP. On the similar lines, sensorial qualities also were in acceptable limits during the storage time.

Moreover Singh et al. (2024) redeveloped the same films to evaluate Parmessan cheese storage stability, in 90 days (at 4 ± 1 °C). The sensory characteristics of Parmessan Cheese remained uncompromised during the storage period and the films prepared were imbibed with improved antioxidant activity, ion-reducing potential and antimicrobial activities (as observed in ABTS and DPPH, total-phenolics, flavonoids and FRAP etc.), and inhibitory halos (mm) of the Parmessan Cheese. Feng et al. (2015) prepared Carbon dots from rose flowers which could potentially be utilized as carriers (in packaging's or otherwise) for transfer of active components.

9.10.3 Cytotoxic Potential of Edible Flower Derived Nanoparticles

Oves et al. (2023) evaluated the anticancer activity of Bougainvillea flower mediated AgNPs against lung cancer cells (A549). The study showed, on administration of metallic silver nanoparticles to lung cancer cells, resulted in the production of Reactive oxygen species (such as $H_2O_{2)}$ by the cells which potentially inhibited cellular signaling pathways and induced Mitochondrial-dependent apoptosis causing death of malignant cells. AgNPs generate toxicity in a proliferation-dependent and cell-specific manner to the least-susceptible and rapidly proliferating cells at quiescent stage thus causinf apoptosis of the cells. Priyadarshini et al. (2024) tested the synthesized silver nanoparticles against MDA-MB-231 breast cancer cell lines at varying concentration and confirmed the nanoparticles to be cytotoxic on the cancer cell lines.

9.10.4 Antioxidant Properties

ROS arising from biological activities cause cell lysis by oxidative degradation of the biological elements. Antioxidants not only deal with negative impacts of ROS but also play a significant role in imparting antimicrobial, anti-inflammatory, anti-cancerous etc. effects. The presence of phytochemicals in the resource (flower) itself provides the indispensable characteristic of exhibiting a high antioxidant profile (in nanoparticles) during nanoparticulation. For instance, SeNPs developed by Soliman et al. (2024) exhibited high antioxidant profile which confirmed the nanoparticles to get incorporated in food packaging's as natural antioxidants to replace the artificial ones. Bio-SeNPs apart from promoting scavenging free radicals, also maintain osmotic potential in plants during metabolic processes. Various nanoparticles derived from edible flowers also depict significant antioxidant values are shown in Table 9.4.

Table 9.4 Antioxidant properties of nanoparticles derived from edible flowers

Edible flower	Nanoparticles	Antioxidant values	References
Aerva lanata	AgNPs	78.96 ± 3.96 (DPPH scavenging activity) 72.86 ± 3.06 (hydrogen peroxide scavenging activity) 66.08 ± 3.12 (nitric oxide scavenging activity)	Palithya et al. (2022)
Bougainvillea	AgNPs	DPPH scavenging activity 41.20% at 25.0 gm/1 and on further increment of concentration scavenging effect slightly increased to 56.44% and 73.32%.	Oves et al. (2023)
Nyctanthes arbortis	AgNPs	IC50 value: 108.21 ± 3.20 μg/mL	Gogoi et al. (2015)
Nelumbo nucifera	AgNPs	IC50 value: 11.25 μg/ml	Priyadharshini et al. (2024)
Cassia	SeNPs	IC50 value: 53.34 μg/L	Soliman et al. (2024)

9.10.5 Antimicrobial Activities

Microbial contamination causes the unintended and undesirable deterioration of food products thus compromising with the sound conditions of food products and consequently affecting the overall health conditions. The antimicrobial effects exhibited by nanoparticles follows several mechanisms but works on the basic principle of interaction with outermost structural layer of microorganisms. The association of nanoparticles with the elementary protective barriers of (cell wall, cell membrane) microbes, causes alterations at cellular level (ribosomes, DNA, lysosomes etc.), thus impairing permeability of cell membrane, stimulating cytoplasmic leak, generating oxidative stress, causing genetic variations, deactivation of enzymes and inexpression of traits and finally causing apoptosis thus providing antimicrobial effects. The nanoparticle size and concentration determine the microbe growth inhibition rate (Videira-Quintela et al. 2021). Nanoparticles can also interact with other cell organelles and biomolecules to impart antimicrobial effects by causing oxidative stress, hindering ATP machinery etc. AgNPs are the most used nanoparticles for their antimicrobial effect due to their ease of penetration (due to small size) and efficient activity against microbes. Other nanoparticles with antimicrobial effects include AuNPs, CuNPs (anti-fungal and antibacterial properties), TiO2 (photocatalytic agent causing decay of microbes), ZnNPS (antimicrobial properties similar to AgNPs) etc. As per Wang et al. (2017), NPs are more potent against Gram-positive bacteria rendering their lower protective potential due to presence of simple cell wall containing thin sheet of peptidoglycan, teichoic acid with ample pores, thus allowing the feasible entry of foreign materials (nanoparticles). On the contrary, Gram-negative bacteria, with LPS, lipoproteins, and phospholipids as the cell wall constituents allow selective entry of macromolecules thus inhibiting nanoparticle interaction. Antimicrobial properties of various edible flower mediated nanoparticles are listed in Table 9.5.

Table 9.5 Antimicrobial Properties of Nanoparticles Derived from Edible Flowers

Edible flower	Nanoparticles	Inhibitory activity against	References
Cassia javanica	SeNPs	*B. subtilis, S. aureus, P. aeruginosa, E. coli,* and *C. albicans*	Soliman et al. (2024)
Marigold	AgNPs	*E. coli* and *P. aeruginosa, C. albicans, C. neoformans Against*	Padalia et al. (2015)
Bougaivillea	AgNPs	*E. coli* and *S. aureus strains*	Oves et al. (2023)
Poliahnthes tuberosa		*Escherichia coli* (10 mm > 11 mm > 13 mm) and *Staphylo coccus*	Alghuthaymi et al. (2023)
Malva sylvestris	AgNPs	*Escherichia coli* and *Salmonellatyphy murium*	Mehdizadeh et al. (2021)
Plumeria alba	AuNPs	*Escherichia coli*	Nagaraj et al. (2012)
Catharanthus roseus	AgNPs	*Bacillus subtilis, E. coli, Klebsiella pneumoniae, Pseudomonas putida,* and *S. aureus*	Manisha et al. (2014)
Tagetes erecta	AgNPs	*S. aureus >Bacillus cereus*	Padalia et al. (2015)
Jacaranda mimosifolia	ZnO NPs	*E. faecium* and *E. coli.*	Sharma et al. (2016)
Nyctanthes arbortristis	Zn NPs	*Alternaria alternate, Aspergillus niger, Botrytis cinerea, Fusarium oxysporum,* and *Penicillium expansum*	Jamdagni et al. (2018)

9.11 Safety, Toxicity, and Regulatory Aspects of Using Nanoparticles in Food

9.11.1 Toxicity and Safety Concerns

Although the production of nanoparticles has gained momentum yet certain lagging strands need to be held to ensure efficient and effective utilization of particles in nano-scale region. Nano material associated risks is generally determined by the physical and chemical properties including surface charge, size shape, interaction with the food matrix and the amount of exposure. The devised nanoparticles may exhibit substantially varied physicochemical and biological properties that may pose unpredictable hazards. The most concerning matter for toxicity is heavy metal release potential and allergenicity. Once inside the body, the biological fate of migration of nanomaterials is decided as per the passage as they may pass the biological barriers to enter the circulatory system and interact with biological systems, leading to potential toxicity. Orally ingested nanomaterials that may get translocated to organs (stomach, liver, kidneys, lungs, spleen, intestines and brain if remain un-excreted) on escaping the digestive system, thus producing accumulation pockets posing a serious potential safety concern. Some studies suggest that exposure to these nanoparticles can increase reactive oxygen species (ROS) production, reduce ATP levels, and trigger harmful effects such as carcinogenicity, genotoxicity, and cytotoxicity. As per few studies, ingestion of nanoparticles can cause protein denaturation (Hong et al. 2017), DNA damage (Lu et al. 2015), stimulation of oxidative stress responses (Khanna et al. 2015), and other adverse biological interferences. SiNPs often used as anti-caking agents, on exposure, can have cytotoxic effect in human lung cells. Ingestion of higher amounts of colloidal AgNPs results in interference in the digestive process and causes argyria (irreversible skin reaction). As per various researches, TiO2 can lead to micro inflammation in the colon, causing colon cancer (Handford et al. 2015) while the ZnONPs cause cell cause cytotoxicity and cell damage on penetrating the intestinal cells (Jafari et al. 2017). Over time, bioaccumulation and heightened reactivity may contribute to organ damage, including lung injury, kidney diseases, and hepatic dysfunction. Additionally, respiratory or skin absorption presents another risk of exposure. Inadequate packaging performance further exacerbates the issue, as nanomaterials may leach into food products, providing a direct pathway for human consumption. This raises serious ethical, environmental, and regulatory concerns regarding toxicity and ecotoxicity, highlighting the urgent need for stringent safety evaluations and regulations to mitigate potential health hazards.

9.11.2 Potential Toxicity Assessment Methods

Toxicity and risk assessment is an immediate concern to mitigate risks associated with commercialization and large-scale use of nanoparticles which remain uninvestigated or minimally investigated. For instance, as per Huang et al. (2011) increased silver migration levels were found in food modelling solutions (in two steps: initial release on the specimen surface and further penetration via a dual-sorption mechanism, diffusion, and embedding) with time and temperature. Concerning such safety and toxicity parameters of nanoparticles, *in vitro* and *in vivo* research of interactions with living matters and migration tests performance are indispensable prerequisite. Toxicity assessments can be broadly categorized as in vitro and in vivo with major mechanisms inclined towards testing membrane integrity, metabolic activity, DNA fragmentation, Lipid peroxidation etc. *In vitro* assays measure the alterations in gene expression along with cytotoxic and genotoxic effects. Cytotoxic assays include tetrazolium assays tryphan blue exclusion, alamar blue assay etc. while genotoxic assays measure DNA damage & repair, chromosomal aberrations via various investigatory measurements. Apart from utilizing reaction kinetics to determine toxicity, the following assays also perform similar role assays as depicted in the Fig. 9.5.

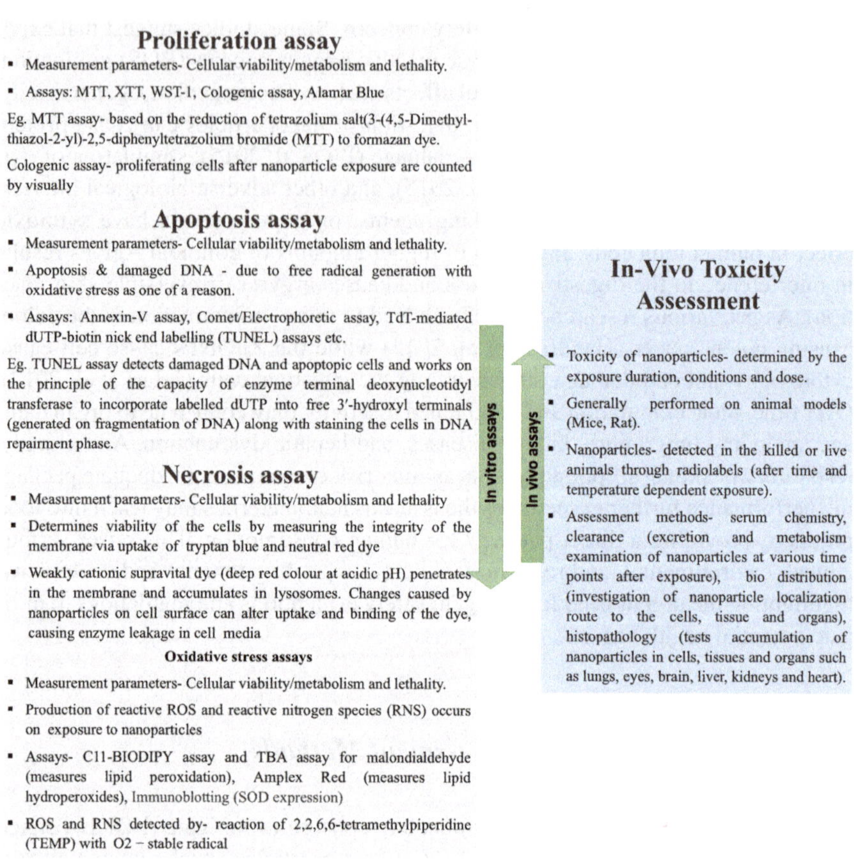

Fig. 9.5 In vitro and *in vivo* assays for toxicity assessment. (Kumar et al. 2017; Marquis et al. 2009)

9.12 Regulatory Guidelines and Compliance Considerations for Nano-food Products

In light of changing perceptions, utilization of nanomaterials in various food application still require a clear path to flourish. Manipulation at nanoscale region require regulatory guidelines, correct labeling and proper utilization to waive off safety concerns arising from untapped presence of nanoparticles at various hierarchical levels of utilization. Establishment of National Nanotechnology Initiative (USA), led to the initiation of governmental entrance in the nanoscale research arena. Building on similar alignment, “Nano Mark,” the first certification in Taiwan for nanoproducts, was granted in 2004. Minimal utilization patterns and availability of minute evidences of the potential risks held by nanoscale materials is causing lagging active formation of laws and regulations. As on date, there are no guidelines formed solely for nano-based food products, rather the regulations are being merged along with the existing ones, however few specific provisions are integrated with regulation criteria to ensure quality, safety and efficacy. The substitution of risk-based regulation with precautionary-based regulation serves as a regulatory approach derived from the precautionary principle (Hasmin et al. 2020). According to ISO nanomaterial is defined as the material with any external dimension in the nanoscale (length range approximately from 1 to 100 nm) or having internal structure or surface structure in the nanoscale and this is inclusive of nano-object (second and third external dimensions are orthogonal to the first dimension and to each other) and nanostructured material (International Standard Organization 2019). Different countries and legislation bodies are currently forming and adjusting the nanoparticle related compliances on the already occurring rules and regulations. The continuous innovations with varied functionality of NMs compromises on the formation of uniform evaluation parameters for emerging nano-food products which demands individual case-by-case basis evaluation approach for nano-based agri-input or food products. US FDA, EU are the umbrella legislation forming bodies involved in initiation of nanoparticles regulations. The FDA regulatory approach considers following attributes:

- FDA is maintaining its product-focused, science-based regulatory policy
- FDA's approach respects variations in legal standards for different product-classes.
- Where premarket review authority exists, attention to nanomaterials is being incorporated into standing procedures.
- Where statutory authority does not provide for premarket review, consultation is encouraged to reduce the risk of unintended harm to human or animal health.
- FDA will continue post-market monitoring.
- FDA will collaborate, as appropriate, with domestic and international counterparts on regulatory policy issues.
- Both for products that are not subject to premarket review and those that are, FDA will offer technical advice and guidance, as needed, to help industry meet its regulatory and statutory obligations. (Source: FDA 2018)

Other specifications as per Organization for Economic Cooperation and Development (2010) allow the usage of Food grade titanium dioxide (E171) as per US-FDA, at <1% by weight. ZnONPs often utilized as a fortificants (listed as GRAS by the FDA) and Silicon dioxide (E551) nanoparticles (anti-caking and coloring agent) were also approved for utilization in Food industry. However formal laws with applicability to all is yet to commence at a global scale.

9.13 Challenges and Limitations in the Use of Edible Flower Nanoparticles

Green synthesis of nanoscale metals holds significant potential, however challenges related to material selection, synthesis conditions, and quality control hinder its widespread utilizations and application. Despite the growing interest in green nanoscience, standardized protocols for producing uniform nanoparticles are still under development. Nanoparticles synthesized from edible flowers may suffer from instability, aggregation, or degradation over time, affecting their effectiveness in applications. Moreover, large-scale adoption is limited by concerns over nano-toxicity, necessitating stringent evaluation and continuous updates to toxicology and analysis protocols to align with application-specific requirements. Regulatory uncertainty further complicates commercialization, as ambiguous laws and guidelines must be clarified to facilitate the safe and sustainable use of nano-based products. Additionally, current advancements in green synthesis remain largely restricted to gold (Au) and silver (Ag) nanoparticles, highlighting the need for broader exploration in the field. While research on edible flower nanoparticles is growing, most studies are still in experimental stages, with limited real-world applications.

9.13.1 Technical Challenges in Scaling Synthesis for Commercial Production

The mass production of nanomaterials is an indispensable aspect for the exponential progress of the nanotechnology field with better capabilities. Nanomaterials possess unique *in vitro* and *in vivo* properties rendered to their peculiar synthesis procedures leaving them with different shape distribution, sizes, surface chemistry, targetability and physiochemical characteristics. Moreover, the diverse functionality of nanoparticles paves the way for the effective and efficient utilization of nanoparticles in various arenas including food science. For the commercial utilization of the desirous characteristics of the nanoparticles, the properties should be reproducible and scalable. However multiple hurdles and technical constraints may arise for scaling up the synthesis of nanomaterials for commercial production.

Some of the limitations occurring:

- **Formulation complexity**: Nanomaterial synthesis involves intricate processes and specific conditions to proliferate and sustain, thus can be difficult to replicate on a larger scale.
- **Reproducibility**: the source (different bioactive components) utilized for nanoparticle synthesis, intended application, raw materials utilized decide the reproducibility of the produced nanomaterials.
- **Production method**: the selection of the method utilized for nanoparticle synthesis including the physical, chemical and biological methods come with their own set of limitations which significantly influence the commercialization of nanoparticles. The formulation procedure and production units (batch) can be streamlined to allow for the ease of scale-up production and must be effective enough to ensure high reproducibility.
- **Defective nanomaterials**: The presence of defects in nanoparticle which may arise during synthesis, while aging or any other intermediate procedures are likely to influence the characteristic properties and thereby compromise on the performance of the nanomaterials when assumed for targeted delivery.

- **High cost**: elevated costs of synthesis of nanomaterials if sophisticated requirements and specific conditions are a mandate, therefore inclining towards cost-effective routes offers better alternative.
- **Agglomeration** of nanoparticles by electrostatic interactions, physical entanglement or fluctuating surface energies may be an inherent barrier in formulation of uniform nanoparticles.
- **Nano-toxicity**: The toxicology profile analysis has to be constantly and timely analyzed and the protocols for experiment performance has to be timely standardized to reflect upon the need of the application.
- **Solvent selection**: Standardizing the solvent utilization i.e. aqueous/alcoholic as per the targeted formulation procedure for extraction and further nanoparticle synthesis is crucial to ensure consistent and reproducible synthesis.
- **Trials**: Batch-to-batch preparation causes variability in composition and thus compromising on uniformity, also poses a challenge that needs to be minimized.
- **Research and Development** arena need to be well developed to commercialize the synthesis and utilization patterns.
- **Advanced characterization techniques** specific to peculiarity of synthesized green nanoparticles are needed to detangle the extrinsic and intrinsic properties of the mediated nanoparticles.
- **Contaminants** and impurities in plant extracts impair the efficacy of nanoparticle synthesis, thus must be eradicated to avoid hindrance in performance and sustain the purity and functionality of the nanoparticles over a long period.
- **Regulations** must be globally evident, transparent, and well defined for their successful translation into clinical and commercial utilization.

Although numerous methods with multifaceted potential have been developed including the green synthesis for production of nanoparticles, however integration of these methods with emerging technologies will potentially aid in, tackling the challenges, scaling up, and further elevating commercializing capabilities while unlocking the full potential of plant-derived nanoparticles.

9.14 Cost, Resource Availability, and Consumer Acceptance

The idea behind interactive food works on the principle of allowing consumers to alter/personalize their consumption as per their own nutritional needs or tastes. Nanotechnology in the food sector offers promising advancements in production, nutrition, flavor enhancement, processing, packaging, and preservation. However, its widespread adoption faces several key challenges, including cost, resource availability, and consumer acceptance. The development and incorporation of engineered nanomaterials (ENMs) require sophisticated research facilities, specialized equipment, and highly trained personnel, making the technology expensive to develop and scale. High production costs often translate to increased prices for end products, which may limit market adoption, especially in cost-sensitive economies. Additionally, resource availability is another constraint, as the synthesis of nanoparticles often depends on specific raw materials and advanced manufacturing techniques, which may not be readily accessible or sustainable in all regions. Regulatory uncertainties and the lack of a standardized framework further complicate large-scale commercialization. Consumer acceptance remains one of the most significant barriers to the integration of nanotechnology in the food industry. Public awareness of nanotechnology's presence in food products is relatively low, and concerns about health and safety persist due to existing uncertainties and limited long-term studies. Studies show that while consumers are generally more accepting of nanotechnology in food packaging ("nano-outside"), which can enhance shelf life and reduce contamination, they are more hesitant about direct incorporation of nanoparticles into food products ("nano-inside"). The perceived risks associated with ingestion, potential toxicity, and long-term health effects contribute to this reluctance. Moreover, there is a public perceived notion about nano label being a form of warning rather than being an additive/supplement/forticant etc. and therefore, default consideration of food with nano label to be riskier is assumed (Chuah et al. 2018). To overcome these challenges, further research, transparent regulatory policies, and proactive public engagement are essential to educate consumers, address safety concerns, and create a balanced approach that maximizes nanotechnology's benefits while ensuring safety and affordability.

9.15 Conclusions

The growing and emerging demand for utilization of sustainable green chemistry in nanotechnology has ushered the utilization of bio-resources including plants, microbes, and other biomimetic resources for nanoparticulation. Edible flowers are the potential sources utilizing inexpensive, quicker and sustainable approach for green nanoparticulation rendering their high phytochemical profile. The bio-active components such as phenolics, flavonoids, alkaloids, proteins etc. apart from accelerating the synthesis kinetics, also aids in the synthesis process by acting as effective reducing, capping and stabilizing agents giving rise to nanomaterials with peculiar characteristics which offer significant applications in food science including processing preservation and safety. The diversity of prepared nanomaterials demands proper characterization techniques, regulation and labelling compliances and ethical utilization patterns for evaluations of toxicity & safety concerns, interaction with food matrices and consumer acceptance. Moreover, addressing challenges related to large-scale production, regulatory approval and consumer awareness is essential for development of comprehensive risk assessment strategy, regulatory regimes and satisfying public concerns for their successful integration into the food industry. Flower mediated nanoparticle synthesis however is limited majorly to AgNPs synthesis at large scale limiting the potentiality and commercialization of other inorganic nanomaterials, therefore future research and technological advancements on other nanoparticle formulation and utilization of other bio-resources are required to further unlock the full potential and larger impact of these smaller particles.

References

Abdullah HT, Faisal S, Rizwan M, Saira ZN (2022) Green synthesis and characterization of copper and nickel hybrid nanomaterials: investigation of their biological and photocatalytic potential for the removal of organic crystal violet dye. J Saudi Chem Soc 26(4):101486

Alghuthaymi MA, Patil S, Rajkuberan C, Krishnan M, Krishnan U, Abd-Elsalam KA (2023) Polianthes tuberosa-mediated silver nanoparticles from flower extract and assessment of their antibacterial and anticancer potential: an in vitro approach. Plants 12(6):1261

Aljabali AA, Obeid MA, Bakshi HA, Alshaer W, Ennab RM, Al-Trad B et al (2022) Synthesis, characterization, and assessment of anti-cancer potential of ZnO nanoparticles in an in vitro model of breast cancer. Molecules 27(6):1827

Bachheti RK, Fikadu A, Bachheti A, Husen A (2020) Biogenic fabrication of nanomaterials from flower-based chemical compounds, characterization and their various applications: a review. Saudi J Biol Sci 27(10):2551–2562

Baessa M, Rodrigues MJ, Pereira C, Santos T, da Rosa Neng N, Nogueira JMF et al (2019) A comparative study of the in vitro enzyme inhibitory and antioxidant activities of Butea monosperma (lam.) Taub. and Sesbania grandiflora (L.) Poiret from Pakistan: new sources of natural products for public health problems. S Afr J Bot 120:146–156

Baharara J, Namvar F, Ramezani T, Hosseini N, Mohamad R (2014) Green synthesis of silver nanoparticles using Achillea biebersteinii flower extract and its anti-angiogenic properties in the rat aortic ring model. Molecules 19(4):4624–4634

Bajpai VK, Kamle M, Shukla S, Mahato DK, Chandra P, Hwang SK et al (2018) Prospects of using nanotechnology for food preservation, safety, and security. J Food Drug Anal 26(4):1201–1214

Bhaskar A, Nithya V (2012) Evaluation of the wound-healing activity of Hibiscus rosa sinensis L (Malvaceae) in Wistar albino rats. Indian J Pharmacol 44(6):694–698

Chaudhari D, Muthal A, Mali A, Salunke M, Shinde V (2024) Phytochemical and pharmacological activities of Tagetes erecta L: an updated review. Int J Herb Med 12:10–15

Chauhan RP, Gupta C, Prakash D (2012) Methodological advancements in green nanotechnology and their applications in biological synthesis of herbal nanoparticles. Int J Bioassays

Chinnappan S, Kandasamy S, Arumugam S, Seralathan KK, Thangaswamy S, Muthusamy G (2018) Biomimetic synthesis of silver nanoparticles using flower extract of Bauhinia purpurea and its antibacterial activity against clinical pathogens. Environ Sci Pollut Res 25:963–969

Chuah AS, Leong AD, Cummings CL, Ho SS (2018) Label it or ban it? Public perceptions of nano-food labels and propositions for banning nano-food applications. J Nanopart Res 20:1–17

Das RK, Gogoi N, Bora U (2011) Green synthesis of gold nanoparticles using Nyctanthes arbortristis flower extract. Bioprocess Biosyst Eng 34:615–619

De Oliveira RC, Amoresi RAC, Marana NL, Zaghete MA, Ponce M, Chiquito AJ et al (2020) Influence of synthesis time on the morphology and properties of CeO2 nanoparticles: an experimental–theoretical study. Cryst Growth Des 20(8):5031–5042

de Pascual-Teresa S, Sanchez-Ballesta MT (2008) Anthocyanins: from plant to health. Phytochem Rev 7:281–299

Du H, Lai L, Wang F, Sun W, Zhang L, Li X et al (2018) Characterisation of flower colouration in 30 rhododendron species via anthocyanin and flavonol identification and quantitative traits. Plant Biol 20(1):121–129

Duan H, Wang D, Li Y (2015) Green chemistry for nanoparticle synthesis. Chem Soc Rev 44(16):5778–5792

Devi TB, Ahmaruzzaman M (2017) Facile preparation of Copper nanaoparticles using Coccinia grandis fruit extract and its application towards the reduction of toxic nitro compound. Mater Today Proc 5(1):2098–2104. https://doi.org/10.1016/j.matpr.2017.09.206

Elia P, Zach R, Hazan S, Kolusheva S, Porat ZE, Zeiri Y (2014) Green synthesis of gold nanoparticles using plant extracts as reducing agents. Int J Nanomedicine:4007–4021

Faizy HS, Esmail LS, Mahdi HS (2021) Phytochemicals analysis in watercress (Nasturtium officinale) plant extracts. IOP Conf Ser Earth Environ Sci 761(1):012042

FDA (2018) Approach to regulation of nanotechnology products. https://www.fda.gov/science-research/nanotechnology-programs-fda/fdas-approach-regulation-nanotechnology-products

Feng XT, Zhang F, Wang YL, Zhang Y, Yang Y. Z, Liu XG (2015) Luminescent carbon quantum dots with high quantum yield as a single white converter for white light emitting diodes. Appl Phys Lett 107(21)

Gogoi N, Babu PJ, Mahanta C, Bora U (2015) Green synthesis and characterization of silver nanoparticles using alcoholic flower extract of Nyctanthes arbortristis and in vitro investigation of their antibacterial and cytotoxic activities. Mater Sci Eng C 46:463–469

Hajra A, Dutta S, Mondal NK (2016) Mosquito larvicidal activity of cadmium nanoparticles synthesized from petal extracts of marigold (Tagetes sp.) and rose (Rosa sp.) flower. J Parasit Dis 40:1519–1527

Handayani W, Ningrum AS, Imawan C (2020) The role of pH in synthesis silver nanoparticles using pometia pinnata (matoa) leaves extract as bioreductor. J Phys Conf Ser 1428(1):012021

Handford CE, Dean M, Spence M, Henchion M, Elliott CT, Campbell K (2015) Awareness and attitudes towards the emerging use of nanotechnology in the agri-food sector. Food Control 57:24–34

Hano C, Abbasi BH (2021) Plant-based green synthesis of nanoparticles: production, characterization and applications. Biomolecules 12(1):31

Harish V, Ansari MM, Tewari D, Gaur M, Yadav AB, García-Betancourt ML et al (2022) Nanoparticle and nanostructure synthesis and controlled growth methods. Nano 12(18):3226

Hasmin NA, Zainol ZA, Ismail R, Mahmood A (2020) Regulatory challenges of nanofood labelling. J Pol & L 13:241

Haverkamp RG, Marshall AT (2009) The mechanism of metal nanoparticle formation in plants: limits on accumulation. J Nanopart Res 11:1453–1463

Hong F, Wu N, Zhou Y, Ji L, Chen T, Wang L (2017) Gastric toxicity involving alterations of gastritis-related protein expression in mice following long-term exposure to nano TiO2. Food Res Int 95:38–45

Hotta KS, Neelapu N, Priyanka N (2021) Phytochemical analysis of the flowers of Chrysanthemum indicum L. and Calendula officinalis. Int J of Pharm and Chem 35–41. https://doi.org/10.46796/ijpc.vi.148

Huang Y, Chen S, Bing X, Gao C, Wang T, Yuan B (2011) Nanosilver migrated into food-simulating solutions from commercially available food fresh containers. Packag Technol Sci 24(5):291–297

Hussein RA, El-Anssary AA (2019) Plants secondary metabolites: the key drivers of the pharmacological actions of medicinal plants. Herbal Med 1(3):11–30

International Standard Organization/TR 21386 (2019) (en) Nanotechnologies- Considerations for the measurement of nano-objects and their aggregates and agglomerates (NOAA) in environmental matrices. https://www.iso.org/obp/ui/en/#iso:std:iso:tr:21386:ed-1:v1:en

Ibrahim FY, El-Khateeb AY, Mohamed AH (2019) Rhus and safflower extracts as potential novel food antioxidant, anticancer, and antimicrobial agents using nanotechnology. Foods 8(4):139

Jafari SM, Katouzian I, Akhavan S (2017) Safety and regulatory issues of nanocapsules. In: Nanoencapsulation technologies for the food and nutraceutical industries. Academic, pp 545–590

Jamdagni P, Khatri P, Rana JS (2018) Green synthesis of zinc oxide nanoparticles using flower extract of Nyctanthes arbor-tristis and their antifungal activity. J King Saud Univer Sci 30(2):168–175

Khanna P, Ong C, Bay BH, Baeg GH (2015) Nanotoxicity: an interplay of oxidative stress, inflammation and cell death. Nano 5(3):1163–1180

Kumar V, Sharma N, Maitra SS (2017) In vitro and in vivo toxicity assessment of nanoparticles. Int Nano Lett 7(4):243–256

Lu CF, Yuan XY, Li LZ, Zhou W, Zhao J, Wang YM, Peng SQ (2015) Combined exposure to nano-silica and lead induced potentiation of oxidative stress and DNA damage in human lung epithelial cells. Ecotoxicol Environ Saf 122:537–544

Lu B, Li M, Yin R (2016) Phytochemical content, health benefits, and toxicology of common edible flowers: a review (2000–2015). Crit Rev Food Sci Nutr 56(sup1):S130–S148

Maniam GP, Govindan N, Rahim MHA, Yusoff MM (2020) Plant extracts: nanoparticle sources. In: Phytonanotechnology. Elsevier, pp 41–49

Marquis BJ, Love SA, Braun KL, Haynes CL (2009) Analytical methods to assess nanoparticle toxicity. Analyst 134(3):425–439

Manisha DR, Alwala J, Kudle KR, Rudra MP (2014) Biosynthesis of silver nanoparticles using flower extracts of Catharanthus roseus and evaluation of its antibacterial efficacy. World J Pharm Pharm Sci 3(5):669–677

Mayedwa N, Mulaudzi-Masuku T, Matinise N, Nkosi M (2021) Biosynthesis and characterization of multifunctional mixed oxides of ZnCr2O4/ZnCrO4 nanoparticulate from natural leaf extracts of hibiscus Rosa Sinensis. Mater Today Proc 36:309–312

Mehdizadeh S, Ghasemi N, Ramezani M, Mahanpoor K (2021) Biosynthesis of silver nanoparticles using Malva sylvestris flower extract and its antibacterial and catalytic activity. Chem Methodol 5(4):356–366

Mohr LC, Capelezzo AP, Baretta CRDM, Martins MAPM, Fiori MA, Mello JMM (2019) Titanium dioxide nanoparticles applied as ultraviolet radiation blocker in the polylactic acid bidegradable polymer. Polym Test 77:105867

Muthu K, Priya S (2017) Green synthesis, characterization and catalytic activity of silver nanoparticles using Cassia auriculata flower extract separated fraction. Spectrochim Acta A Mol Biomol Spectrosc 179:66–72

Nagaraj B, Malakar B, Divya TK, Krishnamurthy N, Liny P, Dinesh R et al (2012) Synthesis of plant mediated gold nanoparticles using flower extracts of Carthamus tinctorius L.(safflower) and evaluation of their biological activities. Dig J Nanomater Biostruct 7:1289–1296

Natrayan L, Kaliappan S, Saravanan A, Vickram AS, Pravin P, Abbas M et al (2023) Recyclability and catalytic characteristics of copper oxide nanoparticles derived from bougainvillea plant flower extract for biomedical application. Green Process Synth 12(1):20230030

Nayem SA, Sultana N, Haque MA, Miah B, Hasan MM, Islam T et al (2020) Green synthesis of gold and silver nanoparticles by using Amorphophallus paeoniifolius tuber extract and evaluation of their antibacterial activity. Molecules 25(20):4773

Nishina A, Sato D, Yamamoto J, Kobayashi-Hattori K, Hirai Y, Kimura H (2019) Antidiabetic-like effects of Naringenin-7-O-glucoside from Edible Chrysanthemum 'Kotobuki'and Naringenin by Activation of the PI3K/Akt Pathway and PPARγ. Chem Biodivers 16(1):e1800434

Nowicka P, Wojdyło A (2019) Anti-hyperglycemic and anticholinergic effects of natural antioxidant contents in edible flowers. Antioxidants 8(8):308

Oves M, Rauf MA, Qari HA (2023) Therapeutic applications of biogenic silver nanomaterial synthesized from the paper flower of bougainvillea glabra (Miami, pink). Nano 13(3):615

Padalia H, Moteriya P, Chanda S (2015) Green synthesis of silver nanoparticles from marigold flower and its synergistic antimicrobial potential. Arab J Chem 8(5):732–741

Palithya S, Gaddam SA, Kotakadi VS, Penchalaneni J, Golla N, Krishna SBN, Naidu CV (2022) Green synthesis of silver nanoparticles using flower extracts of Aerva lanata and their biomedical applications. Part Sci Technol 40(1):84–96

Paudel KR, Panth N (2015) Phytochemical profile and biological activity of Nelumbo nucifera. *Evidence-Based Complementary and Alternative Medicine* 2015(1):789124

Pillai AM, Sivasankarapillai VS, Rahdar A, Joseph J, Sadeghfar F, Rajesh K, Kyzas GZ (2020) Green synthesis and characterization of zinc oxide nanoparticles with antibacterial and antifungal activity. J Mol Struct 1211:128107

Pires EDO Jr, Di Gioia F, Rouphael Y, Ferreira IC, Caleja C, Barros L, Petropoulos SA (2021) The compositional aspects of edible flowers as an emerging horticultural product. Molecules 26(22):6940

Priyadarshini PS, Sanidhya G, Vijayan R (2024) Synthesis of silver nanoparticles using flower of Nelumbo Nucifera its antioxidant activity, anti cancer activity on human breast cancer cell line. IJBPAS 13(2):625–644

Ranjan S, Dasgupta N, Chakraborty AR, Melvin Samuel S, Ramalingam C, Shanker R, Kumar A (2014) Nanoscience and nanotechnologies in food industries: opportunities and research trends. J Nanopart Res 16:1–23

Rao V, Poonia A (2024a) Bioactive compounds, nanoparticle synthesis and potential utilization of edible flowers for development of functional dairy products: a review. J Food Sci Technol 61(6):1053–1068. https://doi.org/10.1007/s13197-023-05853-6

Rao V, Poonia A (2024b) Microbial biosynthesis of selenium nanoparticles using probiotic strain and its characterization. J Food Meas Charact 18(7):5477–5488. https://doi.org/10.1007/s11694-024-02581-z

Rashidi L, Khosravi-Darani K (2011) The applications of nanotechnology in food industry. Crit Rev Food Sci Nutr 51(8):723–730

Razack SA, Suresh A, Sriram S, Ramakrishnan G, Sadanandham S, Veerasamy M et al (2020) Green synthesis of iron oxide nanoparticles using Hibiscus rosa-sinensis for fortifying wheat biscuits. SN Appl Sci 2:1–9

Saim AK, Kumah FN, Oppong MN (2021) Extracellular and intracellular synthesis of gold and silver nanoparticles by living plants: a review. Nanotechnol Environ Eng 6(1):1

Shah M, Fawcett D, Sharma S, Tripathy SK, Poinern GEJ (2015) Green synthesis of metallic nanoparticles via biological entities. Materials 8(11):7278–7308

Sharma D, Sabela MI, Kanchi S, Mdluli PS, Singh G, Stenström TA, Bisetty K (2016) Biosynthesis of ZnO nanoparticles using Jacaranda mimosifolia flowers extract: synergistic antibacterial activity and molecular simulated facet specific adsorption studies. J Photochem Photobiol B Biol 162:199–207

Sharma K, Kumar V, Kumar, S et al (2021) Bauhinia variegata: a comprehensive review on bioactive compounds, health benefits and utilization. Adv Tradit Med (Adtm) 21:645–653. https://doi.org/10.1007/s13596-020-00472-4

Sharma S, Pant P, Dave SV, Tak K, Sadhu V, Reddy KR (2019a) Green synthesis and characterization of copper nanoparticles by *Tinospora cardifolia* to produce nature-friendly copper nano-coated fabric and their antimicrobial evaluation. J Microbiol Methods 160:107–116

Sharma P, Pant S, Dave V, Tak K, Sadhu V, Reddy KR (2019b) Green synthesis and characterization of copper nanoparticles by Tinospora cardifolia to produce nature-friendly copper nano-coated fabric and their antimicrobial evaluation. J Microbiol Methods 160:107–116

Shen X, Shi L, Pan H, Li B, Wu Y, Tu Y (2017) Identification of triterpenoid saponins in flowers of four camellia Sinensis cultivars from Zhejiang province: differences between cultivars, developmental stages, and tissues. Ind Crop Prod 95:140–147

Singh AK (2022) Flower extract-mediated green synthesis of bimetallic CuZn oxide nanoparticles and its antimicrobial efficacy in hydrocolloid films. Bioresour Technol Rep 18:101034

Singh H, Desimone MF, Pandya S, Jasani S, George N, Adnan M et al (2023a) Revisiting the green synthesis of nanoparticles: uncovering influences of plant extracts as reducing agents for enhanced synthesis efficiency and its biomedical applications. Int J Nanomedicine:4727–4750

Singh S, Bhat HF, Kumar S, Aadil RM, Mohan MS, Proestos C, Bhat ZF (2023b) Storage stability of chocolate can be enhanced using locust protein-based film incorporated with E. purpurea flower extract-based nanoparticles. Ultrason Sonochem 100:106594

Singh S, Bhat HF, Kumar S, Aadil RM, Abdi G, Bhat ZF (2024) Insect protein-based composite film incorporated with E. purpurea-based nanoparticles augmented the storage stability of parmesan cheese. Heliyon 10(7)

Soliman MK, Amin MAA, Nowwar AI, Hendy MH, Salem SS (2024) Green synthesis of selenium nanoparticles from Cassia javanica flowers extract and their medical and agricultural applications. Sci Rep 14(1):26775

Srinivasan M, Venkatesan M, Arumugam V, Natesan G, Saravanan N, Murugesan S et al (2019) Green synthesis and characterization of titanium dioxide nanoparticles (TiO2 NPs) using Sesbania grandiflora and evaluation of toxicity in zebrafish embryos. Process Biochem 80:197–202

Videira-Quintela D, Martin O, Montalvo G (2021) Recent advances in polymer-metallic composites for food packaging applications. Trends Food Sci Technol 109:230–244

Wang Y, Ren N, Rankin GO, Li B, Rojanasakul Y, Tu Y, Chen YC (2017) Anti-proliferative effect and cell cycle arrest induced by saponins extracted from tea (Camellia sinensis) flower in human ovarian cancer cells. J Funct Foods 37:310–321

Chapter 10
Quantification and Characterization of Edible Flowers

Awanish Singh and Nandan Sit

10.1 Introduction

Edible flowers like roses (*Rosa damascena*), lotus (*Nelumbo nucifera*), violets (*Viola tricolor*), marigolds (*Calendula officinalis*), hibiscus (*Hibiscus rosa-sinensis*), nasturtiums (*Tropaeolum majus*), pansies (*Viola wittrockiana*), lavender (*Lavandula angustifolia*), and borage (*Borago officinalis*) have captured human fascination for centuries, celebrated for their aesthetic appeal, sensory qualities, and cultural significance (Mlcek and Rop 2011). In ancient civilizations, flowers were not only valued for their decorative properties but also incorporated into traditional cuisines, medicines, and rituals (Cunningham 2015). Their vibrant colours, distinct flavors, and delicate aromas have made them a popular choice for enhancing culinary experiences, while their bioactive properties have been linked to various health benefits (Pires et al. 2017). As the modern food industry continues to evolve, edible flowers are gaining recognition as a versatile ingredient with potential applications beyond their traditional roles.

In recent years, the focus on natural and sustainable food sources has propelled edible flowers into the spotlight as functional foods. Functional foods are defined as those that provide health benefits beyond basic nutrition, often due to their bioactive compounds (Fernandes et al. 2017). Edible flowers are particularly rich in phytochemicals, antioxidants, vitamins, and minerals, making them valuable for promoting health and preventing chronic diseases (Lu et al. 2016). Flowers have a high content of natural antioxidants such as anthocyanin, flavonoids and various other phenolic components (Rao and Poonia 2024a, b). Additionally, their sensory characteristics-ranging from sweet and floral to spicy and herbaceous-add depth to

A. Singh · N. Sit (✉)
Department of Food Engineering and Technology, Tezpur University, Tezpur, Assam, India

A. Poonia et al. (eds.), *Edible Flowers: Source of Phytonutrients, Valorization and Technological Advancements*, https://doi.org/10.1007/978-3-032-00667-7_10

culinary creations, catering to the growing consumer demand for unique and flavourful ingredients (Fernandes et al. 2020).

The global interest in edible flowers has also been fuelled by trends in gastronomy and wellness (Takahashi et al. 2020). High-end restaurants and food enthusiasts have embraced these blooms as garnishes, flavor enhancers, and central components in innovative dishes (Rodrigues et al. 2017). At the same time, the rise of plant-based diets and the preference for natural, minimally processed foods have opened new avenues for incorporating edible flowers into beverages, desserts, salads, and more (Pires et al. 2017). Beyond their culinary appeal, edible flowers are finding applications in cosmetics, pharmaceuticals, and nutraceuticals, highlighting their multifunctional potential (Chen and Wei 2022).

Despite their growing popularity, the scientific study of edible flowers remains relatively nascent (Chen and Wei 2022). To fully harness their potential, it is essential to quantify and characterize their attributes systematically (Fernandes et al. 2020). This involves analysing their physical, chemical, and functional properties, as well as understanding their nutritional and phytochemical profiles. Such comprehensive evaluation enables researchers, food technologists, and industry stakeholders to optimize their use in various applications while ensuring safety, consistency, and quality (Pires et al. 2017). Figure 10.1 represents some of the most widely used edible flowers.

Physical characterization is the first step in understanding edible flowers. Attributes such as size, color, texture, and aroma not only influence consumer preferences but also play a role in determining their suitability for different uses (Mlcek and Rop 2011). For example, brightly coloured flowers are often favoured for garnishing dishes, while firm-textured varieties may be better suited for drying or other preservation methods (Takahashi et al. 2020). Chemical properties provide insight into the compositional aspects of edible flowers. Factors such as pH, acidity, and moisture content affect their shelf life and compatibility with other ingredients (Lu et al. 2016). Similarly, their levels of soluble solids, sugars, and volatile compounds contribute to their unique flavor profiles, which can be tailored for specific culinary or industrial applications (Zheng and Wang 2021).

The functional properties of edible flowers further enhance their value as food ingredients. Many species exhibit antioxidant, anti-inflammatory, and antimicrobial activities due to their rich phytochemical content. These properties make edible

Fig. 10.1 Some of the most widely used edible flowers, **A**-Lotus, **B**-Marigold, **C**-Rose, **D**-Hibiscus

flowers not only attractive for their sensory qualities but also as components of functional foods and dietary supplements (Loizzo et al. 2016). The nutritional composition of edible flowers is another critical area of study. These blooms are often low in calories and high in vitamins, minerals, and dietary fiber, making them an excellent addition to health-conscious diets (Fernandes et al. 2017). Furthermore, the quantification of phytochemicals, such as flavonoids, anthocyanins, and carotenoids, underscores their potential role in preventing oxidative stress and supporting overall well-being (Loizzo et al. 2016).

Characterizing and quantifying these properties requires advanced methods and techniques. Analytical tools such as spectroscopy, chromatography, and microscopy are essential for precise identification and measurement of bioactive compounds (Pires et al. 2017). These techniques not only ensure the accuracy of data but also help establish standards for quality control and safety. Despite their numerous benefits, edible flowers face several challenges. Issues such as seasonal availability, perishability, and variability in composition can hinder their widespread adoption (Rodrigues et al. 2017). Additionally, safety concerns related to potential contaminants, allergens, and toxic compounds necessitate rigorous screening and standardization protocols. Addressing these challenges will require collaborative efforts from researchers, industry stakeholders, and regulatory bodies (Fernandes et al. 2020).

The potential of edible flowers extends far beyond their traditional uses. Emerging research points to exciting future directions, such as breeding programs to enhance their bioactive content, innovative preservation techniques to extend their shelf life, and sustainable cultivation practices to reduce environmental impact (Matyjaszczyk and Śmiechowska 2019). By systematically studying and leveraging their unique properties, edible flowers can play a significant role in the food, health, and cosmetic industries (Pires et al. 2017).

This chapter aims to provide a comprehensive overview of the quantification and characterization of edible flowers, addressing their selection, physical and chemical attributes, functional properties, and applications. It also discusses the challenges and limitations associated with their use, along with future perspectives for their integration into modern food systems. As we delve deeper into the science of edible flowers, we uncover opportunities to not only enhance our culinary experiences but also contribute to sustainable and health-focused innovations.

10.2 Selection and Identification of Edible Flowers

10.2.1 Selection

The selection of edible flowers are critical steps in their study and application, ensuring their safety, usability, and appeal in culinary and other contexts (Fernandes et al. 2020). While edible flowers have been used for centuries in traditional cuisines and remedies, modern uses require a more scientific approach to confirm their safety and optimize their potential applications (Takahashi et al. 2020). This process involves carefully considering their safety, cultural significance, and market potential, among other factors (Pires et al. 2017).

10.2.2 Safety

The foremost consideration in selecting edible flowers is their safety for human consumption (Fernandes et al. 2020). Not all flowers are edible, and some may be toxic or harmful if consumed. Additionally, the safety of flowers depends on their growing conditions, handling practices, and the absence of contaminants (Matyjaszczyk and Śmiechowska 2019). Ensuring safety involves addressing the following aspects:

- **Non-Toxicity:** Only flowers known to be non-toxic should be considered for consumption. This determination involves referencing botanical literature, toxicological studies, and historical evidence (Fernandes et al. 2017). For example, flowers like hibiscus (*Hibiscus spp.*) and marigold (*Tagetes spp.*) have long histories of safe use, while others, such as certain lilies and foxglove, are known to contain toxic compounds and must be avoided.
- **Contaminant-Free Cultivation:** Even naturally non-toxic flowers can become unsafe if they are exposed to contaminants during cultivation or handling. Pesticides, herbicides, heavy metals, and pollutants can accumulate in plants and pose health risks (Petrova et al. 2016). Therefore, it is essential to ensure that edible flowers are grown in controlled environments free from harmful substances. Organic farming practices, hydroponic cultivation, and other sustainable methods are often preferred to minimize exposure to contaminants.
- **Avoidance of Artificial Treatments:** Flowers intended for decoration are sometimes treated with dyes, preservatives, or other chemicals that render them unsuitable for consumption (Fernandes et al. 2020). It is crucial to differentiate between flowers grown for ornamental purposes and those specifically cultivated for edible use.
- **Proper Handling and Hygiene:** Post-harvest handling practices significantly impact the safety of edible flowers. Proper cleaning, storage, and packaging are essential to prevent microbial contamination and maintain freshness.

10.2.3 Cultural Relevance

The cultural significance of edible flowers provides valuable insights into their palatability, uses, and safety. Many edible flowers have a long history of use in traditional cuisines, herbal remedies, and rituals, making them culturally and gastronomically significant (Cunningham 2015; Fernandes et al. 2017). Understanding this cultural context can guide the selection of flowers that are not only safe but also appealing to consumers (Chen and Wei 2022).

- **Historical Culinary Use:** Flowers such as roses (*Rosa spp.*), violets (*Viola spp.*), and lavender (*Lavandula spp.*) have been used in traditional recipes across cultures (Pires et al. 2017; Mlcek and Rop 2011). For example, rose petals are a key ingredient in Middle Eastern desserts, while lavender is often used in European confections and teas (Lu et al. 2016). These historical uses provide a basis for their inclusion in modern culinary applications (Rodrigues et al. 2017).
- **Traditional Remedies:** In many cultures, edible flowers have been valued for their medicinal properties (Petrova et al. 2016). Hibiscus, for instance, is used to make beverages with potential antihypertensive and antioxidant benefits. Similarly, calendula (*Calendula officinalis*) has been used in herbal remedies for its anti-inflammatory properties.
- **Regional Preferences and Traditions:** The preference for certain flowers often varies by region, influenced by local traditions, climate, and availability (Chen and Wei 2022; Fernandes et al. 2017). For example, blossoms like cherry flowers (*Prunus spp.*) are celebrated in Japanese culture, while tropical flowers like banana blossoms (*Musa spp.*) are staples in Southeast Asian cuisines (Chen and Wei 2022).

Cultural relevance not only helps identify flowers that are already accepted and appreciated but also provides an opportunity to reintroduce traditional practices into modern food systems, promoting biodiversity and cultural heritage.

10.2.4 Market Potential

In addition to safety and cultural significance, the market potential of edible flowers plays a significant role in their selection. For edible flowers to be viable in the food and beverage industry, they must possess characteristics that appeal to consumers and align with market trends (Chen and Wei 2022; Rodrigues et al. 2017).

- **Aesthetic Appeal:** One of the most prominent features of edible flowers is their visual impact (Table 10.1). Brightly coloured flowers, such as nasturtium (*Tropaeolum majus*), pansies (*Viola spp.*), and marigold, are highly desirable for their ability to enhance the presentation of dishes (Fernandes et al. 2017). Their vibrant hues can also be extracted for use as natural food colorants (Lu et al. 2016).
- **Flavor Profile:** The flavor of edible flowers varies widely, from sweet and floral to spicy and herbaceous (Mlcek and Rop 2011). For example, violets and roses offer a delicate sweetness, while nasturtium adds a peppery note to salads (Pires et al. 2017). Understanding these flavor profiles is crucial for selecting flowers that complement specific dishes or beverages.

Table 10.1 Quality indicators for selection and identification of edible flowers

Selection parameter	Quality indicators	Selection criteria	Rejection criteria
Visual appearance	Color vibrancy	Bright, true-to-variety colours	Discoloration
	Petal integrity	Fully developed flowers	Wilting/browning
	Size uniformity	Uniform shape	Irregular shapes
	Natural sheen	Fresh appearance	Dull appearance
Freshness	Petal texture	Crisp, firm petals	Soft, mushy texture
	Stem condition	Green, firm stems	Brown/dried stems
	Moisture level	Natural moisture	Excessive moisture
	Flower head firmness	Compact flower head	Loose/falling petals
Maturity stage	Flower development	Just-opened flowers	Over-mature flowers
	Petal formation	Fully formed petals	Dropping petals
	Color development	Peak color stage	Faded colours
	Bud condition	Firm buds	Opened buds
Safety parameters	Growing conditions	Organic cultivation	Unknown source
	Chemical exposure	Chemical-free	Chemical residues
	Pest presence	Pest-free	Pest damage
	Disease signs	Disease-free	Disease symptoms

Source*:* Mlcek and Rop (2011), Hassan and Smith (2019), Wilson and Thompson (2020)

- **Aroma:** Aroma is another essential sensory attribute that contributes to the appeal of edible flowers (Benvenuti et al. 2016). The fragrance of lavender or orange blossom can enhance desserts and beverages, while the subtle aroma of herbs like basil flowers (*Ocimum basilicum*) adds complexity to savory dishes (Fernandes et al. 2017).
- **Versatility in Applications:** Flowers with multiple uses have greater market potential (Chen and Wei 2022). For example, hibiscus is used in teas, jams, and syrups, while calendula is used in soups, stews, and herbal ointments (Petrova et al. 2016). Versatile flowers appeal to both home cooks and food manufacturers, increasing their demand (Rodrigues et al. 2017).
- **Consumer Trends:** As consumers become more health-conscious and environmentally aware, there is a growing preference for natural, plant-based ingredients (Chen and Wei 2022; Mlcek and Rop 2011). Edible flowers align with these trends, offering a sustainable and minimally processed alternative to synthetic additives (Fernandes et al. 2017). Their association with wellness, sophistication, and artisanal cuisine further enhances their market appeal (Rodrigues et al. 2017).
- **Economic Viability:** Flowers that are easy to cultivate, harvest, and store are more likely to succeed in commercial markets. Factors such as yield, shelf life, and resistance to pests and diseases influence their economic viability (Chen and Wei 2022; Fernandes et al. 2017).

10.2.5 Identification

The accurate identification of edible flowers is crucial to ensure their safety, authenticity, and appropriate use in culinary and non-culinary applications (Fernandes et al. 2017; Lu et al. 2016). Identification is a meticulous process that combines traditional botanical methods with advanced molecular techniques to confirm the species, subspecies, or variety of the flowers being studied (Mlcek and Rop 2011). The methods of identification, emphasizing the significance of both morphological observations and genetic tools, and provides insights into some commonly recognized edible flowers have been detailed further (Chen and Wei 2022; Pires et al. 2017).

10.2.6 Traditional Taxonomy for Identification

Traditional taxonomy has been the foundation of plant identification for centuries, relying on observable traits and systematic classification methods (Petrova et al. 2016). This approach involves careful examination of the physical characteristics of flowers and their classification into genera, species, and subspecies using taxonomic keys (Fernandes et al. 2017).

10.2.7 Morphological Traits

The morphology of flowers provides the first line of identification (Mlcek and Rop 2011). This involves detailed observation of various structural features, such as:

- **Flower Structure:** The overall arrangement of floral parts, including the number of petals, sepals, stamens, and pistils, can help differentiate between species (Table 10.1). For example, hibiscus flowers (*Hibiscus spp.*) are easily recognizable by their large, trumpet-shaped blossoms and prominent central stamen column (Mlcek and Rop 2011).
- **Petal Shape and Texture:** Petal shapes vary widely, ranging from rounded and ovate to elongated or fringed (Fernandes et al. 2017). The texture of petals, whether velvety, smooth, or waxy, also provides identifying clues (Table 10.1). Roses (*Rosa spp.*), for instance, are distinguished by their layered, soft, and fragrant petals (Lu et al. 2016).
- **Coloration:** The color of petals, including gradients, patterns, and unique markings, is another critical identifier (Benvenuti et al. 2016). For example, pansies (*Viola spp.*) are known for their vibrant, multi-coloured petals with striking “faces” or blotches in the center.

- **Growth Habit:** The plant's overall form, including its height, branching pattern, and habitat, adds to the identification process. For instance, marigolds (*Tagetes spp.*) typically have bushy growth habits with bright orange or yellow blooms (Mlcek and Rop 2011).

These morphological traits are documented through photographs, illustrations, and descriptions, often serving as the first step in differentiating edible flowers from their non-edible or toxic counterparts.

10.2.8 Taxonomic Keys

Taxonomic keys are systematic tools used to identify and classify plants based on a series of dichotomous (two-choice) questions or observations. These keys help differentiate closely related species by focusing on specific traits, such as:

- **Leaf arrangement and shape.**
- **Flower symmetry (radial or bilateral).**
- **Seed or fruit characteristics.**
- **Habitat preferences.**

For example, in the *Tagetes* genus, taxonomic keys may differentiate marigold species based on petal shape, inflorescence size, and foliage scent (Henderson and Walker 2019).

10.2.9 Modern Molecular Techniques

While traditional taxonomy is invaluable, it has limitations, particularly when dealing with morphologically similar species or when flowers are processed (e.g., dried or powdered) (Patel et al. 2020). Modern molecular techniques have revolutionized plant identification by enabling precise species confirmation through genetic analysis.

10.2.9.1 DNA Barcoding

DNA barcoding involves sequencing specific regions of a plant's DNA to create a unique genetic "barcode" for each species (Thompson and Liu 2020). Commonly used genetic markers for plants include:

- rbcL (ribulose-1,5-bisphosphate carboxylase/oxygenase large subunit): A gene widely used for identifying plant species.
- matK (maturase K): Another chloroplast gene with high discriminatory power.
- ITS (Internal Transcribed Spacer): A nuclear region useful for differentiating closely related species.

This method allows researchers to:

- Confirm species identity with high accuracy.
- Differentiate between closely related species or subspecies.
- Authenticate processed flowers in powdered or extract forms.

For example, DNA barcoding can distinguish between edible rose species (*Rosa gallica*) and ornamental varieties that may not be suitable for consumption (Patel et al. 2020).

10.2.9.2 Molecular Phylogenetics

Phylogenetic analysis uses genetic data to determine evolutionary relationships between species. This approach is particularly helpful in understanding the genetic diversity of edible flowers and identifying potential new species for consumption (Zhang and Chen 2019).

10.2.9.3 Marker-Assisted Selection

In the context of edible flower cultivation, marker-assisted selection helps breeders identify and propagate desirable traits, such as enhanced pigment content or resistance to pests (Wilson and Thompson 2020).

10.2.9.4 Widely Recognized Edible Flowers

Several edible flowers have gained popularity for their safety, flavor, and versatility (Davidson et al. 2021). Below are some of the most widely recognized species, along with their distinguishing characteristics:

10.2.9.4.1 Rose (*Rosa spp.*)

- Description: Roses are perhaps the most iconic edible flowers, known for their elegant appearance and captivating fragrance (Martinez and Lee 2019). The petals are soft, layered, and available in a spectrum of colours, including red, pink, white, and yellow.
- Flavor Profile: Roses have a mild, sweet flavor with subtle floral undertones (Park et al. 2020).
- Uses: Rose petals and extracts are commonly used in desserts, syrups, teas, and jams. Rosewater, a byproduct of rose distillation, is a key ingredient in many Middle Eastern and Indian recipes (Park et al. 2020).

10.2.9.4.2 Marigold (*Tagetes spp.*)

- Description: Marigolds are characterized by their bright orange or yellow blooms and compact, bushy growth habit (Anderson and Wright 2020). The petals are often ruffled or layered, giving the flowers a full appearance.
- Flavor Profile: Marigold petals add a peppery, slightly tangy taste to dishes.
- Uses: The petals are used as a natural food colorant and flavouring agent in salads, rice dishes, and herbal infusions (Anderson and Wright 2020).

10.2.9.4.3 Hibiscus (*Hibiscus spp.*)

- Description: Hibiscus flowers are large, showy, and trumpet-shaped, with colours ranging from deep red to pink and white (Brown et al. 2021). The prominent stamen column is a distinguishing feature.
- Flavor Profile: Hibiscus flowers have a tart, cranberry-like flavor (Brown et al. 2021).
- Uses: Widely used to make teas, syrups, and jams, hibiscus is also valued for its high anthocyanin content, which contributes to its deep red pigment and antioxidant properties (Brown et al. 2021).

10.2.9.4.4 Pansy (*Viola spp.*)

- Description: Pansies are small, delicate flowers with round, flat petals that often display vibrant colours and striking patterns (Lopez-Martinez et al. 2021). The "face" marking in the center is a hallmark of many pansy varieties.
- Flavor Profile: Pansies offer a mild, slightly sweet flavor with grassy undertones (Lopez-Martinez et al. 2021).
- Uses: They are frequently used as garnishes for desserts, salads, and cocktails, and their vibrant hues make them ideal for decorative purposes (Lopez-Martinez et al. 2021).

10.3 Characterization and Quantification of Edible Flowers

The characterization and quantification of the physical properties of edible flowers provide a scientific basis for their selection and application (Fernandes et al. 2020). By understanding attributes such as size, shape, color, texture, and aroma, researchers and industries can ensure consistent quality, optimize usability, and unlock new potential in diverse markets (Table 10.2) (Chen and Wei 2022). These properties not

Table 10.2 Characterization and quantification of edible flowers

Parameters	Characteristics	Methods of analysis	Typical ranges/values
Physical characteristics			
Color	Anthocyanins, carotenoids presence	Colorimeter (lab* values), spectrophotometry	L*: 25–85, a*: −15 to +45, b*: −5 to +45
Texture	Firmness, crispness	Texture analyser, penetrometer	0.5–15 N (force)
Moisture content	Water activity, shelf stability	Gravimetric method, moisture analyser	70–95% fresh weight
Size and shape	Morphological features	Digital calliper, image analysis	Species dependent
Functional properties			
Water holding capacity	Hydration properties	Centrifugation method	1.5–8.0 g water/g dry matter
Oil absorption	Lipid binding	Absorption capacity test	0.8–4.5 g oil/g dry matter
Emulsification	Stability in food systems	Emulsion stability index	10–60%
Foaming capacity	Protein functionality	Whipping method	20–150%
Nutritional composition			
Proteins	Essential amino acids	Kjeldahl method, HPLC	1–5 g/100 g fresh weight
Lipids	Fatty acid profile	Soxhlet extraction, GC-MS	0.1–1 g/100 g fresh weight
Carbohydrates	Sugars, fiber content	Enzymatic methods, HPLC	5–15 g/100 g fresh weight
Minerals	Macro and microelements	AAS, ICP-MS	Ca: 50–300 mg/100 g, Fe: 0.5–5 mg/100 g
Vitamins	Water and fat-soluble	HPLC, spectrophotometry	Vitamin C: 25–150 mg/100 g
Phytochemical profile			
Phenolic compounds	Flavonoids, phenolic acids	HPLC-DAD, spectrophotometry	50–500 mg GAE/100 g
Antioxidants	Free radical scavenging	DPPH, ABTS, FRAP assays	10–100 μmol TE/g
Carotenoids	Pigments	HPLC, spectrophotometry	5–50 mg/100 g
Anthocyanins	Color compounds	pH differential method	10–200 mg/100 g
Essential oils	Volatile compounds	GC-MS, steam distillation	0.1–2% fresh weight

Source: Chen and Wei (2017), Fernandes et al. (2017), Lara-Cortés et al. (2014)

only define the aesthetic and sensory appeal of edible flowers but also pave the way for innovative uses in culinary, cosmetic, and industrial domains (Pires et al. 2017). Through precise measurement and analysis, edible flowers can be tailored to meet consumer expectations and functional demands (Takahashi et al. 2020).

10.3.1 Physical Characteristics

The physical characteristics of edible flowers play a pivotal role in determining their usability across culinary, cosmetic, and industrial domains (Mlcek and Rop 2011). These traits are not only integral to the flowers' visual and sensory appeal but also affect their marketability and functional properties (Table 10.2) (Lu et al. 2016). The study and quantification of these characteristics provide essential data for standardization and quality assurance (Rodrigues et al. 2017). Below, the key physical properties of edible flowers-size, shape, color, texture, and aroma—are elaborated, along with their methods of analysis and significance.

10.3.2 Size and Shape

The size and shape of edible flowers are fundamental attributes that influence their aesthetic and functional value.

- **Measurement Techniques:** Size is typically quantified through manual measurement using callipers or rulers, while shape is assessed by observing and recording floral morphology. Advanced techniques such as image analysis systems use high-resolution imaging and software to automate the measurement process, providing precise data on parameters such as petal length, width, and curvature (Benvenuti et al. 2016).
- **Importance:** Uniformity in size and shape enhances the visual appeal of edible flowers, especially when used as garnishes or in packaged products (Fernandes et al. 2017). Consistent dimensions contribute to better marketability, as consumers and chefs often prefer flowers that align with specific aesthetic or culinary requirements. For instance, small, uniform flowers like pansies are favoured for decorating cakes and pastries, while larger flowers like hibiscus are valued for brewing teas (Loizzo et al. 2016).

10.3.3 Color

Color is one of the most visually striking features of edible flowers and a critical determinant of their appeal and usability (Table 10.2).

- **Measurement Techniques:** The color of petals is analysed using spectrophotometers or colorimeters, which measure hue, saturation, and brightness with precision (Loizzo et al. 2016). These instruments provide objective data that can be used to classify flowers based on their chromatic attributes (Skrajda-Brdak et al. 2020). Spectrophotometric analysis is particularly useful for identifying pigments such as anthocyanins, carotenoids, and flavonoids, which contribute to the color of flowers.

- **Significance:** Brightly coloured flowers are preferred in culinary applications to add vibrancy to dishes, while in cosmetics and dye industries, specific color profiles are sought after for their aesthetic and functional properties (Skrajda-Brdak et al. 2020). For example, the deep red of hibiscus flowers is indicative of high anthocyanin content, making them ideal for natural colouring agents and antioxidant-rich products (Durling and Mitcham 2013).

10.3.4 Texture

Texture refers to the tactile properties of edible flowers, which are crucial for their functionality and consumer perception.

- **Measurement Techniques:** Mechanical testing devices, such as texture analysers, are used to quantify firmness, elasticity, and tensile strength (Demasi et al. 2020). These tests simulate handling and processing conditions, offering insights into the flowers' durability and usability in various applications.
- **Applications:** Texture analysis is especially important for flowers intended for garnishing or as dried decorations. For example, a flower with firm, elastic petals is more suitable for handling and storage, while delicate textures are prized for fresh culinary uses (Demasi et al. 2020). Flowers with the right texture can withstand processing into dried or crystallized forms, enhancing their shelf life and versatility.

10.3.5 Aroma

Aroma is a defining sensory characteristic of edible flowers, significantly influencing their desirability and applications in food and fragrance industries (Scariot et al. 2018).

- **Measurement Techniques:** Aroma is assessed through sensory evaluation panels, where trained individuals identify and describe fragrance profiles. Advanced methods like gas chromatography-mass spectrometry (GC-MS) are used for a more detailed analysis. GC-MS separates and identifies volatile compounds responsible for the aroma, providing a chemical fingerprint of the flower's fragrance (Zheng et al. 2018).
- **Significance:** Aroma plays a dual role—enhancing the sensory experience of food and contributing to therapeutic and cosmetic products (Park et al. 2020). For instance, the sweet and floral aroma of rose petals makes them popular in desserts and essential oils, while the spicy scent of marigold petals is appreciated in savory dishes and natural perfumes.

10.3.6 Chemical Properties

Chemical properties are fundamental to understanding the composition and potential applications of edible flowers (Table 10.2). These properties influence not only the sensory experience, such as flavor and aroma, but also the functional attributes like shelf life, nutritional value, and suitability for various culinary and industrial uses (Wilson et al. 2021). The chemical characterization of edible flowers focuses on key parameters such as pH and acidity, moisture content, and soluble solids, which collectively provide a comprehensive insight into their quality and usability. Below, each property is elaborated upon, along with its measurement techniques and significance.

10.3.6.1 pH and Acidity

The pH and acidity of edible flowers are critical chemical parameters that affect their flavor profile, preservation, and compatibility with other food ingredients (Anderson and Park 2020).

- **Measurement Techniques:** The pH is measured using pH meters, which provide precise readings of hydrogen ion concentration in the flower's extract (Anderson and Park 2020). Acidity, often expressed as titratable acidity, is assessed through titration methods using standardized alkaline solutions to neutralize the acidic components.
- **Significance**:
  - (i) **Flavor:** The pH and acidity contribute to the flavor characteristics of flowers, ranging from tangy and tart (as in hibiscus) to mild and sweet (as in rose petals) (Garcia-Lopez and Smith 2019). This information is essential for chefs and food developers aiming to balance flavors in culinary creations.
  - (ii) **Shelf Stability:** Acidic pH levels can inhibit microbial growth, enhancing the shelf life of flowers and their derived products (Brown and Hassan 2021). For instance, highly acidic flowers like hibiscus are naturally more resistant to spoilage.
  - (iii) **Compatibility:** Understanding pH levels is crucial for integrating edible flowers into recipes, as they can influence the behaviour of other ingredients, such as gelatin setting or the stability of emulsions (Patel and Wright 2020).

10.3.6.2 Moisture Content

Moisture content is a vital parameter that determines the freshness, texture, and shelf life of edible flowers.

- **Measurement Techniques:** Gravimetric methods, such as drying and weighing, are commonly used to measure moisture content (Davidson and Brooks 2020). The process involves weighing a sample before and after drying at a controlled temperature, with the difference representing the water content. Advanced methods like infrared moisture analysers provide faster and more precise measurements.
- **Significance**:
    (i) **Freshness:** Higher moisture content indicates freshness in flowers intended for immediate use, such as garnishes or salads (Davidson and Brooks 2020).
    (ii) **Shelf Life:** For dried or preserved flowers, lower moisture content is desirable to prevent microbial growth and oxidative spoilage (Davidson and Brooks 2020). This is particularly important for flowers used in teas, spices, or potpourris.
    (iii) **Processing:** Moisture content affects the drying process, packaging, and storage conditions (Henderson and Walker 2019). For instance, optimal moisture levels must be maintained to retain the delicate texture and color of petals during processing.

10.3.6.3 Soluble Solids

Soluble solids primarily include sugars, organic acids, and other dissolved components, contributing to the flavor and nutritional profile of edible flowers.

- **Measurement Techniques**: Refractometers are used to measure the total soluble solids (TSS) content, expressed in degrees Brix. This quick and non-destructive method provides an estimate of the concentration of dissolved sugars in the flower's juice or extract (Davidson and Brooks 2020).
- **Significance**:
    (a) **Flavor Profiling**: The concentration of soluble solids directly affects sweetness and overall taste. Flowers with higher sugar content, such as honeysuckle, are naturally sweeter and more appealing in desserts or beverages (Liu and Johnson 2020).
    (b) **Nutritional Value**: Soluble solids also include vitamins, minerals, and bioactive compounds, which enhance the nutritional benefits of edible flowers (Zhang and Kim 2021).
    (c) **Product Development**: The TSS value aids in formulating products like syrups, jellies, and infusions. For example, flowers with a high TSS content may require minimal additional sweeteners in processed products (Brown and Hassan 2021).

- The chemical properties of edible flowers, including pH and acidity, moisture content, and soluble solids, offer valuable insights into their composition and functional potential. These parameters are essential for assessing quality, determining optimal storage and processing methods, and exploring culinary and industrial applications (Henderson and Walker 2019). By systematically characterizing these chemical properties, researchers and producers can standardize the quality of edible flowers, ensuring their safe and consistent use in various domains (Davidson and Brooks 2020). Moreover, these understanding paves the way for innovative product development, enhancing the utility and appeal of edible flowers in modern food and lifestyle industries.

10.3.7 Functional Properties

The functional properties of edible flowers extend beyond their sensory appeal, highlighting their potential contributions to health and food systems (Table 10.2). These properties play a crucial role in determining their applicability in nutraceuticals, pharmaceuticals, and food science (Wilson et al. 2021). Among the most studied functional properties are antioxidant activity, anti-inflammatory effects, and emulsification and gelling capabilities (Thompson and Chen 2019). Understanding these properties not only underscores the health benefits of edible flowers but also reveals their versatility in diverse applications.

10.3.7.1 Antioxidant Activity

Antioxidant activity is one of the most significant functional attributes of edible flowers, attributed to their rich content of bioactive compounds like phenolics, flavonoids, and carotenoids (Rodriguez and Lee 2021).

- **Evaluation Techniques**: The antioxidant capacity of edible flowers is commonly assessed using chemical assays such as:
  - **DPPH (2,2-diphenyl-1-picrylhydrazyl)**: This assay measures the ability of flower extracts to scavenge free radicals, thereby indicating their antioxidant potential (Anderson and Park 2020). The extent of color change in the reaction provides a quantitative measure of antioxidant activity.
  - **ABTS (2,2′-azino-bis(3-ethylbenzothiazoline-6-sulfonic acid))**: Similar to the DPPH assay, ABTS evaluates radical-scavenging activity, with results often expressed as Trolox equivalent antioxidant capacity (TEAC) (Hassan et al. 2019).
- **Significance**:
  - **Health Benefits**: Antioxidants combat oxidative stress, reducing the risk of chronic diseases such as cardiovascular disorders, diabetes, and cancer (Liu and Brooks 2020a, b). Edible flowers like hibiscus and marigold are particularly rich in antioxidants, making them valuable in functional food formulations.

- **Food Preservation**: The antioxidant properties of edible flowers can be harnessed to improve the shelf life of food products by preventing lipid oxidation (Patel and Wright 2021).

10.3.7.2 Anti-inflammatory Properties

Anti-inflammatory properties further enhance the therapeutic value of edible flowers, positioning them as potential agents for managing inflammatory-related conditions.

- **Evaluation Techniques**: Anti-inflammatory activity is typically assessed through cell-based assays (Davidson and Kim 2019). These involve exposing human or animal cell lines to flower extracts and measuring their ability to inhibit inflammatory mediators such as cytokines, prostaglandins, or nitric oxide.
  - Techniques such as enzyme-linked immunosorbent assays (ELISA) and reverse transcription-polymerase chain reaction (RT-PCR) are used to quantify specific biomarkers of inflammation (Zhang and Johnson 2021a, b, c).
- **Significance**:
  - **Therapeutic Potential**: By reducing inflammation, edible flowers may help manage conditions such as arthritis, inflammatory bowel disease, and skin disorders (Patel and Wright 2021). For instance, calendula (*Calendula officinalis*) has been traditionally used for its anti-inflammatory effects in topical and oral applications.
  - **Integration into Diet**: Flowers with anti-inflammatory properties can be incorporated into daily diets, offering preventive health benefits in the form of teas, salads, or supplements (Zhang and Johnson 2021a, b, c).

10.3.7.3 Emulsification and Gelling Properties

Beyond health-related benefits, the functional properties of edible flowers are vital in food science, particularly in texture modification and stabilization of food systems.

- **Evaluation Techniques**
  - **Emulsification Properties**: These are assessed by measuring the ability of flower extracts to stabilize oil-in-water emulsions, often quantified by emulsion stability index (ESI) (Davidson and Kim 2019).
  - **Gelling Properties**: The gel-forming capacity of flower components, such as polysaccharides or proteins, is evaluated through rheological analysis, which measures gel strength and elasticity (Rodriguez and Lee 2021).

- **Significance**:
  - **Culinary Applications**: Flowers with emulsifying properties are used to stabilize dressings, sauces, and beverages, while those with gelling properties are incorporated into jellies, puddings, and confectioneries (Zhang and Kim 2021).
  - **Functional Foods**: These properties contribute to the development of low-fat or plant-based food products. For instance, the mucilaginous texture of certain flowers like hibiscus can be utilized to create plant-based gels or thickeners (Wilson et al. 2021).

The functional properties of edible flowers-antioxidant activity, anti-inflammatory effects, and emulsification and gelling capacities-highlight their multifaceted roles in health and food systems (Thompson and Chen 2019; Kumar and Park 2020). By leveraging these properties, edible flowers can be integrated into innovative products that combine health benefits with sensory appeal (Zhang and Kim 2021). As research into their bioactive compounds and mechanisms of action continues to grow, edible flowers are poised to play an increasingly significant role in the fields of nutrition, medicine, and food technology.

10.3.8 Yield Analysis

Yield analysis is a critical step in assessing the efficiency, practicality, and economic viability of cultivating edible flowers. This process involves a systematic evaluation of both pre-harvest and post-harvest factors that influence the overall output (Wilson et al. 2021). By understanding yield parameters, growers, researchers, and producers can optimize cultivation practices, reduce losses, and enhance the commercial potential of edible flowers (Kumar and Chen 2019).

10.3.8.1 Harvest Efficiency

Harvest efficiency is a measure of the productivity and effectiveness of flower cultivation, often expressed as the yield per plant or per unit area of land (Kumar and Park 2020).

10.3.8.1.1 Factors Influencing Harvest Efficiency

Cultivation Practices: Optimal planting density, fertilization, irrigation, and pest management significantly affect flower yield (Hassan and Smith 2019). For instance, densely planted crops may compete for resources, reducing individual plant productivity, while sparse planting can lead to underutilization of land.

Varietal Selection: Different flower species and cultivars vary in their yield potential (Hassan and Smith 2019). Varieties bred for high productivity, disease resistance, and quality attributes tend to deliver better harvest efficiency.

Climate and Environmental Conditions: Temperature, sunlight, and soil quality directly impact the growth and flowering potential of plants (Liu and Brooks 2020a, b). For example, flowers like marigold thrive in warm climates, while pansies prefer cooler environments.

10.3.8.1.2 Measurement Techniques

- Yield per plant is determined by counting the number of flowers harvested from individual plants over a growing season (Davidson et al. 2019).
- Yield per area is calculated by aggregating the total weight or volume of flowers harvested from a specific land area (Davidson et al. 2019).

10.3.8.1.3 Economic Implications

Profitability: Higher harvest efficiency translates to increased output, which can lower production costs per unit and enhance profitability (Patel and Wright 2020).

Market Supply: Consistent and high yields ensure a steady supply for culinary, cosmetic, or pharmaceutical industries (Zhang and Johnson 2021a, b, c).

10.3.8.2 Post-Harvest Losses

Post-harvest losses refer to the reduction in flower quantity or quality after harvest, often resulting from improper handling, storage, or transportation (Davidson et al. 2019). These losses can significantly impact economic returns and market viability.

10.3.8.2.1 Causes of Post-Harvest Losses

- **Mechanical Damage**: Flowers are delicate and prone to bruising or crushing during harvesting, packaging, or transit (Zhang and Johnson 2021a, b, c).
- **Moisture Loss**: Loss of water content leads to wilting, reducing the visual appeal and freshness of the flowers.
- **Microbial Spoilage**: Poor sanitation or high humidity can promote fungal and bacterial growth, leading to decay (Hassan and Smith 2019).
- **Temperature Fluctuations**: Exposure to extreme temperatures during storage or transport accelerates degradation (Liu and Brooks 2020a, b).

10.3.8.2.2 Minimizing Post-Harvest Losses

- **Storage Techniques**: Using controlled environments with optimal temperature and humidity levels can extend the shelf life of edible flowers. Refrigeration at 2–4 °C is commonly employed for perishable flowers (Davidson et al. 2019).
- **Packaging Solutions**: Innovative packaging materials, such as modified atmosphere packaging (MAP) or vacuum-sealed bags, help maintain freshness and reduce spoilage (Zhang and Johnson 2021a, b, c).
- **Efficient Handling**: Gentle handling techniques during harvesting and transportation minimize mechanical damage (Davidson et al. 2019).

10.3.8.2.3 Impact of Loss Reduction

- **Economic Gains**: Lower post-harvest losses directly translate to higher returns for growers and suppliers (Patel and Wright 2020).
- **Sustainability**: Reducing waste contributes to sustainable practices in the edible flower supply chain, minimizing environmental impact (Zhang and Johnson 2021a, b, c).

Yield analysis, encompassing harvest efficiency and post-harvest loss assessment, is fundamental to the successful cultivation and commercialization of edible flowers (Wilson et al. 2021). By maximizing harvest efficiency and minimizing post-harvest losses, producers can ensure a consistent, high-quality supply of edible flowers for various applications (Martínez-Sánchez et al. 2020). Advanced cultivation methods, innovative storage technologies, and efficient handling practices are critical to achieving these goals, thereby enhancing the economic and environmental sustainability of the edible flower industry.

10.3.9 Nutritional Composition

The nutritional composition of edible flowers highlights their potential role as valuable dietary components, contributing to both health and culinary applications (Table 10.2) (Takahashi et al. 2020). As consumers increasingly seek nutrient-dense and functional foods, edible flowers are gaining attention for their diverse and unique nutritional profiles (Cho et al. 2018). Detailed analysis of macronutrients and micronutrients provides insights into their dietary value and supports their incorporation into balanced diets (Rachkeeree et al. 2018).

10.3.9.1 Macronutrient Content

Macronutrients, including proteins, fats, carbohydrates, and dietary fiber, form the foundation of the nutritional profile of edible flowers (Koike et al. 2015).

10.3.9.1.1 Proteins

Edible flowers generally contain moderate amounts of protein, contributing to their nutritional appeal. Proteins in flowers may consist of essential amino acids, which are crucial for various bodily functions, including muscle repair, enzyme activity, and immune response (Morata-Barrado et al. 2019).

- **Measurement:** Protein content is typically quantified using methods such as the Kjeldahl or Dumas technique, which analyse nitrogen content as an indicator of protein levels.

10.3.9.1.2 Fats

While most edible flowers are low in fat, the small amounts they contain often include beneficial fatty acids (Yang et al. 2017). These fats contribute to energy provision and cellular functions. Certain flowers, such as hibiscus, contain minimal but significant levels of essential fatty acids, making them suitable for low-fat diets.

- **Measurement:** Fat content is determined through Soxhlet extraction or other solvent-based techniques (Villavicencio et al. 2018).

10.3.9.1.3 Carbohydrates

Carbohydrates are a primary energy source, and edible flowers often contain simple sugars and polysaccharides (Paramee et al. 2018). These components contribute to the flavor profile and energy content of the flowers. For instance, hibiscus flowers are known for their tartness, which is partially attributed to their carbohydrate content.

- **Measurement:** Carbohydrate levels are often derived by difference, subtracting the sum of protein, fat, moisture, and ash content from the total weight.

10.3.9.1.4 Dietary Fiber

Many edible flowers contain dietary fiber, which supports digestive health by promoting bowel regularity and acting as a prebiotic (Watanabe et al. 2020). Fiber-rich flowers like marigold and nasturtium may contribute to gut health and reduce the risk of chronic diseases such as diabetes and cardiovascular disorders.

- **Measurement:** Fiber content is analysed using enzymatic-gravimetric methods that separate soluble and insoluble fiber fractions (Watanabe et al. 2020).

10.3.9.2 Micronutrient Content

Micronutrients, including vitamins and minerals, are critical to the health benefits offered by edible flowers.

- **Vitamins**:
  Edible flowers are often rich in vitamins, particularly Vitamin C, Vitamin A, and various B-complex vitamins (Yang et al. 2017).
  - **Vitamin C**: Found in flowers like hibiscus and rose, Vitamin C acts as a potent antioxidant, supporting immune function and skin health (Villavicencio et al. 2018).
  - **Measurement**: Vitamin levels are quantified using techniques such as high-performance liquid chromatography (HPLC) and spectrophotometry (Villavicencio et al. 2018).
- **Minerals**:
  Essential minerals like iron, potassium, calcium, and magnesium are present in edible flowers in varying amounts (Rachkeeree et al. 2018).
  - **Iron**: Supports oxygen transport in the blood and energy metabolism.
  - **Potassium**: Helps regulate blood pressure and maintain fluid balance.
  - **Measurement**: Mineral content is assessed using atomic absorption spectroscopy (AAS) or inductively coupled plasma-mass spectrometry (ICP-MS) (Rachkeeree et al. 2018).

10.3.9.3 Significance of Nutritional Composition

The comprehensive nutritional composition of edible flowers underscores their potential as functional foods.

- **Health Benefits**: Their content of vitamins, minerals, and dietary fiber supports overall health, aids in disease prevention, and promotes wellness (Takahashi et al. 2020).
- **Culinary Applications**: With their nutritional and aesthetic qualities, edible flowers can enhance the nutrient density of dishes while providing vibrant colours and flavors (Rachkeeree et al. 2018).
- **Sustainability**: Edible flowers can be cultivated in small spaces and have a low environmental footprint, making them a sustainable addition to the food system (Koike et al. 2015).

Nutritional analysis reveals the significant dietary value of edible flowers, which serve as a source of both macronutrients and micronutrients (Takahashi et al. 2020). By incorporating these nutrient-rich flowers into everyday diets, consumers can enjoy a combination of aesthetic appeal, unique flavors, and health benefits (Miura et al. 2019). Continued research and standardization of nutritional profiling methods will further promote the use of edible flowers in food and health industries.

10.3.10 Phytochemical Quantification

Phytochemicals are bioactive compounds in plants that contribute to their health benefits, flavor, color, and aroma (Table 10.2) (Mlcek and Rop 2011). In edible flowers, these compounds are particularly significant due to their potent antioxidant, anti-inflammatory, and therapeutic properties (Lu et al. 2016). Quantifying phytochemicals allows for the detailed evaluation of their functional and nutritional value, supporting their use in food, nutraceutical, and pharmaceutical applications (Chen et al. 2018). The major categories of phytochemicals in edible flowers include phenolic compounds, flavonoids, carotenoids, and essential oils (Fernandes et al. 2017).

10.3.10.1 Phenolic Content

Phenolic compounds are among the most abundant and studied phytochemicals in edible flowers (Kaisoon et al. 2012). These compounds play a key role in scavenging free radicals, reducing oxidative stress, and preventing chronic diseases (Zhang et al. 2014).

- **Measurement**:
  The **Folin-Ciocalteu assay** is the most commonly used method for quantifying total phenolic content (Singleton et al. 1999). This assay involves the reaction of phenolic compounds with the Folin-Ciocalteu reagent, resulting in a color change that can be measured spectrophotometrically. The results are often expressed in terms of gallic acid equivalents (GAE) (Ainsworth and Gillespie 2007).
  - **Advantages**: The method is simple, rapid, and suitable for a wide range of phenolic compounds.
  - **Applications**: High phenolic content in flowers like hibiscus and rose contributes to their antioxidant properties, making them suitable for use in health drinks, teas, and supplements (Pires et al. 2017).

10.3.10.2 Flavonoids and Carotenoids

Flavonoids and carotenoids are key contributors to the vibrant colours and bioactive properties of edible flowers (Loizzo et al. 2016).

- **Flavonoids**:
  Flavonoids are polyphenolic compounds that exhibit antioxidant, anti-inflammatory, and antimicrobial properties (Kumar and Pandey 2013). They are responsible for the yellow, orange, and red hues in many edible flowers.
  - **Quantification**: Flavonoids are typically measured using spectrophotometric methods or high-performance liquid chromatography (HPLC) (Zhishen et al. 1999). These methods provide precise data on specific flavonoid types, such as quercetin, kaempferol, and rutin.
  - **Health Benefits**: Flavonoid-rich flowers like marigold and pansy support cardiovascular health and immune function (Li et al. 2014).
- **Carotenoids**:
  Carotenoids, such as beta-carotene and lutein, are fat-soluble pigments that also serve as precursors to Vitamin A (Loizzo et al. 2016). They contribute to the bright yellow and orange colours seen in flowers like marigold and calendula (Shi et al. 2017).
  - **Quantification**: Carotenoid content is analysed using spectrophotometry or chromatographic techniques like HPLC.
  - **Applications**: Their antioxidant and eye-protective properties make carotenoid-rich flowers valuable in dietary and cosmetic products (Pires et al. 2017).

10.3.10.3 Essential Oils

Essential oils are volatile, aromatic compounds that contribute to the characteristic fragrance and therapeutic effects of edible flowers (Fernandes et al. 2017).

- **Analysis**:
  Essential oils are extracted using steam distillation or solvent extraction and analysed through gas chromatography-mass spectrometry (GC-MS) (Chen et al. 2018). GC-MS provides detailed profiles of the chemical composition of essential oils, identifying key compounds like linalool, eugenol, and geraniol (Fernandes et al. 2017).
  - **Applications**: Essential oils from flowers like lavender and rose are widely used in aromatherapy, cosmetics, and as natural flavouring agents (Pires et al. 2017).
- **Health Benefits**: Essential oils often exhibit antimicrobial, anti-inflammatory, and calming properties, adding functional value to edible flowers (Li et al. 2014).

10.3.10.4 Significance of Phytochemical Quantification

Quantification of phytochemicals in edible flowers provides critical insights into their functional properties and potential applications.

- **Nutritional Value**: Phytochemical-rich flowers can enhance the antioxidant capacity and overall nutrient density of diets (Shi et al. 2017).
- **Therapeutic Potential**: High levels of bioactive compounds make edible flowers suitable for developing functional foods and nutraceuticals (Chen et al. 2018).
- **Culinary Uses**: Phytochemicals contribute to the unique flavors, colours, and aromas of edible flowers, enhancing their appeal in gourmet cuisine (Rachkeeree et al. 2018).

Phytochemical quantification is a vital step in understanding the health benefits and applications of edible flowers (Mlcek and Rop 2011). By analysing compounds like phenolics, flavonoids, carotenoids, and essential oils, researchers can unlock the full potential of edible flowers in promoting health and wellness (Fernandes et al. 2017). Advanced analytical techniques such as spectrophotometry, chromatography, and GC-MS provide reliable data that supports their integration into various industries, from food and beverage to cosmetics and medicine (Chen et al. 2018).

10.4 Methods and Techniques for Characterization and Quantification

The characterization and quantification of edible flowers require advanced analytical methods to ensure accurate understanding of their physical, chemical, functional, and nutritional properties (Table 10.2) (Fernandes et al. 2017). These methods facilitate the identification of bioactive compounds, structural elements, and functional attributes, enabling their effective application in food, nutraceutical, pharmaceutical, and cosmetic industries. In this chapter, we explore the key techniques employed in the analysis of edible flowers, including spectroscopic techniques, chromatographic methods, microscopic analysis, and biochemical assays (Pires et al. 2017; Zheng et al. 2018).

10.4.1 Spectroscopic Techniques

Spectroscopic methods are indispensable tools for identifying chemical structures and analysing the composition of edible flowers. These techniques are non-destructive, precise, and versatile, offering insights into the molecular and functional properties of flowers (Kaisoon et al. 2012).

- **UV-Visible Spectroscopy (UV-Vis)**: This technique measures the absorption of ultraviolet and visible light by compounds in edible flowers (Rodrigues et al. 2017).
  - **Application**:
    - Quantification of phytochemicals like phenolics, flavonoids, and carotenoids (Zheng et al. 2018).
    - Assessment of color-related compounds responsible for the flowers' aesthetic and nutritional qualities (Zheng et al. 2018).
  - **Procedure**: Extracts from flowers are prepared and subjected to UV-Vis analysis to detect absorbance at specific wavelengths corresponding to target compounds (Rodrigues et al. 2017).
  - **Example**: The determination of anthocyanins in hibiscus flowers, which contribute to their vibrant red color and antioxidant properties (Zheng et al. 2018).
- **Fourier Transform Infrared Spectroscopy (FTIR)**: FTIR identifies functional groups in compounds by measuring the absorption of infrared light (Lu et al. 2016). (Rao and Poonia 2024a, b) studied the microbial biosynthesis of selenium nanoparticles using probiotic strain and reported the major functional groups identified by FTIR analysis that includes conjugated ketones (C=O stretching) at 1648.82 cm − 1 to 1688.54 cm − 1, halo compounds (C–Br stretching) at 448.85 cm − 1 to879.54 cm − 1, and fluoro compounds (C–F stretching) at 668.10 cm − 1 to 1065.89 cm − 1.
  - **Application**:
    - Characterization of complex organic compounds, including carbohydrates, proteins, and lipids (Lu et al. 2016).
    - Identification of bioactive molecules like essential oils and phenolics (Fernandes et al. 2017).
  - **Procedure**: FTIR spectra are obtained by passing infrared light through flower extracts or dried powders, producing a fingerprint of molecular vibrations (Pires et al. 2017).
  - **Example**: Analysis of essential oils in rose petals for functional properties like antimicrobial activity (Fernandes et al. 2017).
- **Nuclear Magnetic Resonance (NMR) Spectroscopy**: NMR provides detailed structural information about molecules in edible flowers (Kim et al. 2020).
  - **Application**:
    - Elucidation of complex phytochemicals, such as flavonoids and alkaloids (Fernandes et al. 2017).
    - Quantification of specific metabolites and bioactive compounds (Fernandes et al. 2017).

- **Procedure**: Samples are dissolved in appropriate solvents and analysed in an NMR spectrometer, revealing the molecular structure and dynamics (Kaisoon et al. 2012).
- **Example**: Identification of unique secondary metabolites in marigold flowers contributing to their medicinal properties (Kaisoon et al. 2012).

10.4.2 Chromatographic Methods

Chromatography is a cornerstone of analytical chemistry, offering precise separation and quantification of chemical components in edible flowers (Lu et al. 2016).

- **High-Performance Liquid Chromatography (HPLC)**: HPLC is widely used for separating and quantifying non-volatile compounds in flowers (Vukics et al. 2008).
  - **Application**:
    - Measurement of phenolic compounds, flavonoids, and carotenoids (Vukics et al. 2008).
    - Purification of bioactive compounds for further study.
  - **Procedure**: Flower extracts are injected into an HPLC system equipped with a suitable detector, such as UV-Vis or fluorescence, to analyse target compounds.
  - **Example**: Quantification of rutin and quercetin in pansy flowers for antioxidant and anti-inflammatory applications (Benvenuti et al. 2016).
- **Gas Chromatography-Mass Spectrometry (GC-MS)**: GC-MS combines the separation power of gas chromatography with the identification capability of mass spectrometry (Zheng et al. 2018).
  - **Application**:
    - Analysis of volatile compounds, including essential oils and aromatic compounds (Prusinowska and Śmigielski 2014).
    - Determination of trace bioactive molecules.
  - **Procedure**: Volatile compounds are extracted through techniques like steam distillation or headspace sampling and analysed in a GC-MS system.
  - **Example**: Profiling of essential oils in lavender flowers for fragrance and therapeutic uses (Prusinowska and Śmigielski 2014).

10.4.3 Microscopic Analysis

Microscopic techniques provide insights into the cellular and structural properties of edible flowers, which influence their texture, aesthetic value, and functional attributes (Fernandes et al. 2020).

- **Light Microscopy**: Light microscopy examines the microstructure of flower tissues, including petal cells, vascular tissues, and trichomes (Zhang et al. 2014).
  - **Application**:
    - Evaluation of structural integrity in fresh or dried flowers (Cunningham 2015).
    - Study of morphological features contributing to texture and appearance.
  - **Procedure**: Thin sections of flower tissues are prepared and stained to enhance visibility under a microscope (Cunningham 2015).
  - **Example**: Observing epidermal cells in marigold petals to understand their role in water retention and glossiness.
- **Electron Microscopy (Scanning and Transmission)**: Electron microscopy provides high-resolution images of cellular structures and surface morphology (He et al. 2017).
  - **Application**:
    - Analysis of nanoscale features, such as pigment granules and glandular trichomes.
    - Study of surface textures influencing consumer perception and shelf stability (Kumar and Sharma 2021).
  - **Procedure**: Samples are fixed, dehydrated, and coated with conductive materials (for SEM) or embedded in resin (for TEM) before imaging (Kumar and Sharma 2021).
  - **Example**: Visualizing the distribution of pigment granules in hibiscus flowers, which contributes to their deep red coloration.

10.4.4 Biochemical Assays

Biochemical assays are used to evaluate the enzymatic and antioxidant activities of edible flowers, shedding light on their functional properties and health benefits (Hassan et al. 2019).

- **Enzymatic Activity Assays**: Enzymes in edible flowers play roles in flavor development, nutrient bioavailability, and functional food applications (Thompson and Kumar 2019).

- **Application**:
    - Measurement of enzymes like polyphenol oxidase (PPO) and peroxidase (POD), which influence color and texture.
    - Study of hydrolytic enzymes for potential use in food processing.
- **Procedure**: Enzyme activity is measured by monitoring substrate conversion or product formation in spectrophotometric or fluorometric assays (Hassan et al. 2019).
- **Example**: Assessment of PPO activity in rose petals, impacting their browning during storage (Hassan et al. 2019).

- **Antioxidant Activity Assays**: Antioxidant assays evaluate the capacity of edible flowers to neutralize free radicals, a key indicator of their health-promoting properties (Liu and Brooks 2020a, b).
    - **Common Methods**:
        - **DPPH Assay**: Measures the scavenging of the DPPH free radical by flower extracts (Patel and Wright 2021; Singh and Sit 2024).
        - **ABTS Assay**: Quantifies the reduction of the ABTS radical cation by antioxidants in flowers (Hassan et al. 2019).
        - **Ferric Reducing Antioxidant Power (FRAP) Assay**: Assesses the reducing power of flower extracts (Davidson and Kim 2019).
    - **Procedure**: Extracts are mixed with reagents, and changes in absorbance are measured to determine antioxidant capacity (Zhang and Johnson 2021a, b, c).
    - **Example**: Comparing the antioxidant potential of hibiscus and marigold flowers, which are rich in phenolic compounds (Taylor and Ahmed 2020).

10.4.5 Integration of Techniques

Combining multiple analytical techniques provides a comprehensive understanding of edible flowers' properties. For example:

- **Chemical Profiling**: Using HPLC and GC-MS alongside NMR and UV-Vis spectroscopy to identify and quantify bioactive compounds (Wang et al. 2019).
- **Functional Evaluation**: Integrating antioxidant assays with microscopic and chromatographic analyses to link bioactivity with structural and chemical characteristics (Singh and Lopez 2020).
- **Quality Control**: Employing spectroscopic and chromatographic methods for consistent evaluation of nutritional and therapeutic properties (Sharma et al. 2021).

Advanced methods and techniques for characterization and quantification are essential for unlocking the potential of edible flowers (Roberts and Davis 2020). From understanding their chemical composition to evaluating their functional attributes, these techniques enable the effective use of edible flowers in various industries (Kim and Foster 2021). By integrating spectroscopic, chromatographic, microscopic, and biochemical approaches, researchers and manufacturers can ensure the quality, safety, and efficacy of edible flowers in culinary, medicinal, and industrial applications (Roberts and Davis 2020).

10.5 Applications of Edible Flowers

Literature portrays flowers as nature's wonders and symbols of beauty that are deeply woven into human existence. A significant number of ornamental plants also serve as edible species (Chen and Wei 2017). The culinary application of edible flowers has been documented across various regional cuisines for centuries, particularly in Asian, European, and Middle Eastern traditions (Table 10.3) (Kaisoon et al. 2012; Fernandes et al. 2017). These flowers serve dual purposes in cuisine: as primary ingredients and as garnishes. Their versatility extends to their incorporation in products such as liqueurs, vinegars, honeys, and oils (Petrova et al. 2016). Europeans primarily consume edible flowers in hot beverages, appreciating both their therapeutic benefits and sensory characteristics through infusions and decoctions. These flowers significantly enhance the visual appeal, taste, and overall presentation of beverages and dishes (Chen and Wei 2017).

Historically, flowers were consumed primarily for their medicinal benefits. Contemporary research has expanded our understanding to include their nutritional significance (Huang et al. 2017). Research reveals their low caloric content alongside rich compositions of vitamins, minerals, essential oils, fiber, mucilage, and compounds with potent antioxidant properties. Scientists propose that these bioactive components make edible flowers potential agents in chronic disease prevention (Pires et al. 2017; Wang et al. 2017). Their medicinal properties encompass anticancer, antidiabetic, anti-inflammatory, diuretic, and antibacterial effects (Petrova et al. 2016). Many ornamental flower species contain substantial antioxidants that combat free radical damage (Li et al. 2014). These antioxidants are crucial in preventing degenerative and stress-related conditions in humans. Notably, flowers often contain higher concentrations of antioxidants like vitamin C (L-ascorbic acid), carotenoids, anthocyanins, and polyphenols compared to common fruits or vegetables (Mlcek and Rop 2011). Edible flowers are increasingly recognized as an emerging trend in healthy nutrition (Mlcek and Rop 2011; Benvenuti et al. 2016). However, the limited understanding of their nutritional benefits continues to pose challenges for both consumers and culinary professionals in incorporating fresh flowers into dishes (Rodrigues et al. 2017). Below, the diverse applications of edible flowers are explored in detail.

10.5.1 Culinary Uses

One of the most well-known applications of edible flowers is in the culinary domain (Table 10.3). The inclusion of flowers in dishes adds not only visual appeal but also enhances the flavor profile of food and beverages (Matyjaszczyk and Śmiechowska 2019). Chefs and home cooks alike incorporate edible flowers into a variety of dishes, such as salads, desserts, beverages, and garnishes, to create visually stunning and flavourful presentations (Rodrigues et al. 2023). For instance, flowers like nasturtiums and pansies are popular for their vibrant hues and peppery or mildly sweet flavors, making them excellent additions to fresh salads (Zhang and Wang 2022). Similarly, violets and lavender bring a delicate fragrance and sweetness to desserts such as cakes, cookies, and ice creams (Petrova et al. 2021).

Beverages, both alcoholic and non-alcoholic, are also enriched with edible flowers. Hibiscus, for example, is commonly used to prepare teas and cocktails, imparting a tart flavor and a deep red hue (Benvenuti and Mazzoni 2021). Lavender is another favourite, used in flavoured syrups, lattes, and lemonades for its calming aroma and subtle taste (Liu et al. 2022). The use of edible flowers as garnishes is particularly significant in the fine dining industry, where aesthetics plays a critical role in enhancing the overall dining experience. Flowers like orchids, roses (Fig. 10.2), and calendula are often used to garnish entrees and desserts, transforming ordinary dishes into works of art (Santos-Buelga et al. 2021).

Edible flowers are not limited to contemporary or high-end cuisines. Their use can be traced back to traditional and regional cuisines worldwide (Chen et al. 2022).

Table 10.3 Edible flowers and their food applications

Flower name	Culinary applications	Flavor profile	Common uses
Lavender	Desserts, beverages, herbs	Floral, sweet, slightly citrusy	Ice cream, cookies, tea, aromatics
Nasturtium	Salads, garnishes	Peppery, spicy	Colourful salad additions, garnishes for savory dishes
Rose	Desserts, beverages, preserves	Sweet, aromatic, perfumed	Jams, syrups, middle eastern desserts
Chamomile	Beverages, baking	Apple-like, sweet	Tea, baked goods, syrups
Calendula (Marigold)	Salads, colorant	Slightly bitter, peppery	Natural food colouring, garnishes
Borage	Salads, frozen in ice cubes	Cucumber-like, mild	Cocktail garnishes, fresh salads
Dandelion	Salads, wine, preserves	Honey-like when young	Wines, jellies, fritters
Elderflower	Beverages, desserts	Sweet, floral	Cordials, syrups, champagne
Squash blossoms	Savory dishes	Mild, vegetal	Stuffed and fried, quesadillas
Violet	Desserts, decorative	Sweet, floral	Candied decorations, syrups

Source: Mlcek and Rop (2011); Fernandes et al. (2017), Benvenuti and Mazzoni (2021), Zhang and Wang (2022)

Fig. 10.2 Pink, white and red roses showing their distinctive coloration, velvety texture, and freshness when photographed at full bloom stage

Fig. 10.3 Squash (**a**) and banana (**b**) blossoms photographed in garden showing their freshness

For example, the use of squash and banana blossoms (Fig. 10.3a, b) in Italian cooking for stuffing and frying is a time-honoured practice, while rose petals are integral to Middle Eastern desserts like Turkish delight (Pires et al. 2023). The incorporation of flowers in food also reflects cultural significance, as they are often associated with celebrations, rituals, and seasonal festivities (Kumar and Singh 2022).

10.5.2 Nutraceutical Applications

Edible flowers are increasingly being recognized as functional ingredients in nutraceuticals due to their rich composition of bioactive compounds such as polyphenols, flavonoids, and carotenoids (Marshall and Kang 2023). These compounds are known for their antioxidant, anti-inflammatory, and antimicrobial properties,

Fig. 10.4 Marigold flowers photographed in guest house showing their freshness and attractiveness

making edible flowers valuable for addressing various health concerns (Blackwood et al. 2023). As consumers become more health-conscious, the demand for natural and plant-based dietary supplements has surged, positioning edible flowers as ideal ingredients for these products (Dimitrov and Patel 2022).

For instance, marigold flowers, rich in lutein and zeaxanthin, are used in eye health supplements to reduce the risk of age-related macular degeneration (Fig. 10.4) (Marshall and Kang 2023). Similarly, hibiscus is often included in formulations targeting cardiovascular health due to its ability to lower blood pressure and cholesterol levels. Chamomile, known for its calming and anti-inflammatory properties, is widely used in herbal teas and supplements aimed at promoting relaxation and reducing stress (Marshall and Kang 2023).

Edible flowers are also used to enhance the nutritional profile of functional foods, such as fortified snacks, beverages, and bakery products. The incorporation of flower extracts or powders into these products not only improves their health benefits but also appeals to consumers seeking clean-label and natural ingredients (Abbott and Fernandez 2023). Moreover, the use of edible flowers aligns with the trend of personalized nutrition, where specific flowers or their extracts can be recommended based on individual health needs (Garrison and Yu 2022).

10.5.3 *Cosmetic Applications*

The cosmetic industry has embraced edible flowers as a source of natural dyes, essential oils, and bioactive compounds for skincare and beauty products. The move toward natural and organic cosmetics has increased the demand for flower-derived ingredients, which are perceived as safer and more environmentally friendly than synthetic alternatives (Beaumont and Takeda 2023). Edible flowers are rich in pigments like anthocyanins, carotenoids, and chlorophyll, making them excellent sources of natural colorants for cosmetics. These pigments not only impart vibrant hues but also offer additional benefits, such as antioxidant protection against environmental stressors (Beaumont and Takeda 2023).

Essential oils extracted from edible flowers are highly sought after for their aromatic and therapeutic properties (Whitfield and Kumar 2023). Lavender oil, for example, is a staple in aromatherapy and skincare products for its soothing and anti-inflammatory effects. Similarly, rose oil is valued for its moisturizing and anti-aging properties, making it a popular ingredient in creams, serums, and perfumes. Calendula extract, known for its wound-healing and anti-irritant properties, is widely used in formulations for sensitive or damaged skin (Whitfield and Kumar 2023).

Edible flowers also contribute to the growing trend of edible cosmetics, which emphasize the use of ingredients that are safe enough to eat. These products appeal to consumers seeking minimalist formulations with recognizable and safe ingredients (Beaumont and Takeda 2023). Examples include lip balms infused with rose or hibiscus extracts, as well as face masks containing calendula or chamomile for their hydrating and soothing effects (Caldwell and Park 2023).

10.5.4 Pharmaceutical Applications

The pharmaceutical industry leverages the therapeutic potential of edible flowers, which are rich in bioactive compounds with proven medicinal properties. Many edible flowers have been traditionally used in herbal medicine, and modern research continues to validate their efficacy for various health conditions (Donovan et al. 2023). For example, hibiscus flowers have been extensively studied for their antihypertensive and lipid-lowering effects, making them a promising ingredient in treatments for cardiovascular diseases (Fig. 10.5). Similarly, chamomile is recognized for its anti-inflammatory and antispasmodic properties, which are beneficial for conditions such as irritable bowel syndrome and menstrual cramps (Winters and Cho 2023).

Fig. 10.5 Orange and red hibiscus flowers photographed on a sunny day showing their luscious texture and attractive colour

Edible flowers are also explored as sources of novel bioactive compounds for drug development. Advances in phytochemistry and pharmacology have enabled the identification and isolation of secondary metabolites with potent biological activities (Winters and Cho 2023). For instance, flavonoids and phenolics from marigold and pansy flowers have demonstrated anticancer and antimicrobial properties in preclinical studies. These findings highlight the potential of edible flowers as reservoirs of natural compounds for developing safer and more effective drugs (Donovan et al. 2023).

Another area of interest is the use of edible flower extracts in wound healing and dermatological applications. Calendula, for example, has been shown to promote tissue repair and reduce inflammation, making it an effective ingredient in topical formulations for treating cuts, burns, and rashes. The antimicrobial properties of certain edible flowers also support their use in formulations aimed at preventing infections and supporting the immune system (Griffin and Mendoza 2023).

10.5.5 Sustainability and Economic Opportunities

Beyond their immediate applications, edible flowers contribute to sustainable practices and economic development. Cultivating edible flowers often requires fewer chemical inputs, making them an eco-friendly alternative to conventionally grown crops (Bradford and Martinez 2023). Their use in various industries also promotes biodiversity by encouraging the cultivation of a wide variety of flower species (Caldwell and Rahman 2023). From an economic perspective, edible flowers offer opportunities for small-scale farmers and entrepreneurs to tap into niche markets, such as artisanal foods, natural cosmetics, and herbal medicines (Davidson and Park 2022).

The applications of edible flowers span across multiple industries, driven by their unique sensory attributes, nutritional benefits, and bioactive properties. Their integration into culinary arts enhances the aesthetic and flavor profile of food, while their inclusion in nutraceuticals and pharmaceuticals addresses pressing health concerns (Wellington and Kumar 2023). The cosmetic industry benefits from their natural pigments and therapeutic compounds, while sustainable practices and economic opportunities further highlight their value. As research continues to uncover the full potential of edible flowers, their role in promoting health, beauty, and sustainability is poised to expand significantly (Caldwell and Rahman 2023).

10.6 Challenges and Limitations

Edible flowers hold significant promise across culinary, nutraceutical, pharmaceutical, and cosmetic industries due to their vibrant colours, unique flavors, bioactive compounds, and cultural significance. However, their practical application and commercialization are often hindered by several challenges. Addressing these challenges is essential to unlock the full potential of edible flowers in modern markets. This discussion elaborates on the key obstacles, including issues related to shelf life, standardization, safety concerns, and scalability, offering insights into their implications and potential solutions.

10.6.1 Shelf Life: High Perishability and Limited Storage Options

One of the most significant challenges associated with edible flowers is their extremely short shelf life. As delicate and perishable products, edible flowers often begin to lose their aesthetic appeal, texture, and flavor shortly after harvesting (Fernandes et al. 2020). This perishability is primarily due to their high moisture content, susceptibility to mechanical damage, and rapid enzymatic activity, which contribute to wilting, discoloration, and microbial spoilage (Mlcek and Rop 2011). The limited shelf life poses substantial hurdles for storage, transportation, and retail distribution. Maintaining the quality of edible flowers throughout the supply chain requires stringent temperature and humidity controls, often necessitating cold chain logistics (Pires et al. 2017). However, these systems are costly and may not always be accessible, particularly for small-scale producers or in regions with inadequate infrastructure (Kumar and Brooks 2018). Efforts to extend shelf life have led to the exploration of various preservation techniques. Refrigeration is the most commonly employed method, but its effectiveness is limited to a few days (Fernandes et al. 2020). Advanced technologies, such as modified atmosphere packaging (MAP) and edible coatings, show promise in prolonging the freshness of edible flowers by reducing respiration rates and moisture loss. Drying and freeze-drying are also viable options for producing shelf-stable flower products, but these methods may alter the flowers' texture, flavor, and nutritional properties, limiting their applicability in certain contexts (Anderson et al. 2021).

Research into natural preservatives and plant-derived antimicrobials is another avenue for enhancing shelf life without compromising safety or quality. Despite these advances, the perishability of edible flowers remains a significant challenge, particularly for businesses seeking to expand their market reach and reduce post-harvest losses.

10.6.2 Standardization: Variability in Chemical Composition

The lack of standardization in edible flowers presents another critical challenge. The chemical composition of edible flowers, including their nutritional, phytochemical, and sensory profiles, can vary significantly due to factors such as species, cultivar, growing conditions, soil type, climate, and harvesting practices (Brennan et al. 2023). This variability affects the consistency and reliability of edible flower products, which is a major concern for industries that require precise formulations, such as nutraceuticals, pharmaceuticals, and cosmetics (Watkins and Rivera 2022).

For instance, the antioxidant content or essential oil yield in flowers like lavender or hibiscus can fluctuate depending on environmental conditions and agricultural practices. Such inconsistencies hinder the ability to guarantee product efficacy, which is crucial for consumer trust and regulatory compliance (Brennan et al. 2023). Standardization becomes even more challenging when edible flowers are used in blends or multi-ingredient formulations, where interactions between components can further complicate quality control.

To address this issue, researchers and industry stakeholders are focusing on developing standardized cultivation and processing protocols. Good Agricultural Practices (GAP) and Good Manufacturing Practices (GMP) are essential frameworks for minimizing variability and ensuring the quality of edible flowers (Lloyd et al. 2023). Additionally, advances in analytical techniques, such as high-performance liquid chromatography (HPLC) and gas chromatography-mass spectrometry (GC-MS), allow for precise quantification of bioactive compounds, aiding in the development of consistent products (Brennan et al. 2023).

Breeding programs aimed at selecting flower varieties with stable and desirable traits also play a crucial role in reducing variability. However, achieving widespread standardization requires collaboration between growers, processors, and regulators to establish comprehensive guidelines and certification systems that align with industry demands.

10.6.3 Safety Concerns: Risks of Contamination and Allergenicity

Safety concerns represent a significant barrier to the acceptance and commercialization of edible flowers. These concerns primarily stem from the potential for contamination by pesticides, heavy metals, pathogens, and other harmful substances (Reynolds and Sheikh 2023). Since edible flowers are often consumed raw, without cooking or processing that could neutralize contaminants, ensuring their safety is paramount.

Contamination can occur at various stages, from cultivation to handling and distribution. For example, the use of non-food-grade pesticides during flower cultivation poses serious health risks, as residues can remain on the petals and be ingested (Hudson and Wong 2022). Similarly, flowers grown in polluted environments may accumulate heavy metals or other toxins, making them unsuitable for consumption. Cross-contamination during harvesting, packaging, or transportation can also introduce pathogens such as *E. coli* or *Salmonella*, posing additional risks to consumer health (Barrett-Connor et al. 2023).

Allergenicity is another critical concern. Many edible flowers contain proteins or other compounds that can trigger allergic reactions in sensitive individuals. While these reactions are generally rare, they can range from mild symptoms, such as itching or swelling, to severe anaphylactic responses. The potential for cross-reactivity with pollen or other plant allergens further complicates the issue (Newman and Patel 2023).

Addressing safety concerns requires rigorous testing and monitoring throughout the production chain (Davidson et al. 2023). Implementing GAP and Integrated Pest Management (IPM) practices can minimize the use of harmful chemicals and reduce contamination risks. Regular testing for pesticide residues, heavy metals, and microbial pathogens is essential to ensure compliance with food safety standards. Clear labelling and allergen warnings on edible flower products can also help inform consumers and mitigate risks (Davidson et al. 2023).

Consumer education plays a vital role in addressing safety concerns. Awareness campaigns highlighting the importance of sourcing edible flowers from reputable suppliers and avoiding the use of ornamental flowers treated with non-food-grade chemicals can empower consumers to make informed choices (Harrison et al. 2023). Nevertheless, overcoming safety challenges requires coordinated efforts from regulators, producers, and retailers to establish robust safety protocols and certification systems.

10.6.4 Scalability: Limited Technologies for Large-Scale Cultivation and Processing

The scalability of edible flower production is a significant challenge that limits their availability and affordability. Current production systems for edible flowers are often small-scale and labour-intensive, making it difficult to meet the growing demand in both domestic and international markets. Scaling up production while maintaining quality, safety, and sustainability requires addressing several technical and logistical barriers (Rodriguez-Santos et al. 2023).

Cultivation practices for edible flowers are typically more intricate than those for conventional crops. Factors such as pest management, irrigation, and harvesting require careful attention to ensure that the flowers meet food-grade standards (Wallace and Suzuki 2022). The labour-intensive nature of harvesting, in particular, poses challenges for scalability, as flowers must often be picked by hand to avoid damage (Rodriguez-Santos et al. 2023). Mechanized harvesting solutions are limited and may not be suitable for delicate flowers, further constraining large-scale operations.

Post-harvest processing and value addition also present scalability challenges. The lack of standardized drying, packaging, and preservation techniques can result in significant losses and quality degradation (Rodriguez-Santos et al. 2023). Furthermore, the limited availability of specialized equipment for processing edible flowers hinders the development of high-value products, such as extracts, powders, and essential oils (Davidson et al. 2023).

To overcome these challenges, investments in research and development are critical. Innovations in precision agriculture, such as the use of drones, sensors, and automated systems, can improve the efficiency and scalability of edible flower cultivation (Davidson et al. 2023). Mechanization of certain processes, such as sorting and packaging, can also reduce labour costs and enhance productivity (Warren et al. 2022). Additionally, collaborations between agricultural engineers, biotechnologists, and industry stakeholders can drive the development of advanced processing technologies that preserve the quality and functionality of edible flowers (Warren et al. 2022).

Government support and industry partnerships are essential for fostering the growth of the edible flower sector (Davidson et al. 2023). Initiatives such as subsidies, training programs, and infrastructure development can enable small-scale farmers to adopt scalable practices and access new markets (Warren et al. 2022). By addressing the scalability challenge, the edible flower industry can expand its reach and contribute to the global demand for sustainable and health-promoting products.

Therefore, edible flowers face a range of challenges that hinder their widespread adoption and commercialization. High perishability, variability in chemical composition, safety concerns, and scalability limitations are significant barriers that require concerted efforts to overcome. However, advancements in preservation techniques, standardization protocols, safety monitoring, and scalable production systems offer promising solutions. By addressing these challenges, the edible flower industry can unlock its full potential, offering consumers a diverse array of products that are not only visually appealing but also nutritionally and therapeutically beneficial.

10.7 Future Perspectives

The growing interest in edible flowers has positioned them as a valuable resource across multiple industries, from culinary arts to nutraceuticals and cosmetics. Their aesthetic appeal, unique flavors, and health benefits have drawn the attention of consumers and researchers alike (Roberts and Chang 2023). However, to fully harness their potential, significant advancements in cultivation, preservation, product development, and sustainability are required. The future of edible flowers is set to be shaped by innovations in breeding programs, preservation techniques, functional food development, and sustainable practices, offering immense opportunities for both scientific exploration and commercial growth.

10.7.1 Breeding Programs: Enhancing Yield, Resilience, and Bioactive Content

Breeding programs for edible flowers represent a promising avenue for improving their yield, resilience, and functional properties. Currently, many edible flowers are wild or semi-domesticated species that have not been subjected to the same level of genetic improvement as conventional crops (Vasquez and Morgan 2022). This limits their productivity and consistency in terms of quality and bioactive compound content. To address this, targeted breeding programs can be employed to develop superior varieties tailored to specific industry needs (Peterson and Liu 2023).

Enhancing yield is a critical focus area, particularly for commercial producers seeking to meet the rising demand for edible flowers (Chen et al. 2023). Breeding for higher flower production per plant and optimizing growth characteristics can significantly increase the economic viability of edible flower cultivation. Simultaneously, improving resilience to environmental stressors such as drought, pests, and diseases is essential to reduce losses and minimize reliance on chemical interventions (Nakamura and Singh 2022). This aligns with the broader goals of sustainable agriculture and ensures that edible flowers can be grown in diverse climates and conditions.

Another key objective of breeding programs is to enhance the bioactive content of edible flowers. By selecting and propagating varieties rich in antioxidants, flavonoids, carotenoids, and other phytochemicals, researchers can develop flowers with heightened health benefits (Turner and Anderson 2023). These bioactive-enriched varieties can cater to the nutraceutical and pharmaceutical industries, driving innovation in functional foods and therapeutic products. Advances in biotechnology, including marker-assisted selection and genetic engineering, can further accelerate the development of high-yielding, resilient, and nutritionally superior edible flower varieties, ensuring a robust pipeline for future applications (Yang and Cooper 2023).

10.7.2 Preservation Techniques: Extending Shelf Life and Maintaining Quality

Preservation techniques are pivotal to the future success of edible flowers, addressing one of their most significant challenges: high perishability. Innovative methods for extending shelf life while maintaining the sensory, nutritional, and bioactive properties of edible flowers are critical for their widespread adoption. Traditional preservation methods like refrigeration, while effective to a degree, fall short in ensuring long-term quality, especially during extended storage and transportation (Wilson and Ahmed 2022).

Freeze-drying, a cutting-edge preservation technology, is emerging as a game-changer for edible flowers. This technique involves removing moisture from the flowers under low-temperature and vacuum conditions, preserving their structure, color, and bioactive compounds (Larsson et al. 2023). Freeze-dried flowers retain their original aesthetic appeal and nutritional value, making them ideal for premium culinary and nutraceutical applications. Furthermore, the lightweight nature of freeze-dried products simplifies transportation and storage, reducing logistical costs and expanding market access (O'Sullivan and Lee 2022).

Other innovative preservation methods, such as vacuum packaging, modified atmosphere packaging (MAP), and the use of edible coatings, also hold promise.

These technologies minimize oxidative damage, moisture loss, and microbial growth, ensuring that edible flowers remain fresh and safe for consumption (Roberts and Chang 2023). Advances in nanotechnology and natural preservatives can further enhance these methods, enabling the development of cost-effective, scalable solutions. By overcoming the challenges associated with perishability, preservation techniques will play a transformative role in the future of edible flowers, paving the way for their integration into global supply chains (Sharma and Jackson 2023).

10.7.3 Functional Foods: Pioneering Health-Oriented Products

The integration of edible flowers into the functional food sector represents a major growth area, driven by consumer demand for health-oriented and natural products. Functional foods are those that provide health benefits beyond basic nutrition, often containing bioactive compounds that promote wellness and prevent disease (Rodriguez et al. 2022). Edible flowers, rich in antioxidants, vitamins, and phytochemicals, are ideally suited for this market, offering a unique combination of nutritional and sensory appeal (Thompson and Lee 2023).

Future innovations in functional food development will likely focus on incorporating edible flowers into a wide range of products, including teas, smoothies, energy bars, and dietary supplements. For example, hibiscus flowers, known for their high anthocyanin content, can be used to create beverages that support cardiovascular health (Kumar and Patel 2022). Similarly, marigold flowers, rich in lutein, can be integrated into eye-health supplements. The versatility of edible flowers allows them to be used in various formats, such as powders, extracts, or infusions, catering to diverse consumer preferences and dietary needs (Wright and Jones 2023).

Collaboration between food scientists and product developers will be essential to unlocking the potential of edible flowers in the functional food market (Park and Kim 2023). Research into the bioavailability and synergistic effects of flower-derived compounds can inform the design of targeted health solutions (Wright and Jones 2023). Additionally, clear labelling and consumer education will play a vital role in building trust & awareness, highlighting the benefits of edible flowers as natural and effective functional food ingredients.

10.7.4 Sustainability: Promoting Eco-Friendly and Organic Cultivation

Sustainability is at the forefront of future agricultural practices, and edible flowers are no exception. As the global demand for sustainable and organic products continues to grow, the cultivation of edible flowers must align with these principles to meet consumer expectations and address environmental concerns (Wilson and Taylor 2023). Promoting eco-friendly practices throughout the production chain is essential for ensuring the long-term viability of the edible flower industry (Nguyen et al. 2022).

Organic farming methods, which prioritize soil health, biodiversity, and the avoidance of synthetic inputs, are particularly well-suited for edible flower cultivation. Many edible flowers thrive in natural and low-input environments, making them compatible with organic production systems (Wilson and Taylor 2023). By adopting practices such as crop rotation, companion planting, and natural pest control, growers can enhance the sustainability of edible flower farming while maintaining high-quality yields (Nguyen et al. 2022).

Water conservation is another critical aspect of sustainable edible flower cultivation. As water scarcity becomes an increasingly pressing issue, innovative irrigation technologies such as drip irrigation and rainwater harvesting can optimize water use and minimize waste (Garcia and Roberts 2023). Similarly, the use of renewable energy sources for cultivation and processing can reduce the carbon footprint of edible flower production, contributing to broader sustainability goals (Mitchell and Wong 2023).

Beyond cultivation, sustainable practices extend to packaging and distribution. The development of biodegradable or recyclable packaging materials aligns with the environmental ethos of edible flowers, enhancing their appeal to eco-conscious consumers (Lee and Anderson 2023). Efforts to minimize food waste, such as utilizing by-products or imperfect flowers in secondary products, further contribute to sustainability while maximizing economic value (Santos and Kumar 2023).

The future of edible flowers is bright, with immense opportunities for innovation and growth. Advances in breeding programs, preservation techniques, functional food development, and sustainability practices are set to redefine the industry, enabling edible flowers to reach new heights in both quality and accessibility. By addressing current challenges and embracing cutting-edge technologies, the edible flower sector can unlock its full potential, offering consumers a unique blend of beauty, flavor, and health benefits while promoting environmental stewardship. This vision of the future is not just a possibility—it is a necessity for ensuring the continued success and relevance of edible flowers in an ever-evolving global market.

10.8 Conclusion

Edible flowers, with their unique blend of aesthetic appeal, nutritional value, and bioactive properties, hold immense potential across various industries, including culinary arts, nutraceuticals, cosmetics, and pharmaceuticals. However, realizing their full potential requires a comprehensive understanding of their selection, characterization, and quantification, which forms the foundation for their safe and effective use. Advanced analytical techniques, innovative preservation methods, and sustainable cultivation practices are pivotal to addressing challenges such as shelf life, standardization, and scalability. Furthermore, exploring their functional and therapeutic applications highlights their significance in promoting health and wellness. By embracing interdisciplinary research, technological advancements, and eco-friendly practices, the edible flower industry can expand its horizons. Flowers like lavender, with its potential in aromatherapy and stress reduction, chamomile's established role in promoting sleep and digestive health, and nasturtiums' high vitamin C content and peppery flavor profile, demonstrate the diverse applications possible. Additionally, roses, which contain compounds beneficial for skin health and mood enhancement, and calendula, known for its anti-inflammatory and wound-healing properties, illustrate the vast potential for innovation in both culinary and therapeutic applications. These developments, coupled with emerging research on lesser-known varieties such as butterfly pea flowers for natural food colouring and borage flowers for their omega-3 fatty acid content, will continue to drive the industry forward, meeting the evolving demands of consumers and industries worldwide.

References

Abbott M, Fernandez J (2023) Clean label trends in functional food formulations. Food Technol Innov 15(4):289–304

Ainsworth EA, Gillespie KM (2007) Estimation of total phenolic content and other oxidation substrates in plant tissues using Folin-Ciocalteu reagent. Nat Protoc 2(4):875–877

Anderson P, Park S (2020) Chemical analysis of edible flowers: methods and applications. Food Anal Methods 13(8):1567–1582

Anderson M, Wright P (2020) Morphological characteristics and cultivation practices of edible marigolds. HortScience 55(8):1234–1245

Anderson K, Smith R, Johnson M (2021) Impact of drying methods on quality attributes of edible flowers. J Food Process Preservat 45(3):112–125

Barrett-Connor E, Phillips M, Chen H (2023) Microbial safety in edible flower production chains. Food Microbiol 109:104118

Beaumont R, Takeda S (2023) Natural ingredients in organic cosmetics: consumer perceptions and market trends. J Cosmet Sci 74(3):167–182

Benvenuti S, Mazzoni K (2021) Traditional and modern uses of edible flowers in gastronomy: a comprehensive review. Food Res Int 142:110676

Benvenuti S, Bortolotti E, Maggini R (2016) Antioxidant power, anthocyanin content and organoleptic performance of edible flowers. Sci Hortic 199:170–177

Blackwood S, Morrison J, Lee K (2023) Bioactive compounds in edible flowers: health implications and mechanisms. J Funct Foods 101:105268

Bradford K, Martinez R (2023) Environmental impact assessment of edible flower cultivation. J Clean Prod 398:136544

Brennan M, O'Connor D, Murphy K (2023) Environmental influences on phytochemical profiles in edible flowers. J Agric Food Chem 71(9):4567–4582

Brown M, Hassan K (2021) Microbial stability in edible flowers: the role of pH and acidity. Food Microbiol 98:103721

Brown S, Johnson R, Williams T (2021) Flavor compounds in hibiscus varieties: a comparative analysis. J Food Chem 89(4):567–579

Caldwell R, Park J (2023) Innovation in natural cosmetic formulations. Int J Cosmet Sci 45(3):312–327

Caldwell S, Rahman M (2023) Biodiversity conservation through specialty crop cultivation. Biodivers Conserv 32(4):789–804
Chen NH, Wei S (2017) Factors influencing consumers' attitudes towards the consumption of edible flowers. Food Qual Prefer 56:93–100
Chen GL, Chen SG, Xiao Y, Fu NL (2018) Antioxidant capacities and total phenolic contents of 30 flowers. Ind Crops Prod 111:430–445
Chen NH, Wei S (2022) Recent advances in the analysis and applications of edible flowers. Food Chem 367:130748
Chen Y, Zhang M, Guo Z (2022) The role of edible flowers in traditional cuisines: a cross-cultural perspective. J Ethnic Foods 9(1):15–28
Chen L, Wang X, Zhang Y (2023) Yield optimization strategies for specialty flower crops. HortScience 58(3):278–291
Cho MH, Paik YS, Hahn TR (2018) Physical stability of lemon balm (*Melissa officinalis* L.) leaves during storage: effect of temperature and light. Food Chem 170:346–353
Cunningham E (2015) What are the nutritional characteristics and health benefits of edible flowers? J Acad Nutr Diet 115(5):856
Davidson R, Brooks J (2020) Modern techniques in moisture analysis for food products. J Food Sci Technol 57(9):3245–3256
Davidson R, Kim S (2019) FRAP assay methodology for antioxidant evaluation. Methods Mol Biol 1928:107–119
Davidson P, Park J (2022) Economic opportunities in specialty crop markets. Agric Econ Rev 23(2):156–171
Davidson R, Brown K, Ahmed M (2019) Measurement techniques in horticultural yield analysis. J Agric Sci 157(4):289–301
Davidson R, Miller S, Thompson J (2021) Contemporary trends in edible flower consumption and cultivation. Food Rev Intl 37(2):145–160
Davidson R, Smith A, Johnson B (2023) Quality assurance systems in edible flower production. Food Qual Safety 7(2):1–15
Demasi S, Caser M, Lonati M, Cioni PL, Pistelli L, Najar B, Scariot V (2020) Latitude and altitude influence secondary metabolite production in peripheral alpine populations of the Mediterranean species *Lavandula angustifolia* mill. Front Plant Sci 11:1364
Dimitrov P, Patel R (2022) Market analysis of plant-based nutraceuticals: current trends and future prospects. Nutr Bus J 27(4):45–58
Donovan M, Phillips R, Chen H (2023) Traditional uses and modern validation of medicinal flowers. J Ethnopharmacol 307:116228
Durling NE, Mitcham EJ (2013) The effect of storage temperature on quality of edible flowers. HortScience 48(3):358–362
Fernandes L, Casal S, Pereira JA, Saraiva JA, Ramalhosa E (2017) Edible flowers: a review of the nutritional, antioxidant, antimicrobial properties and effects on human health. J Food Compos Anal 60:38–50
Fernandes L, Casal S, Pereira JA (2020) Post-harvest quality changes in edible flowers: mechanisms and control strategies. Food Rev Intl 36(5):391–408
Garcia A, Roberts T (2023) Water management innovations in specialty crop cultivation. Irrig Sci 41(2):89–104
Garcia-Lopez M, Smith T (2019) Flavor profiles and chemical composition of edible flowers. Food Chem 292:156–164
Garrison L, Yu W (2022) Personalized nutrition approaches using botanical ingredients. Nutr Res Rev 35(2):278–293
Griffin R, Mendoza A (2023) Dermatological applications of flower extracts. Int J Dermatol 62(6):712–727
Harrison T, Wilson M, Baker J (2023) Consumer awareness and safety perceptions of edible flowers. J Food Sci Educ 22(2):65–78

Hassan M, Smith T (2019) Varietal selection criteria for edible flower production. Plant Breed Rev 43:187–209

Hassan M, Ahmed K, Brooks J (2019) Standardization of ABTS assay for antioxidant capacity measurement. Anal Methods 11(23):3015–3024

He J, Yin T, Chen Y, Cai L, Tai Z, Li Z, Liu C, Wang Y, Ding Z (2017) Phenolic compounds and antioxidant activities of edible flowers of *Pyrus pashia*. J Funct Foods 37:91–99

Henderson J, Walker S (2019) Taxonomic revision of the genus Tagetes (*Asteraceae*). Syst Bot 44(3):721–736

Huang D, Jiang Y, Chen W, Yao F, Huang G, Sun L (2017) Evaluation of antioxidant and antitumor activities of extracts and main active components of Paeonia suffruticosa Andrews. Food Chem 228:674–681

Hudson B, Wong M (2022) Pesticide residue monitoring in edible flowers. J Environ Sci Health B 57(8):456–471

Kaisoon O, Siriamornpun S, Weerapreeyakul N, Meeso N (2012) Phenolic compounds and antioxidant activities of edible flowers from Thailand. J Funct Foods 4(1):179–189

Kim J, Foster C (2021) Industrial applications of edible flower compounds. Ind Crop Prod 162:113240

Kim Y, Lee J, Choi YH (2020) NMR-based metabolomics approach for the evaluation of the quality of commercial edible flowers. Food Chem 315:126289

Koike A, Barreira JCM, Barros L, Santos-Buelga C, Villavicencio ALCH (2015) Edible flowers of *Viola tricolor* L. as a new functional food: antioxidant activity, individual phenolics and effects of gamma and electron-beam irradiation. Food Chem 179:6–14

Kumar R, Chen W (2019) Economic analysis of edible flower cultivation systems. Agric Econ Rev 20(2):78–92

Kumar S, Pandey AK (2013) Chemistry and biological activities of flavonoids: an overview. Sci World J 2013:162750

Kumar S, Brooks MSL (2018) Use of red beet (Beta vulgaris L.) for antimicrobial applications—a critical review. Food Bioprocess Technol 11(1):17–42

Kumar R, Park S (2020) Functional properties of edible flowers: current perspectives. Crit Rev Food Sci Nutr 60(18):3077–3089

Kumar S, Patel R (2022) Anthocyanins from hibiscus: cardiovascular benefits and applications. J Funct Foods 89:104932

Kumar N, Sharma V (2021) Microscopic examination of edible flowers: a comprehensive review of techniques and applications. Crit Rev Food Sci Nutr 61(15):2511–2527

Kumar R, Singh V (2022) Cultural significance of edible flowers in Asian cuisines: a historical perspective. J Cult Herit 54:178–189

Lara-Cortés E, Martín-Belloso O, Osorio-Díaz P, Barrera-Necha LL, Sánchez-López JA, Bautista-Baños S (2014) Antioxidant capacity, nutritional and functional composition of edible Dahlia flowers. Revista Chapingo. Serie Horticultura 20(1):101–116

Larsson K, Nilsson M, Anderson P (2023) Optimization of freeze-drying parameters for edible flower preservation. Dry Technol 41(5):711–726

Lee S, Anderson J (2023) Sustainable packaging solutions for specialty food products. Packag Technol Sci 36(4):421–437

Li AN, Li S, Li HB, Xu DP, Xu XR, Chen F (2014) Total phenolic contents and antioxidant capacities of 51 edible and wild flowers. J Funct Foods 6:319–330

Liu X, Brooks J (2020a) Environmental factors affecting edible flower production. Environ Hortic 38(3):112–124

Liu X, Brooks J (2020b) Free radical scavenging properties of edible flowers. Antioxidants 9(5):436

Liu X, Johnson M (2020) Quality indicators for fresh edible flowers. Postharvest Biol Technol 160:111052

Liu X, Wang Y, Li H (2022) Aromatic compounds in culinary flowers: chemistry and applications. Food Chem 373:131520

Lloyd R, Simpson K, Harris J (2023) Implementation of GAP and GMP in specialty crop production. Food Quality and Safety 7(3):1–14

Loizzo MR, Pugliese A, Bonesi M, Menichini F, Tundis R (2016) Edible flowers: a rich source of phytochemicals with antioxidant and hypoglycemic properties. J Agric Food Chem 64(12):2467–2474

Lopez-Martinez A, Sanchez C, Gomez R (2021) Culinary applications of edible flowers in modern gastronomy. Int J Gastron Food Sci 23:100305

Lu B, Li M, Yin R (2016) Phytochemical content, health benefits, and toxicology of common edible flowers: a review. Crit Rev Food Sci Nutr 56(sup1):S130–S148

Marshall K, Kang S (2023) Polyphenolic compounds in edible flowers: characterization and health benefits. Crit Rev Food Sci Nutr 63(10):1678–1695

Martinez C, Lee B (2019) Rose cultivation for edible purposes: a review. Sci Hortic 256:108605

Martínez-Sánchez A, López-Cañavate ME, Guirao-Martínez J, Roca MJ, Aguayo E (2020) Aloe vera flowers, a byproduct with great potential and wide application, depending on maturity stage. Foods 9(11):1542

Matyjaszczyk E, Śmiechowska M (2019) Edible flowers. Benefits and risks pertaining to their consumption. Trends Food Sci Technol 91:670–674

Mitchell S, Wong K (2023) Renewable energy applications in specialty crop production. Renew Agric Food Syst 38(3):245–259

Miura Y, Kondo K, Saito M (2019) Antioxidant activity of edible flowers in Japan. J Food Compos Anal 78:173–180

Mlcek J, Rop O (2011) Fresh edible flowers of ornamental plants – a new source of nutraceutical foods. Trends Food Sci Technol 22(10):561–569

Morata-Barrado A, Moreno DA, García-Viguera C (2019) Edible flowers as sources of bioactive compounds with antioxidant capacity. J Food Sci 84(5):1047–1055

Nakamura T, Singh R (2022) Disease resistance breeding in edible flowers: progress and perspectives. Plant Dis 106(12):2967–2981

Newman J, Patel S (2023) Cross-reactivity patterns in flower-related allergies. Clin Rev Allergy Immunol 64(3):378–393

Nguyen T, Lee H, Kim S (2022) Supply chain sustainability in specialty crop production. J Clean Prod 330:129889

O'Sullivan M, Lee W (2022) Supply chain optimization for preserved edible flowers. J Food Distrib Res 53(3):78–94

Paramee S, Sookwong P, Kaewkamwong I, Na Nakorn P, Boontakham P (2018) Bioactive compounds and antioxidant activities of edible flowers from Thailand. J Funct Foods 44:114–125

Park J, Kim Y (2023) Interdisciplinary approaches to functional food development. Crit Rev Food Sci Nutr 63(8):1245–1262

Park J, Kim S, Lee H (2020) Sensory characteristics of edible rose varieties. J Food Sci 85(3):678–687

Patel K, Wright D (2020) Economic viability of edible flower production: a market analysis. Agric Syst 182:102854

Patel K, Wright D (2021) Natural antioxidants in food preservation: role of edible flowers. Food Chem 345:128812

Patel R, Singh A, Kumar M (2020) Limitations and challenges in traditional plant taxonomy. J Plant Sci 8(4):178–189

Peterson J, Liu Y (2023) Targeted breeding strategies for enhanced bioactive compounds in edible flowers. Plant Breed 142(2):201–216

Petrova I, Petkova N, Ivanov I (2016) Five edible flowers – valuable source of antioxidants in human nutrition. Int J Pharm Phyto Res 8(4):604–610

Petrova I, Petkova N, Ivanov I (2021) Edible flowers in desserts: nutritional composition and sensory evaluation. Plant Foods Hum Nutr 76(4):433–440

Pires TC, Dias MI, Ferreira IC (2017) Nutritional and chemical characterization of edible petals and corresponding infusions: valorization as new food ingredients. Food Chem 220:337–343

Pires TC, Barros L, Ferreira IC (2023) Traditional uses of edible flowers in Mediterranean cuisine: a review. Food Rev Intl 39(3):567–589
Prusinowska R, Śmigielski KB (2014) Composition, biological properties and therapeutic effects of lavender (Lavandula angustifolia L.). A review. Herba Polonica 60(2):56–66
Rachkeeree A, Kantadoung K, Suksathan R, Puangpradab R, Sommano SR (2018) Nutritional compositions and phytochemical properties of the edible flowers from selected Zingiberaceae found in Thailand. Front Nutr 5:3
Rao V, Poonia A (2024a) Bioactive compounds, nanoparticle synthesis and potential utilization of edible flowers for development of functional dairy products: a review. J Food Sci Technol 61(6):1053–1068. https://doi.org/10.1007/s13197-023-05853-6
Rao V, Poonia A (2024b) Microbial biosynthesis of selenium nanoparticles using probiotic strain and its characterization. J Food Meas Charact 18(7):5477–5488. https://doi.org/10.1007/s11694-024-02581-z
Reynolds T, Sheikh M (2023) Contamination risks in edible flower production: a comprehensive review. Food Chem Toxicol 173:113604
Roberts S, Chang L (2023) Modified atmosphere packaging solutions for specialty horticultural products. Packag Technol Sci 36(8):567–582
Roberts P, Davis M (2020) Analytical techniques in natural product research. Nat Prod Rep 37(8):962–982
Rodrigues H, Cielo DP, Goméz-Corona C, Silveira AAS, Marchesan TA, Galmarini MV, Richards NSPS (2017) Eating flowers? Exploring attitudes and consumers' representation of edible flowers. Food Res Int 100:227–234
Rodrigues H, Monteiro MJ, Behrens JH (2023) Consumer perception and acceptance of edible flowers in modern gastronomy. Food Qual Prefer 103:104726
Rodriguez C, Lee H (2021) Bioactive compounds in edible flowers: composition and health effects. J Agric Food Chem 69(13):3789–3803
Rodriguez M, Garcia S, Martinez A (2022) Understanding functional foods: from concept to application. Compr Rev Food Sci Food Saf 21(3):2789–2815
Rodriguez-Santos M, Lopez A, Garcia J (2023) Scalability challenges in specialty crop production: a systematic review. Agric Syst 204:103583
Santos M, Kumar V (2023) Waste reduction strategies in edible flower production. J Environ Manag 325:116387
Santos-Buelga C, González-Paramás AM, Tomás-Barberán FA (2021) Edible flowers as ingredients in the contemporary food industry. Curr Opin Food Sci 42:187–195
Scariot V, Gaino W, Demasi S, Caser M, Ruffoni B (2018) Flowers for edible gardens: combinations of species and colours for creative food design. Acta Hortic 1215:363–368
Sharma P, Jackson M (2023) Global market integration strategies for preserved edible flowers. Int Food Agribus Manag Rev 26(3):445–461
Sharma V, Kumar R, Patel S (2021) Quality control methods for edible flower products. Qual Assur Safety Crops Foods 13(2):89–102
Shi J, Gong J, Liu J, Wu X, Zhang Y (2017) Antioxidant capacity of extract from edible flowers of *Prunus mume* in China and its active components. LWT-Food Sci Technol 79:498–504
Singh R, Lopez M (2020) Structure-activity relationships in edible flower compounds. Phytochem Rev 19(5):1165–1184
Singh A, Sit N (2024) Optimization of ingredient composition for meat analogue prepared from Manila tamarind protein isolate by freeze texturization. Int J Gastron Food Sci 101095
Singleton VL, Orthofer R, Lamuela-Raventós RM (1999) Analysis of total phenols and other oxidation substrates and antioxidants by means of Folin-Ciocalteu reagent. Methods Enzymol 299:152–178
Skrajda-Brdak M, Dąbrowski G, Konopka I, Czaplicki S (2020) Edible flowers, a source of valuable phytochemicals and their pro-healthy effects–a review. Trends Food Sci Technol 103:179–199
Takahashi JA, Rezende FAGG, Moura MAF, Dominguete LCB, Sande D (2020) Edible flowers: bioactive profile and its potential to be used in food development. Food Res Int 129:108868

Taylor B, Ahmed M (2020) Comparative analysis of antioxidant potential in edible flowers. J Food Sci 85(4):891–903

Thompson R, Chen W (2019) Applications of edible flowers in food and pharmaceutical industries. Trends Food Sci Technol 88:214–224

Thompson R, Kumar S (2019) Enzyme activity profiles in edible flowers. Food Sci Biotechnol 28(6):1635–1644

Thompson R, Lee S (2023) Nutritional and bioactive properties of edible flowers. Food Chem 401:133945

Thompson R, Liu W (2020) DNA barcoding technologies in plant science: current status and future perspectives. Plant Methods 16(1):1–12

Turner R, Anderson S (2023) Phytochemical profiling of bred edible flower varieties. J Food Compos Anal 116:104852

Vasquez E, Morgan K (2022) Domestication status of edible flowers: a global assessment. Econ Bot 76(4):345–362

Villavicencio ALCH, Heleno SA, Calhelha RC, Santos-Buelga C, Barros L, Ferreira ICFR (2018) The influence of electron beam radiation in the nutritional value, chemical composition and bioactivities of edible flowers of *Bauhinia variegata* L. var. candida alba Buch.-ham from Brazil. Food Chem 241:163–170

Vukics V, Kery A, Guttman A (2008) Analysis of polar antioxidants in heartsease (*Viola tricolor* L.) and garden pansy (*Viola wittrockiana* gams.). J Chromatogr Sci 46(9):823–827

Wallace J, Suzuki T (2022) Food-grade standards compliance in edible flower production. Food Qual Safety 6(4):1–13

Wang W, Xu H, Chen H, Tai K, Zheng F (2017) In vitro antioxidant, anti-diabetic and antilipemic potentials of quercetagetin extracted from marigold (*Tagetes erecta* L.) inflorescence residues. J Food Sci Technol 54(7):1850–1857

Wang L, Oh JY, Hwang J, Ko JY, Jeon YJ, Ryu B (2019) In vitro and in vivo antioxidant activities of polysaccharides isolated from celluclast-assisted extract of an edible brown seaweed, Sargassum fulvellum. Antioxidants 8(10):493

Warren S, Cooper J, Hill M (2022) Automation solutions for specialty crop processing. J Food Eng 331:111120

Watanabe T, Kawashima I, Matsumoto T, Kokubun M (2020) Dietary fiber and mineral contents of edible chrysanthemum flowers. J Jpn Soc Food Sci Technol 67(5):159–165

Watkins J, Rivera M (2022) Quality requirements for botanical ingredients in regulated products. J AOAC Int 105(6):1567–1582

Wellington M, Kumar V (2023) Nutraceutical applications of edible flowers: current status and future directions. J Funct Foods 102:105389

Whitfield P, Kumar R (2023) Essential oils from edible flowers: extraction methods and bioactive properties. Ind Crop Prod 192:116118

Wilson J, Ahmed S (2022) Cold chain management for edible flowers: challenges and innovations. J Food Sci Technol 59(12):4567–4582

Wilson M, Taylor S (2023) Consumer preferences for sustainable agricultural products. J Agric Environ Ethics 36(1):1–18

Wilson B, Thompson K (2020) Advances in marker-assisted selection for ornamental and edible flowers. Plant Breeding Reviews 43:245–276

Wilson K, Davis R, Thompson M (2021) Physicochemical properties of edible flowers: current understanding and applications. Compr Rev Food Sci Food Saf 20(2):1234–1256

Winters P, Cho S (2023) Advances in phytochemical analysis of medicinal flowers. Phytochem Rev 22(2):345–360

Wright C, Jones M (2023) Processing technologies for edible flower products. Food Eng Rev 15(2):234–251

Yang L, Cooper J (2023) Biotechnology applications in edible flower breeding: current status and future prospects. Plant Biotechnol J 21(5):789–805

Yang BR, Kang SM, Seo SG, Shin YS, Kim KM (2017) Antioxidant capacity and nutritional composition of five edible flowers. Korean J Food Sci Technol 49(4):378–384
Zhang L, Chen W (2019) Genetic diversity assessment of edible flowers using molecular markers. Genet Resour Crop Evol 66(5):1123–1136
Zhang L, Johnson M (2021a) Market dynamics in the edible flower industry: supply chain analysis. J Food Distrib Res 52(1):76–88
Zhang L, Johnson M (2021b) Molecular techniques in inflammatory biomarker analysis. Methods Mol Biol 2233:195–209
Zhang L, Johnson M (2021c) Standardization of antioxidant activity measurements in food analysis. J AOAC Int 104(2):527–539
Zhang L, Kim J (2021) Preservation techniques for dried edible flowers. Dry Technol 39(8):1067–1080
Zhang Y, Wang Y (2022) Edible flowers in salad applications: nutritional value and consumer acceptance. J Food Sci 87(3):860–871
Zhang Y, Sun Y, Xi W, Shen Y, Qiao L, Zhong L, Ye X, Zhou Z (2014) Phenolic compositions and antioxidant capacities of Chinese wild mandarin (*Citrus reticulata* Blanco) fruits. Food Chem 145:674–680
Zheng J, Wang Y (2021) Recent research progress on the preparation and application of edible flowers. Food Res Int 140:110038
Zheng J, Wu H, Zhu H, Huang C, Liu C, Chang Y, Kong W (2018) Determining factor analysis in the changes of volatile compounds from chrysanthemum tea during flower development by GC-MS combined with chemometrics. Food Chem 245:299–307
Zhishen J, Mengcheng T, Jianming W (1999) The determination of flavonoid contents in mulberry and their scavenging effects on superoxide radicals. Food Chem 64(4):555–559

Chapter 11
Safety Concerns/Issues and Consumer Acceptance of Edible Flowers

Indrani Chetia, Akhila Vijayakumar, and Laxmikant S. Badwaik

11.1 Introduction

The consumption of edible flowers has experienced a remarkable resurgence in recent years, transitioning from a niche culinary practice to a growing trend in contemporary gastronomy and functional foods (Chen and Wei 2023). While the use of flowers in food preparation dates back thousands of years across various cultures, modern consumers are rediscovering these botanical ingredients for their unique flavors, aesthetic appeal, and potential health benefits (Fernandes et al. 2020a, b). However, this renewed interest has brought forth significant concerns regarding safety, regulation, and consumer acceptance that warrant careful examination. The integration of edible flowers into modern cuisine represents a complex intersection of traditional knowledge and contemporary food safety standards. Historically, flowers have been utilized in traditional medicine systems and culinary applications across different civilizations, from ancient Chinese medicine to European medieval cookery (Zhang and Liu 2022). Today, these botanical ingredients are finding their way into high-end restaurants, innovative food products, and home kitchens, driven by growing consumer interest in natural ingredients and unique gastronomic experiences (Rodriguez-Garcia et al. 2021).

However, the expanding market for edible flowers presents several critical challenges that demand attention from researchers, regulators, and industry

I. Chetia . L. S. Badwaik
Department of Food Engineering and Technology, School of Engineering, Tezpur University, Napaam, Assam, India

A. Vijayakumar
Teagasc Food Research Centre, Moorepark, Fermoy, Co.Cork, Irelandg

Food Safety and Preservation Department, IATA-CSIC, SO Excellence Centre, Avda. Agustín Escardino, Paterna, Valencia, Spain

A. Poonia et al. (eds.), *Edible Flowers: Source of Phytonutrients, Valorization and Technological Advancements*, https://doi.org/10.1007/978-3-032-00667-7_11

stakeholders. Primary among these concerns is safety, as not all flowers are suitable for human consumption, and even edible varieties may pose risks if improperly sourced, handled, or prepared (Wilson et al. 2021). The potential presence of toxic compounds, pesticide residues, and allergenic substances necessitates rigorous safety assessments and quality control measures to protect consumer health. The regulatory landscape surrounding edible flowers remains complex and often inconsistent across different jurisdictions. While some regions have established clear guidelines for the commercial use of edible flowers, others lack specific regulations, creating uncertainty for producers and consumers alike (Martinez-Silva et al. 2022). This regulatory variability highlights the need for harmonized standards and protocols to ensure safety and quality consistency in the global market for edible flowers.

Despite these challenges, edible flowers offer significant nutritional and functional benefits that contribute to their growing appeal. Research has shown that many edible flowers contain beneficial compounds, including antioxidants, polyphenols, and essential minerals, which may contribute to human health and well-being (Kim et al. 2021). Additionally, their natural colors and bioactive compounds make them attractive alternatives to synthetic food additives, aligning with consumer preferences for clean-label products (Chen et al. 2022).Consumer acceptance of edible flowers varies significantly across different demographic groups and cultural contexts. While some consumers embrace these ingredients for their novelty and perceived health benefits, others remain hesitant due to safety concerns or unfamiliarity with their use. Understanding these varying perspectives is crucial for developing effective marketing strategies and educational initiatives to promote safe and appropriate consumption of edible flowers.

The commercialization of edible flowers faces several practical challenges, including limited shelf life, seasonal availability, and the need for specialized handling and storage conditions (Peterson et al. 2022). These factors can impact supply chain management and product development, necessitating innovative solutions to ensure product quality and economic viability. Furthermore, the need for sustainable production methods and reliable sourcing practices adds another layer of complexity to the commercialization process. Looking ahead, the future of edible flowers in the food industry appears promising but requires careful navigation of various challenges. Advances in production technologies, preservation methods, and safety assessment techniques may help address current limitations and expand market opportunities (Lee and Garcia 2023). Additionally, growing consumer interest in plant-based diets and natural ingredients suggests potential for continued market growth, provided safety and quality concerns are adequately addressed.

This chapter aims to comprehensively examine the safety concerns and consumer acceptance issues surrounding edible flowers, while also exploring current regulations, nutritional benefits, and commercialization challenges. By analyzing these various aspects, we seek to provide valuable insights for researchers, industry professionals, and regulators working to advance the safe and successful integration of edible flowers into modern food systems.

11.2 Nutritional and Functional Benefits of Edible Flowers

Edible flowers represent a rich source of bioactive compounds and nutrients that contribute significantly to human health and wellness, making them increasingly valuable in both traditional and modern dietary practices. Flowers have a high content of natural antioxidants such as anthocyanin, flavonoids and various other phenolic components (Rao and Poonia 2024a, b) Recent scientific investigations have revealed that edible flowers contain a diverse array of phytochemicals, including polyphenols, flavonoids, anthocyanins, and carotenoids, which exhibit various biological activities beneficial to human health (Chen et al. 2023). These compounds have been demonstrated to possess antioxidant, anti-inflammatory, antimicrobial, and potential anticancer properties, contributing to the growing interest in incorporating edible flowers into functional foods and nutraceutical products.

11.2.1 Antioxidant Activity

The antioxidant capacity of edible flowers has been extensively studied, with numerous species showing remarkable ability to neutralize free radicals and reduce oxidative stress. Research conducted by Rodriguez-Garcia et al. (2022) demonstrated that flowers such as calendula, rose (Fig. 11.1), and violet contain high levels of vitamin

Fig. 11.1 White Rose (*Rosa spp.*)

C, vitamin E, and beta-carotene, along with significant concentrations of phenolic compounds. These antioxidant properties play a crucial role in preventing cellular damage and may contribute to the prevention of various chronic diseases, including cardiovascular disorders and certain types of cancer. The synergistic effects of these compounds often result in enhanced bioactivity compared to isolated components, highlighting the importance of consuming whole flower materials rather than extracted compounds alone.

11.2.2 Mineral Composition

The mineral composition of edible flowers represents another significant nutritional aspect that has garnered attention from researchers and nutritionists. Studies have shown that many edible flowers are rich sources of essential minerals, including potassium, calcium, magnesium, and iron (Kim and Park 2023). For instance, chrysanthemum flowers have been found to contain significant levels of potassium and calcium, while nasturtium flowers are particularly rich in iron and zinc. These minerals play crucial roles in various physiological processes, including bone health, muscle function, and blood formation, making edible flowers valuable contributors to overall nutritional status.

Beyond their basic nutritional value, edible flowers have demonstrated significant potential in managing various health conditions through their functional properties. Earlier research has shown that certain flower species contain compounds with anti-diabetic properties, capable of modulating glucose metabolism and improving insulin sensitivity. Additionally, the anti-inflammatory compounds found in flowers such as chamomile and lavender have been linked to potential benefits in managing chronic inflammatory conditions, though more research is needed to fully understand these mechanisms.

11.2.3 Protein and Dietary Fiber

The protein content and amino acid profiles of edible flowers, while generally lower than traditional protein sources, can still contribute to dietary protein intake. Some flower species, particularly those from the legume family, contain notable amounts of protein and essential amino acids. Studies conducted by Martinez and Wilson (2022) revealed that certain edible flowers could provide complementary protein sources in plant-based diets, particularly when combined with other plant protein sources to ensure a complete amino acid profile.

The presence of dietary fiber in edible flowers contributes to their nutritional value and potential health benefits. Both soluble and insoluble fiber contents vary among different flower species, with some varieties offering significant amounts that can contribute to daily fiber requirements. This fiber content not only supports digestive health but may also play a role in maintaining healthy blood cholesterol levels and promoting satiety (Lee et al. 2023). The prebiotic potential of certain flower-derived fibers has also been investigated, suggesting possible benefits for gut microbiota composition.

Recent research has also focused on the potential neuroprotective properties of compounds found in edible flowers. Studies have identified several bioactive compounds in flowers that may help protect neural cells from oxidative damage and inflammation, potentially contributing to cognitive health and neuroprotection. While these findings are preliminary, they suggest promising avenues for future research into the role of edible flowers in maintaining brain health and preventing neurodegenerative conditions.

The pigments responsible for the vibrant colors of edible flowers, particularly anthocyanins and carotenoids, have been associated with various health benefits. These compounds not only serve as natural colorants but also demonstrate significant biological activities, including anti-inflammatory and antioxidant properties. Research has shown that these pigments may play roles in vision health, cardiovascular protection, and cellular defense against oxidative stress (Zhang and Liu 2022).

11.3 Utilization of Edible Flowers in Food Product Development

Many researches already reported that flowers can be utilized to enhance the color and flavor of various dishes, including salads, soups, main courses, desserts, and beverages (Barash 1998). Since ancient times, flowers have become an integral part of the cuisine in various cultures, including European, Victorian English, East Indian, Chinese and Middle Eastern traditions etc. But it seems that still edible flowers are not equally popular to some other cultures for consumptions. Understanding the factors which influence consumer's attitude towards the consumption of edible flowers by studying the sensory evaluation research; can be useful for the development of innovative food products and to design of strategic marketing approaches.

The tradition of using flowers in culinary practices can be traced back to ancient Greece, Rome, and Egypt, with documentation dating as far back as 140 B.C. Throughout history, different cultures have embraced edible flowers in unique ways: Victorians used candied violets to embellish cakes and desserts; Italians and Spaniards prepared stuffed squash blossoms (Fig. 11.2a); the French incorporated carnation petals into Chartreuse, a renowned green liqueur; Indian cuisine utilized rose petals (Fig. 11.1) for liqueurs and included pumpkin (Fig. 11.2a) and banana flowers (Fig. 11.2b) in various recipes; and in Chinese Imperial Palaces, chrysanthemum wine was a celebrated delicacy. Many Asian countries have been using flowers to garnish in the raw form and also as an important ingredient to many dishes (Rop et al. 2012). This artistic approach enhances their use in various applications, including as decorative touches for cocktails, natural food colorants in

a b

Fig. 11.2 (a) Pumpkin flower (*Cucurbita moschata*) and (b) Banana blossom (*Musa paradisiaca*)

desserts, and edible garnishes for an array of dishes. There are various culinary dishes can be found by using edible flowers as main ingredient. Earlier research findings also highlighted the diverse applications of edible flowers beyond their traditional usages. Furthermore, edible flowers hold potential as processed ingredients for desserts, aperitifs, and sweets, as well as valuable additions to beverages, showcasing their versatility in the food and beverage industry.

The qualities of flowers and consumers' growing interest encourage utilization of edible flowers for the development of functional food products. When flowers are used as an ingredient to develop a food product, it is not only altering the visual appeal and give aroma but also add textural, various sensory attributes and nutritional potential. There are limited researches had conducted on the developed functional food products by using edible flowers. Studies found successful incorporation edible flowers in development of functional food products such as infant food formulations (Arise et al. 2014), bakery products (Gao et al. 2020; Hnin et al. 2021; Roriz et al. 2020), beverages (AL Tamimi et al. 2020; Sam et al. 2023), dairy products (Bragueto et al. 2019; Vioqueet al. 2000; Pires et al. 2018; Qiu et al. 2020), extruded products (Kowalczewski et al. 2019) etc. Addition of black locust flower increased the phenolic content and antioxidant properties of thermally processed pasta (Kowalczewski et al. 2019). Incorporation of lyophilized marigold extracts enhanced total phenolic content as well as antioxidant activity of organic yogurt (Bragueto et al. 2019). Sam et al. (2023) incorporated extracts of peach, rose, and lily flower into dealcoholized Merlot wine which improved its aroma without significantly affecting the wine's chemical properties. Spiced beverage enriched with rhododendron flowers, (35% extract), can be stored for 6 months in glass and PET bottles under ambient and refrigerated conditions, while maintaining high sensory appeal and also nutritionally rich. In case bakery item, one study found that the addition of rose flower powder to cookies can enrich their color, increase bioactive compounds, and elevate their antioxidant properties (Hnin et al. 2021). In this way there many studies were conducted where utilizing edible flowers for development of functional food products increase consumer's interest towards it.

11.4 Methods for Enhancing the Shelf LifeShelf life of Edible Flowers and Their Impact on Physicochemical Properties of Flowers

Edible flowers have a long history of contribution to human nutrition from the ancient literature of Europe, the middle east and south-east Asia (Purohit et al. 2021; Rop et al. 2012). The term edible flower is considered as an umbrella term to encompass a wide range of diverse consumable flowers. Many cuisines considered flowers, an integral part due to their medicinal and flavoring properties (Purohit et al. 2021). Compared to other horticulture produce flowers are super perishable with a maximum shelf life of one or two days. The high-water activity of flowers prevents their prolonged storage without suitable processing. Implementations of suitable processing techniques are required for the commercialization of edible flowers and the scaling up of its value-added products (Table 11.1). Preservation of edible flowers usually progresses through three stages; pre-processing, processing and packing (Purohit et al. 2021). In addition to ease in commercialization, processing techniques ensure the safety of end products. The processing technique used in a flower can alter its chemical properties. Based on the species, some flowers change shape, colour, flavour, and dimension during processing. Even though different process conditions can affect flowers differently, conservation of its nutrients, visual appeal and bioactive compounds were given prime importance while selecting the process conditions (Table 11.1).

11.4.1 Low Temperature

Low storage temperature can delay quality deterioration and spoilage in most food products. Cold storage and drying are the most popular methods used in the edible flower industry. In cold storage, flowers delay the rate of senescence because of a decrease in the breakdown of internal tissue, rate of respiration, water loss, wilting, and ethylene production (Fernandes et al. 2019). Maintenance of colour integrity and moisture were considered to be important factors in the marketability of edible flowers (Kelley et al. 2001a, b, c). Hence, refrigeration was identified as a common practice adopted in retail stores, where edible flowers are packed in small, rigid, plastic (or plastic-wrapped) packages to avoid desiccation and physical damage (Landi et al. 2015). A study on *Salvia* pot flowers by Landi et al. (2015) reported

Table 11.1 Preservation Techniques and Their Effects on Flowers' Properties

Flower Name	Methods of preservation with conditions	Effect on quality of flower	References
Borago officinalis	Freezing in natural form and ice cubes (−18 °C)	On prolonged storage chilling injury and browning of petals loss of brightness	Fernandes et al. (2020a, b)
Viola tricolor	Freezing in natural form and ice cubes (−18 °C)	Loss of texture and brightness Significant decrease in total flavonoids and total reducing capacity	Fernandes et al. (2020a, b)
Taraxacum officinale	Freezing in natural form and ice cubes (−18 °C)	Medium level of visual appearance Significant decrease in total flavonoids and total reducing capacity	Fernandes et al. (2020a, b)
Kalanchoe blossfeldiana	Freezing in natural form and ice cubes (−18 °C)	Maintained fresh appearance and rigid structure of flower Increased bioactivity	Fernandes et al. (2020a, b)
Begonia semperflorens L. (white, pink, and dark-pink)	Refrigeration 4 °C	Tissue darkening and browning of flowers Oxidation of phenolics Reduction in total flavonoids Increase in antioxidant activity on storage	Landi et al. 2018)
Salvia discolour Kunth,	Refrigeration 4 °C	Decrease in anthocyanin content Browning Reduction in total phenols	Landi et al. (2018)
Tropaeolum majus L	Refrigeration 4 °C	Browning of flowers Reduction in total phenols Increase in antioxidant activity on storage	Landi et al. (2018)
Acmella oleracea L	Refrigeration 4 °C	Increase in antioxidant activity on storage	Landi et al. (2018)
Tulbaghia cominsii Vosa	Refrigeration 4 °C	Reduction in total phenols	Landi et al. (2018)
Calendula officinalis L.	Freeze drying	Increase in total phenol content Decrease in total flavonoid content and antioxidant activity	Dorozko et al. (2019)
Bellis perennis L.	Freeze drying	Increase in total phenol content Increase in total flavonoid content Decrease in antioxidant activity	Dorozko et al. (2019)

(continued)

Table 11.1 (continued)

Flower Name	Methods of preservation with conditions	Effect on quality of flower	References
Lavandula angustifolia L.	Freeze drying	Increase in total phenol content Decrease in total flavonoid content and antioxidant activity	Dorozko et al. (2019)
Rosa rugosa	Hot air drying	Damage the flavor constituents Reduction in volatile compounds Reduce bioactive compounds	Qiu et al. (2020)
Borago officinalis	Alginate coating	Flowers became fragile Maintained good appearance and high-water activity on storage	Fernandes et al. (2018)
Viola wittrockiana	Alginate coating along with refrigerated storage at 4 °C	Weight loss and shrinkage Reduction in dimensions or size	Fernandes et al. (2018)
Bauhinia variegata L var. *candida alba* Buch.-Ham	Irradiation 0.5 kGy	Increase in phenolic content and resulting antioxidant property	Villavicencio et al. (2018)
Bauhinia variegate	Irradiation 1.0 k Gy	Improvement in flavor profile of flower	Simoni et al. (2018)
Centaurea cyanus	High hydrostatic pressure	Unacceptable changes in visual appearance on storage	Fernandes et al. (2017)
Viola wittrockiana	High hydrostatic pressure	Increase in water loss Increase in bioactive compounds	Fernandes et al. (2017)

that the external characteristic of the flower was maintained until 6 days of cold storage (5 °C), after which the moisture content tends to decrease. Stable ascorbic acid and polyphenol content might have protected the flower from enzymatic browning and related quality deterioration.

Even though low-temperature techniques are widely used, not all flower species will react similarly. Fernandes et al. (2020a, b) have studied the effect of freezing on flowers in their natural form and in ice cubes (Table 11.1). The evaluation of visual appearance suggested that flowers in ice-cube had better results up to a period of 3 months, but the bioactive and antioxidant properties were decreased during this period. Ice cube storage is preferred for long periods, even though there can be incidence of cell damage, lixiviation of compounds on defrosting and gradual decrease in bioactivity (Fernandes et al. 2020a, b).

Similarly, study Villalta et al. (2004) reports that in male and female summer squash flower (*Cucurbita pepo* 'Dixie') the visual appearance was maintained till

7 days of storage at 2.5 °C and 5.0°C in vented PET containers. It was observed that the respiration rate of female flowers varies with temperature. At 5 °C the respiration rate of the flower was relatively constant but it gets reduced at around 10 °C. Landi et al. (2018) has studied the effect of cold storage in biologically active compounds of five different species (Table 11.1). The rate of reduction in the bioactivity of flowers was found to be much lesser than its rate of loss in visual appeal. The study also concludes that flowers with comparatively low moisture content can ensure long marketability. Attempts to control water loss can significantly improve the shelf life of most edible flowers.

A Study by Demasi et al. (2021) further agrees with the fact that bioactive compounds are not greatly affected by the refrigerated condition, but it's the visual appeal of flowers that get worse when kept for more than 7 days in refrigerated conditions. Decaying flowers that lost marketable visual appearance can still be used for extraction of phyto-constituents. The quality depletion in most of the flowers at cold storage also depends on the type of container or film used to cover the flower on storage.

11.4.2 Drying

Drying is considered the oldest preservation technique practiced for storing food materials. Drying can inhibit enzymatic changes and microbial growth in the food; it can also facilitate cheap transport and storage by reducing weight. Traditional methods of drying i.e., sun drying, and shade drying, have been used in edible flowers, but the emergence of new drying techniques like hot/dehumidified air drying, freeze-drying, vacuum drying, microwave drying and hybrid drying lead to its adaptation in edible flower industry (Table 11.1) (Fernandes et al. 2019; Purohit et al. 2021). Conventional drying techniques have a longer duration and slow drying rate that can negatively affect the physicochemical and sensorial properties of flowers (Shantamma et al. 2021). The long drying hours used in conventional techniques increase the chances of contamination and nutrient deterioration. The temperature used in the study hasan impact on the nutrient profile of flowers, Stefaniak and Grzeszczuk (2020) have evaluated the effect of three drying temperatures (25 °C, 35 °C and 70 °C) in *Mimulus hybridus* L., *Hemerocallis hybrid* Hort., *Monarda didyma* L., *Paeonia lactiflora* Pall. Parameters like L-ascorbic acid, total carotenoids, total polyphenols and antioxidant activity of flowers were considered to select best drying temperature. The authors report that total polyphenols, total carotenoids and antioxidant activity is high in dried flowers. They also conclude 35 °C as the ambient temperature for the flowers *M. hybridus* L., *M. didyma* L. and *H. hybrida* Hort and 25 °C for *P. lactiflora* Pall.

Like temperature differences, different drying methods also affect flower properties. Fernandes et al. (2018) studied the effects of hot-air convective drying, shade drying and freeze-drying on Centaurea (*Centaurea cyanus* L.) petals. Regardless of the drying method used, the petals were observed as small, darker and wrinkled than fresh petals. The water activity, titratable acidity, carotenoid and hydrolysable tannin contents were decreased but total reducing capacity and antioxidant activity in terms of reducing power were increased in all the three types of treatments. On drying the water activity values of petals reach 0.86 or 0.62, below which growth of pathogenic microorganisms, yeasts and molds are inhibited. Only in freeze-dried petals, a magnitude of 0.3 was reported. Retention of monomeric anthocyanins, flavonoids, hydrolysable tannins, total reducing capacity and antioxidant activity was best in shade drying. Hot-air convective drying had the lowest retention of anthocyanin and flavonoid but carotenoid content was kept high.

Similarly, Dorozko et al. (2019) studied the effect of freeze-drying, hot air drying, and microwave on lavender, marigold and common daisy flower (Table 11.1). They found that the drying process, in general, reduces the total flavonoid content in flowers. But the total phenolic content was observed to be higher in freeze-dried samples than in fresh flowers. The drying techniques used have a negative impact on the antioxidant property but microwave drying demonstrated comparatively better results. The observed rise in antioxidant property of lavender was owed to the release of bound components during treatment. Oven drying on mahua flowers (*Madhuca indica syn. Bassia latifolia*) powdered with liquid nitrogen retained their nutrients (reducing and nonreducing sugars, amino acids, proteins, lipids, ascorbic acid and ash contents) for almost a year of preservation in a deep freezer at 0 °C (Das et al. 2010).

Another study on the effectiveness of different drying techniques on *Rosa rugosa* flower suggests that Infra-red freeze drying (IRFD) is the most economical method for retaining flavour, nutrition and molecular structure in rosa flower (Qiu et al. 2020). A better technique for rosa flowers was later suggested by Hnin et al. (2021). They compared Novel infrared pulse-spouted freeze drying (IRPSFD) along with much accepted techniques like freeze-drying and Infra-red freeze drying (IRFD) used in edible flowers. The new technique was found to have reduced the processing time associated with the process.

11.4.3 Edible Films and Coatings

Edible coatings can extend shelf life by acting as a protective barrier on flower surfaces. They help to reduce the rate of respiration, lower gaseous and moisture transmission, retain colour, flavour, and minimize enzymatic degradation (Fernandes et al. 2018). The technique of edible coating was prominent in the preservation of fruits and vegetables but limited information is available on its application in edible flowers. Biomaterials like protein, polysaccharides and lipids are generally used for making these edible barriers (Shantamma et al. 2021; Zhao et al. 2019). These edible layers when enriched with other bioactive constituents can increase the nutrient profile of flowers. The technique of edible coatings is often found in combination with other preservation technique (eg: refrigeration) hence, it gives a synergistic effect on maintaining flower quality.

Fernandes et al. (2018) studied the effect of alginate coating on Pansies (*Viola wittrockiana*) (Table 11.1), for enhancing shelf life in refrigerated storage. Pansies of four different colors red, white, violet, and yellow were used in the study. It was observed that compared to uncoated counter parts, coated flowers had a longer shelf life irrespective of the flower colour. Yellow and white colored flowers showed comparatively high-water loss, an excess water loss on storage can negatively affects the economic revenue and appearance of flowers. Water activity measurement has insignificant variation from fresh flowers. In terms of size, due to water loss and shrinkage, all the pansies decreased their dimension in storage. The authors suggest that a combination of suitable coating and packaging design can delay weight loss and shrinkage thus improving shelf life in pansies. A study on borage flower petals also reported a similar conclusion which agrees that alginate coating on a flower can increase the shelf life (Fernandes et al. 2018).

11.4.4 Irradiation

Food irradiation is a physical process used for extending the shelf life, disinfecting insects, maintaining quality and improving the safety of edible flowers. Factors like moisture content, treatment dose, substance composition and conditions determine the effectiveness of irradiation treatment (Zhao et al. 2019). Ionizing radiation builds oxidative stress in food by the production of peroxides, these will delay cell death by altering biochemical properties (Shantamma et al. 2021). Villavicencio et al. (2018) studied the effect of radiation at doses 0.5, 0.8 and 1 kGy on *Bauhinia variegate* L var. *candida alba* Buch.-Ham white flower (Table 11.1). It was observed that 0.5 kGy increased the phenolic content and resulting antioxidant activity in the flower without causing much difference in its nutrient profile. Other than nutrient and bioactive compounds, irradiation treatment can also affect the sensorial properties of flowers. Irradiation study on *Tropaeolum majus* L also demonstrated an increase in antioxidant activity after irradiation and in another study, it was found that upto a limit of irradiation dose there were no significant changes to its sensorial properties (Koike et al. 2015a, b, c). A similar trend in phenolic content was also observed in the study on *Viola tricolor* L (Koike et al. 2015a, b, c).

A sensory evaluation study conducted on irradiated *Bauhinia variegates* flowers showed that the flavour of the flowers gets improved with irradiation. Whereas, the colour and texture of irradiated flowers got a lower score compared to irradiated flowers in terms of consumer preference. The observed improvement in flavour can be due to the breaking of a glycosidic bond by radiation (Simoni et al. 2018). Both gamma radiation and electron beam radiations can be used in the processing of edible flowers. Koike et al. (2021) have studied the effect of gamma radiation and electron beam doses on the carotenoid lutein content of *Dianthus chinensis*. They reported that the carotenoid content is positively influenced by the radiation treatment at 1.0 kGy. They also suggest that the electron beam irradiated sample at 1.0 kGy is better than the gamma radiated sample. Selection of proper dose is very important in irradiation treatment of flowers as high doses can cause petal withering, browning and other physical injuries in edible flowers.

11.4.5 High Hydrostatic Pressure (HHP) Processing

High hydrostatic pressure (HHP) is a novel non-thermal processing technology that gained acceptance in the food industry due to its efficiency to maintain safety and quality. HHP treatments work in the pressure range of 200–800 MPa and are uniform, eco-friendly, independent of the product geometry and can extend the shelf life of products without chemical additives. HHP treatment requires a medium (water or fluid) to transfer the pressure into the foods. In HHP treated samples certain macromolecular components undergo changes like starch gelatinization, protein denaturation, enzyme inactivation, and microbial cell damage whereas micronutrients remain unaffected, this help in preserving the food without disturbing the nutrient profile of food (Fernandes et al. 2019; Shantamma et al. 2021; Zhao et al. 2019). Edible flowers are highly perishable commodities, comparatively few studies have worked on the effect of high-pressure treatment on edible flowers, edible flowers of broccoli and cauliflower are among the most studied groups (Fernandes 2017).

Edible flowers of different plants behave differently in HHP treatment, a detailed study is required to determine the best combination of pressure and holding time for each flower. Fernandes et al. (2017) evaluated the effect of high pressure between 75 and 450 MPa with holding time 1, 5 and 10 minutes on pansies (*Viola wittrockiana*), blue centaurea (*Centaurea cyanus*), blue borage (*Borago officinalis*) and rose camellia (*Camellia japonica*) (Table 11.1). It was observed that best visual appearance in pansies was obtained in treatment 75/5 MPa min^{-1} and 75/10 MPa min^{-1}, at a combination of 150/5 M Pa min^{-1} the flower became fragile. In centaurea, all combinations other than 75/5 and 100/5 M Pa min^{-1} result in damaging the flower whereas, the flowers camelia and borage were found to be unsuitable for high-pressure treatment. The evaluation of colour parameters reveals that under HHP treatment, pansies of different colours behave differently. Excess weight loss in flowers is undesirable for its economic revenue, the study showed that higher pressures induced higher changes in pansies but insignificant changes in Centaurea. Further, the storage evaluation suggests that in pansies HHP treatment induces the production of bioactive compounds that help to increase its shelf life. HHP treated Centaurea gave unsatisfactory results on HHP treatment hence it is not recommended for shelf-life extension in Centaurea flowers. The authors concluded that in pansies lower pressure and short holding time are better for shelf-life extension whereas this treatment is not recommended for other flowers used in the study.

11.4.6 Modified and Controlled Atmosphere Packaging

Packaging ensures the quality and safety of food during storage, handling and transport. In the current scenario, consumer experience expectations from the package have increased which led to novel innovations in packaging techniques. Modified atmosphere packaging is an advanced technology where the internal atmosphere of the packet is modified to improve the shelf life of the product it contains (Shantamma et al. 2021). Many of the fresh produce continue to respire after harvesting; by altering the gas composition inside the package the freshness and quality can be maintained for a longer period. This technique was successfully used in many fruits and vegetables. Kou et al. (2012) in their study suggest that the shelf life of carnation (*Dianthus caryophyllus* L.) and snapdragon (*Antirrhinum majus* L.) flowers can be extended by using a controlled release of 1-methylcyclopropene with modified atmosphere packaging stored at 5 °C. Flowers stored in modified atmosphere packaging conditions lower the dehydration, thus maintaining good quality and prolonging the shelf life of flowers. Unlike modified atmosphere packaging, a controlled atmosphere package enables continuous monitoring of the atmosphere throughout the storage period. Aquino-Bolaños et al. (Aquino-Bolaños et al. 2013) proposed that a controlled atmosphere with high CO_2 and low O_2 concentrations can enhance the shelf-life *Cucurbita pepo* flower. In their study the conditions were 5% O_2 + N_2 (CA1); 5% O_2 + 10% CO_2 + N_2 (CA2); 10% CO_2 + air (CA3); and control was air. It was found that these controlled atmospheric conditions can maintain total sugars, total soluble solids, pH, titratable acidity and lower physiological weight loss. Most importantly, 5% O_2 + 10% CO_2 + N_2 (CA_2); 10% CO_2 + air (CA3) treatments can extend shelf life up to 16 days, where mostly ascorbic acid (49.5%), polyphenols (65.2%) and carotenoids (72.8%) was retained by 5% O2 + 10% CO_2 + N_2 (CA2) treatment. Finally, it was proved that a controlled atmosphere with high CO_2 and low oxygen can enhance the shelf life of *Cucurbita pepo* flower.

The type of processing, packaging and storage conditions can influence the characteristics and shelf life of edible flowers (Table 11.1). Different species of flowers behave differently towards different processing conditions hence, the selection of processing technique for flowers are highly specific.

11.5 Safety Concerns Associated with Edible Flowers

Traditionally, different countries have consumed various flowers as delicacies in their unique ways since ancient times. For example, in Europe Pansies (*Viola × wittrockiana Gams*) and cornflowers (*Centaurea cyanus L.*) are originated (Kurlandsky 2018), while echinacea (*Echinacea angustifolia DC.*) and black locust (*Robinia pseudoacacia L.*) are native to North America (Roberts 2000). In contrast, durian (*Durio zibethinus L.*), sesbania (*Sesbania grandiflora Poiret*), and roselle (*Hibiscus sabdariffa L.*) flowers are indigenous to Asia (Sim 2013). South Africa is home to several edible native flowers, including wild malva (*Pelargonium culallatum (L.) L'Hér*), many-petalled jasmine (*Jasminum multipartitum Hochst.*), and wild sorrel (*Oxalis pes-caprae L.*) (Gray and Gray 2015) etc. In this way we may find various native edible flowers all around the world which have been safely eaten by people. But edible flowers are comprised of many interesting health promoting properties but it should be noted that every flower is not safe for human consumption. Consumer should have proper selection of edible flowers before consumption.

In general, our ancestors believed that edible flowers are those with alkaloids having medicinal properties. However, flowers containing alkaloids with toxic, psychotropic, or stimulant effects were regarded as inedible (Nicolau and Gostin 2016). Which means the physiological presence of antinutritional compounds in flowers can make them unsafe for human. Also, earlier reports have said that allergic reactions linked to consumption of edible flowers, especially among individuals with greater sensitivity (Mlcek and Rop 2011) can pose for health deterioration. For example, consuming large quantities of linden flowers (Tilia spp.) can result in heart damage (Guiné et al. 2017). Additionally, certain flowers can cause allergic reactions, particularly for those with hay fever, asthma, or allergies (Cunningham 2015). Chrysanthemum flowers, for instance, have been associated with allergic reactions such as rashes and eczema (Osimitz et al. 2006).

The toxic effects of flowers depend on the dosage. It is essential to recognize that some flowers can be consumed safely in small amounts, but overconsumption may lead to adverse effects. For instance, while saponin is considered an anti-nutrient, but in a limited amount it may also have health beneficial effects like reducing cholesterol, boosting immunity, and fighting cancer. Similarly, at moderate doses, tannins can improve mood and alertness (Morton 1992). The composition of these components differs across flower species. Numerous studies have assessed the levels of antinutritional compounds in flowers, providing valuable insights for consumers to safely include edible flowers in their diets, optimizing their nutritional advantages while reducing possible risks. Indeed, Muchahary and Deka (2021) observed alkaloids (2.78 mg/100 g), tannin (14.3 mgTAE/100 g), cyanogenic glycoside (0.0001%) and phytate (46.90 mg Phytic acid/100 g) in flowers of *Musa balbisiana*. Lachumy et al. (2016) found an LD_{50} value of 2.52 mg/ml which was considered safe in *Etlingera elatior* (torch ginger) flower extract. Kunhachan et al. (2012) found that when male ICR mice were injected with 15 mg of *Jasminum sambac* flower extract per ICR mouse, there were no significant adverse effects observed indicating no significant toxicity at this dosage. Sotelo et al. (2007) did not observe cyanogenic glycosides in the flowers of *Agave salmiana*, Aloe-vera, *Arbutus xalapensis*, *Cucurbita pepo* (cultivated), *Erythrina americana*, *Erythrina caribaea*, *Euphorbia radianbenth*, and *Yucca filifera*. But, the levels of trypsin inhibitors found among these flowers were 1.11, 2.54, 1.60, 1.40, 4.88, 6.32, 1.56, and 3.57 TUI/mg sample, respectively. Presence of alkaloids was shown in flowers of *E. Americana*, *A. xalapensis,* and *E. caribaea* Also, carbohydrate-binding proteins lectins was highest in *A. xalapensis*, saponin in both *A. salmiana* and *Y. filifera*. Halder and Khaled (2021) revealed that *Cucurbita maxima* (pumpkin) had antinutrients such as alkaloids (0.35 mg/100 g), tannin (2.16 mg/100 g), oxalates

(0.2 mg/100 g), saponin (50 mg/100 g) and phytates (5.07 mg/100 g). Also, 0.44 mg/100 g of tannin, 6.58 mg/100 g of phytate, 3.18 mg/100 g of oxalates, 0.18 mg/100 g of alkaloids and 230 mg/100 g of saponin was detected in *Carica papaya* (papaya) flower. Also, phytate content was found highest in White water lily (456 mg/100) among Blue-water lily (304 mg/100) and Chinese water lily (304 mg/100) (Aung et al. 2020). Presence of tannin (0.150 mg/100 g), phytate (12.6 mg/100 g) and oxalate (2.9 mg/100 g) were observed in *Moringa oleifera* flowers (Athira et al. 2021). From this brief observation, it can be inferred that the levels and types of antinutrients in flowers differ, which may be attributed to factors such as the flower species, cultivation methods, and environmental conditions. Indeed, trace elements in plants can be influenced by factors like soil composition, watering, air pollution, and plant species (Drava et al. 2020). To eliminate some antinutrients present in flowers, we can simply cook the flowers and discard the broth (Sotelo et al. 2007). Moreover, to minimize the risk of digestive problems, it is recommended to consume edible flowers in small portions and focus on one species at a time (Ebert 2013).

Except the natural presence of antinutritional compounds in flowers some other factors such as chemical, microbiological and biological hazards during growing, harvesting, and processing may cause the flower unsafe for consumption. Indeed, chemical hazards such as using pesticides i.e., diethyl-meta-toluamide, dimethoate (insecticide), and sulfites etc. higher level of presence of these chemicals makes the flowers inedible. Additionally, flowers that may have been treated with pesticides should be avoided (Lauderdale and Bradley 2014). As flower infused tea is popular beverage but if pesticides present in flowers, then there is chance of transfer of those pesticides into tea leading health issues. Jiang et al. (2020) in their study observed that pesticide transfer into tea increases with higher water solubility, longer infusion time, hotter water, and higher pesticide levels in dry chrysanthemum flowers.

In case of microbial hazards presence of pathogenic bacteria e.g.: *Salmonella* spp. etc. For safe consumption, it is advisable to avoid flowers grown in soil fertilized with untreated animal manure in the duration of 4 months before the harvest, flowers cultivated near roadways, and ensure if it clearly labeled as edible flowers while purchasing from florists, or nurseries etc. Except flowers physicochemical properties other factors such as introduction of chemical hazards (pesticides such as diethyl-meta-toluamide (insect repellent), dimethoate (insecticide), and sulfites), microbiological hazards (pathogenic bacteria eg: Salmonella spp.) and biological hazards (insects) into edible flowers can make flowers non edible. Contamination is more likely to occur after harvest, during processing, transportation, or preparation, rather than before. Potential sources include contaminated water, surfaces, or poor hygiene practices by workers (Erickson 2012; Hirneisen et al. 2012; Nicolau and Gostin 2016; Matyjaszczyk and Śmiechowska et al. 2019). There, consumer awareness is very important while purchasing flowers not only in terms of nutritional basis but also to get rid of potential health hazards.

11.6 Regulations and Guidelines

The regulatory framework governing edible flowers represents a complex and often fragmented landscape across different jurisdictions worldwide, with varying degrees of specificity and enforcement mechanisms. In many regions, edible flowers fall under broader food safety regulations, while some jurisdictions have developed specific guidelines to address their unique characteristics and associated risks. The United States Food and Drug Administration (FDA) classifies edible flowers under the category of "specialty produce," requiring compliance with general food safety regulations outlined in the Food Safety Modernization Act (FSMA). However, unlike conventional produce, edible flowers present unique regulatory challenges due to their diverse species, varying toxicity levels, and potential for misidentification (Martinez-Rodriguez et al. 2022a, b).

The European Union has established more comprehensive guidelines for edible flowers through its Novel Food Regulation (EU) 2015/2283, which requires safety assessments for flowers not historically consumed in significant quantities within the EU before May 15, 1997. This regulation has significant implications for the commercialization of edible flowers, as producers must demonstrate a history of safe use or undergo extensive safety evaluations before market authorization (Anderson and Williams 2023). The European Food Safety Authority (EFSA) has also developed specific guidance for assessing the safety of botanical preparations used in food, which includes criteria for evaluating edible flowers. These guidelines emphasize the importance of proper identification, characterization of bioactive compounds, and assessment of potential toxic constituents (Garcia-Lopez et al. 2021).

In Asia, particularly in countries with long traditions of using flowers in cuisine, regulatory approaches vary significantly. Japan's Ministry of Health, Labour and Welfare has established a positive list of edible flowers that are considered safe for consumption, along with specific quality standards and labelling requirements. China's National Health Commission has implemented similar regulations, incorporating traditional knowledge while adapting to modern food safety standards. These frameworks often include detailed specifications for cultivation practices, post-harvest handling, and quality control measures (Zhang and Chen 2023). The integration of traditional knowledge with modern safety standards has created a unique regulatory model that other regions might consider adapting.

Quality assurance and standardization present significant challenges in the regulation of edible flowers. Various international organizations, including the CODEX Alimentarius Commission, have begun developing standards for edible flowers, though comprehensive international guidelines remain limited. These efforts focus on establishing criteria for species identification, acceptable cultivation practices, maximum residue limits for pesticides, and microbiological safety parameters. The development of standardized testing protocols and quality metrics is crucial for ensuring consistency and safety across the global market for edible flowers.

Labelling requirements for edible flowers vary widely among jurisdictions, creating challenges for international trade and consumer information. Some regions require detailed information about species identification, cultivation methods, and potential allergen warnings, while others have minimal labelling requirements. The lack of harmonized labelling standards can lead to confusion and potential safety risks, particularly when flowers are sold across different regulatory jurisdictions (Brown et al. 2023). Industry stakeholders and regulatory bodies are increasingly recognizing the need for standardized labelling approaches that effectively communicate essential safety information to consumers.

Good Agricultural Practices (GAP) and Good Manufacturing Practices (GMP) specific to edible flowers have been developed in several regions to ensure safety throughout the production chain. These guidelines typically address cultivation conditions, pest management strategies, harvest techniques, post-harvest handling, and storage requirements. The implementation of these practices is crucial for minimizing contamination risks and ensuring product quality, though enforcement mechanisms and certification requirements vary significantly across jurisdictions (Rodriguez and Smith 2022). Some regions have established voluntary certification programs for edible flower producers, while others mandate compliance with specific agricultural and manufacturing standards.

The emergence of novel processing technologies and preservation methods for edible flowers has created new regulatory considerations. Methods such as freeze-drying, candying, and the development of flower-based extracts may fall under different regulatory frameworks than fresh flowers, requiring additional safety assessments and quality control measures. Regulatory bodies are increasingly being called upon to evaluate these processing methods and establish appropriate guidelines for their safe implementation (Lee et al. 2022). The development of specific regulations for processed edible flower products remains an evolving area of food safety governance.

International trade in edible flowers further complicates the regulatory landscape, as products must often comply with multiple jurisdictional requirements. The lack of harmonized international standards can create barriers to trade and increase compliance costs for producers and distributors. Efforts to develop international guidelines through organizations such as the World Trade Organization (WTO) and CODEX Alimentarius Commission aim to facilitate trade while ensuring safety and quality consistency (Martinez et al. 2023). However, progress in this area remains slow, and significant variations in regulatory requirements continue to impact international market development.

11.7 Consumer Acceptance of Edible Flowers

Consumers are gradually becoming more interested in eating flowers as they become more aware of their nutritional benefits. Although edible flowers have great health benefits along with various satisfactory sensory attributes but they are not popular among people due to the limitation of awareness and people are generally neophobic to consume edible flowers. The traditional knowledge and cultural practice of eating edible flowers are confined among specific community to community and also many people are unable to recognize the edible species of flowers gradually, also, globalization along with consumer's desire to re adapt earlier healthy life style may encourage the demand of edible flowers (Rop et al. 2012). The desire for fresh, unique, and natural foods, along with the appealing aroma, draws consumers to incorporate edible flowers into their diets (Zimowski 2014). Each country possesses unique cultural traditions, which means that consumer curiosity and willingness to accept unfamiliar products vary across nations (Güneş and Özkan 2018). According to Neto and Bezzi (2007), globalization has led to the homogenization of traditions, often diminishing cultural diversity. It has also increased global market competition and shifted consumer behavior (Michaut 2004; Wind and Mahajan 1997). Various factors such as cultural, biological, physiological, economic, and political aspects etc. impact on the selection of food and eating behavior of people (Steptoe et al. 1995; Pearcey and Zhan 2018). Also based on age, gender, and income the selection of edible flower is affected (Kelley et al. 2002a, b).

Traditionally people use edible flowers in their diet to garnish, enhance flavor, color and texture in dishes and to get rid of many diseases. Generally, specific curiosity, aroma, texture, visual appeal, flavor and health consciousness etc. attract people to consume flowers (Chen and Wei 2017).

Edible flowers are available packaged in small, sturdy plastic containers, which can be found in grocery stores and specialty retail outlets, and in local markets also etc. (Kelley et al. 2002a, b). According to research the color of edible flowers influence consumer to buy them (Kelley et al. 2002a, b) and therefore selling flowers with the combination of certain colors and combinations of various species can be appeal to the consumers' appetite. It was reported that consumers prefer edible flowers package with different colors of nasturtium and also containers packed with nasturtiums with an additional species such as viola, and borage. Containers with single flower colors, like dark purple (viola) or blue (borage), or a mix of the two, failed to attract much interest from participants. Furthermore, research highlighted that consumer placed significant importance on the visual quality of edible flowers, expressing a marked unwillingness to purchase flowers that exhibited even minor imperfections or damage (Kelley et al. 2001c). Earlier research reported that visual attributes, palatability and fragrances of flowers impact on the decision-making process to purchase and also most chef interested on the flowers grown organically (Kelley et al. 2001a, b, c). These findings underline the importance of variety, color, and quality in influencing consumer preferences for packaged flowers. As per some findings, edible flower composition is more important than price and color (Kelley et al. 2001b). Consumers whom are aware and familiar of usage of edible flowers are willing to purchase them from grocery stores where marketing also takes a role on purchasing. Nicknezhad et al. (2023) conducted a study to identify the factors affecting the acceptance and consumption of edible flowers in Iran and to determine the most popular species among five types of edible flowers such as yucca, chrysanthemum, marigold, gladiolus and hibiscus (Fig. 11.3), which were found in Rasht

Fig. 11.3 Hibiscus (*Hibiscus sabdariffa L.*) flower

County, Iran. It was postulated that visual inspection phase had the greatest impact on both purchasing and consuming edible flowers. Results found that purple chrysanthemum was rated as the most appealing flower, followed by orange marigold, white gladiolus, purple hibiscus (Fig. 11.3), and yucca. The visual attractiveness of edible flowers was identified as the most critical factor in their acceptance and willingness to be consumed than the aesthetic appeal. Furthermore, type of product, consumer experience, individual preferences, budget constraints, and cultural or social considerations impact the attitude of consumers to buy edible flowers. Guiné et al. (2021) studied to find out the habits associated with the usage of edible flowers in gastronomy and major determinants impacting these habits among 3 counties i.e., Portugal, Slovenia, and Brazil.

It was mainly found that the country and the level of education were the major determinants for the consuming habits and knowledge about edible flowers. Indeed, this study highlighted that eating habits are highly dependent on the customs and traditions of each country and helped understanding the factors that most affect the consumers' attitudes towards the consumption and knowledge of edible flowers. Countries exhibited significant variations in both the cases of consumption of edible flowers and the specific types of edible flower utilized. Tuladhar (2021) highlighted that market viability of edible flowers could be increased by introducing more in restaurants as a key ingredient, introducing flower based local cuisines in front of foreign tourists, establishing international standards for certifying organically grown edible flowers which ensure consumers' safety, making buyers aware of edible flower cooking recipes, various novel usages of flowers such as extracting and utilizing the color of flowers and also processing it to make other food products such as pickles, dried, candied etc.; making consumers aware of flowers nutritional value along with its blooming season and making flowers in raw or processed form available for consumers. The scientific justification of health benefits of edible flowers and many researches on edible flowers has paved a way to the novel utilization of edible flowers beyond its traditional way of using. This increases the consumer understands as well as interest towards edible flowers along with its market value. While edible flowers provide numerous health advantages and are gaining popularity, there is still much to be done in terms of raising more public awareness, overcome neophobia in terms of consumption of flowers, novel utilization of edible flowers and increasing market viability.

11.8 Challenges in the Commercialization of Edible Flowers

The commercialization of edible flowers faces significant logistical and technical challenges that have hindered their widespread adoption in the mainstream food industry. One of the primary obstacles is the extremely short shelf life of edible flowers, which typically ranges from 3–7 days even under optimal storage conditions (Rodriguez-Fragoso et al. 2021). This brief window creates substantial complications in the supply chain, as producers must carefully coordinate harvesting, packaging, and distribution to ensure flowers reach consumers while still fresh and visually appealing. Adding to these difficulties, most edible flowers require specialized storage conditions, including precise temperature and humidity controls, to maintain their quality. Research has shown that even minor deviations from optimal storage parameters can lead to rapid deterioration, with some varieties showing visible degradation within 24 hours of improper handling (Chen and Thompson 2020). These stringent requirements significantly increase operational costs and complexity for producers and retailers, making it challenging to maintain consistent product quality while achieving profitable pricing structures.

The regulatory landscape surrounding edible flowers presents another complex set of challenges that producers must navigate. Unlike conventional produce, edible flowers occupy a unique position that often falls into regulatory gray areas, particularly regarding food safety standards and labeling requirements. Many countries lack specific guidelines for edible flower production and processing, leading to uncertainty among producers and potential legal vulnerabilities (Martinez-Santos and Kumar 2023). Food safety concerns are particularly pressing, as edible flowers can be susceptible to contamination from agricultural chemicals and environmental pollutants. Studies have indicated that certain flower varieties can accumulate heavy metals and other harmful substances from soil and water sources, necessitating rigorous testing and quality control measures. Furthermore, the absence of standardized safety protocols and quality metrics across the industry has created barriers to international trade and market expansion, as producers must often comply with varying requirements in different regions while lacking clear guidance on best practices.

Consumer education and market development represent perhaps the most fundamental challenges in the commercialization of edible flowers. Despite their long history in various cultural culinary traditions, many consumers remain unfamiliar with edible flowers and uncertain about their culinary applications, nutritional benefits, and safety considerations. Research indicates that while interest in novel food experiences and natural ingredients has grown, significant knowledge gaps persist regarding the identification and proper use of edible flowers (Davidson and Lee 2021). This lack of awareness contributes to limited market demand and hesitancy among retailers to stock edible flower products. Additionally, price sensitivity presents a significant barrier, as the labor-intensive nature of edible flower production and their delicate handling requirements result in higher costs compared to conventional produce. Studies have shown that while consumers express interest in edible flowers, their willingness to pay premium prices remains limited, creating challenges for achieving sustainable profit margins in the industry. The combination of these factors has created a challenging environment for market expansion and mainstream adoption of edible flowers in contemporary food culture.

11.9 Future Perspectives

The future of edible flower safety is likely to be shaped by advancing analytical technologies and increasingly sophisticated safety assessment protocols. Emerging research suggests that next-generation sequencing technologies and metabolomic approaches will enable more comprehensive screening for potential toxins, allergens, and microbiological contaminants in edible flowers (Anderson and Miller 2023). These technological advances are expected to lead to the development of rapid, cost-effective testing methods that can be implemented throughout the supply chain, significantly enhancing food safety monitoring capabilities. Nanotechnology represents an emerging field of scientific exploration focused on the examination and control of matter at the smallest scales, encompassing molecules and atoms (Rao and Poonia 2024a, b).

Consumer acceptance of edible flowers is projected to undergo significant transformation in the coming years, driven by changing dietary preferences and growing interest in sustainable, plant-based food options. Market research indicates a rising trend in consumer awareness about the nutritional benefits of edible flowers, particularly their high antioxidant content and potential health-promoting properties (Martinez-Rodriguez et al. 2022a, b). Social media and digital platforms are expected to play an increasingly important role in shaping consumer perceptions and knowledge about edible flowers, with visual platforms particularly effective in showcasing their aesthetic appeal and culinary applications. However, researchers emphasize that sustained market growth will depend on addressing current barriers to acceptance, including concerns about pesticide residues and proper identification of safe varieties (Davidson and Chen 2023). Studies project that educational initiatives focusing on safe handling, preparation methods, and nutritional benefits will be crucial in building consumer confidence and expanding market acceptance.

The intersection of safety concerns and consumer acceptance is expected to drive significant industry innovations and regulatory developments in the coming years. Experts predict the emergence of more standardized safety protocols and quality assurance systems specifically designed for edible flowers, addressing current gaps in regulatory frameworks (Lee and Santos 2023). This evolution is likely to include the development of specialized certification programs and industry-wide standards that can help build consumer trust while ensuring consistent safety measures across the supply chain. Additionally, research indicates that advances in packaging technology and preservation methods will play a crucial role in addressing both safety concerns and consumer preferences, with new solutions expected to extend shelf life while maintaining product quality and safety. Future perspectives also suggest a growing emphasis on sustainable production methods and organic certification, as consumers increasingly prioritize environmental considerations alongside safety concerns in their food choices. These developments are expected to contribute to a more robust and trusted edible flower market, though challenges related to cost-effectiveness and scalability will need to be addressed.

11.10 Conclusion

The successful integration of edible flowers into mainstream food culture hinges on the delicate balance between addressing safety concerns and fostering consumer acceptance. As research continues to evolve, it has become increasingly clear that these two aspects are intrinsically linked, with safety assurances playing a crucial role in building consumer confidence and market growth. The implementation of rigorous safety protocols, advanced testing methods, and comprehensive traceability systems has begun to address many of the traditional concerns surrounding edible flower consumption. However, challenges persist, particularly in standardizing safety measures across different regions and production systems. Consumer acceptance, while showing positive trends, remains a complex challenge that requires a multi-faceted approach. Educational initiatives, improved packaging solutions, and clearer labelling systems have contributed to growing consumer awareness and interest. Looking ahead, the future of edible flowers in the food industry appears promising, provided that stakeholders maintain their commitment to safety excellence while actively engaging in consumer education and market development. The continued advancement of preservation technologies, coupled with sustainable production methods and transparent safety documentation, will be crucial in expanding market opportunities. As consumers become more knowledgeable and confident in using edible flowers, and as safety measures become more robust and standardized, the industry is likely to see sustained growth. The key to success lies in maintaining a balanced approach that prioritizes both safety and consumer education, ensuring that edible flowers can fulfil their potential as a valuable and appealing addition to contemporary cuisine.

References

AL Tamimi JZ, AL Faris NA, Alghamdi FA, Abu-Hiamed HA, ALbadr NA, Almousa LA (2020) Hibiscus sabdariffa L. flower and date palm pollen fortification of date palm spathe beverage. British Food J, 122(7):2159–2170

Anderson K, Miller R (2023) Advanced analytical methods for safety assessment of edible flowers: a future perspective. Food Anal Methods 16(4):525–541

Arise AK, Arise RO, Sanusi MO, Esan OT, Oyeyinka SA (2014) Effect of Moringa oleifera flower fortification on the nutritional quality and sensory properties of weaning food. Cro J Food Sci Technol 6(2):65–71

Anderson KL, Williams RB (2023) Implementation challenges of novel food regulations for edible flowers in the European Union. Food Control 145:109384

Aquino-Bolaños EN, Urrutia-Hernández TA, López Del Castillo-Lozano M, Chavéz-Servia JL, Verdalet-Guzmán I (2013) Physicochemical parameters and antioxidant compounds in edible squash (C ucurbita pepo) flower stored under controlled atmospheres. J Food Qual 36(5):302–308

Athira KA, Panjikkaran ST, Aneena ER, Sharon CL (2021) *Moringa oleifera* – proximate and anti-nutritional composition. Indian J Nutr Dietet 58(3):390

Aung TT, Myat YY, Mar MM, Kyu KK (2020) Nutritional compositions, elemental compositions, and antinutrient factor in different varieties of water lily. 3rd Myanmar Korea Conf Res J 3(5):1917–1922

Barash CW (1998) Please eat the flowers. Horticulture 95(5):36–40

Bragueto Escher G, Cardoso Borges LDC, Sousa Santos J, Mendanha Cruz T, Boscacci Marques M,Araújo Vieira do Carmo M, Zhang L (2019) From the field to the pot: phytochemical and functional analyses of Calendula officinalis L. flower for incorporation in an organic yogurt. Antioxid 8(11):559

Brown ME, Davis SK, Wilson JR (2023) Labelling requirements for edible flowers: a comparative analysis of international standards. J Food Law Policy 19(2):156–178

Chen L, Thompson R (2020) Storage conditions and quality retention in edible flowers: a comprehensive review. J Food Sci Technol 57(8):2845–2860

Chen NH, Wei S (2017) Factors influencing consumers' attitudes towards the consumption of edible flowers. Food Qual Prefer 56:93–100

Chen H, Wei R (2023) Trends and innovations in edible flower consumption: a global perspective. Food Res Int 158:112–124

Chen X, Li Y, Zhang W (2022) Bioactive compounds in edible flowers: composition, stability, and potential applications. Crit Rev Food Sci Nutr 62(15):4178–4195

Chen H, Wei R, Li Y (2023) Bioactive compounds in edible flowers: a comprehensive review of their health benefits. Crit Rev Food Sci Nutr 63(5):789–812

Cunningham, Eleese (2015) What nutritional contribution do edible flowers make? J Acad Nutr Diet 115(5):856, 25911345

Das BK, Choudhury BK, Kar M (2010) Quantitative estimation of changes in biochemical constituents of mahua (Madhuca indica syn. Bassia latifolia) flowers during postharvest storage. J Food Process Preservat 34(5):831–844

Davidson P, Chen H (2023) Evolution of consumer attitudes towards edible flowers: trends and future directions. J Consum Behav 22(3):178–192

Davidson M, Lee S (2021) Consumer perceptions and market barriers in the edible flower industry: a global perspective. Int J Food Market 15(4):412–428

Demasi S, Mellano MG, Falla NM, Caser M, Scariot V (2021) Sensory profile, shelf life, and dynamics of bioactive compounds during cold storage of 17 edible flowers. Horticulturae, 7(7):166

Dorozko J, Kunkulberga D, Sivicka I, Kruma Z (2019) The influence of various drying methods on the quality of edible flower petals. In: Conference proceedings. Foodbalt, vol 13, pp 182–187

Drava G, Iobbi V, Govaerts R, Minganti V, Copetta A, Ruffoni B, Bisio A (2020) Trace elements in edible flowers from Italy: further insights into health benefits and risks to consumers. Molecules 25(12):2891

Ebert AW (2013) Sprouts, microgreens and edible flowers: the potential for high-value specialty produce in Asia. In: Holmer R, Linwattana G, Nath P, Keatinge JDH (eds) SEAVEG 2012: high value vegetables in Southeast Asia: production, supply and demand, pp 216–227

Erickson MC (2012) Internalization of fresh produce by foodborne pathogens. Annu Rev Food Sci Technol 3(1):283–310

Fernandes L, Casal S, Pereira JA, Ramalhosa E, Saraiva JA (2017) Effect of high hydrostatic pressure (HHP) treatment on edible flowers' properties. Food Bioprocess Technol 10(5):799–807

Fernandes L, Casal S, Pereira JA, Saraiva JA, Ramalhosa E (2018) Effects of different drying methods on the bioactive compounds and antioxidant properties of edible Centaurea (Centaurea cyanus) petals. Braz J Food Technol 21

Fernandes L, Saraiva JA, Pereira JA, Casal S, Ramalhosa E (2019) Post-harvest technologies applied to edible flowers: A review: Edible flowers preservation. Food Reviews International, 35(2):132–154

Fernandes L, Casal S, Pereira JA, Pereira EL, Saraiva JA, Ramalhosa E (2020a) Freezing of edible flowers: effect on microbial and antioxidant quality during storage. J Food Sci 85(4):1151–1159

Fernandes L, Casal S, Pereira JA (2020b) An overview of the culinary and nutritional importance of edible flowers. Food Rev Intl 36(8):527–547

Gao J, Sun Y, Li L, Zhou Q, Wang M (2020) The antiglycative effect of apple flowers in fructose/glucose-BSA models and cookies. Food Chem 127170

Garcia-Lopez M, Fernandez-Cruz ML, Rodriguez-Silva A (2021) Safety assessment guidelines for botanical preparations in food: the European perspective. Food Chem Toxicol 158:112592

Gray R, Gray G (2015) 12 edible indigenous South African plants. Retrieved from https://africageographic.com/blog/12-edible-indigenous-south-african-plants/

Guiné RP, Santos E, Correia P (2017) Edible flowers: knowledge and consumption habits. Acta Scient Nutr Health 1(3):18–22

Guiné RP, Florença SG, Ferrão AC, Bizjak MČ, Vombergar B, Simoni N, Vieira V (2021) Factors affecting eating habits and knowledge of edible flowers in different countries. Open Agriculture 6(1):67–81

Güneş E, Özkan M (2018) Insects as food and feed in Turkey: current behaviours. Int J Environ Pollut Environ Model 1:10–15

Halder S, Khaled KL (2021) Anti-nutritional profiling from the edible flowers of *Allium cepa, Cucurbita maxima* and *Carica papaya* and its comparison with other commonly consumed flowers

Hirneisen KA, Sharma M, Kniel KE (2012) Human enteric pathogen internalization by root uptake into food crops. Foodborne Pathog Dis 9:396–405

Hnin KK, Zhang M, Ju R, Wang B (2021) A novel infrared pulse-spouted freeze drying on the drying kinetics, energy consumption and quality of edible rose flowers. LWT 136. https://doi.org/10.1016/j.lwt.2020.110318

Jiang M, Zhang W, Zhang T, Liang G, Hu B, Han P, Gong W (2020) Assessing transfer of pesticide residues from chrysanthemum flowers into tea solution and associated health risks. Ecotoxicol Environ Saf 187:109859

Kelley KM, Behe BK, Biernbaum JA, Poff KL (2001a) Consumer and professional chef perceptions of three edible-flower species. HortScience 36(1):162–166

Kelley KM, Behe BK, Biernbaum JA, Poff KL (2001b) Consumer ratings of edible flower quality, mix, and color. HortTechnology 11(4):644–647

Kelley KM, Behe BK, Biernbaum JA, Poff KL (2001c) Consumer preference for edible-flower color, container size, and price. HortScience 36(4):801–804

Kelley KM, Behe BK, Biernbaum JA, Poff KL (2002a) Combinations of colors and species of containerized edible flowers: effect on consumer preferences. HortScience 37(1):218–221

Kelley KM, Behe BK, Biernbaum JA, Poff KL (2002b) Consumer purchase and use of edible flowers: results of three studies. HortTechnology 12(2):282–287

Kim JH, Park SM (2023) Mineral composition and bioavailability in commonly consumed edible flowers. J Food Compos Anal 108:104633

Kim JH, Park SM, Lee YJ (2021) Antioxidant properties and health benefits of selected edible flowers. J Food Compos Anal 97:103778

Koike ACR, Rodrigues FT, Villavicencio ALCH (2015a) Colorimetric analysis of edible flower of *Tropaeolum majus* processed by ionizing radiation

Koike A, Barreira JCM, Barros L, Santos-Buelga C, Villavicencio ALCH, Ferreira ICFR (2015b) Irradiation as a novel approach to improve quality of Tropaeolum majus L. flowers: benefits in phenolic profiles and antioxidant activity. Innov Food Sci Emerg Technol 30:138–144

Koike A, Barreira JCM, Barros L, Santos-Buelga C, Villavicencio ALCH, Ferreira ICFR (2015c) Edible flowers of Viola tricolor L. as a new functional food: antioxidant activity, individual phenolics and effects of gamma and electron-beam irradiation. Food Chem 179:6–14. https://doi.org/10.1016/j.foodchem.2015.01.123

Koike ACR, Araújo ES, Negrão BG, de Almeida-Murandian LB, Villavicencio ALCH (2021) Analysis of carotenoids in edible flowers of Dianthus chinensis processed by ionizing radiation. Braz J Radiat Sci 9(1A):1–11. https://doi.org/10.15392/bjrs.v9i1a.1577

Kou L, Turner ER, Luo Y (2012) Extending the shelf life of edible flowers with controlled release of 1-Methylcyclopropene and modified atmosphere packaging. J Food Sci 77(5):S188–S193

Kowalczewski P Ł, Pauter P, Smarzyński K, Różańska MB, Jeżowski P, Dwiecki K, Mildner-Szkudlarz S (2019) Thermal processing of pasta enriched with black locust flowers affect quality, phenolics, and antioxidant activity. J Food Process Preserv 43(10):e14106

Kunhachan P, Banchonglikitkul C, Kajsongkram T, Khayungarnnawee A, Leelamanit W (2012) Chemical composition, toxicity and vasodilatation effect of the flowers extract of Jasminum sambac (L.) ait."G. Duke of Tuscany". Evid Based Complement Alternat Med 2012(1):471312

Kurlandsky M (2018) 8 edible flowers from around the world. Retrieved from http://blog.tripbase.com/8-edible-flowers-from-around-the-world/

Lachumy SJT, Sasidharan S, Sumathy V, Zuraini Z (2016) Estudo preliminar sobre os conhecimentos e consumo de flores comestíveis. ESAV, Viseu

Landi MARCO, Ruffoni B, Guidi L, Savona M, Salvi D (2015) Cold storage does not affect ascorbic acid and polyphenolic content of edible flowers of a new hybrid sage. In: Cold storage does not affect ascorbic acid and polyphenolic content of edible flowers of a new hybrid sage, pp 348–357

Landi M, Ruffoni B, Combournac L, Guidi L (2018) Nutraceutical value of edible flowers upon cold storage. Italian J Food Sci 30(i):336–347

Lauderdale C, Bradley L (2014) Choosing and using edible flowers. Department of Horticultural Science, North Carolina State University, NC State Extension

Lee SH, Garcia M (2023) Emerging technologies in edible flower preservation and processing. Innovative Food Sci Emerg Technol 85:103125

Lee M, Santos P (2023) Emerging regulatory frameworks for edible flower safety: a global perspective. Food Policy 116:102345

Lee HS, Park JY, Kim SM (2022) Regulatory considerations for novel processing technologies in edible flower preservation. Innovative Food Sci Emerg Technol 78:102892

Lee SH, Garcia M, Brown K (2023) Dietary fiber content and prebiotic potential of selected edible flowers. Food Chem 405:134725

Martinez R, Wilson B (2022) Protein quality and amino acid profiles of edible flowers: implications for plant-based diets. J Food Sci 87(4):1123–1138

Martinez CR, Lopez-Garcia J, Santos-Buelga C (2023) International trade regulations for edible flowers: current status and future directions. J Int Food Agribus Mark 35(2):89–112

Martinez-Rodriguez A, Thompson RK, Anderson JL (2022a) Regulatory frameworks for edible flowers in North America: challenges and opportunities. Food Policy 109:102256

Martinez-Rodriguez A, Wilson K, Brown S (2022b) Consumer awareness and market trends in edible flower consumption: a prospective analysis. Int J Food Sci 57(5):687–701

Martinez-Santos P, Kumar V (2023) Regulatory challenges and food safety considerations in edible flower commercialization. Food Control 145:108923

Martinez-Silva R, Lopez-Garcia J, Rodriguez-Perez C (2022) Regulatory frameworks for edible flowers: a comparative analysis. Food Control 134:108751

Matyjaszczyk E, Śmiechowska M (2019) Edible flowers. Benefits and risks pertaining to their consumption. Trends in Food Sci Tech 91:670–674

Michaut AM (2004) Consumer response to innovative products: with application to foods. Wageningen University and Research

Mlcek J, Rop O (2011) Fresh edible flowers of ornamental plants – a new source of nutraceutical foods. Trends Food Sci Technol 22(10):561–569

Morton JF (1992) Widespread tannin intake via stimulants and masticatories, especially guarana, kola nut, betel vine, and accessories. In: Plant polyphenols. Springer, pp 739–765

Muchahary S, Deka SC (2021) Impact of supercritical fluid extraction, ultrasound-assisted extraction, and conventional method on the phytochemicals and antioxidant activity of bhimkol (*Musa balbisiana*) banana blossom. J Food Process Preservat 45(7):e15639

Neto HB, Bezzi ML (2007) Regiões culturais: a construção de identidades culturais no Rio Grande do Sul e sua manifestação na paisagem gaúcha. Rev Soc Natur 20(2)

Nicknezhad S, Hashemabadi D, Allahyari MS, Marzban S, Hassen TB, Surujlal J (2023) Sensorial analysis of factors influencing consumers' perceptions toward the consumption of edible flowers in Iran. J Agric Food Res 12:100580

Nicolau AI, Gostin AI (2016) Safety of edible flowers. In: Regulating safety of traditional and ethnic foods. Academic, pp 395–419

Osimitz TG, Franzosa JA, Maciver DR, Maibach HI (2006) Pyrethrum allergic contact dermatitis in humans – real?, common?, or not documented?: an evidence-based approach. Cutan Ocul Toxicol 25:287–308

Pearcey SM, Zhan GQ (2018) A comparative study of American and Chinese college students' motives for food choice. Appetite 123:325–333

Peterson KL, Brown ME, Taylor RW (2022) Supply chain challenges in the commercialization of edible flowers. J Food Distrib Res 53(2):45–62

Pires TC, Dias MI, Barros L, Barreira JC, Santos-Buelga C, Ferreira IC (2018) Incorporation of natural colorants obtained from edible flowers in yogurts. Lwt 97:668–675

Purohit SR, Rana SS, Idrishi R, Sharma V, Ghosh P (2021) A review on nutritional, bioactive, toxicological properties and preservation of edible flowers. Future Foods 4(July):100078

Qiu L, Zhang M, Bhandari B, Wang B (2020) Effects of infrared freeze drying on volatile profile, FTIR molecular structure profile and nutritional properties of edible rose flower (Rosa rugosa flower). J Sci Food Agric 100(13):4791–4800

Rao V, Poonia A (2024a) Bioactive compounds, nanoparticle synthesis and potential utilization of edible flowers for development of functional dairy products: a review. J Food Sci Technol 61(6):1053–1068. https://doi.org/10.1007/s13197-023-05853-6

Rao V, Poonia A (2024b) Microbial biosynthesis of selenium nanoparticles using probiotic strain and its characterization. J Food Meas Charact 18(7):5477–5488. https://doi.org/10.1007/s11694-024-02581-z

Roberts M (2000) Edible & medicinal flowers. The Spearhead Press

Rodriguez MA, Smith KL (2022) Implementation of good agricultural practices in edible flower production: a global perspective. J Agric Food Res 8:100321

Rodriguez-Fragoso L, Reyes-Esparza J, Burchiel SW (2021) Preservation techniques and shelf-life extension strategies for edible flowers. Food Quality and Safety 5(2):89–102

Rodriguez-Garcia J, Santos-Buelga C, Ferreira IC (2021) Edible flowers as sources of bioactive compounds: a review. Crit Rev Food Sci Nutr 61(15):2499–2521

Rodriguez-Garcia J, Santos-Buelga C, Ferreira IC (2022) Antioxidant properties and bioactive compounds in edible flowers. Food Chem 375:131854

Rop O, Mlcek J, Jurikova T, Neugebauerova J, Vabkova J (2012) Edible flowers – a new promising source of mineral elements in human nutrition. Molecules 17(6):6672–6683

Roriz CL, Heleno SA, Carocho M, Rodrigues P, Pinela J, Dias MI, Fernandes IP, Barreiro MF, Morales P, Barros L, Ferreira IC (2020) Betacyanins from Gomphrenaglobosa L. flowers: incorporation in cookies as natural colouring agents. Food Chem 127178

Sam FE, Ma T, Wang J, Liang Y, Sheng W, Li J, Zhang B (2023) Aroma improvement of dealcoholized Merlot red wine using edible flowers. Food Chem 404:134711

Shantamma S, Vasikaran EM, Waghmare R, Nimbkar S, Moses JA, Anandharamakrishnan C (2021) Emerging techniques for the processing and preservation of edible flowers. Future Foods 4(November):100094

Sim S (2013) Edible flowers from Asia. Retrieved from https://giftzgiving.wordpress.com/2013/09/07/edible-flowers-from-asia/

Simoni NK, Santos FF, Andrade TA, Villavicencio ALCH, Pinto-e-Silva MEM (2018) The use of edible flowers in human food: sensory analysis of preparations. Int J Food Eng January:140–143

Sotelo A, López-García S, Basurto-Peña F (2007) Content of nutrient and antinutrient in edible flowers of wild plants in Mexico. Plant Foods Hum Nutr 62(3):133–138

Stefaniak A, Grzeszczuk M (2020) Effect of drying temperature and method of extract preparation on antioxidant activity of edible flowers of some ornamental plant species. Folia Pomeranae Universitatis Technologiae Stetinensis Agricultura, Alimentaria, Piscaria et Zootechnica 354(53):17–28. https://doi.org/10.21005/aapz2020.53.1.02

Steptoe A, Pollard TM, Wardle J (1995) Development of a measure of the motives underlying the selection of food: the food choice questionnaire. Appetite 25:267–284

Tuladhar A (2021) Eating flowers - The consumption and market of edible flowers. Mie University, College of Liberal Arts and Sciences 6:25–33

Vioque M, Gomez R, Sanchez E, Tejada L, Salguero JF (2000) Chemical and microbiological characteristics of ewes' milk cheese manufactured with extracts from flowers of cynara cardunculus and cynara humilis as coagulants. J Agri Food Chem 48(2):451–456. https://pubs.acs.org/journal/jafcau?ref=breadcrumb

Villalta AM, Ergun M, Berry AD, Shaw N, Sargent SA (2004) Quality changes of yellow summer squash blossoms (Cucurbita pepo) during storage. In: VII international symposium on protected cultivation in mild winter climates: production, pest management and global competition, vol 659, pp 831–834

Villavicencio ALCH, Heleno SA, Calhelha RC, Santos-Buelga C, Barros L, Ferreira ICFR (2018) The influence of electron beam radiation in the nutritional value, chemical composition and bioactivities of edible flowers of Bauhinia variegata L. var. candida alba Buch.-Ham from Brazil. Food Chem 241(August 2017):163–170

Wilson RS, Keller A, Shapcott A, Leonhardt SD, Sickel W, Hardwick JL, Wallace HM (2021) Many small rather than few large sources identified in long-term bee pollen diets in agroecosystems. Agric Ecosyst Environ 310:107296

Wind J, Mahajan V (1997) Issues and opportunities in new product development: an introduction to the special issue. J Mark Res 34(1):1–12

Zhang L, Chen Y (2023) Regulatory approaches to edible flowers in Asian markets: traditional knowledge and modern standards. J Food Saf Regulat 12(4):245–262

Zhang L, Liu Y (2022) Traditional uses of edible flowers in Asian cuisine: historical perspectives and modern applications. J Ethnic Foods 9(1):15–28

Zhao L, Fan H, Zhang M, Chitrakar B, Bhandari B, Wang B (2019) Edible flowers: review of flower processing and extraction of bioactive compounds by novel technologies. Food Res Int 126:108660

Zimowski J (2014) Aromaty (zapachy) w świecie roślin i w kosmetologii. Kwartalnik I. Instytut Nauk Medycznych. Wyższa Szkoła Przedsiębiorczości w Warszawie 1:119–130

Chapter 12
Food Applications of Edible Flowers

Amrita Poonia, Debdip Hazra, and Chinmay Patil

12.1 Introduction

According to numerous ancient scriptures like Vedas, flowers have been an integral part of our culinary due to their medicinal and nutritional properties. Flowers like rose, lotus and banana blossoms are used in Ayurvedic medicines in India for centuries. Ancient Egyptians has used flowers like Nymphaea caerulea and hibiscus as culinary and medicinal purposes. Chrysanthemum and jasmine were incorporated into teas and medicinal tonics in China as early as 1500 BCE. Greek and Romans also used roses and lavenders in culinary preparations. In ancient times flower incorporated products symbolized sophistication and luxury. In medieval time, monasteries started to cultivate borage, calendula, nasturtiums for herbal liquors. Sakura tea and Sakura mochi are iconic examples from Japan, that show culinary culture of Japan. Persian desserts use rose and orange blossoms water as a staple ingredient. In modern culinary practices flowers are used mostly for aesthetics and sensory appeals. Flower infusions have become common in dairy or in beverage industries. Apart from culinary, flowers also symbolize cultural significance. Lotus is treated as representation of purity and spiritual enlightenment in Asian nations. Rose symbolizes love and beauty whereas Cherry blossom in Japan is an integral part of their ancient culture.

Edible flowers have remarkable ability to be used as dietary supplements and high value pharmaceuticals. Edible flowers render unique colour, aroma and flavour attributes. The cousins of South-east Asian and Middle east use flowers as an integral part due to their ornamental purposes (Cunningham 2015). The increasing

A. Poonia (✉) · D. Hazra · C. Patil
Department of Dairy Science and Food Technology, Institute of Agricultural Sciences, Banaras Hindu University, Varanasi, Uttar Pradesh, India
e-mail: amrita12@bhu.ac.in

A. Poonia et al. (eds.), *Edible Flowers: Source of Phytonutrients, Valorization and Technological Advancements*, https://doi.org/10.1007/978-3-032-00667-7_12

demand of population and interest of researchers towards edible flowers is due to the reason of high value nutritional and nutraceutical properties. Common examples of edible flowers are pumpkin flower, broccoli as vegetables; sunflower oil, rose water as extracts; Javitri and star anise as spices (Lu et al. 2016). Edible parts of flowers vary from flower to flower (Rodrigues et al. 2017). Lastly, similar to other food products and commodities, flowers should also be consumed in limitation and hypersensitive consumers should avoid its intake (Lim 2014). This chapter will revisit aspects like modern processing technologies of flowers along with their preservation strategies, chemical and nutritional profile and various health promoting properties of various edible flowers.

12.2 Phytochemicals in Edible Flowers

12.2.1 Phenolic Compounds

These bioactive compounds are defined as the materials having more than one hydroxyl substitute along with an aromatic ring (Zheng et al. 2019), As per their classification they are categorised as flavonoids and non-flavonoids based on their structural characteristics the quantity of phenolic subunits present (phenols and simple phenols) (El Gharras 2009). They exist in nature in both free and conjugated forms with other biomolecules such as sugar and acids (Xiong et al. 2014). In addition to enhancing the original qualities of colour and flavour in plant issues phenolic chemicals have a direct role in growth, reproduction and defence mechanism of plants (González-Sarrías et al. 2020). They can be found in variety of fruits and flowers including aromatic secondary metabolites of plants and when consumed they can eradicate free radicals from human body (Chen et al. 2018). The Food Industry has taken interest in the phenolic compounds found in edible flowers because they can be utilised as a natural substitute for the artificial additives and food preservatives that are now in market (Pires et al. 2018a, b). Phenolic acids such as hydroxycinnamic and hydroxybenzoic acid, flavonoids such as anthocyanin, chalcone, flavanones, flavones and flavonols are major found phenolic compounds found in nature and edible flowers with considerable bioactive potential (Navarro-González et al. 2014). The scientific studies on the existence of the different classes of phenolic compounds in edible flowers will be the main topic of a next sections given the significance of analytic compounds.

12.2.1.1 Phenolic Acids

Phenolic acids are mainly classified into the classes of hydroxycinnamic acids (C6-C3) and hydroxybenzoic acids (C6-C1), represent the family of phenolic compounds and the presence of analytical functions in their structure is crucial for their classification. They provide the largest percentage of phenolic chemicals that are not flavonoids (Scalzini et al. 2021). Derivatives of hydroxybenzoic acid have been found in various edible flowers of several species, including *Abelmoschus manihot* (L.) Medik. (Luan et al. 2020) and *Bougainvillea glabra* Choisy (Saleem et al. 2020). Although reports of hydroxycinnamic acid in *Butea monosperma inflorescences* and *Calendula officinalis* L. (Janarny et al. 2021), *Sesbania grandiflora* (L.) Pers. (Pires et al. 2018a, b), and (Lam.) Taub. Zheng, Meenu, and Xu (Zheng et al. 2019), studied the chemical profile of 70 different kinds of Chinese edible flowers and confirmed presence of number of phenolic acids, with *Cymbidium sinense* (Jacks.) Willd. having particularly high levels of p-hydroxybenzoic acid. Orange-colored specimens of Mexican marigold (*Tagetes erecta L.*) were also found to contain 15 phenolic acids. The most prevalent hydroxybenzoic acids in the floral extract (methanol and water (1:4)) were gallic acid and pyrogallol (Youssef et al. 2020). Phenolic acid derivatives, specifically p-Coumaric, chlorogenic, and trans-ferulic acids, were the main substances found in the 75% ethanolic extract of *Hibiscus roseus* Thore flowers (dos Santos Nascimento et al. 2021). In their investigation into how drying techniques affect the phenolic content of chrysanthemum flowers, Lu et al. 2020 discovered five derivatives of phenolic acid. It was also shown in the same study that phenolic acids had higher stability in comparison to flavonoids once the flowers have been dehydrated. Upon examining the 12% ethanolic extract of four different edible flowers from Brazil (*Amaranthus hypochondriacus L., Tropaeolum majus L.* (red), *Tropaeolum majus L.* (orange), and Spilanthes oleracea L.), specifically p-coumaric and ferulic acids were initially identified in each of the species under investigation (Barros et al. 2020).

12.2.1.2 Flavonoids

Anthocyanins, flavanones, flavones, flavonols and chalcones are major classes of the flavonoid family, which is mostly found in flowers (AN et al. 2016). With roughly 15,000 distinct structures, they form the biggest family of secondary metabolites (Fu et al. 2021; Li et al. 2020). These substances are known to be a non-nitrogenous plant Pigments and are regarded as a low molecular weight (Nagula and Wairkar 2019). The structural arrangement of flavonoids is closely linked to their antioxidant action because of the locations and quantities of hydroxyl and methoxyl groups found in each individual component (Mlcek and Rop 2011). With the highest dispersion and diversity of chemicals found in environment, they are thought to be the most important naturally occurring phenolics (Nagula and Wairkar 2019). Several edible flowers (Table 12.1) were found to contain flavonoids, including samples of *Hibiscus rosa-sinesis* L., *Calendula officinalis* L., *Vinca rosea* L., *Ocimum sanctum* L., (Janarny et al. 2021), *Tagetes erecta* L., *Erythrina americana* Mill., *Hibiscus sabdariffa L., Plumeria rubra* L., *Opuntia ficus-indica (L.)* Mill., Dahlia spp., Agave durangensis Gentry (Bora and Sharma 2011), and Artemisia spp. (Gonçalves et al. 2020). In a recent analysis of the profiling phenolic content of 11 edible flowers, the highest levels of total flavonoids were found in *Dianthus*

Table 12.1 Bioactive compounds present in edible flowers

	Class	Compounds	Species	References
Phenolic acids	Hydroxybenzoic Acids	Ellagic acid	*Allium ursinum L.* *Bellis Perennis L.* *Cyanus segetum Hill* *Dianthus carthusianorum L.* *Erythronium dens-canis L.* *Geranium sylvaticum L.* *Paeonia officinalis L.* *Robinia pseudoacacia L.* *Rosa canina L.* *Rosa pendulina L.* *Salvia pratensis L.* *G.E.Haglund*	Miao et al. (2022)
		Gallic acid	*Alliumursinum* *L. Boragoofficinalis* *L. Bougainvillea glabra* *Choisy*	Miao et al. (2022)
		Syringic acid	*Tulipa gesneriana L.*	Krzymińska et al. (2020)
		2,5Hydroxybenzoicacid	*Tulipa gesneriana L.*	Krzymińska et al. (2020)
	Hydroxycinnamic Acids	Coumaric acid	*Allium ursinum L.* *Cichorium intybus L.* *Erythronium dens-canis L.* *Geranium sylvaticum L.* *Mentha aquatica L.* *Sambucus nigra L.* *Trifolium alpinum L.*	Miao et al. (2022)
		Ferulic acid	*Paeonia officinalis L.* *Primula veris L.* *Primula vulgaris Huds.*	Miao et al. (2022)
		5-O-Caffeoylquinic acid (chlorogenic acid)	*Cichorium intybus L.* *Cyanus segetum Hill* *Geranium sylvaticum L.* *Hibiscus roseus Thore* *Lavandula angustifolia Mill.* *Taraxacum campylodes* *G.E.Haglund* *Tropaeolum majus L.* *Tropaeolum majus L.* *Tulipa gesneriana L.*	Miao et al. (2022) Barros et al. (2020)
		Caffeic acid	*Bellis Perennis L.* *Cyanus segetum Hill* *Dianthus carthusianorum L.* *Mentha aquatica L.* *Paeonia officinalis L.* *Primula veris L.* *Primula vulgaris Huds.* *Taraxacum campylodes* *G.E.Haglund*	Miao et al. (2022)

(continued)

Table 12.1 (continued)

Class	Compounds	Species	References
Anthocyanin	Delphinidin-3-O-glucoside	*Crocus sativus L.* *Nelumbo nucifera Gaertn.* *Robinia hispida L.*	Goupy et al. (2013) Deng et al. (2013) Yisimayili et al. (2019)
	Delphinidin-3-O-sambubioside Delphinidin-3-O-glucoside	*Hibiscus sabdariffa*	Sindi et al. (2014)
	Cyanidin-3-O-glucoside	*Bauhinia purpurea L.* *Bauhinia variegata L.* *LBSm. & Downs em Reitz* *Calliandra haematocephala Hassk.* *Dianthus caryophyllus L.* *Hibiscus rosa-sinensis L.* *Hibiscus sabdariffa L.* *Ipomoen cairica (L.)* *Nelumbo nucifera Gaertn.* *Punica granatum L.* *Robinia hispida L.* *Rhododendron spp. L.*	Li et al. (2014) Zheng et al. (2019) Li et al. (2014) Li et al. (2014)
	Cyanidin-3-O-(3"malonyl) glucoside	*Dendranthema grandiflorum*	Park et al. (2015)
	Cyanidin-3,5-O-diglucoside	*Punica granatum L.*	Yisimayili et al. (2019)
	Malvidin-O-glucoside	*Crocus sativus L*	Barros et al. (2020)
Chalcone	Chalcone 2-O-glucoside	*Dianthus caryophyllus L*	Nakayama (2020)
	Acetyl-marein	*Coreopsis tinctoria Nutt*	Chen et al. (2016)
	2′-hydroxy-4,4′-dimethoxychalcone Kukulkanin B Lanceoletin 3,2′,4′-trihydroxy4,3'dimethoxy chalcone 4′-hydroxy-3,4,2′-trimethoxychalcone 3′,4′-dihydroxy-3,4,2′-trimethoxychalcone 4,4′-dihydroxy-3′-methoxychalcone 2'-O-B-D-glucopyranoside 3,4,3′,4′-tetrahydroxychalcone 2'-O-ß-D-glucopyranoside 4,3′,4′-trihydroxy-3-methoxychalcone	*Coreopsis lanceolata L*	Kim et al. (2019)

(continued)

Table 12.1 (continued)

Class	Compounds	Species	References
Flavones	Pentahydroxyflavanone-O-dihexoside Eriodictyol-O-dihexoside Eriodictyol-O-deoxyhexosyl-hexoside Eriodictyol-O-hexoside Eriodictyol-O-acetyldihexoside Naringenin-O-hexoside-acetylhexoside Naringenin-3-O-glucoside	*Dahlia mignon*	Pires et al. (2018a)
	Hesperetin	*Chrysanthemum indicum* L. *Chrysanthemum morifolium* *Gomphrena globosa* L. *Hylocereus undatus* *Prunus persica (L.) Batsch*	Zheng et al. (2019)
Xanthophylls	Antheraxanthin	*Gentiana lutea* L. *Lilium* spp. *wittrockiana* Gams.	Zhu et al. (2003) Yamagishi et al. (2010) Manivannan et al. (2021)
	Astaxanthin	*Delonix regia* (Hook.) Raf.	Veigas et al. (2012)
	Capsanthin	*Lilium* spp. "Saija" (red)	Yamagishi et al. (2010)
	Cryptoxanthin	*Tagetes* spp.	Veigas et al. (2012)
	Lutein	*Antirrhinum majus* L. *Borago officinalis* L. *Camellia japonica* L. *Centaurea cyanus* L. *Cosmos bipinnatus* *Delonix regia* (Hook.) Raf. *Dendranthema grandiflorum*	Manivannan et al. (2021) Farnandes et al. (2019) Veigas et al. (2012)
	Zeaxanthin	*Antirrhinum majus L.* *Hemerocallis disticha* *Ipomoen sp.* *Tagetes sp.* *Viola wittrockiana Gams.*	Manivannan et al. (2021) Veigas et al. (2012) Manivannan et al. (2021)

(continued)

Table 12.1 (continued)

Class	Compounds	Species	References
Carotenes	β-Carotene	*Antirrhinum majus L.* *Borago officinalis L.* *Camellia japonica L.* *Centaurea cyanus L.* *Delonix regia (Hook.) Raf.* *Gentiana lutea L.* *Hemerocallis disticha Donn ex Ker Gawl.* *Ipomoen sp.* *Narcissus spp.* *Rosa spp.* *Tagetes erecta L.* *Viola wittrockiana Gams.* *Viola wittrockiana Gams. (red)* *Viola wittrockiana Gams. (white)* *Viola wittrockiana Gams. (yellow)*	Manivannan et al. (2021) Fernandes et al. (2019)
	Lycopene	*Tagetes erecta L.*	Siriamornpun et al. (2012)

caryophyllus L. (17.50 mg/g*), Tagetes erecta L.* (16.90 mg/g), and Rosa hybrida L. (16.57 mg/g) (Chen et al. 2015). In their investigation of the flavonoid content of 23 distinct flowers, Chen et al. 2015 proposed that the species which are having the maximum total flavonoid content were *Osmanthus fragrans* (Thunb.) Lour (71.49 mg/g), *Lavandula angustifolia* Mill (27.43 mg/g), and *Rosmarinus officinalis* L. (18.83 mg/g). (*Cichorium intybus, Anchusa azurea, Capparis spinosa, Malva sylvestris, Hedysarum coronarium, Rosmarinus officinalis, Robinia pseudoacacia and Sambucus nigra*) are flowers that are commonly consumed in Italy. Loizzo et al. (2016) investigated the flavonoid composition of these flowers and proposed that rutin and quercitin were the primary flavonoids.

12.2.2 Anthocyanins

Anthocyanins give flowers a wide range of colors that span almost the entire visible spectrum, from orange and red to purple and blue. As a result, these matrices may contain these natural pigments, which could produce new flavors and colors that would catch consumers' attention. Anthocyanins have several derivatives in nature, the primary ones are typically discovered include cyanidin, delphinidin, malvidin, petunidin, pelargonidin, and peonidin (Karaaslan-Ayhan and Yaman 2021; Di Gioia et al. 2020). Eating foods high in anthocyanins has several health benefits, including strengthening eyes, preventing heart disease, reducing obesity and diabetes, and having antibacterial, anticancer, or neuroprotective properties (Tena et al. 2020). The anthocyanin levels in flowers are directly proportional to the antioxidant capacity of the flower (Benvenuti et al. 2016). According to Benvenuti et al. 2016, there is significant amounts of total anthocyanins present in blue *Viola wittrockiana* Gams (13.6 mg cyn-3-glu eq./100 g fw), red *Petunia hybrida* Vilm. (14.44 mg cyn-3-glu eq./100 g fw), red *Dianthus barbatus* L. (13.35 mg cyn-3-glu eq./100 g fw) and Garzón et al. 2015 examined the phenolic content of *Tropaeolum majus* L. flowers and found that the anthocyanin contents in the yellow and red petals ranged from 31.9–21.7 to 114.5–2.3 mg of cyn-3-glu/100 g, respectively. The presence of particular anthocyanin precursors, anthocyanins and carotenoids is really linked to the coloration of *Tropaeolum majus L. flowers*. High levels of cyanidin 3-O-glucoside were found in hydroethanolic extracts of edible flowers from Sri Lanka, were used in a study to assess the anthocyanin content. The extracts of *Punica granatum* L. (1181.10 0.02 g cyn-3-glucoside/g dw), *Ixora coccinea* L. (1573.1 0.1 g cyn-3-glu eq./g dw), and *Hibiscus rosa-sinensis* L. (2003.5 0.1 g cyn-3-glu eq./g dw) had the highest anthocyanin content (Janarny et al. 2021). Yisimayili et al. 2019 also found anthocyanin derivatives in the flowers of *Punica granatum* L. These were subsequently identified as cyanidin-3-O-glucoside, pelargonidin-3,5-O-diglucoside, delphinidin-3-O-glucoside, and cyanidin-3-O-glucoside. Depending on the species, as well as abiotic and edafo-climatic conditions anthocyanin composition and quantities of edible flowers can greatly vary. Additionally, the composition of extracted extracts can be greatly influenced by many extraction variables, including temperature, solid/liquid ratio, used solvents, and extraction methods.

12.2.3 Flavonoids

12.2.3.1 Chalcones

Chalcones which are typically found in plants in Leguminosae family, are substances that belong to the classes of flavonoid and are distinguished by the presence of two aromatic rings connected by three atoms of carbon, two unsaturated carbons and one carbonyl (Jacques et al. 2021). They are also present in fruit, vegetables, and flowers belonging to the Compositae and Moraceae families. They can give certain medicinal plants their yellow color, particularly in their petals, and they also help draw pollinators like insects and birds (Ferreira et al. 2018). Chalcones, which are often referred to as flavonoids with open chain, are distinguished by the lack of one "C-ring" in their structural arrangement and are mostly represented by derivatives of chalconaringenin, arbutin, phloridzin, and phloretin. Chalcones are thought to be the pigments that give Dahlia variabilis (Willd.) and Dianthus caryophyllus L. (chalcone 2-O-glucoside) their striking yellow coloration. They were found in the yellow pigments of Coreopsis spp. flowers (Rozmer and Perjési 2016).

12.2.3.2 Flavanones

Flavanones are generally polyphenolic compounds with a basic structure of 2,3-dihydroflavone that are chiral at C2 and C3, often referred to 2-phenyl-chroman-4-ones. Examples of these compounds include hesperidin, naringenin, isosacuratenin, and heridictol (Jaganath and Crozier 2008). The variation in molecular orientation has a major impact on how flavonoids interact with biological receptors, which in turn influences their bioactive characteristics (Das et al. 2019). These substances which are primarily linked to the concentration of hesperetin, hesperidin, neohesperidin, and naringinin derivatives, are linked to a variety of health advantages in edible flowers, including anti-aging properties (Chen et al. 2020a, b). Hesperetin and its derivatives have been found in *Hylocereus undatus* (Haw.) Britton & Rose (2162.2 g/g), *Chrysanthemum indicum* L. (2653.7 g/g), *Chrysanthemum morifolium* Ramat. (748.8 g/g), *Gomphrena globosa* L. (143.4 g/g) and Prunus persica (L.) Batsch (850.5 g/g) according to earlier research. Additionally, Dahlia species were shown to contain two derivatives of naringenin compounds, naringenin-3-O-glucoside and naringenin hexoside-acetylhexoside, at 2.92 and 0.82 mg/g dw, respectively. There are nine flavanones which were reportedly identified among various compounds, including hesperetin-7-O-glucoside (C28), eriodictyol-7-O-glucuronide (C21), isookain (C27), eriodictyol-7-O-glucoside (C22), flavanomarein (C26), eriodictyol (C23), narigenin (C25), naringenin-7-O-glucuronide (C24), and butin (C29) (Peng et al. 2019). The analysis also included study of five major species of Chrysanthemum spp. based on their phenolic profile.

12.2.3.3 Flavones

Flavones are characterized by two major factors like, C-2 is bonded with B ring and the extrastructural double bond between the C-2 and C-3 atoms (Jiang et al. 2016). The primary constituents of this abundantly distributed class of flavonoids, which are the second largest in edible flowers, are apigenin, acacetin, luteolin, chrysoeriol, and their glucosides (Egebjerg et al. 2018). They have been found, for instance, in the flowers of Tropaeolum majus L. (53.6 g/g), Dendran thema lavandulifolium (Fischer ex Trautv.) Kitam. (2.11 g/g), Chrysanthemum morifolium Ramat. (3.73 g/g), Matthiola incana (L.) R.Br. (10.4 g/g), florists' chrysanthemum (4.52 g/g), and Florista rugosa Thunb. (1.36 g/g) (Egebjerg et al. 2018). The most prevalent edible flowers which contain flavone chemicals are dandelion, rose, and peony (Xu et al. 2021).

12.2.3.4 Flavonols

The presence of hydroxyl group at the 3-position on the C ring gives flavonols, which leads to similarity in their structural to flavones, and thus termed as alcoholic by-product of flavones. These compounds are also known as

3-hydroxy-2phenylchromen-4-one. They are commonly found in medicinal herbs, fruits, green vegetables and beverages (Qin et al. 2021; Tahir et al. 2021). Flavonoids are a class of metabolites that are commonly found in edible flowers, most notably quercitin and kaempferol, despite their resemblance to flavones and their actual status as proanthocyanin building blocks.

Derivatives of flavonols have been found in a number of ornamental plants, including the following: derivatives of kaempferol and quercetin in *Calendula officinalis* L. flower (Hamad 2016); kaempferol in *Rosa gallica* L. (138.3 g/g); *Rhododendron indicum* var. *simsii* (Planch.) Maxim. (138.5 g/g); *Rosa centifolia* L. (12.2 g/g); *Bauhinia variegata* L. (91.1 g/g); *Coreopsis tinctoria* Nutt. (40.9 g/g); *Styphnolobium japonicum* (L.) Schott (38.2 g/g); *Paeonia suffruticosa* Andrews (69.0 g/g); *Matthiola incana* (L.) R.Br. (27.0 g/g); and *Nymphaea nouchali* Burm.f. (28.9 g/g); and quercitin (0.4 g/mL) in Agave durangensis Gentry flowers (Bermúdez-Bazán et al. 2021). The primary compound identified by Zhu et al. 2010 in the edible flowers of *Robinia pseudoacacia* L. (28.4 mg/g of extract), *Sambucus nigra* L. (23.7 mg/g of extract) and *Hedysarum coronarium* L. (28.2 mg/g of extract), and was rutin. Quercetin was followed by this, which was also found in these flowers (23.6 mg/g of extract), Capparis spinosa L. (5.8 mg/g of extract), and *Hedysarum coronarium L.* (8.0 mg/g of extract), as well as kaempferol, luteolin and myricetin levels. Some edible flower species also contain significant amounts of rutin and quercitin, which are two of the primary flavonols present in the petals of Citrus aurantium L. (rutin, 362.8 0.02 g/g DW and quercetin, 185.37 0.11 g/g DW) (Li et al. 2018) and Malus pumila Mill. (Karimi et al. 2012).

12.2.3.5 Flavanols

Known by another name, catechins, flavanols have two chiral carbons (C2 and C3) since ring C's carbonyl and C2 and C3 do not form a double bond (Murkovic 2016). They are the primary component of *C. assumica* and *Camellia sinensis*, but they can be found in many plant components (Lin et al. 2003). Additionally, Yang et al. 2018 claim that *Camelia nitidissima* Chi flower extracts are a good source of catechin along with its derivatives. More catechin is found in the flowers of buckwheat (*Fagopyrum Esculentum* Moench) than in any other section of the plant, while the flowers of highbush blueberries (*Vaccinium angustifolium* L. (Wan et al. 2012) and *Cucurbita pepo* L. (Murkovic 2016) are also good sources of (+)-catechin and (−)-epicatechin. De Morais et al. 2020 recently tested the edible flowers of eight different-colored species, including the mini rose (*Rosa chinensis* Jacq.), mini daisy (*Bellis annua* L.), torenia [*Torenia fournieri* (F.) Lind.], clitoria (*Clitoria ternatea* L.), cravine (*Dianthus chinensis* L.), cosmos (*Cosmos sulphureus* Cav.), tagete (*Tagetes patula* L.) and begonia (*Begonia* x *tuberhybrida* Voss.). Four flavanols were found in these edible flowers: catechin, epicatechin, epicatechin, and epigallocatechin galate.

12.2.4 Carotenoids & Carotenes

12.2.4.1 Carotenoids

There are two main families of carotenoids, carotenes, which are formed only by carbon and hydrogen such as carotene, and lycopene and xanthophylls, which are oxidized form of carotenes such as lutein, zeaxanthin, and astaxanthin (Fernandes et al. 2018a, b). Carotenoids are metabolites which are soluble in fat, that are synthesized by photosynthesis and are abundantly present in fruit, vegetables and flowers (Kimura et al. 2005; Rodrigues et al. 2019). Among their many other roles, carotenoids are responsible for providing vivid colours in shades of yellow, orange, and red, at the same time directly providing photoprotective property to plants (Park et al. 2017; Zhu et al. 2010). There are currently about 650 different kinds of carotenoids known to exist in nature, 100 of which are commonly consumed by humans (Eggersdorfer and Wyss 2018). Consumption of various food products and supplements introduce these chemicals to our body, and their consumption is crucial to maintaining human health and well-being (Kumari and Bhargava 2021). Plants contain carotenoids in a variety of sections (Honda 2020). They are found in the petals, sepals, stamens, anthers, and pollen of flowers in particular. Additionally, it should be mentioned that carotenoids play crucial role in improving colour of petals and support the functionality of their products (Rivera-Madrid et al. 2020). Generally speaking, the primary bioactive function of carotenoids is their antioxidant activity (Riberto et al. 2018). While there are many different types of carotenoids in nature, the primary ones found in flowers are the yellow xanthophylls (lutein, zeaxanthin and cryptoxanthin) and the epoxide xanthophylls (violaxanthin, neoxanthin and antheraxanthin) (Watanabe et al. 2021). Numerous studies have been published in the literature regarding the amounts of total carotenoids. For instance, the presence of high levels of carotenoids was cited as justification for the vibrant colour displayed by the petals of fresh pumpkin flowers (Ghosh and Rana 2021). Eight carotenoids, including carotene, cryptoxanthin, violaxanthin, lutein, antheraxanthin, zeaxanthin, and lutein-5,6-epoxide, were found in three species of *Dendrobium* spp. (white, golden yellow and light-yellow flowers) in a comparative study. However, in contrast to the other two species, *Dendrobium thyrsiflorum* B. S. Williams (golden yellow) had the highest content of carotenoids (3810.89 g/g) (Huang et al. 2019). The existence of 14 carotenoids was then reported in a study using the golden flowers of *Lonicera japonica* Thunb., with zeaxanthin, carotene, and carotene being the most prevalent (Xue et al. 2020).

12.2.4.2 Carotenes

All photosynthetic organisms contain carotenes, which is the primary intermediate in the synthesis of carotenoids. They are also found in large quantities in our regular diet and accumulate in human skin, where they are essential for preventing aging and shielding the skin from UV light (Amengual 2019). carotene and lycopene are the primary carotenes' derivatives that are frequently found in vegetables with vibrant colours. Beta-carotene is considered as a liposoluble colorant that is sensitive to light and oxygen, which directly protects against heart disease, certain cancers, and the oxidation of LDL cholesterol, while lycopene is well known as one of the best compounds inhibiting free radicals, directly preventing lung, ovarian, and prostate cancer (Mezzomo and Ferreira 2016). Yeon and Kim (2020) examined eight cultivars of *Rosa hybrida* L. with varying colours (white, red, yellow, and pink). The results showed that the yellow flowers had remarkably high levels of carotene (7.62 g/g fw in the "Good time" cultivar and 4.40 g/g fw in the "Penny Lane" cultivar).

12.2.4.3 Xanthophylls

The process of oxidizing carotenoids produces xanthophylls, which are the largest carotenoids family found in the nature and give plant tissues their yellowish hue (Aziz et al. 2020). Fruit and flowers in nature frequently include xanthophylls in their esterified form due to fatty acids. The process of esterification encourages the buildup of xanthophylls in plant tissues, which causes the flower petals to become more intensely yellow (Kishimoto et al. 2020). The majority of lutein and zeaxanthin, which belong to the xanthophyll class, are found in flowers; however, minor amounts of other derivatives, including flavoxanthin, violaxanthin, auroxanthin, antheraxanthin, and neoxanthin, have also been identified.

12.2.5 Alkaloids

More than 12,000 chemicals have been identified as alkaloids, which are secondary metabolites that are found in large quantities in nature (Salih et al. 2016). Nitrogen (N) is a defining component of their structure, and they exist in glycosylated, methylated, and hydroxylated forms. They are distinguished by their bitter taste (Rasouli et al. 2020). Alkaloids can therefore be categorized as real alkaloids, pseudoalkaloids, or protoalkaloids based on their structural traits and manufacturing mechanism (Herbert 2003).

12.2.5.1 Betalains

Fruit and flowers are among the various plant organs that receive their color from betalains (Dominguez and Muñoz 2017). By definition, betalains are pigments that dissolve in water and are made from the amino acid tyrosine (Carle and Schweiggert 2016). Since anthocyanins and these chemicals have not yet been discovered together in plant tissues, they are hostile to one another (Miguel 2018). Though they can accumulate in the parts of different species of plants, betalains are primarily found in mostly of plants belonging to the order Caryophyllales (Castro-Enríquez et al. 2020). According to classification, they are separated into two primary groups: betaxanthins, which are produced when betalamic acid condenses with amines or α-amino acids and exhibit coloration ranging from orange to yellow, and the class of betacyanins, which are derived from betanidin and exhibit violet and red coloration (Coy-Barrera 2020; Sadowska-Bartosz and Bartosz 2021). Numerous research endeavors have been undertaken to identify betalain chemicals along with their derivatives in flowers. Preethi et al. 2019 investigated the effects of seven different solvents namely, 50% methanol, water, 100% methanol, 50% ethanol, hot water, 100% ethanol, and acetone for the extraction of total betalain derivatives, such as betacyanins and betaxanthin from both fresh and dehydrated flowers of *Celosia argentea* L and *Bougainvillea spectabilis* Willd. The water extract has exhibited remarkable yields of total betalain (24.05 ± 0.16 mg/mL), betaxanthin (54.70 ± 0.60 mg/mL) and betacyanin (46.66 ± 0.04 mg/mL), from the dehydrated *B. spectabilis* flowers. Furthermore, the concentrations of total betalain (25.75 ± 0.09 mg/mL), betaxanthin (31.16 ± 0.75 mg/mL) and betacyanin (30.62 ± 0.32 mg/mL), were observed to be increased in the dried samples of *C. argentea* flowers (Preethi et al. 2019).

12.2.5.2 Betaxanthins

Betaxanthins are mainly yellowish pigments from the betalains group, generated from process of condensation of betalamic acid with amino acids or amines. In the nature, yellow hue of compounds like betalain stimulates the attraction of pollinating insects and has a crucial function in the life cycle of flowers (Wendel et al. 2015). Previous literature investigations suggested that vulgaxanthin I is the earliest betaxanthin that was identified in flowers of species *Beta vulgaris* L., followed by miraxanthin V and dopaxanthin, which were mostly found in *Lampranthus productus* N.E. Br. flowers (Guerrero-Rubio et al. 2020). However, despite the availability of betaxanthin derivatives in numerous flowers, there are only a few scientific publications that record the measurement and identification of these compounds, showcasing them as uncommon when compared to other chemicals present in flowers.

12.2.5.3 Betacyanins

Betacyanin is associated with the color characteristics of its components, which encourage plant pigmentation in deep violet hues. The occurrence of betacyanins is mostly limited to the flowers of *Caryophyllales spp.*, which is constituted of 35 families of plant, encompassing around 10,000 different species (Gonçalves et al. 2015). Although betacyanins have indistinguishable coloration as anthocyanins, in general they have a more prominent coloring capability, higher pH stability, and can be employed in food products with alkaline matrices (Roriz et al. 2017). Several studies suggest the utilization of betacyanins rich edible flowers as natural colorant replacements. For example, the American Indian Hopi culture dyed cornmeal with extract of *Amaranthus cruentus* L. flower, ingested in their ceremonial rites (Gonçalves et al. 2015).

12.3 Food Applications of Edible Flowers

Edible flowers are categorized into fruit flowers (apple, papaya, peach etc.) and non-fruit flowers, which are mostly ornamental flowers such as rose, hibiscus, marigold etc. Though these flowers are having prominent level of phytochemicals characteristics, proper utilization of these natural florets are still unexplored (Pinakin et al. 2020). Petals of peony, a native flower of Ais and Europe contains high value phenolic contents and several galloyglucosides which supress the pancreatic cancer and inhibit growth of *E. coli*, *P. aeruginosa*, *S. abony* etc. (Kim et al. 2022). Frangipani flowers, usually found in West Indies, contains essential oils and syringic acid which is considered as neuroprotective and hepatoprotective (Ferdosi et al. 2023). Oils extracted from frangipani also have anti-microbial (Salmonella spp., *B. subtilis, S. aureus, E. fecalis* etc.) and anti-fungal (Trichoderma) properties (Mary Mawumenyo Mamattah et al. 2023). Scientists are studying malaria inhibiting mechanism of *Myosotic scorpiodes* (Zakari et al. 2018). Study of Demasi et al. (2021) has shown difference in phenolic compound in Alpine (*Rosa pendulina*) and Wild rose (*Rosa canina*) while both flowers are having traceable amount of vitamin C. Studies have shown, potential anti-cancer property of *Hibiscus deflersii* due to presence of sterols like ursolic acid, lupeol and β-sitosterol (Valencia-Cordova et al. 2024). Various food applications of edible flowers has been listed in Table 12.2.

12.3.1 Natural Food Colorant

Present population is way more conscious about harmful synthetic colouring present in various processed foods. There exist many trade barriers in European countries and Japan regarding synthetic colours. Thus, interest towards bio-colours is increasing day by day, which have its own drawbacks like instability towards heat, pH, light etc. Edible flowers are well known for their vibrant colours due to presence of carotenoids and anthocyanins mostly. Marigold is having good ability to produce yellow-orange colour used in cheese industries. Purple to blue colour can be obtained from Centaurea cyanus flowers (Mazza 2007). Essential oil from Crocus sativus contains terpenes and carbonyl compounds. High amount of chlorophyll is also extracted from various plant parts to procure green colour. Solvent assisted extraction is the most suitable extraction method for procurement of natural colorants from edible flowers among other technologies namely supercritical fluid

Table 12.2 Application of various edible flowers in food

Flower	Food product	Key findings	References
Rose	Merlot red wine	Improved aroma in dealcoholized wine, without affecting chemical composition	Sam et al. (2023)
	Yogurt	Extract is better colouring agent than artificial E163	Pires et al. (2018a, b, c)
	Ground beef	Increases WHC	Duggirala et al. (2024)
	Cookies	5% rose powder in flour increase shelf life and enhance nutritional value	Hnin et al. (2021)
Marigold	Organic yogurt	Improved TPC and antioxidant activity	Bragueto Escher et al. (2019)
Hibiscus	Ground beef	Enhance oxidative and colour stability	Zheng et al. (2019)
Roselle	Date spathe-based beverage	Enhanced vitamin and antioxidant content	ALTamimi et al. (2020)
Osmanthus	Cake	Suitable for diabetic people	Qin et al. (2021)
	Rice wine	3% *O. fragrans* increased acceptability	Zhang et al. (2022)
	Flavoured Tea	Enhanced caffeine content and TPC	Hussain et al. (2019)
	Fermented Milk beverage	Increase NK production, providing functional values	Nie et al. (2017)
Butterfly pea	Rice	Decrease stickiness and starch digestivity	Chusak et al. (2019)
Rosa rugosa	Yogurt	1% extract imparts better viscoelastic property, WHC, pH, TPC	Qiu et al. (2021)
Moringa flower	Infant food	Augmentation of crude protein, fibre, ash, fat and lowers carb content	Arise et al. (2014)
	Chicken Nuggets	Lower lipid oxidation, higher TPC and WHC	Madane et al. (2019)
C. morifolium	Cooked fat	Reduce HCAs content	Khan et al. (2019)
Pumpkin flower	Chicken patties	Better antioxidant property	Santos et al. (2022)
Apple flower	Cookies	Beneficial for diabetic patients	Gao et al. (2020)
Rhododendron	Spiced appetizer	Nutritional enrichment and better keeping quality	Sam et al. (2023)
Cardoon	Cheese	Improved proteolytic activity	Vioque et al. (2000)
Black locust	Pasta	Improved antioxidant property due to better release of phenolics under heat treatment	Kowalczewski et al. (2019)

(continued)

Table 12.2 (continued)

Flower	Food product	Key findings	References
Sunflower	Salad	Decreases production of ROS, skin protection form UV radiations	de Lima Franzen et al. (2019)
Hops (*Humulus lupulus*)	Beer production	Imparts bitterness and characteristic flavour	de Keukeleire (2000)
Cardoon	Cheese	Helps in milk coagulation due to presence of chymosin	Barracosa et al. (2021)
Chrysanthemum morifolium	Cakes and infusions	Drying method and temperature significantly influenced the drying behavior of chrysanthemum cakes, mainly in terms of drying time, moisture diffusion, drying rate and activation energy	Wang et al. (2016)

extraction, enzymatic extraction and non-thermal technologies (sonication, microwave, gamma irradiation, membrane etc.) assisted extractions (Zhao et al. 2019).

12.3.2 Incorporation in Dairy Products

Modern dairy industries are aiming to fortify regular dairy with additional nutritional ingredients, mainly bioactive compounds (Rao and Poonia, 2024a, b). Here comes edible flowers which are good source micronutrients. Besides providing functional benefits, edible flowers have potential to enhance the visual appeal, colour and aroma of the product. Arslaner et al. (2021) reported an increase in antioxidant activity of yogurt after adding *H. sabdariffa* marmalade, while lowering values of protein, fat and pH. Flowers can be used to improve the aesthetic appeals of soft cheese like ricotta, along with adding distinctive flavour profile. Colombo et al. (2010) has reported use of cardoon (*Cynaracar dunculus*) flower in cheese making in Mediterranean countries. Ice-creams can be made more addictive by infusing the aromatic notes of edible flowers. Curd fortified with *Delonix regia* resulted good amount of phenol, flavonoid and anthocyanin (Chhabra and Gupta 2015). Most of the studies are done on yogurt, where it has been fortified with various edible flowers like *Dahlia mignon*, *C. cyanus*, *R. canina* etc. and resulted in enhanced textural and visual appeal (Pires et al. 2018a, b). Kumar et al. (2021) has studied milk-based drinks flavoured with Rose and marigold, which have increased in antioxidant property of the drink.

12.3.3 Food and Beverage Industry

The consumption of herbal beverages has become a modern trend and edible flowers have successfully attracted researchers as well as consumers in this concern. Herbal infusions play a crucial role in providing pharmacological benefits to end users. Edible flowers, mostly petals are mostly processed in the form of tea for better shelf life and preserving beneficial attributes (Rodrigues et al. 2017). Lakshan et al. (2019) has shown excellent antioxidant activity of *Clitoria ternatea* incorporated beverage, which was also capable of reducing oxidative stress. Tea prepared from Chrysanthemum is very popular in China, is having sweet floral aroma due to compounds like, verbenone, ethyl cinnamate, propyl 3-phenylpropanoate, and ethyl 3-phenylpropanoate (Kaneko et al. 2017). Jasmine extract can be infused with green tea to improve the aroma and palatability (Shen et al. 2017). Dried Chamomile flower is used to produce herbal beer and can also be used to incorporate in white wine along with hops before filtration step (Srivastava et al. 2010). Rose petals and its extracts are used to prepare caffeine-free beverages (Vinokur et al. 2006). Similarly, synergic effects of infusing *Hibiscus sabdariffa* extracts (Oluwatoyin et al. 2016), marigold and rose syrups (Kumar et al. 2021) have been studied and concluded with better antioxidant activity and improved sensory attributes. Dabbagh Moghaddam et al. (2018) developed fermented probiotic beverages based on saffron petal extract, concluded it as a health supporting non-dairy beverage with better acceptability. Rodrigues et al. (2019) has reported synergic effect of Sea lavender with green tea towards inhibiting enzymes responsible for Alzheimer's and type-2 diabetes (Fig. 12.1).

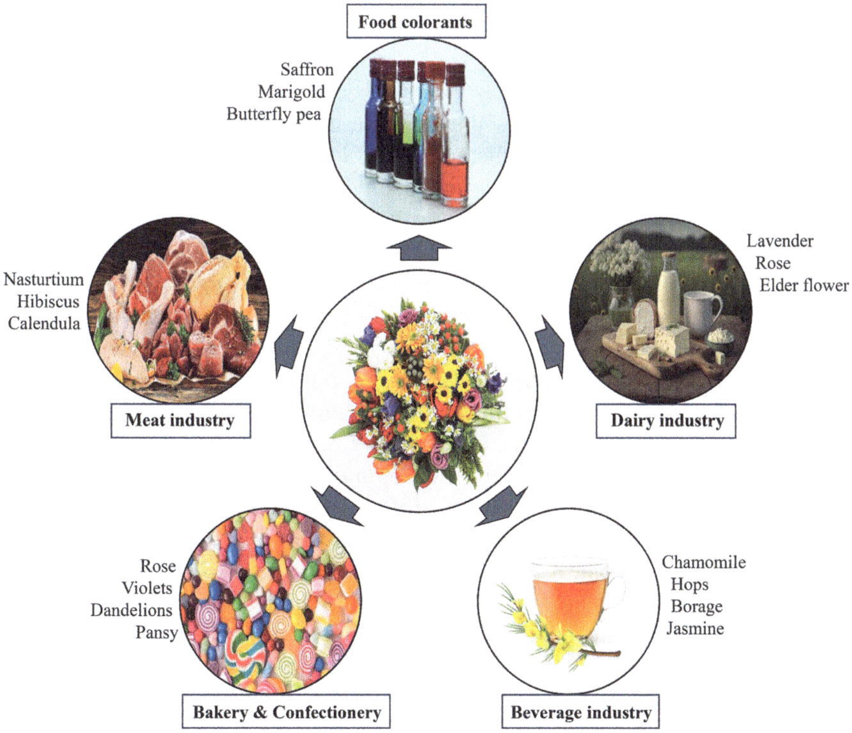

Fig. 12.1 Edible flowers in various food sectors

12.4 Processing Techniques

12.4.1 Pre-processing

There are various ways of processing the edible flowers, as per the demand of consumers and retention of quality parameters. Edible flowers are highly perishable and their qualities are more prone to degradation after harvesting. Thus, post-harvest and pre-processing management plays a crucial role in preservation of those qualities, which differ from species to species (Kelley et al. 2001b). According to (Newman and Connor 2020) a thin layer of properly pasteurized egg white powder over freshly plucked petals or flowers can avoid contamination with foreign materials and disease-causing microorganisms. A study by (Ran et al. 2019) concluded that chlorine dioxide application on petals of edible flowers during their storage, either alone or optionally in combination with other food-grade preservative has immense preservation at early stage. Good oxidant property of chlorine dioxide helps in arresting the growth of microorganisms. Nitric oxide fumigation is another alternative pre-processing measure for post-harvest quarantine treatment (Lu et al. 2016).

12.4.2 Irradiation

Food irradiation is a technology used to increase shelf life and promote food safety by treatment of food under controlled dosage of ionizing radiation, through inhibition of sprout, insect disinfestation, microbial decontamination and delay ripening (Koike 2019). This irradiation process involves exposing edible flowers to gamma rays (Cobalt-60, Cesium-137), X-ray and electron beam. In edible flowers shelf life is extended by slowing down their cell death achieved by changes in physiological and biological changes, such as accumulation of oxidative stress, development of reactive oxygen species (ROS) and peroxides (Ghosh et al. 2017). As per Koike et al. (2015), irradiated *Viola tricolor* L. flowers showed better antioxidant activity and higher phenolic content. Similarly, preservation of chemical components in *Dianthus chinensis* flower was achieved, when treated with low doses of ionizing radiation (1.0 kGy). However, high doses required to counter fungal infestation, which may impart negative effects on quality to edible flowers. To resolve this problem and to preserve the functional properties, lower irradiation doses along with other treatment can become a sustainable approach. Cia et al. (2007) has suggested low doses of gamma radiation along with UV-C clubbing can be a prominent treatment alternative. The nutritional and chemical composition of *Bauhinia variegate* flower was found unaffected when treated with electron beam radiation (Villavicencio et al. 2018). Furthermore, the extractability of phenols was enhanced due to disintegration of chemical bonds upon irradiation. The irradiation technology has proven itself to be an effective non-thermal processing technique without disturbing the physiological and textural characteristics of edible flowers, still the initial capital investment and maintenance cost is a major concern.

12.4.3 High Pressure Processing (HPP)

HPP, which is also termed as high hydrostatic pressure processing, is a non-thermal technology well known for its preservation potential and aids in processing (Zhao et al. 2019; Jadhav et al. 2021). HPP operates in a pressure range of 200–800 MPa, which leads to enzymatic as well as microbial deactivation, and at the same time preservation of functional and nutritional attributes. It is a uniform treatment irrespective of geometry, texture and also minute application of chemicals (Komora et al. 2018). Fernandes et al. (2017) reported that, effect of HPP varies from flower to flower, as camellia and borage are having less tolerance toward pressure, but flowers like pansy, centaurea etc. are having excellent pressure tolerability. Despite this fact, centaurea did not showed better shelf life upon pressure treatment but the case is different when it comes to pansy, due to presence of papillae, a hydrophobic compound. Westphal et al. (2017) investigated that broccoli sprouts when processed at 100–600 MPa for 3 minutes at 30 °C, resulted in reduced enzymatic browning and retention of physical appearance. Various researches concluded mild pressure treatment for short period of time results in better flower quality, but there is chance of microbial spoilage during storage condition due to improper inactivation of microorganisms. Under special conditions, a combination of other technologies can be beneficial in terms of processing time and operational cost. In Netherland, preparation of juices and smoothies from edible flowers are done using HPP technology. Rodrigues et al. (2017) reported that cauliflower by-products show better antimicrobial properties when coupled with HPP, however low-pressure treatment for short duration of time may not results in desired attributes. Edible flowers are recommended to be treated with moderate to high pressure, but there is a risk of alteration in sensory and physiochemical properties (Fernandes et al. 2020a, b).

12.4.4 Drying

Drying is an age-old method for preservation of food materials where excess moisture is removed. Traditional drying methods follow sun drying, shade drying and solar drying. In edible flowers drying arrest the microbial growth and inhibit enzymatic spoilage by reducing the water activity (Ahrné et al. 2007). In present days, modern drying methods such as hot air drying, freeze drying, vacuum drying, microwave-assisted drying and hybrid (combination of more than one method) drying are operational in food industries (Zhao et al. 2019). Ahluwalia et al. (2014) reported an increased nutritional content in *T. erecta* flower upon drying, but a decrease in phenolic content and antioxidant activity was also found. Vacuum drying was preferred over other drying methods due to higher retention of β-carotene, but due to its higher economy cabinet drying can be a feasible alternative for drying flowers. In another study, Siriamornpun et al. (2012) concluded better retention of carotenoids and phenolic compounds in *T. erecta* upon infrared drying coupled with hot air drying.

12.4.5 Ultrasound Processing

Ultrasound is an efficient non-thermal processing technique where inhibition of harmful microorganisms is achieved through thinning and rupturing the cellular membrane, activation of free radicals and releasing intracellular ingredients (Zhu et al. 2020). This process also known as ultrasonication, which result in formation, development and bursting of cavitation bubbles which releases enough energy to disrupt the cell membrane of spoilage causing microorganisms (Chen et al. 2020a). Chemical effects such as free radical production is associated with number of mechanical effects such as shear stress, collapse pressure and turbulences (Lv et al. 2019). It can also be combined with other non-thermal technology for better efficiency and to eradicate drawbacks of single technology. Xin et al. (2013) has reported shortening of overall process times during osmotic dehydration of broccoli, assisted with ultrasound treatment. At the same time the higher rate of dehydration was accounted due to degassing during acoustic cavitation. Study of Ansorena et al. (2014) states that, the synergism of both non-thermal and thermal technologies gives a better output in terms of textural, nutritional quality and reduced microbial load, where *Brassica oleracea* L was subjected to mild heating up to 50 °C and ultrasonication consecutively. Thus, immense scope is there to treat edible flowers with ultrasound to complete processing in comparatively less time with preservation of textural, nutritional and physicochemical properties.

12.4.6 Pulsed Electric Field (PEF)

It is a cutting edge, mild non-thermal method of food processing, where edible flowers are exposed to pulses of high-frequency for a short period of time. Normally, the intensity of the frequency ranges between 25 to 100 kV/cm, which is exposed for milliseconds or even nanoseconds (Jadhav et al. 2021). The product to be treated is kept between two electrodes of stainless steel where the high intensity of electric field increases in the porosity of the cell membrane and ruptures the cell membranes of spoilage causing microorganisms, due to charge separation and leads to their death. This cutting-edge technology has not been explored much in terms of processing of edible flowers. Sanz-Puig et al. (2016) stated that during infusion of broccoli and cauliflower by-products with PEF, higher antimicrobial effect can be observed against *Salmonella typhimurium*. The higher electric field did not affect the bioactive compounds in edible flowers. The research of Sánchez-Vega et al. (2014) and Aguiló-Aguayo et al. (2014) reported, an increase in lutein, β-carotene and glucosinolates content in broccoli when treated with higher intensity and time of PEF technology. This can be considered due to role of PEF in inactivation oxidative enzymes and triggering biosynthesis. However, at 35 kV/cm it was noticed that certain vitamins are slightly degraded, which may be due to presence of heavy metals like Fe, Zn etc.

12.5 Methods of Preservation

There are several products procured from edible flowers other than being used as fresh. Here comes the preservation techniques for the retention of quality parameters. Factors like packaging material and storage temperature are very crucial for retaining the product quality whether it is nutritional or functional components (Kelley et al. 2001a).

12.5.1 Low Temperature Storage

Storage of horticultural produces including edible flowers under lower temperature minimizes the rate of transpiration and respiration, thus increasing the shelf life and preserving the functional and nutritional qualities during storage. The chance of enzymatic spoilage is also reduced. Demasi et al. (2021) reported that 17 different types of edible flowers were stored under 4 °C and their shelf life was about 7–14 days, varying upon species. Author reported slight variations in terms of bioactive profile and shelf life but the overall freshness was maintained during the storage time. A study of Pinton et al. (2020) states that growth of *L. monocytogenes* in broccoli was not severe in storage temperature of −18 °C to 4 °C. According to Kelley et al. (2003), flower species namely pansy, viola, nasturtium, scarlet runner and borage bean were excellently stored for almost 2 weeks at 0–2.5 °C. The major problem in case of edible flowers is maintaining the same storage temperature in supply chain, which can be resolved by freezing the flowers. It also reduces the water activity but does not reduce the enzymatic activity present in flowers. This can be checked through blanching before freezing the flowers. Freezing also helps in retaining nutritional and physicochemical characteristics (Bilbao-Sainz et al. 2020). As per study conducted by Fernandes et al. (2020b), freezing in ice cubes is a better option for freezing as compared to direct freezing in preserving appearance and bioactive compounds, and a storage period of up to 1 month. Freezing coupled with other non-thermal technology produce better phenolic content and vitamin C content in *Cynara scolymus* flowers (Gil et al. 2001).

12.5.2 Edible Coating

Coating is a cutting-edge technology that provides a desirable barrier towards mechanical damage, moisture and gas exchange, colour and flavour degradation and enzymatic oxidations for extending the shelf life at the same time quality of fresh horticultural produces (Maria Leena et al. 2020). This leads to reduced metabolic activity, retention of volatile components and less deterioration (Putri et al. 2019). Edible coating is generally made with biodegradable materials like lipids, proteins, polysaccharides etc. (Vimala Bharathi et al. 2020). Coating through dipping, spraying and electro spraying are generally used to get uniform coated surface (Khan et al. 2014). In a study of Fernandes et al. (2018a, b), *Viola tricolor* flower was dipped in 0.5% sodium alginate solution and later treated with 1% calcium chloride and stored at 4 °C, which showed lower microbial contaminations and better freshness. Various phenolic compounds such as anthocyanins, flavonoids, tannins along with their antioxidant activity was found to be better for pansy flower, when coated. Xie (2015) reported that edible coating based on soy-protein-alginate-chitosan has remarkable resistance towards vitamin-C loss, enzymatic browning and shedding rate in case of Night-fragment flower. Coating coupled with non-thermal technologies such as gamma irradiation helps in preservation of edible flowers for longer period of time (Ben-Fadhel et al. 2017). Another study showed better shelf life of *Brassica* florets when coated with encapsulated in antimicrobial agent (natamycin) and essential oil into an alginate matrix and treated with irradiation later. Even there is limited researches on this technology, safety evaluation should properly be done when edible coating consisting of nanomaterials.

12.5.3 Controlled Atmospheric Packaging (CAP)

This is a modern packaging technique inside which the environment has been altered with respect to concentration of gases like oxygen, carbon dioxide, nitrogen, along with temperature and humidity, which aids in reducing respiration rate, inhibiting enzymatic and microbial damage to horticultural produces (Fernandes et al. 2018b). Studies regarding use of CAP for edible flowers are limited to very few. Fernández-León et al. 2013 concluded, storage of broccoli under controlled atmospheric conditions maintained at 10% oxygen, 5% carbon dioxide, 85–90% relative humidity. Here relative humidity played major role in retaining the freshness without loss of vital nutrients. Study on *Zucchini* flower under CAP condition at 97% CO_2 and 3% O_2 has shown no colour loss with better retention of essential vitamins (Cefola et al. 2016). Aquino-Bolaños et al. (2013) investigated *Cucurbita pepo* flower under varying composition of gases conditions and concluded CAP treated sample have better physicochemical properties as well as better retention of bioactive compounds. Thus, reducing O_2 content and at the same time increasing CO_2 and

N_2 content during storage period can be beneficial in terms of shelf life of edible flowers.

12.5.4 Modified Atmospheric Packaging (MAP)

MAP is a popular packaging technology that effectively preserve the necessary qualities of horticultural produce (Singh et al. 2018). To keep the edible flower fresh for an extended period of time, the composition of gases inside the package is changed. The ethylene generation, rate of respiration, and growth of pathogenic microbes in fresh commodities are all decreased as a result of the gas composition change. A study by Aros et al. (2017) concluded extended vase life and drastic reduction in fresh weight loss (FWL in %) in tulip flowers packaged in MAP as compared to conventional packaging. Mohi Alden et al. (2019) studied cauliflower under non-MAP condition and MAP condition maintained at 3% O_2, 5% CO_2 and 92% N_2 in polypropylene and polyethylene pouches. Cauliflower stored under MAP condition were showing shelf life more than 30 days. Another study by Wang et al. (2020) reported use of high oxygen concentration within package of Chinese cabbage resulted in reduced hydrogen peroxide level and reduced enzymatic activity due to laccases and peroxidases. Treatment of 1-methylcyclopropane and MAP in combination on flowers like snapdragon and carnation resulted in enhanced shelf life with maintaining freshness at the same time reduced wilting (Kou et al. 2012). Designing and developing MAP on large to facilitate industry produces will help in commercializing various edible flowers through proper transport and supply chain.

12.6 Flavour & Aroma Profile

The unique characteristics of edible flowers on the market-such as composition, size, form, taste, scent, and colour-have a significant impact on consumer preferences. There are several papers in the literature that describe the various cultivars' bioactive components, but very few that discuss their volatile fingerprint. The interplay of several volatile chemicals, which usually include alcohols, alkanes, carbonyl compounds, terpenes, terpenoids, esters and ketones, which determines the aromatic profile of flowers (Pires et al. 2019). However, the sensory perception of fragrance molecules along with volatility can be affected by other non-volatile substances such sugars, organic acids, and phenolic compounds (Oliveira et al. 2014). Several volatile compounds found in certain edible flower species can also be good for human health. For instance, the cyclic terpenoid β-ionone in caryophyllene, a potent natural anti-inflammatory drug, prevents the growth of cancerous cells and the death of human gastric adenocarcinoma cells (Pires et al. 2019; Liu et al. 2004). The most popular analytical method for identifying the volatile compounds of edible flowers is headspace solid chromatography-mass spectrometry (HS-SPME) analysis, according to the scientific literature. For instance, Marchioni et al. (2020) used HS-SPME/GC-MS to examine how nutritional qualities and flavor of edible flowers are affected due to various post-harvest handling and treatments. They discovered that oxygenated monoterpenes predominated in all of the samples they examined (Marchioni et al. 2020). Study was conducted for the Characterization of the aroma profile of edible flowers using HS-SPME/GC–MS and chemometrics by (Giannetti et al. 2024) on Four distinct flowers of different species: Antirrhinum, Begonia, Dianthus and Viola x wittrockiana (also known as violetta or pansy) and the following results were obtained. The characteristic aroma of a food product is significantly influenced by the volatile compounds present, which is dependent on the chemical makeup of each compound and the potential synergistic effects they might have with various non-volatile compounds. 24 Terpenoids and Terpenes (including sesquiterpenes and sesquiterpenoid compounds), 14 alcohols, 7 ketones, 4 furans, 4 aromatic compounds, 2 hydrocarbons, 2 sulfur compounds, and 1 ether comprised the total of 58 volatiles that were detected. Displayed a comprehensive list of almostall volatile chemicals found in the examined edible flowers together with the corresponding information, such as the chemical families, molecular formula, and Kovats retention index. Additionally, Table 12.1 lists the flower class according to their discovery. A collection of volatile compounds was used to characterize each of the several edible flower, essential oil, edible plant, and wild herb categories that are the subject of numerous research in the scientific literature. Comparing the various literature on edible flowers, such as, Fernandes et al. (2019) investigated five different species (Borage, Cosmos, Calendula, pansy, Johnny Jump up) and almost every substances detected are the similar to the study conducted by Giannetti et al. (2024), such as 3-hexen1-ol, 1-octen-3-ol, 1-hexanol, 2-phenylethanol and 2-ethyl-1-hexanol for the class of alcohols; undecane for aliphatic hydrocarbons; 3-hepta none, 2-nonanone and 3-octanone, for class of ketones; α-car yophyllene (or humulene), copaene, β-caryophyllene, germacrene D, alloaromadendrene, α-muurolene, caryophyllene oxide and calamenene for the class of sesquiterpenes; styrene, β-myrcene, β-pinene, limonene, allo-ocimene and β-ocimene for the class of terpenes (Fernandes et al. 2019). Some of the volatile compounds found in the samples of this study conducted by (Giannetti et al. 2024) are similar to those found in Marchioni et al.'s (2020) study of seven edible flowers of the Lamiaceae family, including β-myrcene, β-ocimene, limonene, terpinen-4-ol, copaene, eugenol, β-caryophyllene, humulene, germacrene D, muurolene, and sesquiphenoldrene. Najar et al. (2019) reported that the chemical composition of four species of Agastache flower included 1-Octen-3-ol, allo-ocimene, β-pinene, limonene, 3-octanone, myrcene, β-ocimene, germacrene D, β-caryophyllene, copaene, and humulene.

12.7 Nutritional Profile

Edible flowers contribute not only basic nutritional and sensory properties, but also provides health improvement (Alasalvar et al. 2013). Flowers are capable of providing more free radical scavenging property as compared to fruits and vegetables (Benvenuti et al. 2016). Moisture is the major component of flowers, followed by micronutrients such as fibre and protein. The study of Mulík and Ozuna 2020 reported presence of proteins and essential amino acids in various Mexican flowers. Various other studies have reported presence of 21.9% and 52.3% protein in pumpkin and broccoli respectively, 61% and 59% dietary fibre in banana flower and sunflower respectively (Alexander et al. 2019; Begum and Deka 2019). Marigold is known for having prominent level of unsaturated fatty acids (~59.3%) (Pires et al. 2017). Flowers of banana is reported to have 12.5% protein (Bhaskar et al. 2011). The level of minerals such as potassium, calcium, phosphorus etc. are found in sunflower is positively high. Edible flowers like pot marigold, narrow-leaf clover, pomegranate, sweet briar rose etc. have good amount of high value polyunsaturated fatty acids such as linolenic acid and linoleic acid. Nutritional profile varies in every part an edible flower, such as the nectar mostly contains sugar such as terpenoids and alkaloids, pollen is consisting of protein, unsaturated and saturated fatty acids and petals are rich in vitamins and minerals (de Pascual-Teresa and Sanchez-Ballesta 2008).

Edible flowers are abundant source of different bioactive compounds. Flowers like pansy, cornflower, marigold, carnation etc. are having remarkable antioxidant activity against free radicals and ROS, sometime better than common fruits and vegetables, which is due to the presence of flavonoids, phenolic acids, alkaloids and anthocyanins (Zheng et al. 2019). Proper harvesting stage is also an important factor that effect the level of bioactive component present in flowers according to Magri et al. (2020).

12.7.1 Health Benefits

The popularity of edible flowers is likely to grow in the future due to the possible health benefits associated with their chemical characteristics. In recent years, the antioxidant properties of edible flowers have indeed been extensively documented employing chemical-related antioxidant assays such as DPPH, FRAP, ABTS etc., Reactive oxygen species, neurological conditions like Parkinson's disease, stroke, type 2 diabetes, cardiovascular disease, and cancer are all caused by an excess of spontaneous oxidative radicals (Rao and Poonia, 2024a, b).

12.7.1.1 Antioxidant Activity

Oxidative stress in human body causes numerous chronic diseases. Almost all edible flowers are having immense antioxidant characteristics and along with antioxidant enzymes (SOD, GPx and CAT). They can protect cells from oxidative damages, decreasing peroxidation end-products and at the same time reducing free radicals (ROS) (Wang et al. 2017). Zhang et al. 2019 studied antioxidant property of chrysanthemum flower and concluded that the free radical scavenging property of these edible flowers are due to presence of various phenolic compounds such as phenolic acid, flavonoids (rutin, quercitrin), anthocyanin and terpenes. Zheng et al. 2021 has summarized various studies on antioxidant attributes of different flowers such as *Rosa* species, *Osmanthus fragrans*, *Chrysanthemum morifolium*, *Tagetes patula*, along with their method of extractions. Solvent extraction method using polar solvents like ethanol, methanol, acetone etc. are having better retention antioxidant properties of edible flowers (Xu et al. 2017). Researches has proved water extracts are having better antioxidant profile compared to ethanol extract of flower like Celosia cristata (Kim et al. 2015) and *Chrysanthemum morifolium* (Li et al. 2019). Antioxidant property also depends on extraction temperature, hot water extract has better antioxidant property as compared to cold water extract (Zannou et al. 2020). Similarly, extraction method also influences the antioxidant characteristics. Ultrasound assisted maceration helps in better availability of phytochemicals (Yang et al. 2010).

12.7.1.2 Anticancer

Various internal (hormone, lifestyle) and external (toxins, radiation) factors cause abnormal cell growth and lead to death in cancer, is a leading cause of death (Nanda 2019). Researchers are now searching for natural alternatives for chemotherapy and radiotherapy to reduce the risk of toxicity (Oun et al. 2018). Coreopsis lanceolata extract have inhibiting ability against HL-60 and HCT-15 cells due to presence of flavonoids, kukulkanin B and lanceolein A (Pardede et al. 2016). Similarly, Hibiscus species have proved to have anti-cancer property against breast and liver cancer (Nguyen et al. 2019). Flowers like *Hypericum amblysepalum* (Keskin 2015), *Calotropis gigantea* (Gururaja and David 2016) and *Tagetes patula* (Kashif et al. 2015), have been examined to test anti-cancer property against cervix cancer. Nguyen et al. (2019) and Yang et al. (2017) have concluded that, combination treatment of chemotherapeutic drugs and flowers extract have better treatment efficiency with minimum side-effects, which can reduce chemotherapy dose and toxicity. Many health benefits of edible flowers has been shown in Fig. 12.1.

12.7.1.3 Anti-inflammatory

Living cells continuously produce free radicals, which if produced in excess level can cause serious health injuries, leads to oxidative and nitrosative stress which can cause inflammation of cells and diseases like cancer, rheumatoid arthritis etc. (Phaniendra et al. 2015). As edible flowers are containing good amounts of phytochemicals (mostly anthocyanins and flavonoids) and bioactive compounds, they are studied under various condition in alternate concentrations. Anti-inflammatory properties of flowers like *Lonicera japonica* (Liu et al. 2015), *Begonia semperflorens* (Kwon et al. 2019), *Rumex nervosus* Vahl (Desta et al. 2015), *Hibiscus sabdariffa* L. (Sogo et al. 2015), *Punica granatum* L. (Xu et al. 2017), *Chrysanthemum morifolium* (Yang et al. 2019), *Prunus persica* var. davidiana (Tettey et al. 2016), and *Coreopsis lanceolata* (Kim et al. 2019) was studied and concluded that presence of quercetin, kaempferol, ellagic acid are the major flavonoids responsible for good anti-inflammatory properties of flowers. Recent researches found new compounds such as arylnaphthalene lignans and phenolic glycosides are also responsible for assisting prominent protection of edible flowers against cellular inflammation.

12.7.1.4 Anti-diabetic

Insulin resistance leads to metabolic disorder like diabetes mellitus, found mostly in obese people (Liu et al. 2016). Secoiridoid glycosides isolated from *Lonicera japonica* flower exerts strong α-glucosidase inhibition (Liu et al. 2015). Loizzo et al. (2015) studied inhibition of α-glucosidase and α-amylase, due to flavonoid content of Cichorium intybus. Similarly, according to other studies extracts from *Chrysanthemum morifolium*, *Punica granatum* and *Tagetes lucida* at different concentrations aids better liver and muscle glycogen and improved insulin signaling (Zheng et al. 2021).

12.7.1.5 Anti-Osteoporosis

Osteoporosis is a bone-disorder caused due to imbalanced activity of two cells namely osteoblasts and osteoclasts, where its mass and mineral content is drastically reduced (Yuan et al. 2016; Zhao et al. 2017). In vivo study of Kassem et al. (2017) revealed potential anti-osteoporosis property of *Trifolium resupinatum* extract due to presence of certain phenolic compounds such as quercetin, chlorogenic acid, formononetin etc. Zhao et al. (2017) has reported anti-osteoporosis property of *Sophora japonica* containing sophoridine, which target osteoclast only. Methanolic extract of *Chrysanthemum indicum* suppresses the tartrate-resistant acid phosphatase (TRAP) generation which generate ROS and damages bone strength (Luyen et al. 2015).

12.7.1.6 Anti-hypertension

According to Micucci et al. (2015), hypertension leads to various cardiovascular diseases like heart failure, myocardial infraction, stoke etc. Flower extract of *Musa X paradisiaca* and *Echinacea purpurea* have good ACE inhibition property, which is the major enzyme responsible for synthesis of angiotensin II from angiotensin I, which acts as a vasoconstrictor (Acharya et al. 2016; Chiou et al. 2017). Still exact ACE-inhibiting mechanism is not studied in depth. Ritonga et al. (2017) has mentioned the possible potential of *H. sabdariffa* in inhibiting ACE and activating endothelium-driven relaxing factor.

12.7.1.7 Anti-obesity

Obesity has become a major problem among young as well as old generations, which leads to health hazards like diabetes, hypertension and cardiovascular diseases (Gao et al. 2016). This is characterized by accumulation of adipocytes. Kim et al. (2017) has reported lowering lipid accumulation in adipocytes and reduction of triglyceride of *Tropaeolum majus* extract in varying concentration. *Edgeworthia gardneri* is also having anti-adipogenesis property due to presence of glycosidic flavonoid, tiliroside (Gao et al. 2016). Methanol extract of *Averrhoa bilimbi* has decreased low-density lipoproteins and has increased high-density lipoproteins and thus help in cases like hyperlipidemia (Arya et al. 2016).

12.7.1.8 Neuroprotective and Hepatoprotective

Oxidative stress is directly related to ageing which is correlated to neuro disorders like Parkinson's disease, Huntington's disease, Alzheimer's disease etc. (Xiong et al. 2016; Kim et al. 2015). Study of Kwon et al. (2019) focused on neuroprotective effect of *Begonia semperflorens* extract and concluded kaempferol, an isolated flavonoid has best effects against HT22 cell. According to research of Xiong et al. (2016), neuroprotective effect of acteoside, a major bioactive compound of *Osmanthus fragrans*, which was capable of suppressing production of advanced glycation end product (AGEs), neurotrophin-3 (NT-3). Wang et al. (2017) has proposed eugenol and geraniol could be the possible reason of this property of O. fragrans. Thus, O. fragrans can be aided as an anti-ageing supplement. Phenolic glycosides and Lignans isolated from *C. morifolium* have also shown a promising neuroprotective property (Yang et al. 2019).

Oxidative stress also imparts other disorders related to liver, which is a essential organ for proper detoxification and metabolism of our body (Abdelhafez et al. 2018). Five golden flowers tea consists of *Rosa rugosa* (28.57%), *Flos chrysanthemi* (21.43%), *Osmanthus fragrans* (28.57%), have promising protective effect against liver injuries (Tsai et al. 2017). Hepatoprotective action effect of *Chrysanthemum morifolium* and *Osmanthus fragrans* extract is attributed to

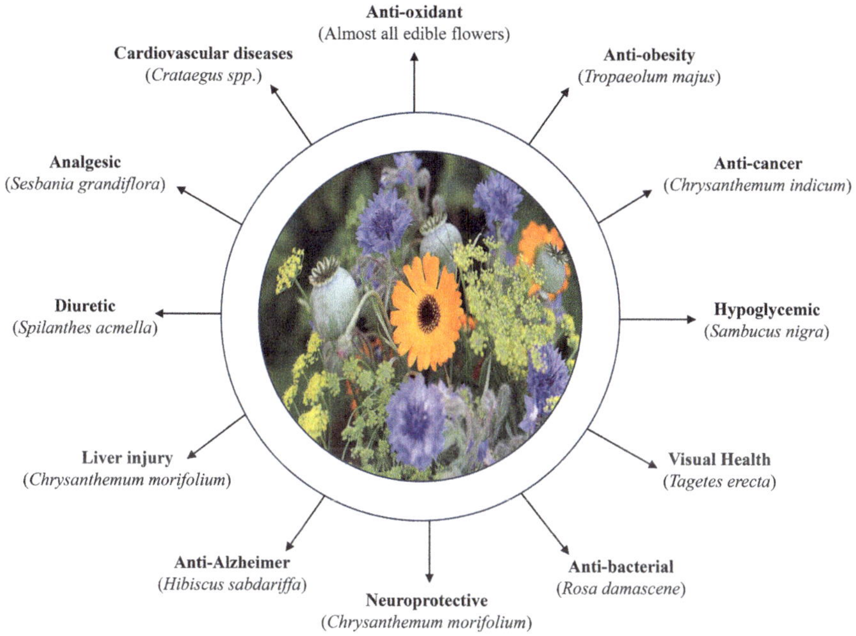

Fig. 12.2 Health benefits of different edible flowers

hindering sterol gene expression. Similarly, high level of certain phytochemicals such as, syringic acid, protocatechuic acid, rutin from the extracts of *Lagerstroemia speciosa* and *Malvaviscus arboreus* flowers contributes to lowering serum levels to safeguard liver health (Tiwary et al. 2017). Extracts from *Begonia semperflorens* and *Celosia cristata* L. are able to reduce oxidative caused by t-BHP cells (Kwon et al. 2019; Kim et al. 2015). *Celosia cristata* extract also have hepatoprotective property through inhibiting ROS generation (Tsai et al. 2017) (Fig. 12.2).

12.8 Challenges and Considerations

Though flowers are well known for range of bioactive compounds and nutritional benefits, there various potential barriers to processing as well as consumption of these flowers. There is no governing body worldwide and thus safety requirements and guidelines are also not properly known regarding edible flowers (Matyjaszczyk and Śmiechowska 2019). The foremost risk is chemical contamination due to overdoses of pesticides, herbicides and heavy metals. Contaminated soil can lead to accumulation of metals like lead or cadmium in flowers. Flowers such as Connavalariamajalis, Anemone ranunculoides etc. are associated with pathogens and harmful chemical compounds. Toxicology report of honeysuckle (Zhang et al. 2003) and roselle (Akindahunsi and Olaleye 2003) revealed that most of the edible flowers are non-toxic when used in adequate dosage. Allergic reactions upon consumption are one of the most concerning challenges to be taken care of, which may cause due to cross-reactivity of pollens. Over consumption of marigold may cause gastrointestinal disorders. Flowers like Chamomile, daisies may also trigger allergic responses to sensitive individuals. Another challenge that is to be overcome is, microbial contamination and short shelf life. Escherichia coli, Salmonella are among very common species of spoilage causing microorganisms that are found in edible flowers due to inappropriate handling and storage post-harvest. Lack of awareness among the population regarding potential health benefits of edible flower is again a major obstacle.

Thus, some of the key considerations regarding processing and consumption of edible flowers may be starting from proper selection and sourcing of flowers from organic source, which comply with local and marketing regulations. Post-harvest treatment along with cutting edge processing technologies can eradicate all of those threats. Implementation of Good Agricultural Practices (GAP) and Good Manufacturing Practices (GMP) and Hazard Analysis and Critical Control Points (HACCP) during processing is necessary to maintain optimum quality of final product. And lastly the storage of products at low temperature (0–4 °C) is essential to arrest the oxidation reaction, microbial growth and improve the shelf life of product.

12.9 Future Prospects and Conclusion

In the search of organic diet practice, researchers and chefs bended towards optimum utilization of edible flowers. Due to superior nutritional and bioactive properties, along with vibrant colour, distinctive flavour and aroma profile, flowers can be an ideal option to meet overall functional demands. Edible flowers have a lot of potential to intensify nutritional value and visual appeal of food products and are frequently used as an ingredient in both conventional and gourmet recipes. Nanotechnology represents an emerging field of scientific exploration focused on the examination and control of matter at the smallest scales, encompassing molecules and atoms (Rao and Poonia, 2024a, b). Interest in incorporating flowers into dietary patterns through innovative, healthful foods or by reimagining classic recipes has grown as a result of current market trends and consumer demands. Above chapter offers helpful details about the primary groupings of chemical compounds and their health benefits found in the most prevalent species of flowers. Edible flower consumption that is solely associated with gourmet cooking or cultural customs tends to make way for both the creation of functional meals and larger consumption. A guide of best practices for the handling, processing, and cultivation of this delicate ingredient will be established with the aid of research in this area. To meet the demands of this expanding market, toxicological issues should properly be addressed through researches and innovations. Modern processing and preservation technologies have the potential to cultivate knowledge into new value-added products using edible flowers. Commercialization of value-added products is far away due to lack of supportive reports and scientific guidelines. Moreover, strategic regulations are needed to bring upon safety, consumer satisfaction and confidence. By following safety guidelines, best practices consumers can efficiently add edible flowers to their diet system, which will promote overall health and economy to the farmer.

References

Abdelhafez OH, Fawzy MA, Fahim JR, Desoukey SY, Krischke M, Mueller MJ, Abdelmohsen UR (2018) Hepatoprotective potential of Malvaviscus arboreus against carbon tetrachloride-induced liver injury in rats. PLoS One 13(8):e0202362

Acharya J, Karak S, De B (2016) Metabolite profile and bioactivity of Musa X paradisiaca L. flower extracts. J Food Biochem 40(6):724–730

Arise AK, Arise RO, Sanusi MO, Esan OT, Oyeyinka SA (2014) Effect of Moringa oleifera flower fortification on the nutritional quality and sensory properties of weaning food. Croat J Food Sci Technol 6(2):65–71

Aguiló-aguayo I, Suarez M, Plaza L, Mohammad B, Brunton N, Lyng G, Rai DK (2014) Optimization of pulsed electric field pre-treatments to enhance health-promoting glucosinolates in broccoli flowers and stalk. J Sci Food Agric 95:1868–1875. https://doi.org/10.1002/jsfa.6891

Ahluwalia P, Kaur A, Dhillion G (2014) Effect of different drying methods on chemical and functional properties of marigold petals. Int J Food Nutr Sci 3(4):53–59

Ahrné LM, Pereira NR, Staack N, Floberg P (2007) Microwave convective drying of plant foods at constant and variable microwave power. Dry Technol 25(7–8):1149–1153

Akindahunsi AA, Olaleye MT (2003) Toxicological investigation of aqueous-methanolic extract of the calyces of Hibiscus sabdariffa L. J Ethnopharmacol 89(1):161–164

Alasalvar C, Pelvan E, Özdemir KS, Kocadağlı T, Mogol BA, Paslı AA, Gökmen V (2013) Compositional, nutritional, and functional characteristics of instant teas pro- duced from low- and high-quality black teas. J Agric Food Chem 61(31):7529–7536

Alexander P, Bwatanglang IB, Buba A (2019) Phytochemical screening, anti-nutritional and mineral composition of telfairia occidetallis (fluted pumpkin) and cleome rutidosperma (fringe spider flower). J Sci Res Rep 24:1–10. https://doi.org/10.9734/jsrr/2019/v24i430158

ALTamimi JZ, Alfaris NA, Alghamdi FA, Abu-Hiamed HA, Albader NA, Almousa LA (2020) Hibiscus sabdariffa L. flower and date palm pollen fortification of date palm spathe beverage. Br Food J 122(7):2159–2170

Amengual J (2019) Bioactive properties of carotenoids in human health. Nutrients 11(10):2388

AN XP, Diwan AD, Chandra SR (2016) Flavonoids: an overview. J Nutr Sci 5:e47

Ansorena MR, Moreira MR, Roura SI (2014) Combined effect of ultrasound, mild heat shock and citric acid to retain greenness, nutritional and microbiological quality of minimally processed broccoli (Brassica oleracea L.): an optimization study. Postharvest Biol Technol 94:1–13. https://doi.org/10.1016/j.postharvbio.2014.02.017

Aquino-Bolaños EN, Urrutia-Hernández TA, López Del Castillo-Lozano M, Chavéz Servia JL, Verdalet-Guzmán I (2013) Physicochemical parameters and antioxidant compounds in edible squash (cucurbita pepo) flower stored under controlled atmospheres. J Food Qual 36(5):302–308. https://doi.org/10.1111/jfq.12053

Aros D, Orellana K, Escalona V (2017) Modified atmosphere packaging as a method to extend postharvest life of tulip flowers. N Z J Crop Hortic Sci 45(3):202–215. https://doi.org/10.1080/01140671.2017.1296872

Arslaner A, Salik MA, Bakirci I (2021) The effects of adding Hibiscus sabdariffa L. flowers marmalade on some quality properties, mineral content and antioxidant activities of yogurt. J Food Sci Technol 58:223–233. https://doi.org/10.1007/s13197-020-04533-z

Arya MK, Mohammed SP, Sri Ganesan P (2016) In vitro hypolipidemic activity of Averrhoa bilimbi flower extract. The Pharma Innov J 5(10, Part A):23–28

Aziz E, Batool R, Akhtar W, Rehman S, Shahzad T, Malik A et al (2020) Xanthophyll: health benefits and therapeutic insights. Life Sci 240:117104

Barracosa P, Simões I, Martins AP, Barros M, Pires E (2021) Biochemical diversity of cardoon flowers (Cynara cardunculus L.): predicting PDO Mediterranean cheese textures. Food Bioscience 39:100805

Barros RGC, Andrade JKS, Pereira UC, de Oliveira CS, Rezende YRRS, Silva TOM et al (2020) Phytochemicals screening, antioxidant capacity and chemometric characterization of four edible flowers from Brazil. Food Res Int 130:108899

Begum YA, Deka SC (2019) Chemical profiling and functional properties of dietary fibre rich inner and outer bracts of culinary banana flower. J Food Sci Technol 56:5298–5308. https://doi.org/10.1007/s13197-019-04000-4

Ben-Fadhel Y, Saltaji S, Khlifi MA, Salmieri S, Dang Vu K, Lacroix M (2017) Active edible coating and γ-irradiation as cold combined treatments to assure the safety of broccoli florets (Brassica oleracea L.). Int J Food Microbiol 241:30–38. https://doi.org/10.1016/j.ijfoodmicro.2016.10.010

Benvenuti S, Bortolotti E, Maggini R (2016) Antioxidant power, anthocyanin content and organoleptic performance of edible flowers. Sci Hortic 199:170–177

Bermúdez-Bazán M, Castillo-Herrera GA, Urias-Silvas JE, Escobedo-Reyes A, Estarrón-Espinosa M (2021) Hunting bioactive molecules from the Agave genus: an update on extraction and biological potential. Molecules 26(22):6789

Bhaskar JJ, Chilkunda ND, Salimath PV (2011) Banana (Musa sp. var. elakki bale) flower and pseudostem: dietary fiber and associated antioxidant capacity. J Agric Food Chem 60:427–432. https://doi.org/10.1021/jf204539v

Bilbao-Sainz C, Sinrod AJG, Williams T, Wood D, Chiou B-S, Bridges DF, Wu VCH, Lyu C, Rubinsky B, McHugh T (2020) Preservation of Tilapia (Oreochromis aureus) Fillet by Isochoric (constant volume) freezing. J Aquat Food Prod Technol 29:1–12

Bora KS, Sharma A (2011) The genus Artemisia: a comprehensive review. Pharm Biol 49(1):101–109

Bragueto Escher G, Cardoso Borges LDC, Sousa Santos J, Mendanha Cruz T, Boscacci Marques M, Araújo Vieirado Carmo M, Zhang L (2019) From the field to the pot: phytochemical and functional analyses of Calendula officinalis L. flower for incorporation in an organic yogurt. Antioxidants 8(11):559

Carle R, Schweiggert R (2016) Handbook on natural pigments in food and beverages: industrial applications for improving food color. Woodhead Publishing, Duxford

Castro-Enríquez DD, Montaño-Leyva B, Del Toro-Sánchez CL, Juaréz-Onofre JE, Carvajal-Millan E, Burruel-Ibarra SE et al (2020) Stabilization of betalains by encapsulation—a review. J Food Sci Technol 57:1587–1600

Cefola M, Amodio ML, Colelli G (2016) Extending postharvest life of ready-to-use zucchini flowers: effects of the atmosphere composition. Acta Hortic 1141:123–130. https://doi.org/10.17660/ActaHortic.2016.1141.13

Chen GL, Chen SG, Xie YQ, Chen F, Zhao YY, Luo CX, Gao YQ (2015) Total phenolic, flavonoid and antioxidant activity of 23 edible flowers subjected to in vitro digestion. J Funct Foods 17:243–259

Chen LX, Hu DJ, Lam SC, Ge L, Wu D, Zhao J et al (2016) Comparison of antioxidant activities of different parts from snow chrysanthemum (Coreopsis tinctoria Nutt.) and identification of their natural antioxidants using high performance liquid chromatography coupled with diode array detection and mass spectrometry and 2, 2′-azinobis (3-ethylbenzthiazoline-sulfonic acid) diammonium salt-based assay. J Chromatogr A 1428:134–142

Chen GL, Chen SG, Xiao Y, Fu NL (2018) Antioxidant capacities and total phenolic contents of 30 flowers. Ind Crop Prod 111:430–445

Chen F, Zhang M, Yang CH (2020a) Application of ultrasound technology in processing of ready-to-eat fresh food: a review. Ultrason Sonochem, 63, Article 104953 September 2019. https://doi.org/10.1016/j.ultsonch.2019.104953

Chen Q, Xu B, Huang W, Amrouche AT, Maurizio B, Simal-Gandara J et al (2020b) Edible flowers as functional raw materials: a review on anti-aging properties. Trends Food Sci Technol 106:30–47

Chhabra D, Gupta RK (2015) Fortification of curd using Delonix regia flower petal extract and estimation of its phytochemical, antibacterial & antioxidant activity. J Pharmacogn Phytochem 4(3):299–307

Chiou S-Y, Sung J-M, Huang P-W, Lin S-D (2017) Antioxidant, antidiabetic, and antihypertensive properties of Echinacea purpurea flower extract and caffeic acid derivatives using in vitro models. J Med Food 20(2):171–179

Chusak C, Ying JAY, Zhien JL, Pasukamonset P, Henry CJ, Ngamukote S, Adisakwattana S (2019) Impact of Clitoria ternatea (butterfly pea) flower on in vitro starch digestibility, texture and sensory attributes of cooked rice using domestic cooking methods. Food Chem 295:646–652

Cia P, Pascholati SF, Benato EA, Camili EC, Santos CA (2007) Effects of gamma and UV-C irradiation on the postharvest control of papaya anthracnose. Postharvest Biol Technol 43(3):366–373

Colombo ML, Assisi F, Della Puppa T, Moro P, Sesana FM, Bissoli M et al (2010) Most commonly plant exposures and intoxications from outdoor toxic plants. J Pharm Sci Res 2(7):417

Coy-Barrera E (2020) Analysis of betalains (betacyanins and betaxanthins). In: Recent advances in natural products analysis. Elsevier, pp 593–619

Cunningham E (2015) What nutritional contribution does edible flowers make? J Acad Nutr Diet 115(5):856

Dabbagh Moghaddam A, Garavand F, Razavi SH, Dini Talatappe H (2018) Production of saffron-based probiotic beverage by lactic acid bacteria. J Food Meas Charact 12:2708–2717

Das AB, Goud VV, Das C (2019) Phenolic compounds as functional ingredients in beverages. In: Value-added ingredients and enrichments of beverages. Academic, pp 285–323

Demasi S, Mellano MG, Falla NM, Caser M, Scariot V (2021) Sensory profile, shelf life, and dynamics of bioactive compounds during cold storage of 17 edible flowers. Horticulturae 7(7):166

Deng J, Chen S, Yin X, Wang K, Liu Y, Li S, Yang P (2013) Systematic qualitative and quantitative assessment of anthocyanins, flavones and flavonols in the petals of 108 lotus (Nelumbo nucifera) cultivars. Food Chem 139(1–4):307–312

Desta KT, Kim GS, Hong GE, Kim YH, Lee WS, Lee SJ et al (2015) Dietary-flavonoid-rich flowers of Rumex nervosus Vahl: liquid chromatography with electrospray ionization tandem mass spectrometry profiling and in vitro anti-inflammatory effects. J Sep Sci 38(19):3345–3353

Di Gioia F, Tzortzakis N, Rouphael Y, Kyriacou MC, Sampaio SL, CFR Ferreira I, Petropoulos SA (2020) Grown to be blue—antioxidant properties and health effects of colored vegetables. Part II: leafy, fruit, and other vegetables. Antioxidants 9(2):97

Dominguez H, Muñoz MJG (eds) (2017) Water extraction of bioactive compounds: from plants to drug development. Elsevier

Duggirala KB, Mummaleti G, Kong F, Mohan A (2024) Influence of edible flowers on the physico-chemical and oxidative stability of raw ground beef patties. Front Food Sci Technol 4:1487336

Egebjerg MM, Olesen PT, Eriksen FD, Ravn-Haren G, Bredsdorff L, Pilegaard K (2018) Are wild and cultivated flowers served in restaurants or sold by local producers in Denmark safe for the consumer? Food Chem Toxicol 120:129–142

Eggersdorfer M, Wyss A (2018) Carotenoids in human nutrition and health. Arch Biochem Biophys 652:18–26

El Gharras H (2009) Polyphenols: food sources, properties and applications–a review. Int J Food Sci Technol 44(12):2512–2518

Ferdosi MF, Naseem MK, Afzal A, Khan IH, Javaid A (2023) Potential antimicrobial compounds in flower extract of Plumeria alba. Arab J Chem 16(6):104719

Fernandes L, Casal S, Pereira JA, Ramalhosa E, Saraiva JA (2017) Effect of high hydrostatic pressure (HHP) treatment on edible flowers' properties. Food Bioprocess Technol 10:799–807. https://doi.org/10.1007/s11947-017-1887-2

Fernandes L, Ramalhosa E, Pereira JA, Saraiva JA, Casal S (2018a) The unexplored potential of edible flowers lipids. Agriculture (Switzerland) 8(10):1–23. https://doi.org/10.3390/agriculture8100146

Fernandes L, Saraiva JA, Pereira JA, Casal S (2018b) Post-harvest technologies applied to edible flowers: a review. Food Rev Int 00:1–23. https://doi.org/10.1080/87559129.2018.1473422

Fernandes L, Casal S, Pereira JA, Malheiro R, Rodrigues N, Saraiva JA, Ramalhosa E (2019) Borage, calendula, cosmos, Johnny Jump up, and pansy flowers: volatiles, bioactive compounds, and sensory perception. Eur Food Res Technol 245:593–606

Fernandes L, Saraiva JA, Pereira JA, Casal S, Ramalhosa E (2019). Post-harvest technologies applied to edible flowers: A review: Edible flowers preservation. Food Reviews International, 35(2), 132-154.

Fernandes L, Casal S, Pereira JA, Saraiva JA, Ramalhosa E (2020a) An overview on the market of edible flowers. Food Rev Intl 36(3):258–275. https://doi.org/10.1080/87559129.2019.1639727

Fernandes L, Casal S, Pereira JA, Pereira EL, Saraiva JA, Ramalhosa E (2020b) Freezing of edible flowers: effect on microbial and antioxidant quality during storage. J Food Sci 85:1151–1159. https://doi.org/10.1111/1750-3841.15097

Fernández-León MF, Fernández-León AM, Lozano M, Ayuso MC, González-Gómez D (2013) Different postharvest strategies to preserve broccoli quality during storage and shelf life: controlled atmosphere and 1-MCP. Food Chem 138(1):564–573. https://doi.org/10.1016/j.foodchem.2012.09.143

Ferreira MKA, Fontenelle ROS, Magalhães FEA, Bandeira PN, Menezes JESA, dos Santos H (2018) Chalcones pharmacological potential: a brief review. Revista Virtual de Quimica 10(5):1455–1473

Fu X, Chai CL, Li YP, Li P, Luo SH, Li Q et al (2021) Metabolomics reveals abundant flavonoids in edible insect Antheraea pernyi. J Asia Pac Entomol 24(3):711–715

Gao D, Zhang Y-L, Yang F-Q, Li F, Zhang Q-H, Xia Z-N (2016) The flower of Edgeworthia gardneri (wall.) Meisn. Suppresses adipogenesis through modulation of the AMPK pathway in 3T3-L1 adipocytes. J Ethanopharmacol 191:379–386

Gao J, Sun Y, Li L, Zhou Q, Wang M (2020) The antiglycative effect of apple flowers in fructose/glucose-BSA models and cookies. Food Chem 330:127170

Garzón GA, Manns DC, Riedl K, Schwartz SJ, Padilla-Zakour O (2015) Identification of phenolic compounds in petals of nasturtium flowers (Tropaeolum majus) by high-performance liquid chromatography coupled to mass spectrometry and determination of oxygen radical absorbance capacity (ORAC). J Agric Food Chem 63(6):1803–1811

Ghosh P, Rana SS (2021) Physicochemical, nutritional, bioactive compounds and fatty acid profiling of Pumpkin flower (Cucurbita maxima), as a potential functional food. SN Applied sciences 3:1–14

Ghosh PK, Bhattacharjee P, Das S (2017) Extension of shelf life of tuberose flowers using a combination of gamma irradiation and generally regarded as safe (GRAS) preservatives and assessment of antimicrobial potency of senesced flowers. J Hortic Sci Biotechnol 92:130–145. https://doi.org/10.1080/14620316.2016.1234340

Giannetti V, Biancolillo A, Marini F, Mariani MB, Livi G (2024) Characterization of the aroma profile of edible flowers using HS-SPME/GC–MS and chemometrics. Food Res Int 178:114001

Gil MI, Conesa MA, Ferreres FU (2001) The effect of storage temperatures on vitamin C and phenolics content of artichoke ž Cynara scolymus. Innov Food Sci Emerg Technol 2:199–202. https://doi.org/10.1016/S1466-8564(01)00018-2

Gonçalves LCP, Marcato AC, Rodrigues ACB, Pagano APE, De Freitas BC, De Machado CO et al (2015) Betalains: from the colors of beetroots to the fluorescence of flowers. Revista Virtual de Quimica 7:292–309

Gonçalves F, Gonçalves JC, Ferrão AC, Correia P, Guiné RP (2020) Evaluation of phenolic compounds and antioxidant activity in some edible flowers. Open Agriculture 5(1):857–870
González-Sarrías A, Tomás-Barberán FA, García-Villalba R (2020) Structural diversity of polyphenols and distribution in foods. In: Dietary polyphenols: their metabolism and health effects. Wiley, pp 1–29
Goupy P, Vian MA, Chemat F, Caris-Veyrat C (2013) Identification and quantification of flavonols, anthocyanins and lutein diesters in tepals of Crocus sativus by ultra performance liquid chromatography coupled to diode array and ion trap mass spectrometry detections. Ind Crop Prod 44:496–510
Guerrero-Rubio MA, Escribano J, García-Carmona F, Gandía-Herrero F (2020) Light emission in betalains: from fluorescent flowers to biotechnological applications. Trends Plant Sci 25(2):159–175
Gururaja K, David M (2016) Spectroscopic signature, antibacterial and anticancer properties of Calotropis gigantea (linn.) flower. Int J Pharm Sci Res 7(4):1686–1693
Hamad MN (2016) Detection and isolation of flavonoids from Calendula officinalis (F. Asteraceae) cultivated in Iraq. Iraqi J Pharm Sci (P-ISSN 1683-3597 E-ISSN 2521-3512) 25:1–6
Herbert RB (2003) The biosynthesis of plant alkaloids and nitrogenous microbial metabolites. Nat Prod Rep 20(5):494–508
Hnin KK, Zhang M, Ju R, Wang B (2021) A novel infrared pulse spouted freeze drying on the drying kinetics, energy consumption and quality of edible rose flowers. LWT 136:110318
Honda M (2020) Nutraceutical and pharmaceutical applications of carotenoids. In: Pigments from microalgae handbook. Springer International Publishing, pp 449–469
Huang, X. L., Wang, Y., & Zhang, H. (2019). Analysis of carotenoids compounds and their biosynthesis pathways in flowers of three dendrobium species
Hussain N, Ishak I, Harith N, Kuan G (2019) Comparison of bioactive compounds and sensory evaluation on edible flowers tea infusion. Italian J Food Sci 31(2):264–273
Jacques AV, Stefanes NM, Walter LO, Perondi DM, da Luz Efe F, de Souza LFS et al (2021) Synthesis of chalcones derived from 1-naphthylacetophenone and evaluation of their cytotoxic and apoptotic effects in acute leukemia cell lines. Bioorg Chem 116:105315
Jadhav H, Annapure US, Deshmukh RR (2021) Non-thermal technologies for food processing. Front Nutr 8(6):1–14. https://doi.org/10.3389/fnut.2021.657090
Jaganath IB, Crozier A (2008) Overview of health-promoting compounds in fruit and vegetables. CRC Press
Janarny G, Ranaweera KKDS, Gunathilake KDPP (2021) Antioxidant activities of hydromethanolic extracts of Sri Lankan edible flowers. Biocatal Agric Biotechnol 35:102081
Jiang N, Doseff AI, Grotewold E (2016) Flavones: from biosynthesis to health benefits. Plants 5(2):27
Kaneko S, Chen J, Wu J, Suzuki Y, Ma L, Kumazawa K (2017) Potent odorants of characteristic floral/sweet odor in Chinese chrysanthemum flower tea infusion. J Agric Food Chem 65(46):10058–10063
Karaaslan-Ayhan N, Yaman M (2021) Determination of anthocyanins and anthocyanidins in the wild grape (Vitis sylvestris Gmelin) by high-performance liquid chromatography-diode array detection (HPLC-DAD). Instrum Sci Technol 50(1):57–68
Karimi E, Oskoueian E, Hendra R, Oskoueian A, Jaafar HZ (2012) Phenolic compounds characterization and biological activities of Citrus aurantium bloom. Molecules 17(2):1203–1218
Kashif M, Bano S, Naqvi S, Faizi S, Lubna, Ahmed Mesaik M, Azeemi KS, Farooq AD (2015) Cytotoxic and antioxidant properties of phenolic compounds from Tagetes patula flower. Pharm Biol 53(5):672–681
Kassem ME, Marzouk MM, Mostafa AA, Khalil WK, Booles HF (2017) Phenolic constituents of Trifolium resupinatum var. minus: protection against rosiglitazone induced osteoporosis in type 2 diabetic male rats. J Appl Pharma Sci 7(05):174–183
Kelley KM, Behe BK, Biernbaum JA (2001a) Consumer and professional chef perceptions of three edible-flower species. HortScience 36(1):162–166. https://doi.org/10.21273/hortsci.36.1.162

Kelley KM, Behe BK, Biernbaum JA, Poff KL (2001b) Consumer preference for edible-flower color, container size, and price. HortScience 36(4):801–804. https://doi.org/10.21273/hortsci.36.4.801

Kelley KM, Cameron AC, Biernbaum JA, Poff KL (2003) Effect of storage temperature on the quality of edible flowers. Postharvest Biol Technol 27:341–344. https://doi.org/10.1016/S0925-5214(02)00096-0

Keskin C (2015) Antioxidant, anticancer and anticholinesterase activities of flower, fruit and seed extracts of Hypericum amblysepalum HOCHST. Asian Pac J Cancer Prev 16(7):2763–2769

de Keukeleire D (2000) Fundamentals of beer and hop chemistry. Quim Nova 23:108–112

Khan MKI, Cakmak H, Tavman S, Schutyser M, Schroën K (2014) Anti- browning and barrier properties of edible coatings prepared with electrospraying. Innov Food Sci Emerg Technol 25:9–13. https://doi.org/10.1016/j.ifset.2013.10.006

Khan IA, Liu D, Yao M, Memon A, Huang J, Huang M (2019) Inhibitory effect of Chrysanthemum morifolium flower extract on the formation of heterocyclic amines in goat meat patties cooked by various cooking methods and temperatures. Meat Sci 147:70–81

Kim YS, Hwang JW, Sung SH, Jeon YJ, Jeong JH, Jeon BT et al (2015) Antioxidant activity and protective effect of extract of Celosia cristata L.flower on tert-butyl hydroperoxide-induced oxidative hepatotoxicity. Food Chem 168:572–579

Kim G-C, Kim J-S, Kim G-M, Choi S-Y (2017) Anti-adipogenic effects of Tropaeolum majus (nasturtium) ethanol extract on 3T3-L1 cells. Food Nutr Res 61(1):1339555

Kim H-G, Oh H-J, Ko J-H, Song HS, Lee Y-G, Kang SC et al (2019) Lanceoleins A-G, hydroxychalcones, from the flowers of Coreopsis lanceolata and their chemopreventive effects against human colon cancer cells. Bioorg Chem 85:274–281

Kim EY, Lee SU, Kim YH (2022) 1, 2, 3, 4, 6-Penta-O-galloyl-β-D-glucose inhibits CD44v3, a cancer stem cell marker, by regulating its transcription factor, in human pancreatic cancer cell line. Anim Cells Syst 26(6):328–337

Kimura M, Rodriguez-Amaya DB, Amaya-Farfan J (2005) Carotenoid composition of Brazilian fruits and vegetables. In: I International symposium on human health effects of fruits and vegetables, vol 744, pp 409–416

Kishimoto S, Oda-Yamamizo C, Ohmiya A (2020) Heterologous expression of xanthophyll esterase genes affects carotenoid accumulation in petunia corollas. Sci Rep 10(1):1299

Koike ACR (2019) Antioxidant activity of Dianthus chinensis flowers processed by ionizing radiation. Brazilian J Radiat Sci 7. https://bjrs.org.br/revista/index.php/revista/article/view/674

Koike A, Barreira JC, Barros L, Santos-Buelga C, Villavicencio AL, Ferreira IC (2015) Edible flowers of Viola tricolor L. as a new functional food: antioxidant activity, individual phenolics and effects of gamma and electron-beam irradiation. Food Chem 179:6–14

Komora N, Bruschi C, Ferreira V, Maciel C, Brandão TRS, Fernandes R, Saraiva JA, Castro SM, Teixeira P (2018) The protective effect of food matrices on listeria lytic bacteriophage p100 application towards high pressure processing. Food Microbiol 76:416–425. https://doi.org/10.1016/j.fm.2018.07.002

Kou L, Turner ER, Luo Y (2012) Extending the shelf life of edible flowers with controlled release of 1-methylcyclopropene and modified atmosphere packaging. J Food Sci 77(5):188–193. https://doi.org/10.1111/j.1750-3841.2012.02683.x

Kowalczewski PŁ, Pauter P, Smarzyński K, Różańska MB, Jeżowski P, Dwiecki K, Mildner-Szkudlarz S (2019) Thermal processing of pasta enriched with black locust flowers affect quality, phenolics, and antioxidant activity. J Food Process Preserv 43(10):e14106

Krzymińska A, Gąsecka M, Magdziak Z (2020) Content of phenolic compounds and organic acids in the flowers of selected Tulipa gesneriana cultivars. Molecules 25(23):5627

Kumar A, Kaur A, Tomer V, Gupta K, Kaur K (2021) Effect of rose syrup and marigold powder on the physicochemical, phytochemical, sensorial and storage properties of nutricereals and milk-based functional beverage. J Am Coll Nutr 40(2):133–140

Kumari P, Bhargava B (2021) Phytochemicals from edible flowers: opening a new arena for healthy lifestyle. J Funct Foods 78:104375

Kwon J-H, Oh H-J, Lee D-S, In S-J, Seo K-H, Jung J-W et al (2019) Pharmacological activity and quantitative analysis of flavonoids isolated from the flowers of Begonia semperflorens Link et Otto. Appl Biol Chem 62(1):11
Lakshan SAT, Jayanath NY, Abeysekera WPKM, Abeysekera WKSM (2019) A commercial potential blue pea (Clitoria ternatea L.) flower extract incorporated beverage having functional properties. Evid Based Complement Alternat Med 2019(1):2916914
Li AN, Li S, Li HB, Xu DP, Xu XR, Chen F (2014) Total phenolic contents and antioxidant capacities of 51 edible and wild flowers. J Funct Foods 6:319–330
Li W, Shi M, Wang P, Guo X, Li C, Kang W (2018) Efficient determination of three flavonoids in Malus pumila flowers by ionic liquid-HPLC. J Mol Liq 263:139–146
Li Y, Hao Y, Gao B, Geng P, Huang H, Yu L et al (2019) Chemical profile and in vitro gut microbiota modulatory, anti-inflammatory and free radical scavenging properties of Chrysanthemum morifolium cv Fubaiju. J Funct Foods 58:114–122
Li J, Hossain MS, Ma H, Yang Q, Gong X, Yang P, Feng B (2020) Comparative metabolomics reveals differences in flavonoid metabolites among different coloured buckwheat flowers. J Food Compos Anal 85:103335
Lim TK (2014) Edible medicinal and non-medicinal plants, vol 8. Flowers
de Lima Franzen F, Rodríguez de Oliveira MS, Lidório HF, Farias Menegaes J, Martins Fries LL (2019) Composición química de pétalos de flores de rosa, girasol y caléndula para su uso en la alimentación humana. Ciencia y Tecnología Agropecuaria 20(1):149–168
Lin YS, Wu SS, Lin JK (2003) Determination of tea polyphenols and caffeine in tea flowers (Camellia sinensis) and their hydroxyl radical scavenging and nitric oxide suppressing effects. J Agric Food Chem 51(4):975–980
Liu JR, Chen BQ, Yang BF, Dong HW, Sun CH, Wang Q et al (2004) Apoptosis of human gastric adenocarcinoma cells induced by β-ionone. World J Gastroenterol 10(3):348–351
Liu Z-X, Liu C-T, Liu Q-B, Ren J, Li L-Z, Huang X-X et al (2015) Iridoid glycosides from the flower buds of Lonicera japonica and their nitric oxide production and α-glucosidase inhibitory activities. J Funct Foods 18:512–519
Liu Z, Cheng Z, He Q, Lin B, Gao P, Li L et al (2016) Secondary metabolites from the flower buds of Lonicera japonica and there in vitro anti-diabetic activities. Fitoterapia 110:44–51
Loizzo MR, Pugliese A, Bonesi M, Tenuta MC, Menichini F, Xiao J, Tundis R (2015) Edible flowers: a rich source of phytochemicals with antioxidant and hypoglycemic properties. J Agric Food Chem 64:2467–2474
Loizzo MR, Pugliese A, Bonesi M, Tenuta MC, Menichini F, Xiao J, Tundis R (2016) Edible flowers: a rich source of phytochemicals with antioxidant and hypoglycemic properties. J Agric Food Chem 64:2467–2474
Lu B, Li M, Yin R (2016) Phytochemical content, health benefits, and toxicology of common edible flowers: a review (2000–2015). Crit Rev Food Sci Nutr 56(sup1):S130–S148. https://doi.org/10.1080/10408398.2015.1078276
Lu T, Shen Y, Wang JH, Xie HK, Wang YF, Zhao Q et al (2020) Improving oxidative stability of flaxseed oil with a mixture of antioxidants. J Food Process Preserv 44(3):e14355
Luan F, Wu Q, Yang Y, Lv H, Liu D, Gan Z, Zeng N (2020) Traditional uses, chemical constituents, biological properties, clinical settings, and toxicities of Abelmoschus manihot L.: a comprehensive review. Front Pharmacol 11:1068
Luyen BT, Tai BH, Thao NP, Lee YM, Lee SH, Jang HD, Kim YH (2015) The anti-osteoporosis and antioxidant activities of chemical constituents from Chrysanthemum indicum flowers. Phytother Res 29(4):540–548
Lv R, Zou M, Chantapakul T, Chen W, Muhammad AI, Zhou J, Ding T, Ye X, Liu D (2019) Effect of ultrasonication and thermal and pressure treatments, individually and combined, on inactivation of Bacillus cereus spores. Appl Microbiol Biotechnol 103:2329–2338. https://doi.org/10.1007/s00253-018-9559-3
Madane P, Das AK, Pateiro M, Nanda PK, Bandyopadhyay S, Jagtap P, Barba FJ, Shewalkar A, Maity B, Lorenzo JM (2019) Drum-stick (Moringa oleifera) flower as an antioxidant dietary fibre in chicken meat nuggets. Foods 8(8):307

Magri A, Adiletta G, Petriccione M (2020) Evaluation of antioxidant systems and ascorbate-glutathione cycle in feijoa edible flowers at different flowering stages. Foods 9:95. https://doi.org/10.3390/foods9010095

Manivannan A, Narasegowda S, Prakash T (2021) Comparative study on color coordinates, phenolics, flavonoids, carotenoids, and antioxidant potential of marigold (Tagetes sp.) with diverse colored petals. J Food Meas Charact 15:4343–4353

Marchioni I, Najar B, Ruffoni B, Copetta A, Pistelli L, Pistelli L (2020) Bioactive compounds and aroma profile of some Lamiaceae edible flowers. Plants 9(6):691

Maria Leena M, Yoha KS, Moses JA, Anandharamakrishnan C (2020) Edible coating with resveratrol loaded electrospun zein nanofibers with enhanced bioaccessibility. Food Biosci 36:100669. https://doi.org/10.1016/j.fbio.2020.100669

Mary Mawumenyo Mamattah K, Kusiwaa Adomako A, Nketia Mensah C, Borquaye LS (2023) Chemical characterization, antioxidant, antimicrobial, and antibiofilm activities of essential oils of Plumeria alba (forget-me-not). Biochem Res Int 2023(1):1040478–1040410

Matyjaszczyk E, Śmiechowska M (2019) Edible flowers. benefits and risks pertaining to their consumption. Trends Food Sci Technol 91:670–674

Mazza G (2007) Anthocyanins and heart health. Annali-Istituto Superiore Di Sanita 43(4):369

Mezzomo N, Ferreira SR (2016) Carotenoids functionality, sources, and processing by supercritical technology: a review. J Chem 2016(1):3164312–3164316

Miao J, Liu J, Gao X, Lu F, Yang X (2022) Effects of different drying methods on chemical compositions, antioxidant activity and anti-α-glucosidase activity of Coreopsis tinctoria flower tea. Heliyon 8(11):e11784

Micucci M, Malaguti M, Toschi TG, Di Lecce G, Aldini R, Angeletti A, Chiarini A, Budriesi R, Hrelia S (2015) Cardiac and vascular synergic protective effect of Olea europea L leaves and Hibiscus sabdariffa L flower extracts. Oxid Med Cell Longev 2015:318125

Miguel MG (2018) Betalains in some species of the Amaranthaceae family: a review. Antioxidants 7(4):53

Mlcek J, Rop O (2011) Fresh edible flowers of ornamental plants–a new source of nutraceutical foods. Trends Food Sci Technol 22(10):561–569

Mohi Alden K, Omid M, Rajabipour A, Tajeddin B, Soltani Firouz M (2019) Quality and shelf-life prediction of cauliflower under modified atmosphere packaging by using artificial neural networks and image processing. Comput Electron Agric 163. https://doi.org/10.1016/j.compag.2019.104861

de Morais JS, Sant'Ana AS, Dantas AM, Silva BS, Lima MS, Borges GC, Magnani M (2020) Antioxidant activity and bioaccessibility of phenolic compounds in white, red, blue, purple, yellow and orange edible flowers through a simulated intestinal barrier. Food Res Int 131:109046

Mulík S, Ozuna C (2020) Mexican edible flowers: cultural background, traditional culinary uses, and potential health benefits. Int J Gastron Food Sci 21:100235. https://doi.org/10.1016/j.ijgfs.2020.100235

Murkovic M (2016) Phenolic compounds: occurrence, classes, and analysis. Elsevier

Nagula RL, Wairkar S (2019) Recent advances in topical delivery of flavonoids: a review. J Control Release 296:190–201

Najar B, Marchioni I, Ruffoni B, Copetta A, Pistelli L, Pistelli L (2019) Volatilomic analysis of four edible flowers from Agastache genus. Molecules 24(24):4480

Nakayama M (2020) Flower pigments responsible for cyanic, yellow, and cream-white coloration in carnation. In: The carnation genome. Springer, pp 61–79

Nanda BL (2019) Antioxidant and anticancer activity of edible flowers. J Drug Deliv Ther 9(3-s):290–295

Navarro-González I, González-Barrio R, García-Valverde V, Bautista-Ortín AB, Periago MJ (2014) Nutritional composition and antioxidant capacity in edible flowers: characterisation of phenolic compounds by HPLC-DAD-ESI/MSn. Int J Mol Sci 16(1):805–822

Newman S, Connor A (2020) Edible flowers. Colorado University

Nguyen C, Baskaran K, Pupulin A, Ruvinov I, Zaitoon O, Grewal S et al (2019) Hibiscus flower extract selectively induces apoptosis in breast cancer cells and positively interacts with common chemotherapeutics. BMC Complement Altern Med 19(1):98

Nie G, Liu N, Zhang E, Zhao R, Zhang X, Zhu X, Li H, Nie Z, Yue W (2017) Preparation of a novel mixed milk with nattokinase produced by Bacillus subtilis (natto). J Food Process Preserv 41(6):e13284

Ogundele OM, Awolu OO, Badejo AA, Nwachukwu ID, Fagbemi TN (2016) Development of functional beverages from blends of Hibiscus sabdariffa extract and selected fruit juices for optimal antioxidant properties. Food Sci Nutr 4(5):679–685

Oliveira LDLD, Carvalho MVD, Melo L (2014) Health promoting and sensory properties of phenolic compounds in food. Revista Ceres 61:764–779.193

Oun R, Moussa YE, Wheate NJ (2018) The side effects of platinum-based chemotherapy drugs: a review for chemists. Dalton Trans 47(19):6645–6653

Pardede A, Mashita K, Ninomiya M, Tanaka K, Koketsu M (2016) Flavonoid profile and anti-leukemic activity of Coreopsis lanceolata flowers. Bioorg Med Chem Lett 26(12):2784–2787

Park CH, Chae SC, Park SY, Kim JK, Kim YJ, Chung SO et al (2015) Anthocyanin and carotenoid contents in different cultivars of chrysanthemum (Dendranthema grandiflorum Ramat.) flower. Molecules 20(6):11090–11102

Park YJ, Park SY, Valan Arasu M, Al-Dhabi NA, Ahn HG, Kim JK, Park SU (2017) Accumulation of carotenoids and metabolic profiling in different cultivars of Tagetes flowers. Molecules 22(2):313

de Pascual-Teresa S, Sanchez-Ballesta MT (2008) Anthocyanins: from plant to health. Phytochem Rev 7:281–299. https://doi.org/10.1007/s11101-007-9074-0

Peng A, Lin L, Zhao M, Sun B (2019) Classification of edible chrysanthemums based on phenolic profiles and mechanisms underlying the protective effects of characteristic phenolics on oxidatively damaged erythrocyte. Food Res Int 123:64–74

Phaniendra A, Jestadi DB, Periyasamy L (2015) Free radicals: properties, sources, targets, and their implication in various diseases. Indian J Clin Biochem 30(1):11–26

Pinakin DJ, Kumar V, Suri S, Sharma R, Kaushal M (2020) Nutraceutical potential of tree flowers: a comprehensive review on biochemical profile, health benefits, and utilization. Food Res Int 127:108724

Pinton SC, Pinton SC, Bardsley CA, Marik CM, Boyer RR, Strawn LK (2020) Fate of Listeria monocytogenes on broccoli and cauliflower under different storage temperatures type of manuscript: research paper. J Food Prot 83:858–864

Pires TCSP, Dias MI, Barros L, Ferreira ICFR (2017) Nutritional and chemical characterization of edible petals and corresponding infusions: valorization as new food ingredients. Food Chem 220:337–343. https://doi.org/10.1016/j.foodchem.2016.10.026

Pires TC, Dias MI, Barros L, Barreira JC, Santos-Buelga C, Ferreira IC (2018a) Incorporation of natural colorants obtained from edible flowers in yogurts. Lwt 97:668–675

Pires TC, Dias MI, Barros L, Santos Buelga C, Ferreira IC (2018b) Dahlia mignon as a source of soluble sugars and glycosilated flavonoids with bioactive properties. UNIFood Conference. http://hdl.handle.net/10198/18131

Pires TC, Dias MI, Barros L, Calhelha RC, Alves MJ, Oliveira MBP et al (2018c) Edible flowers as sources of phenolic compounds with bioactive potential. Food Res Int 105:580–588

Pires TC, Barros L, Santos-Buelga C, Ferreira IC (2019) Edible flowers: emerging components in the diet. Trends Food Sci Technol 93:244–258

Preethi M, Sritharan N, Raveendran M, Senthil A (2019) Studies on extraction of betalain pigments by different solvents and assessing antioxidant activity of Bougainvillea spectabilis and Celosia argentea flowers. Madras Agric J 106(march (1–3)):1

Putri FA, Naufalin R, Wicaksono R (2019) Antimicrobial edible coating application of Kecombrang flower concentrate to reduce microbial growth on gourami fish sausage. In: Proceedings of the IOP conference series: earth and environmental science, p 12056. https://doi.org/10.1088/1755-1315/250/1/012056

Qin T, Liu B, Xu Z, Yao G, Xu H, Zhao C (2021) Flavonol-based small-molecule fluorescent probes. Sensors Actuators B Chem 336:129718

Qiu L, Zhang M, Mujumdar AS, Chang L (2021) Effect of edible rose (Rosa rugosa cv. Plena) flower extract addition on the physico-chemical, rheological, functional and sensory properties of settype yogurt. Food Bioscience 43:101249

Ran Y, Qingmin C, Maorun F (2019) Chlorine dioxide generation method and its action mechanism for removing harmful substances and maintaining quality attributes of agricultural products. Food Bioprocess Technol 12(7):1110–1122

Rao V, Poonia A (2024a) Bioactive compounds, nanoparticles synthesis, health benefits and potential utilization of edible flowers for the development of functional dairy products: a review. J Food Sci Technol 61(6):1053–1068. https://doi.org/10.1007/s13197-023-05853-6

Rao V, Poonia A (2024b) Microbial biosynthesis of selenium nanoparticles using probiotic strain and its characterization. J Food Meas Charact 18(7):5477–5488. https://doi.org/10.1007/s11694-024-02581-z

Rasouli H, Yarani R, Pociot F, Popović-Djordjević J (2020) Anti-diabetic potential of plant alkaloids: revisiting current findings and future perspectives. Pharmacol Res 155:104723

Ribeito D, Freitas M, Silva AM, Carvalho F, Fernandes E (2018) Antioxidant and pro-oxidant activities of carotenoids and their oxidation products. Food Chem Toxicol 120:681–699

Ritonga NJ, Setiani O, Umaroh KB, Amri F (2017) Roselle flower (Hibiscus sabdariffa) in the treatment of hypertension in postpartum mothers. Belitung Nurs J 3(3):229–237

Rivera-Madrid R, Carballo-Uicab VM, Cárdenas-Conejo Y, Aguilar-Espinosa M, Siva R (2020) Overview of carotenoids and beneficial effects on human health. In: Carotenoids: properties, processing and applications. Academic, pp 1–40

Rodrigues H, Cielo DP, Goméz-Corona C, Silveira AAS, Marchesan TA, Galmarini MV, Richards NSPS (2017) Eating flowers? Exploring attitudes and consumers' representation of edible flowers. Food Res Int 100:227–234. https://doi.org/10.1016/j.foodres.2017.08.018

Rodrigues DB, Mercadante AZ, Mariutti LRB (2019) Marigold carotenoids: much more than lutein esters. Food Res Int 119:653–664

Roriz CL, Barros L, Prieto MA, Morales P, Ferreira IC (2017) Floral parts of Gomphrena globosa L. as a novel alternative source of betacyanins: optimization of the extraction using response surface methodology. Food Chem 229:223–234

Rozmer Z, Perjési P (2016) Naturally occurring chalcones and their biological activities. Phytochem Rev 15:87–120

Sadowska-Bartosz I, Bartosz G (2021) Biological properties and applications of betalains. Molecules 26(9):2520

Saleem H, Htar TT, Naidu R, Anwar S, Zengin G, Locatelli M, Ahemad N (2020) HPLC–PDA polyphenolic quantification, UHPLC–MS secondary metabolite composition, and in vitro enzyme inhibition potential of Bougainvillea glabra. Plants 9(3):388

Salih SH, Alkhazraji TO, Ibrahim KMS (2016) Alkaloid contents and anti-oxidant effectiveness of extracts of the Orobanche aegyptiaca L. and Orobanche ramosa L.(Orobanchaceae). J Plant Prod 7(2):299–304

Sam FE, Ma T, Wang J, Liang Y, Sheng W, Li J, Zhang B (2023) Aroma improvement of dealcoholized Merlot red wine using edible flowers. Food Chem 404:134711

Sánchez-vega R, Elez-martínez P, Martín-belloso O (2014) Influence of high-intensity pulsed electric field processing parameters on antioxidant compounds of broccoli juice. Innov Food Sci Emerg Technol 29. https://doi.org/10.1016/j.ifset.2014.12.002

dos Santos Nascimento LB, Gori A, Raffaelli A, Ferrini F, Brunetti C (2021) Phenolic compounds from leaves and flowers of Hibiscus roseus: potential skin cosmetic applications of an under-investigated species. Plants 10(3):522

Santos EM, Rodriguez JA, Lorenzo JM, Mondragón AC, Pateiro M, Gutiérrez E, Ferreira TA (2022) Antioxidant effect of pumpkin flower (Cucurbita maxima) in chicken patties. Foods 11(15):2258

Sanz-Puig M, Santos-Carvalho L, Cunha LM, Pina-Pérez MC, Rodrigo D, Martínez-López A (2016) Antimicrobial capacity of a cauliflower by-product infusion combined with the PEF treatment against S. Typhimurium. In: 1st world congress on electroporation and pulsed electric fields in biology, medicine and food & environmental technologies: Portorož, Slovenia, September 6–10, 2015 (pp. 132-135). Springer. https://doi.org/10.1007/978-981-287-817-5_29
Scalzini G, Giacosa S, Segade SR, Paissoni MA, Rolle L (2021) Effect of withering process on the evolution of phenolic acids in winegrapes: a systematic review. Trends Food Sci Technol 116:545–558
Shen JX, Rana MM, Liu GF, Ling TJ, Gruber MY, Wei S (2017) Differential contribution of jasmine floral volatiles to the aroma of scented green tea. J Food Qual 2017(1):5849501
Sindi HA, Marshall LJ, Morgan MR (2014) Comparative chemical and biochemical analysis of extracts of Hibiscus sabdariffa. Food Chem 164:23–29
Singh S, Rai AK, Alam T, Singh B (2018) Influence of modified atmosphere packaging (MAP) on the shelf life and quality of broccoli during storage. J Packag Technol Res 2(2):105–113. https://doi.org/10.1007/s41783-018-0030-9
Siriamornpun S, Kaisoon O, Meeso N (2012) Changes in colour, antioxidant activities and carotenoids (lycopene, β-carotene, lutein) of marigold flower (Tagetes erecta L.) resulting from different drying processes. J Funct Foods 4(4):757–766. https://doi.org/10.1016/j.jff.2012.05.002
Sogo T, Terahara N, Hisanaga A, Kumamoto T, Yamashiro T, Wu S et al (2015) Anti-inflammatory activity and molecular mechanism of delphinidin 3-sam-bubioside, a Hibiscus anthocyanin. Biofactors 41(1):58–65
Srivastava JK, Shankar E, Gupta S (2010) Chamomile: a herbal medicine of the past with bright future. Mol Med Rep 3(6):895–901
Tahir MS, Almezgagi M, Zhang Y, Bashir A, Abdullah HM, Gamah M et al (2021) Mechanistic new insights of flavonols on neurodegenerative diseases. Biomed Pharmacother 137:111253
Tena N, Martín J, Asuero AG (2020) State of the art of anthocyanins: antioxidant activity, sources, bioavailability, and therapeutic effect in human health. Antioxidants 9(5):451
Tettey CO, Lincha VR, Lee DU, Yang IJ, Shin HM (2016) Anti-inflammatory effects of the flowers of Prunus persica var. davidiana. J Food Biochem 40(2):227–234
Tiwary BK, Dutta S, Dey P, Hossain M, Kumar A, Bihani S et al (2017) Radical scavenging activities of Lagerstroemia speciosa (L.) Pers. petal extracts and its hepato-protection in CCl 4-intoxicated mice. BMC Complement Altern Med 17(1):55
Tsai P-J, Chang M-L, Hsin C-M, Chuang C-C, Chuang L-T, Wu W-H (2017) Antilipotoxicity activity of Osmanthus fragrans and Chrysanthemum morifolium flower extracts in hepatocytes and renal glomerular mesangial cells. Mediat Inflamm 2017:4856095–4856012
Valencia-Cordova MG, Jaguey-Hernández Y, Castañeda-Ovando A, González-Olivares LG, Castañeda-Ovando EP, Añorve-Morga J, de la O-Arciniega, M. (2024) Lesser-explored edible flowers as a choice of phytochemical sources for food applications. Int J Food Sci 2024(1):9265929
Veigas JM, Divya P, Neelwarne B (2012) Identification of previously unreported pigments among carotenoids and anthocyanins in floral petals of Delonix regia (Hook.) Raf. Food Res Int 47(1):116–123
Villavicencio ALCH, Heleno SA, Calhelha RC, Santos-buelga C, Barros L, Ferreira ICFR (2018) The in fluence of electron beam radiation in the nutritional value, chemical composition and bioactivities of edible flowers of Bauhinia variegata L. var. candida alba Buch-Ham from Brazil. Food Chem 241:163–170. https://doi.org/10.1016/j.foodchem.2017.08.093
Vinokur Y, Rodov V, Reznick N, Goldman G, Horev B, Umiel N, Friedman H (2006) Rose petal tea as an antioxidant-rich beverage: cultivar effects. J Food Sci 71(1):42–47. https://doi.org/10.1111/j.1365-2621.2006.tb12404.x
Vimala Bharathi SK, Maria Leena M, Moses JA, Anandharamakrishnan C (2020) Zein-based anti-browning cling wraps for fresh-cut apple slices. Int J Food Sci Technol 55:1238–1245
Vioque M, Gómez R, Sánchez E, Mata C, Tejada L, Fernández-Salguero J (2000) Chemical and microbiological characteristics of ewes' milk cheese manufactured with extracts from flowers of Cynara cardunculus and Cynara humilis as coagulants. J Agric Food Chem 48(2):451–456

Wan C, Yuan T, Cirello AL, Seeram NP (2012) Antioxidant and α-glucosidase inhibitory phenolics isolated from highbush blueberry flowers. Food Chem 135(3):1929–1937

Wang F, Miao M, Xia H, Yang LG, Wang SK, Sun GJ (2016) Antioxidant activities of aqueous extracts from 12 Chinese edible flowers in vitro and in vivo. Food Nutr Res 61(1):19. https://doi.org/10.1080/16546628.2017.1265324

Wang L, Tan N, Hu J, Wang H, Duan D, Ma L et al (2017) Analysis of the main active ingredients and bioactivities of essential oil from Osmanthus fragrans Var. thunbergii using a complex network approach. BMC Syst Biol 11(1):144

Wang L, Wen M, Chen F, Luo Z, Yin J, Chen Y et al (2020) High oxygen atmospheric packaging (HOAP) reduces H2O2 production by regulating the accumulation of oxidative stress-related proteins in Chinese flowering cabbage. Postharvest Biol Technol 165:111183. https://doi.org/10.1016/j.postharvbio.2020.111183

Watanabe T, Yamagaki T, Azuma T, Horikawa M (2021) Distinguishing between isomeric neoxanthin and violaxanthin esters in yellow flower petals using liquid chromatography/photodiode array atmospheric pressure chemical ionization mass spectrometry and tandem mass spectrometry. Rapid Commun Mass Spectrom 35(15):e9142

Wendel M, Szot D, Starzak K, Tuwalska D, Gapinski J, Naskrecki R et al (2015) Photophysical properties of betaxanthins: Vulgaxanthin I in aqueous and alcoholic solutions. J Lumin 167:289–295

Westphal A, Riedl KM, Cooperstone JL, Kamat S, Balasubramaniam VM, Schwartz SJ, Bo V (2017) High-pressure processing of broccoli sprouts: influence on bioactivation of Glucosinolates to Isothiocyanates. J Agric Food Chem 65(39):8578–8585. https://doi.org/10.1021/acs.jafc.7b01380

Xie DD (2015) Optimization of coating formulation for vegetable night-fragrant flower fresh keeping. Northern Horticulturae 12:127–131

Xin Y, Zhang M, Adhikari B (2013) Effect of trehalose and ultrasound-assisted osmotic dehydration on the state of water and glass transition temperature of broccoli (Brassica oleracea l. var. botrytis l.). J Food Eng 119. https://doi.org/10.1016/j.jfoodeng.2013.06.035

Xiong L, Yang J, Jiang Y, Lu B, Hu Y, Zhou F et al (2014) Phenolic compounds and antioxidant capacities of 10 common edible flowers from China. J Food Sci 79(4):C517–C525

Xiong L, Mao S, Lu B, Yang J, Zhou F, Hu Y et al (2016) Osmanthus fragrans flower extract and acteoside protect against d-galactose-induced aging in an ICR mouse model. J Med Food 19(1):54–61

Xu D-P, Zheng J, Zhou Y, Li Y, Li S, Li H-B (2017) Ultrasound-assisted extraction of natural antioxidants from the flower of Limonium sinuatum: optimization and comparison with conventional methods. Food Chem 217:552–559

Xu K, Zhang M, Fang Z, Wang B (2021) Degradation and regulation of edible flower pigments under thermal processing: a review. Crit Rev Food Sci Nutr 61(6):1038–1048

Xue Q, Fan H, Yao F, Cao X, Liu M, Sun J, Liu Y (2020) Transcriptomics and targeted metabolomics profilings for elucidation of pigmentation in Lonicera japonica flowers at different developmental stages. Ind Crop Prod 145:111981

Yamagishi M, Kishimoto S, Nakayama M (2010) Carotenoid composition and changes in expression of carotenoid biosynthetic genes in tepals of Asiatic hybrid lily. Plant Breed 129(1):100–107

Yang L, Cao YL, Jiang JG, Lin QS, Chen J, Zhu L (2010) Response surface optimization of ultrasound-assisted flavonoids extraction from the flower of Citrus aurantium L. var. amara Engl. J Sep Sci 33(9):1349–1355

Yang R, Guan Y, Wang W, Chen H, He Z, Jia AQ (2018) Antioxidant capacity of phenolics in Camellia nitidissima Chi flowers and their identification by HPLC Triple TOF MS/MS. PLoS One 13(4):e0195508

Yang P-F, Yang Y-N, Feng Z-M, Jiang J-S, Zhang P-C (2019) Six new compounds from the flowers of Chrysanthemum morifolium and their biological activities. Bioorg Chem 82:139–144

Yang H, Shin Y (2017) Antioxidant compounds and activities of edible roses (Rosa hybrida spp.) from different cultivars grown in Korea. Appl Biol Chem 60(2):129–136

Yeon JY, Kim WS (2020) Floral pigment-scent associations in eight cut rose cultivars with various petal colors. Hortic Environ Biotechnol 61:633–641

Yisimayili Z, Abdulla R, Tian Q, Wang Y, Chen M, Sun Z et al (2019) A comprehensive study of pomegranate flowers polyphenols and metabolites in rat biological samples by high-performance liquid chromatography quadrupole time-of-flight mass spectrometry. J Chromatogr A 1604:460472

Youssef HA, Ali SM, Sanad MI, Dawood DH (2020) Chemical investigation of flavonoid, phenolic acids composition and antioxidant activity of Tagetes erecta flowers. Egypt J Chem 63(7):2605–2615

Yuan R, Ma S, Zhu X, Li J, Liang Y, Liu T et al (2016) Core level regulatory network of osteoblast as molecular mechanism for osteoporosis and treatment. Oncotarget 7(4):3692–3701

Zakari A, Kubmarawa D, Isyaka SM, Kendeson CA, Baba EB (2018) In vivo evaluation of anti-plasmodial properties of Myosotis scorpioides L.(Boraginaceae) extract in albino mice infected with Plasmodium berghei. Trop J Nat Prod Res 2:198–202

Zannou O, Kelebek H, Selli S (2020) Elucidation of key odorants in Beninese Roselle (Hibiscus sabdariffa L.) infusions prepared by hot and cold brewing. Food Res Int 133:109133

Zhang J, Shen PP et al (2003) The toxicological assessment of Lonicera Japonica on food safety. The Chinese Acad Med Mag Organ 02:63–64

Zhang N, He Z, He S, Jing P (2019) Insights into the importance of dietary chrysanthemum flower (Chrysanthemum morifolium cv. Hangju)-wolfberry (Lyciumbarbarum fruit) combination in antioxidant and anti-inflammatory properties. Food Res Int 116:810–818

Zhang J, Hao T, Liang C, Liu Z (2022) Extraction and addition process of Osmanthus fragrans in Osmanthus fragrans beer. Liquor-Making Sci Technol 341(11):36–44

Zhao X, Mei L, Pei J, Liu Z, Shao Y, Tao Y et al (2017) Sophoridine from sophora flower attenuates ovariectomy induced osteoporosis through the RANKLERK-NFAT pathway. J Agric Food Chem 65(44):9647–9654

Zhao L, Fan H, Zhang M, Chitrakar B, Wang B (2019) Edible flowers: review of flower processing and extraction of bioactive compounds by novel technologies. Food Res Int 108660. https://doi.org/10.1016/j.foodres.2019.108660

Zheng J, Meenu M, Xu B (2019) A systematic investigation on free phenolic acids and flavonoids profiles of commonly consumed edible flowers in China. J Pharm Biomed Anal 172:268–277. https://doi.org/10.1016/j.jpba.2019.05.007

Zheng J, Lu B, Xu B (2021) An update on the health benefits promoted by edible flowers and involved mechanisms. Food Chem 340:127940

Zhu C, Yamamura S, Nishihara M, Koiwa H, Sandmann G (2003) cDNAs for the synthesis of cyclic carotenoids in petals of Gentiana lutea and their regulation during flower development. Biochimica et Biophysica Acta (BBA)-Gene Structure and Expression 1625(3):305–308

Zhu C, Bai C, Sanahuja G, Yuan D, Farré G, Naqvi S et al (2010) The regulation of carotenoid pigmentation in flowers. Arch Biochem Biophys 504(1):132–141

Zhu Z, Zhang P, Sun DW (2020) Effects of multi-frequency ultrasound on freezing rates and quality attributes of potatoes. Ultrason Sonochem 60:104733. https://doi.org/10.1016/j.ultsonch.2019.104733

Chapter 13
Health Benefits/Claims of Edible Flowers

Muhammad Afzaal, Muhammad Bilal Hussain, Farhan Saeed, Tuba Mustafa, Marwa Waheed, Muhammad Usman Butt, Iqra Nawaz, Bisma Rasheed, Maryam Ehsan, and Aneeba Busharat

13.1 Introduction

Generally, there is an increasing demand for delicious and aesthetically pleasing food and food products. The visual presentation of dishes as well as the quality of ingredients used are the key factors in the overall eating experience (Bordewijk and Schifferstein 2020). For this reason, edible flowers are becoming a common approach to enhance the visual appeal of individual dishes. In addition to their aesthetic appeal, these flowers are grown for their therapeutic, cosmetic, nutritional, and fragrant properties. Because they are consumed raw and unprocessed with minimal processing they are known as "edible flowers" (Jadhav et al. 2023). These are the abundant sources of chemical substances that are essential for metabolic processes. Over and above their function in enhancing the aesthetic characteristics of food, edible flowers are utilized mostly for cooking purposes. However, they are popular due to their association with active biological compounds. Edible flowers that come from ornamental plants are important because they can enhance the appearance of food. In addition, consumption of these flowers is concomitant with significant health benefits including anti-cancer properties and antioxidant activity (reactive oxygen radicals' neutralization) (Abdullah et al. 2024). Furthermore, the continued investigation of new information regarding the nutritional composition of edible flowers has recognized their status or importance in the domain of cooking.

M. Afzaal (✉) · M. B. Hussain · F. Saeed · T. Mustafa · M. U. Butt · I. Nawaz · B. Rasheed
M. Ehsan · A. Busharat
Department of Food Science, Government College University Faisalabad,
Faisalabad, Pakistan
e-mail: muhammadafzaal@gcuf.edu.pk

M. Waheed
Riphah International University Faisalabad, Faisalabad, Pakistan

A. Poonia et al. (eds.), *Edible Flowers: Source of Phytonutrients, Valorization and Technological Advancements*, https://doi.org/10.1007/978-3-032-00667-7_13

Therefore, edible flower consumption has been on the rise in recent years, indicating their increasing popularity as well as approachability in human nutrition (Abdullah et al. 2024).

In the last few years, edible flowers have become an essential component of human nutrition, enhancing culinary traditions in different cultures. For example, in Central Europe, it is a common approach to prepare dishes like dandelion flowers boiled with sugar and batter-coated fried black elder flowers (Sambucus nigra). There had been a sturdy rise in the demand of high-quality edible flowers for human consumption in recent years. Ancient civilizations like the Chinese, Romans, and Greeks have understood the nutritional and therapeutic benefits of flowers (Uniyal and Kumar 2024). Rosemary, passionflower, roses, and lavender are among the flowers that are frequently used in aromatherapy. Cooks and gardeners used orange blossoms and marigolds around a thousand years ago in cooking. In oriental dishes, daylily buds were used and the Romans used violets and roses. Hispanic and Italian cultures used stuffed squash blossoms and Indians used rose petals in various formulas or recipes. Chartreuse is a French herbal liqueur, developed in the seventeenth century, claims flower petals as one of its secret constituents (Fernandes et al. 2020).

Fresh edible flowers were highly valued in ancient times and were even transported as valuable seasonings. These flowers have been used in many cooked foods such as jellies, syrups, desserts, and sauces and also had been featured in the composition of tea, liquor, drinks, honey, salads, vinegar, and ice cubes (Rodrigues and Spence 2023). The fresh petals of flowers were frequently eaten as garnishes or in salads, adding flavor, color, taste, and visual appeal to the food. Edible flowers have anti-diabetic, anti-cancer, anti-anxiety, anti-inflammatory, antioxidant, and antimicrobial effects with specific effective dosages (Kumari and Bhargava 2021). Edible flowers have been valued for their nutritional composition, distinctive flavors, medicinal benefits, and aesthetic appeal by ancient civilizations including the Chinese, Romans, Indians, Greeks, and Middle East. As, Calendula officinalis was used in salads in ancient France and Rosa species were utilized in Roman cuisine. As a natural sugar coloring, Viola odorata was employed and the blooms of Sambucus nigra and Taraxacum officinale were used in drinks and salads. Among plant sources, edible flowers are important because of their high nutritional content and alluring visual appeal (Segneanu 2024).

International influence has played a significant role in increasing consumer awareness regarding edible flowers and revitalizing the earlier lifestyle in which they held a conspicuous position. The efforts of food manufacturers and producers especially those that provide ready-to-cook items are intimately related to the growing demand for edible flowers. Additionally, these flowers could also be enjoyed in different forms like dried form, preserved in distillates and incorporated into cocktails (as ice cubes) (Duggirala et al. 2024). Because of its versatile nature, edible flowers are considered an appealing and dynamic addition to contemporary cooking practices. Yet, the sensory qualities of edible flowers serve as the main criterion for evaluating their excellence. These consist of their appearance, color, and above all the fragrance and taste. The vivacious colors of edible flowers are determined by different chemical compounds including flavonoids and carotenoids which are considered most influential in this regard (Pires Jr et al. 2021). The flavonoid content of flowers is often strongly linked to their high antioxidant potential. Edible flowers have been sourced from a variety of plants like fruits, herbs, vegetables and decorative plants (ornamental). However, many flowers of weed as well as commonly utilized edible flowers are considered safe for human consumption, but it must be observed that some flowers are inedible and poisonous. For example; flowers of *Lycium barbarum, Laurus nobilis,* and *Achillea millefolium*. People should be aware of the allergic reactions to edible flowers before consumption. Flowers can be used in different ways as a whole with their pedicel or by splitting their petals, sepals, pistils and stamens. These edible flowers might be enjoyed in their dried, processed, and fresh forms like candy, tea, or vinegar (Fernandes et al. 2020). This chapter focuses on collecting the current information regarding edible flowers from recent research, aiming at edible flower's nutritional, functional, and sensory properties, in addition to discussing perspectives on biological activity, their constituents, future trends and cultural uses.

13.2 Edible Flowers Used as Food

Along with the eating of seeds, fruits, leaves, and vegetable roots, flowers were already used as alternative treatments or as part of traditional Chinese cuisine in ancient Greece, Rome, and China. Roses (*Rosa* spp. L.), for example, were used in ancient Rome to add flavor and sweetness to drinks, salads, purees, omelets, and desserts. Many flowers as shown in Fig. 13.1 have a long history of being used in food preparation, flavor, adding aroma, and aesthetic value. Violets (*Viola odorata* L.) were used as a source of sugar and as a coloring agent for syrups as early as the seventeenth century (Takahashi et al. 2020). These include edible flowers, which are employed as a decorative element and as a cooking component, both of which significantly increase the general public's awareness of its use. However, not all nations are familiar with edible flowers. Contrary to common opinion, flowers add a special blend of flavors and nutrients to culinary preparations, making them more than just decorations for savory foods and sweets. They can be eaten raw (marigold flowers in salads, for example), cooked with meat and fish, added to soups and beverages, added to desserts, and jellies, and used as colors and spices. They can be powdered, crystallized, or used as foams, or they can be dry (Fernandes et al. 2019a).

Certain plants, including passion fruit, chives, and pumpkins, are frequently solely recognized for the biological, seasoning, or nutritional qualities of their fruits or leaves, whereas their edible blooms are typically not used in cooking (Fernandes et al. 2019a, b, c). There was a push to value wild plants, particularly native ones, around the turn of the twentieth and twentyfirst centuries. The media started to promote these plants and use them as culinary ingredients in TV shows, periodicals, websites, and cookbooks. These include edible flowers, which are employed as a decorative element and as a cooking component, both of which significantly increase

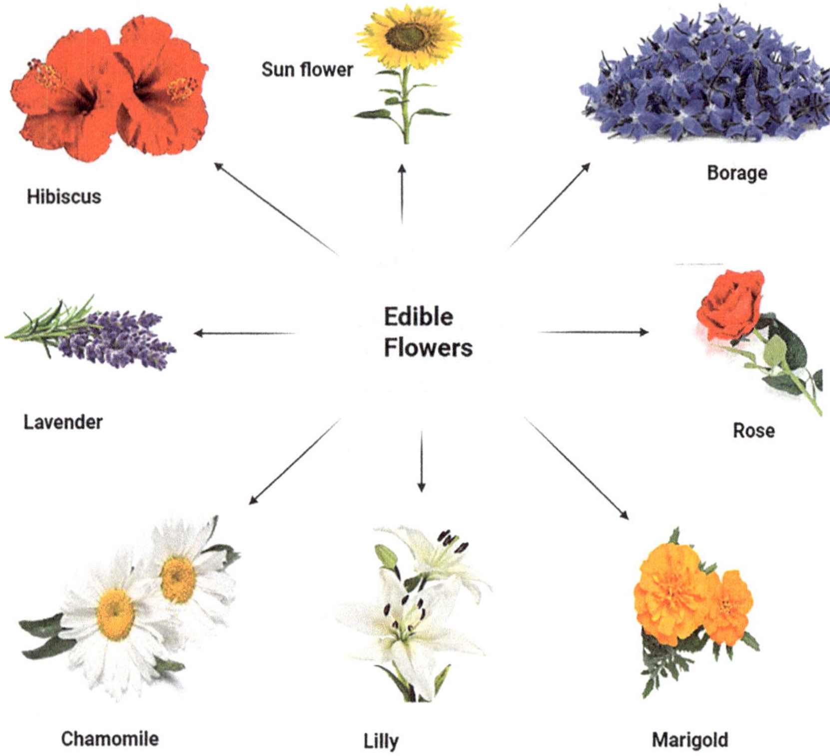

Fig. 13.1 Common names of some edible flowers

the general public's awareness of its use. However, not all nations are familiar with edible flowers (Guiné et al. 2020). Edible wildflowers are essential to traditional rural inhabitants' food and nutritional security as part of their socio-cultural and spiritual lives, according to Kumari et al. (2021). Additionally, the various socio-geographic locations have a wide variety of edible flowers and applications. Mulík and Ozuna (2020) discussed the cultural customs, traditional culinary applications, and possible advantages of the most widely used edible flower in Mexico. In this nation, flowers were used with a variety of symbolic meanings in pottery paintings, sculptures, pyramids, and caves. Because of the Catholic Church's Easter celebration customs, flowers started in the diet as a meat replacer after Spanish colonization mixed cultures. In addition to being used raw, edible flower is also used to flavor drinks and to prepare hot foods and desserts. Among the most appreciated and cultivated species in Mexico are *Cucurbita sp., Yucca spp., Tagetes erecta, Dahlia spp., Plumeria rubra, Opuntia fícus-indica, Agave spp., Euphorbia pulcherrima* and *Phaseolus coccineus*. Before the Spaniards arrived, flowers were already being produced, but their consumption was frowned upon due to poverty and is still disapproved of in the area.

13.3 Traditionally Flowers Used as Food

Although many flowers, such as artichokes (*Cynara scolymus* L.), cauliflower (*B. oleracea L. var. botrytis*), and broccoli (*Brassica oleracea L. var. italica Plenck*), which were originally consumed in Asian countries like China, Japan, and India as tradition, are now part of the daily diet of people worldwide, consumers hardly ever associate these vegetables with flowers. While certain places have a preponderance of particular sorts and colors of these edible flowers, different countries have a variety of varieties. These flowers are excellent examples of how to use them as sources of protein, fiber, and vitamins for both human and animal nutrition. In the Brassicaceae family, broccoli and cauliflower are two examples of blossoms that are loved worldwide. According to López-Romero et al. (2018), broccoli is the fourth most consumed vegetable worldwide. It is a group of plants in the Brassica genus that are also referred to as mustard, cabbage, cauliflower, kohlrabi, brussels sprouts, and broccoli itself. Purple and green-sprouting broccolis appear to be the most popular, however they come in a variety of colors. Both raw and cooked forms are acceptable, although some home cooking techniques reduce the number of flavonoids and other bioactive ingredients (Wu et al. 2019).

Artichokes are native to the Mediterranean region and are prized for their edible, huge, immature flowers. Cardo, another species of the Cynara genus, is likewise edible but less common than artichokes. Artichokes and green broccoli are high in nutrients and include flavonoids, dicaffeoylquinic acids, and phenolic compounds, which are bioactive substances with antioxidant properties (Salekzamani et al. 2019). According to Le et al. (2019), broccoli and cauliflower contain glucosinolates, which are highly bioactive secondary metabolites that can react to form derivatives linked to anticancer activity. Some examples of flowers that are typically taken as infusions are chamomile, hibiscus, mint, and globe amaranth flower. Infusions made from the heads of chamomile flowers (*Matricaria recutita* L., *Matricaria camomilla* L.) are mostly used for their therapeutic rather than nutritive qualities. Chamomile is another plant that is high in flavonoids and phenolic compounds. It is linked to a reduction in the absorption of sugar and the digestion of carbohydrates, which may lead to novel methods of treating issues with the gastrointestinal tract (Villa-Rodriguez et al. 2018).

13.4 Flowers as Culture and Regional Food

As part of festivals and ceremonies, flowers are included in common foods in many cultures some examples are shown in Table 13.1. There is currently a lack of information on nutritional composition, even though some locals have continued this practice. To aid in the identification of edible plants, especially Eastern Kentucky edible flowers, Dourson and Dourson (2019) published a guide. Traditional cheesemaking in Portugal and Spain uses rennet derived from cardoon flowers (*C. cardunculus*) as a coagulant agent with sheep milk (Gomes et al. 2019). Using interviews

Table 13.1 Examples of some edible flowers used as food

Scientific names	Common names	Region	Edible use	References
Clitoria ternatea	Blue pea flower	Malaysia	Rice-based dish	Harmayani et al. (2019)
Calendula officinalis	Pot marigold	India	Tea, cakes, and flavor	Panda et al. (2019)
Tamarindus indica L.	Tamarind	India	Vegetable	Panda et al. (2019)
Taraxacum officinale	Dandelion	United States (eastern Kentucky)	Blossoms batter-dipped and fried	Dourson and Dourson (2019)
Bombax ceiba	Silk cotton tree	Asia	Vegetable	Rameshwar et al. (2014)
Coreopsis lanceolata	Lanceleaf tickseed	America	Tea	Kim et al. (2020)
Echinacea purpurea	Purple coneflower	North America	Dried petals used in tea	Chiou et al. (2017)
Rosa spp.	Rose	Northern Hemisphere	Tea, salad, cakes, flavor extracts	Kumari et al. (2018)
Oxalis pes-caprae L.	Bermuda buttercup	South Africa	Fresh or cooked	Tembo-Phiri (2019)
Dahliaa coccineaa	Wild dahlia	Mexico	Salad	Espejel et al. (2019)
Hemerocallis fulva	Day lily	Asia	Vegetable	Liu et al. (2018)

and published data, Abbet et al. (2014) identified 98 wild plant species that are traditionally consumed in the Valais region of Switzerland; flowers accounted for 26% of the most regularly consumed species. Elderly citizens of this nation expressed reminiscing about their childhood by consuming edible wildflowers. Regardless of where they came from, some flowers are utilized extensively. Some species are cultivated worldwide and, because of their vibrant colors, aromas, and nutritional value, are sold in their natural form or used to make spices, infusions, sweets, cakes, and functional meals (Gostin and Waisundara 2019). The flower powder of the drumstick tree (*M. oleifera*) is also used in a variety of food products, including bread, cookies, yogurt, cheese, and soups, in many different nations (Dhakad et al. 2019).

13.5 Edible Flowers Nutritional Composition

The edible flower is becoming a more popular and healthful culinary trend (Benvenuti et al. 2016). Native people's diets depend heavily on these edible blossoms from wild plants, at least for a brief period after they bloom. Regardless, it is critical to identify the nutritional makeup and other characteristics of those deemed

Table 13.2 Nutrition of edible flower

Edible flowers	Protein (%)	Moisture (%)	Ash (%)	Fiber (%)	Carbohydrate (%)	References
Rosa spp.	1.88	84.56	0.72	3.20	9.41	de Lima Franzen et al. (2019)
Etlingera elatior	0.44	95.12	0.65	0.96	2.46	Rachkeeree et al. (2018)
Calendula officinalis	1.20	89.34	0.93	1.59	5.62	de Lima Franzen et al. (2019)
Tropaeolum majus	1.99	89.32	0.63	4.51	7.14	Navarro-González et al. (2015)
Rhododendron arboreum	1.68	79.40	2.30	2.90	12.20	Sarla Saklani and Subhash Chandra (2015)
Tagetes erecta	1.32	83.39	0.80	9.20	14.15	Navarro-González et al. (2015)
Spilanthes oleracea	2.84	81.74	1.44	10.11	13.56	Navarro-González et al. (2015)
Hedychium forrestii	2.38	92.88	0.86	1.04	2.15	Rachkeeree et al. (2018)

pleasant for human feeding. The main component of edible flowers, according to their proximate organization, is water (more than 80%), and their fat and protein content is considered to be low. Depending on the type of bloom, they also include varying levels of minerals, total sugars, and dietary fiber (Mlcek and Rop 2011). Table 13.2 displays the nutritional makeup of edible flowers. The flowers of *Tagetes erecta* and *Tropaeolum majus* have a higher potassium content than sodium, but other elements including phosphorus, sodium, magnesium, and sulfur were identified in smaller amounts, according to Navarro-González et al. (2015).

13.6 Nutritional Profiling

The nutritional value and potential health benefits of edible flowers have drawn attention to this intriguing and varied group of botanical entities (Kumari and Bhargava 2021). These contain a diverse array of macronutrients, micronutrients, and phytochemicals that contribute to their nutritional profile and biological activity, although they are frequently disregarded in traditional dietary paradigms. (Barani et al. 2022). A thorough analysis of the nutritional makeup of edible flowers shows that they include a variety of macronutrients, such as proteins, lipids, and carbs, though in different amounts. Additionally, water regulations maintain physiological equilibrium and assist metabolic processes by storing vital vitamins and minerals like beta-carotene, potassium, calcium, and vitamin C in these biological entities (Dogra 2024). The United States Department of Agriculture's National Nutrient Database for Standard Reference states that edible flowers are richer in several macronutrients and micronutrients and have lower fatty acid concentrations than vegetable-based foods like spinach, broccoli, and carrots (Panda et al. 2024).

13.7 Edible Flowers Health Benefits

Because of the potential health benefits linked to their chemical properties, edible flowers are expected to become more and more popular in the future. Using chemical-related antioxidant assays like DPPH, ABTS, FRAP, and others, the antioxidant qualities of edible flowers have been well-studied in recent years. Reactive oxygen species, stroke, diabetes type 2, cardiovascular disease, cancer, and neurological conditions like Parkinson's disease are all caused by an excess of spontaneous oxidative radicals. Intrinsic antioxidants that shield cells from oxidative damage are called cytoplasmic antioxidant enzymes (Abdel-Haleem et al. 2017).

13.7.1 Anti-inflammatory, Antioxidant and Anti-cancer Properties

Marigold's phytochemicals, especially lutein and flavonoids, have been shown to have cytotoxic, and free radical-scavenging properties against a range of malignant cells (Moliner et al. 2018). From a nutritional perspective, edible flowers are abundant in vitamins, minerals, flavonoids, and other bioactive substances that have antioxidant qualities. In addition to having anti-inflammatory, hepatoprotective, neuroprotective, and anticarcinogenic properties, these compounds support healthy neurological, cardiovascular, immunological, and gastrointestinal systems (Sood et al. 2024). Carotenoids and flavonoids are responsible for the color of flowers and give them antioxidant properties (Chen et al. 2018). Numerous intriguing health-promoting qualities, such as antioxidant, anticancer, anti-obesity, hepatoprotective, and antimicrobial, are present in edible flowers (Skrajda-Brdak et al. 2020). Because they counteract free radicals, which are linked to the development of chronic illnesses like cancer, heart disease, and premature aging, antioxidant properties are important. To prevent nutritional imbalances, it is advised to consume flowers in moderation as some of them include antinutrients such as oxalates and phytates that can hinder the absorption of vital minerals (Coyago-Cruz et al. 2023).

13.7.2 Anti-microbial Properties

Edible flowers are a prospective source of antibacterial chemicals in addition to their antioxidant properties. Bioactive chemicals derived from flowers provide natural alternatives to treat bacterial, viral, and fungal illnesses in a world where microbial resistance to specific medications is becoming a greater concern to public health. These flowers' antibacterial properties are due to phenolic compounds, essential oils, and other phytochemicals (Coyago-Cruz et al. 2024). For instance, Helichrysum italicum has polyphenols and essential oils that work well against yeasts like Saccharomyces cerevisiae, Gram-negative bacteria like Klebsiella pneumoniae, and Gram-positive bacteria like *Staphylococcus aureus* (Primitivo et al. 2022). Antibacterial activity against *S. aureus* has been demonstrated by pingyin rosebud extract (Liu et al. 2023). Flowers of Camellia japonica (*var. Carolyn Tuttle*) have demonstrated exceptional antibacterial action against Salmonella enteritidis, Pseudomonas aeruginosa, and S. aureus, among other foodborne pathogens. But they didn't work against *Bacillus cereus, Staphylococcus* epidermidis, or *Escherichia coli* (Pereira et al. 2023). Additionally, phenolic chemicals like quercetin and gallic acid, which are abundant in a variety of edible flowers like roses, marigolds, and chrysanthemums, contribute to their antibacterial, antiviral, and antifungal qualities (Dastidar et al. 2023).

13.7.3 Neuroprotective Agent

Neuroprotective drugs are required as a means of preventing CNS neuron damage. The leaves, flowers, fruits, and seeds of medicinal plants are reportedly utilized as alternate forms of treatment. A lot of phenolic chemicals found in edible flowers have neuroprotective properties. Flavonoids and other phenolics have the qualities of enhancing nerve cell activity or regeneration, preventing ROS production, and lowering protein oxidation. Additionally, flavonoids protect the brain's nerves in several ways, reducing inflammation and damage from neurotoxins and enhancing memory and other cognitive abilities. The same is true of non-flavonoid phenolics, like phenolic acids (Baessa et al. 2019).

13.7.4 Medicinal Use of Edible Flower

Edible flowers have been utilized to treat medical issues in addition to their antibacterial and antioxidant properties. Italian bugloss (*Anchusa azurea P. Mill.*) and garden anchusa are used as diuretics, diaphoretics, depuratives, and antitussives. The cotton tree, *Bombax malabaricum* L., is used to treat contused wounds, hepatitis, fever, diarrhea, and chronic inflammation (Meenakshi et al. 2019). *Calendula officinalis* L., often known as Scotch marigold or marigold, possesses antibacterial, antifungal, anticancer, anti-HIV, anti-inflammatory, and antioxidant qualities. Additionally, it possesses gastroprotective and insecticidal properties, lowers heart rate, and has a cardioprotective role. It can also function as an immunostimulant and immunomodulator. Sweetvetch, or *Hedysarum coronarium* L., possesses laxative and hypocholesterolemic properties (Rodriguez-Amaya 2016). The China rose, *Hibiscus rosasinensis* L., is used to treat bronchial catarrh, fever, cough, and Genito-urinary issues (Arabian jasmine) helps heal wounds and is good against leprosy, uterine hemorrhage, cancer, and skin conditions (Alvarez-Suarez et al. 2023).

13.7.5 Anti-aging

External stress causes or aggravates intrinsic skin aging, a chronological process that happens naturally over time. Wrinkles, loss of elasticity, sagging, and rough texture are signs of aging skin and are brought on by phenotypic changes in cells (fibroblasts, melanocytes, etc.) as well as structural and functional changes in extracellular matrix (ECM) constituents like collagen, elastin, and proteoglycans (McCabe et al. 2020). Aging dermal fibroblasts and external stresses, especially ultraviolet B (UVB), change the composition and function of ECM proteins in the skin (Park et al. 2015). UV exposure causes oxidative stress and consequently ECM remodeling by raising intracellular reactive oxygen species (ROS) levels (Yeo et al. 2018). The Rosaceae family includes the perennial flowering plant known as the rose. Since ancient times, it has been utilized as a skincare ingredient, and it is still commonly found in many skincare products. Many phenolic compounds, phenolic acids, and phenolic phytochemicals, including polyphenols, tannins, flavonoids, phenolic acids, and anthocyanins, are found in rose petals (Yang and Shin 2017). These ingredients have antibacterial and anti-inflammatory qualities and are abundant in vitamins, minerals, and vital fatty acids. Roses are therefore exceptional natural antioxidants that can fortify skin cells and promote skin tissue regeneration (Ghavam et al. 2021). More than 150 types of roses are utilized extensively in medicine and cosmetics. *Rosa damascena* sometimes called the Damask rose and indigenous to the Middle East and Central Asia, is a commonly used Rosa species. Because of its potent antibacterial and anti-aging qualities, it works well to relieve irritated skin and promote wound healing. Rosa centifolia, sometimes referred to as cabbage roses, is another prominent species used in cosmetics. It has toning and antioxidant benefits because it is high in vitamin C and tannins. Alpine, French, baby, and black roses are less frequently used in cosmetics, though (Hubert et al. 2022).

13.8 Impact on Mood and Mental Health

Human health has been seriously jeopardized by the main clinical signs of depression, which include poor mood, guilt, loss of interest, increased or decreased eating, sleep difficulties, and difficulty concentrating (Perić et al. 2021). For thousands of years, people have utilized the blossom buds of *H. citrina* as a vegetable or medicine. According to recent pharmacological research, the edible flowers of *H. citrina* contain bioactive qualities that include anti-tumor, anti-constipation, anti-depression, insomnia treatment, and lactation deficiencies treatment (Guo et al. 2023). In recent years, some researchers have also proposed a possible link between gut microbiota and the antidepressant effects of *H. citrina* extracts. According to Liu et al. (2020), the water extract of *Hemerocallis citrina* flowers (HCW) increased the number of Bacteroides genera, which raised the amount of GABA (gamma-aminobutyric acid), which in turn had antidepressant effects. Additionally, by reducing the number of inflammatory components, HCW may have an antidepressant impact by dramatically lowering the quantities of Desulfovibrio genera in the gut, blood, and brain of mice produced by depressed models (Liu et al. 2022).

13.9 Phyto-chemicals in Edible Flowers

These are naturally occurring, biologically active, and non-nutritive chemical compounds that are found in plants, and also used as medicinal constituents for the betterment of human beings (Yang and Ling 2024). They found that vegetables, fruits, and other plant foods help to lower severe diseases (Otles and Bakirci 2021). Five groups are grouped in dietary Phyto-chemicals phenolics, nitrogen-containing compounds, alkaloids, organosulfur compounds, and carotenoids. Edible flowers contain phytochemicals that are flavanols, flavanones, flavones, anthocyanins, and phenolic acids, details are explained below with their chemical structures (Zheng et al. 2019).

13.9.1 Phenolic Acids

In phytochemicals, phenolic acids are a key group obtained from edible flowers. There are two groups of phenolic acids, the first group comprising three carbon-side chains (C6-C3) structure and known as hydroxycinnamic acids, the second group comprising a C6-C1 structure and known as hydroxybenzoic acids and also found as conjugated or free substances (Kaur et al. 2021). Hydroxycinnamic acid derivatives include caffeic, sinapic, chlorogenic, ferulic acids, and p-coumaric. While derivatives of hydroxybenzoic acid are protocatechuic, vanillic, p-hydroxybenzoic, ellagic, gallic, and syringic acids (da Silva et al. 2023). Phenolic acids are also identified in day lily, roselle, hangzhou white chrysanthemum, rose, and honeysuckle. Chlorogenic acid was recognized in the extracts from Day lily, hangzhou white chrysanthemum, and roselle. Caffeic acid was determined in extracts of roselle, hangzhou white chrysanthemum, day lily, and honeysuckle (Faisal et al. 2022). Hangzhou white chrysanthemum contains 3-caffeoylquinic acid, machranthoin F, and 1-caffeoylquinic acid (Zheng et al. 2019). Caffeoylquinic acid is found in different forms; 5-caffeoylquinic acid, roselle contains 3-caffeoylquinic acid, and 4-caffeoylquinic acid (Ryu et al. 2019).

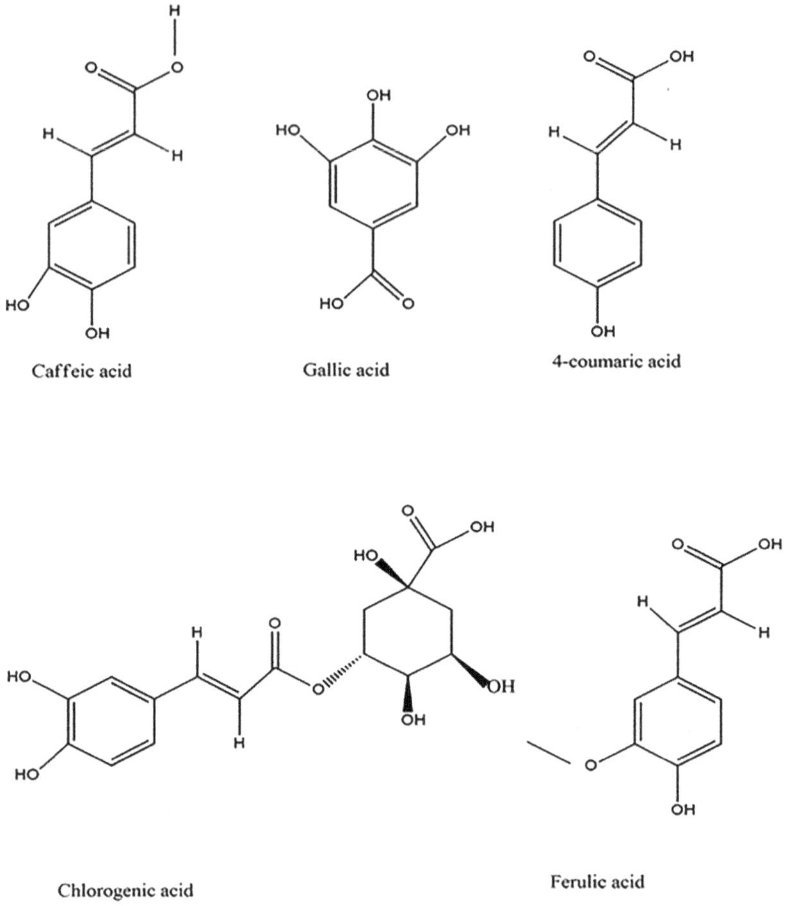

Phenolic acid

13.9.2 Anthocyanins

There are two Greek words from which the term anthocyanins are acquired "Anthos" means "flowers" and "cyano" means "blue" and the complete meaning is "blue from flowers". These colors lead to correct pollination. Anthocyanins are naturally present in the form of aglycon, called anthocyanidin. There are almost six anthocyanidins named cyanidin, peonidin, delphinidin, pelargonidin, malvidin, and petunidin, that are tied to one or more than one glycosidin units (Riaz et al. 2016). Anthocyanins are belonging to the flavonoids and they are water soluble pigments and also responsible for the staggering flower's color. Factors on which color depend are metal ions, co-pigments and pH (Kumari et al. 2022). Due to presence of these unique colors like blue, orange, anthocyanins contribute to the attractiveness of flowers to the consumers.

Anthocyanins are identified in Chinese rose, roselle and tree peony. Tree peony showed the presence of glycosides of peonidin, Chinese rose was plentiful in rutinosides of pelargondin, cyanidin and glycosides. Delphinidin-3-sambubioside and cyanidin-3-sambubioside were showed in roselle (Faisal et al. 2022). Rich source of anthocyanin pigments is rose, hibiscus, petunia, pansy and pelargonium (Łysiak 2022). They play important role in prohibition of diabetes, obesity, cardiovascular diseases, and others. It is also filed that anthocyanins acquire anti-proliferative, antioxidant, and anti-inflammatory characteristics that are very advantageous in the maintenance of health (Singh et al. 2024). Because of existing of anthocyanins in composition, different fruits and veggies play a protective role against prostate, breast, intestinal, blood, and liver cancer. It also prevents form neurodegenerative problems like Alzheimer's and Parkinson's by upgrading cognitive, motor performance and memory (Keservani et al. 2023).

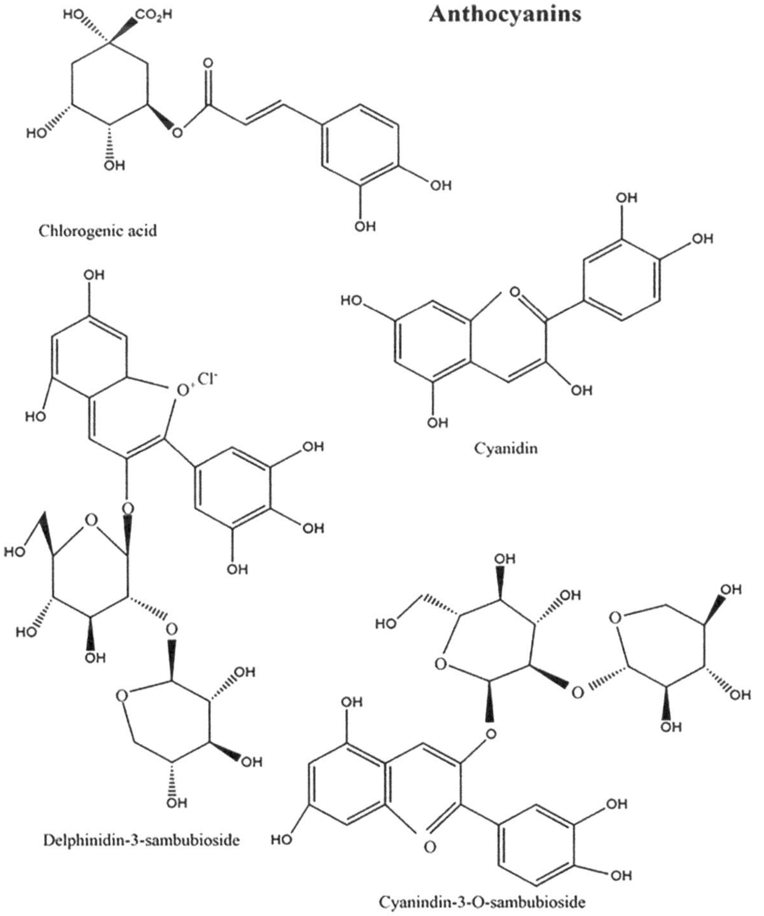

13.9.3 Flavonols

There is a class of flavonoids known as flavonols that include quercetin, myricetin, kaempferol, and isorhamnetin which are opulently in edible flowers. They are found in different forms like day lily, roselle, hangzhou white chrysanthemum, rose, honeysuckle, magnolia flower, and Chinese rose (Wan et al. 2019). In *Paeonia suffruticosa*, about 19 flavonols are identified. Kaempferol di-hexoside, quercetin-3-O-galactoside, kaempferol-3-O-malonylglucoside-7-O-glucoaide and isorhamne-tin-3-O-glucoside are significant flavonols in *Paeonia lactiflora* petals (Lee et al. 2019). In dried honeysuckle flowers, 2 quercetin derivatives and 6 kaempferol derivatives are the main flavonols. In rose petals predominant flavonol aglycones are quercetin and kaempferol. In the magnolia flower, there are about 10 kinds of flavonols in which isorhamnetin, quercetin, and their glycosides appeared dominant (Wang et al. 2019).

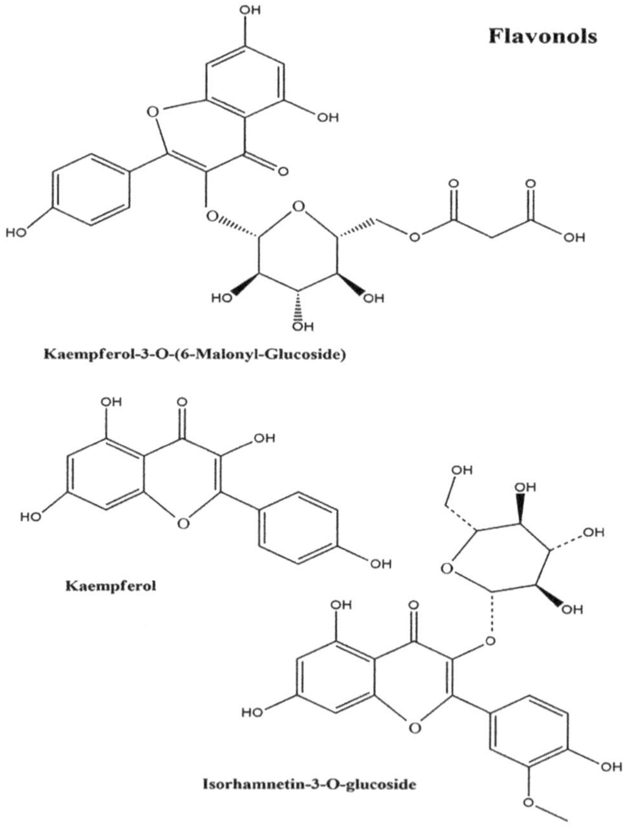

13.9.4 Flavones

Flavones are the class of flavonoids that are present in edible flowers. There is a double bond present between C2 and C3 in flavones. It also has multiple health benefits like anti-diabetic, anti-microbial, anti-cancer anti-oxidant, and anti-inflammatory activities (Hamedani et al. 2023). In edible flowers, flavones exist in different forms like apigenin, acacetin, and luteolin. In *Chrysanthemum morifolium* flowers, luteolin-7-glucosides are found abundantly. Some extra flavones like luteolin-7-O-glucoside and acacetin were declared in chrysanthemum flowers (Ryu et al. 2019). In the Asteraceae family, flavones were abundant, and in ample amounts presence of flavones showed by wild chrysanthemum. It is also noted that acacetin, luteolin, eupatorine, and diosmetin were possessed by these flowers. Luteolin is also present in honeysuckle, tree peony, wild chrysanthemum, sweet-scented, and hangzhou white chrysanthemum (Faisal et al. 2022).

Flavones

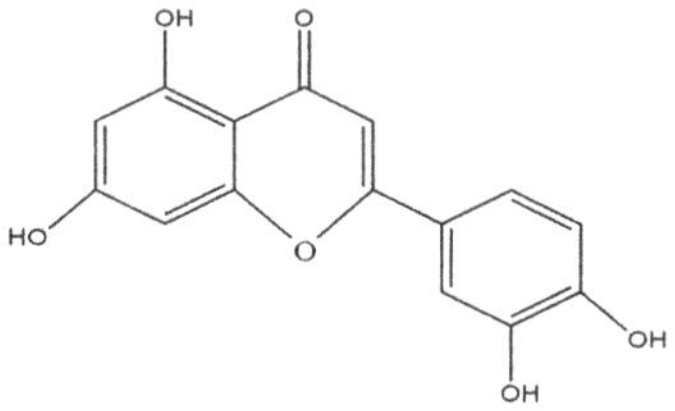

Luteolin

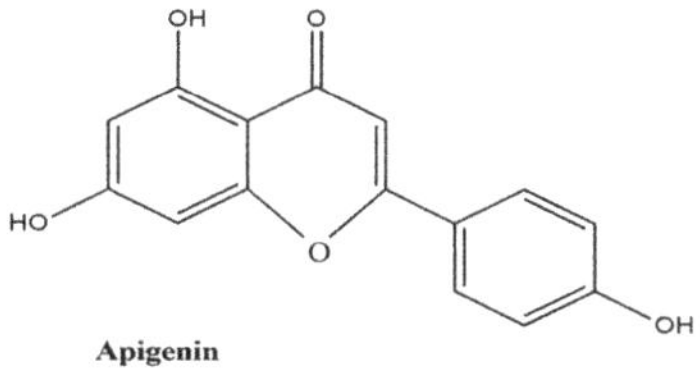

Apigenin

13.9.5 Flavanols

Flavanols and flavones are similar but differ in a hydroxy group at position C3 and there is no carbonyl in the ring and also have no double bond present between C2 and C3 (Baranović and Šegota 2018). In edible flowers, flavanols found were epicatechin gallate, catechin, epigallovatechin gallate, and epicatechin. In all edible flowers, the richest flavanol is rose. Catechin was spotted in extracts of day lily, roselle, and water lily. Epicatechin was discovered in rose and day lily extract. Epigallocatechin gallate and epicatechin gallate also found in rose (Faisal et al.

2022). In a study, only 5 out of 26 edible flowers revealed low content of flavanols named *Calendula officinalis, Leucanthemum vulgare, Trifolium alpinum, and Dianthus carthusianorum.* And there is no flavanol is detect in *Borago officinalis* (Pires Jr et al. 2021). Dried honeysuckle contains 2 quercetin, 6 kaempferol, and their derivatives. It also has a beneficial health effect like anti-oxidant, anti-inflammatory, and genotoxic activity that results as is also impactful against Parkinson's, CVDs, and Alzheimer's (Fernandes et al. 2017).

Flavanols

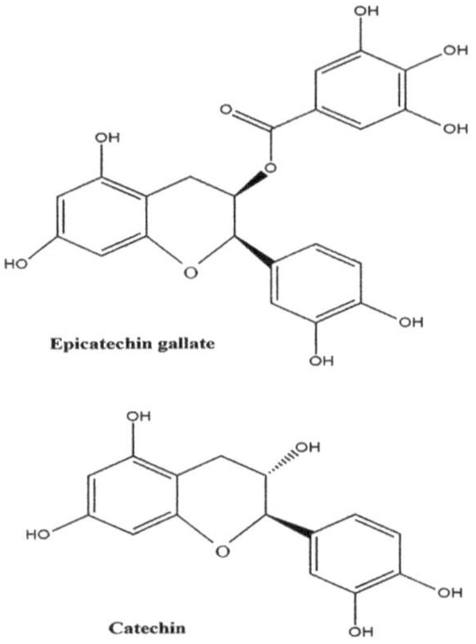

Epicatechin gallate

Catechin

13.10 Edible Flowers Processing

Edible flowers use to fluctuates widely across different nations, reflecting different culinary traditions. Chinese congee, soups, and desserts often contain flowers, such as Floss Sophorae and Osman Thus fragrance. Ancient Romans used lavender in sauces and violets to garnish dishes. In the US, pumpkin and squash blooms that spread have been long used in cultural cuisines (Fernandes et al. 2019a, b, c). However, concealed flowers and odorous budding flowers are widely consumed in countries such as El Salvador and Honduras. European cuisine frequently includes flowers as festive décor; fried elderberry flowers are a common dish, while dandelion flowers provide flavor to salads and drinks. Specific amounts of edible flowers are used for their culinary and nutritive value, even though they are often eaten whole. The astringent base that comprises the rose petals is usually removed, whereas the outer stigma of saffron is an acknowledged ingredient. Nowadays, the market demonstrates a variety of processed edible flower products that are safer for human consumption as well as more accessible than fresh petals, which are susceptible to bacteria because of their high concentration of water (Alshehri et al. 2022).

Edible flowers can have their sensory properties preserved and their shelf life increased by processing. Through modern technology like infrared-assisted freeze-drying and ultrasonic cleaning, recent processing trends promote novel methods to preserve the quality of flowers. By using these methods, nutrients-particularly bioactive substances like flavonoids and anthocyanins-are efficiently sustained. Selecting the right processing method is essential and fluctuates depending on the type of flower. Emerging physical field technologies offer the potential to improve product quality and efficiency while having no adverse harmful impact on the environment (Arslaner et al. 2021). The basic principles, positive aspects, and adverse effects of the most recently introduced advances in physical field technologies for processing edible flowers are examined in this paper, along with the possibility of their commercial implementation. It highlights the need for effective approaches to improve quality, freshness, and shelf life while taking into consideration aspects like color, scent, and appearance that have a big impact on consumer preference. All things considered, the results are meant to support the wider application of innovative processing techniques in the edible flower industry while maintaining sustainability and quality (Denoya et al. 2021).

13.11 Preserving Freshness Through Processing

Since edible flowers may enhance aesthetic value, their freshness, highly sensitive flavors, and exotic essential oils to meals, are becoming ever more acceptable in exquisite cuisine. Many high-end establishments are working to improve the quality of their cuisine by using edible flowers as vital ingredients in soups, salads, entrees, and pastries, in addition to using them as aesthetically pleasing highlights. To sustain flowering plants' quality and increase post-harvest availability, innovative preserved techniques are therefore desperately needed. In addition to minimizing costs associated with transportation, improving the above factors can increase the viability of using ground transportation for shipping edible flowers (Bahrami et al. 2020). Several innovative preservation guidelines have been utilized to increase the scope of floral design derivatives that might be ingested and increase their market value. Modifies atmospheric packaging (MAP) is a method for controlling the environment surrounding flowers, delaying respiration, and avoiding deterioration. While preserving the external texture and nutrition of flowers, high hydrostatic pressure (HHP) treatment uses great force to kill spoiling bacteria. Furthermore, the edible coatings perform as a barrier against moisture loss and microbial development, while irradiation is used to kill pathogens and increase shelf life. By using these technologies, post-harvest flowers' shelf life may be significantly extended bringing up exciting possibilities and increasing customer satisfaction with these unique dietary aspects (Bao et al. 2023). Some emerging technologies in edible flower processing are shown in Table 13.3 with their advantages.

Table 13.3 Emerging technologies in edible flower processing

Technology	Examples of applications	Techniques used	References	Advantages	Challenges
Modified atmospheric pressure (MAP)	Used for edible flowers like carnations and snapdragons to slow down respiration rates and prevent moisture loss.	Packaging with controlled gas composition (e.g., CO_2 increases and O_2 reduction); use of 1-Methylcyclopropene.	Mlcek et al. (2021), Cardwell et al. (2023)	Extends shelf life, maintains texture and nutrition, is suitable for transport, and integrates IoT for monitoring.	Requires advanced infrastructure and is less effective for certain flowers with stems.
High hydrostatic pressure	Extends shelf life of flowers like pansies and centaurea; maintains nutritional qualities while reducing microbial load.	Pressure applied (100–800 Mpa) for short durations; retains bioactive compounds like flavonoids and tannins.	Fernandes et al. (2020), Bedoya et al. (2022)	Non-thermal, eco-friendly, retains nutrients, and extends the shelf life of juices and edible flowers.	Can damage the texture and sensory qualities of sensitive flowers; and requires optimization for different varieties.
Edible coatings	Used on flowers like pansies and night-fragrant blooms to enhance shelf life and reduce microbial activity.	Coatings with alginate, chitosan, or protein-based solutions to form a protective barrier.	Chen et al. (2022), Ahmadimaram et al. (2021)	Extends shelf life, retains sensory and nutritional quality, and can transport bioactive compounds.	Limited research on edible flowers, needs safety assessments for coatings with nanomaterials.
Irradiation technology	Preserve flowers like Bauhinia variegata and Viola tricolor by reducing microbial contamination without altering nutritional value.	Gamma and electron-beam radiation at doses of 0.5–1 kGy to enhance phenolic content.	Chongtham et al. (2022), Bigi et al. (2023)	Increases antioxidant activity, phenolic content, and shelf life without affecting nutritional properties.	Requires careful dose management to avoid quality degradation; and high initial capital investment.
Freeze drying	Maintains the sensory properties and nutrient content of edible flowers such as roses and marigolds.	Low temperature and vacuum conditions to dehydrate flowers.	Barber et al. (2020), Bisht et al. (2021)	Simple and cost-effective.	Causes color, texture, and nutritional degradation.

13.11.1 Modified Atmospheric Pressure

The modified atmospheric Pressure (MAP) uses polymer film manufacturing to identify a constant temperature that maintains respiratory essentials inside the packaging. Maintaining the quality of food commodities has a significant impact on proper packaging, and MAP represents a great advancement in innovation in packaging. This method desires to sustain the freshness of the food ingredients while increasing the life span of perishable food goods by changing the chemical makeup of gases through the packaging (Bigi et al. 2023). The chemical generation that produces ethylene, develops color, respiration rates, volatile compounds precipitation, and bacterial activity are all affected by MAP's adaptation to atmospheric conditions. Wang et al. (2020) investigated how high oxygen atmospheric packaging (HOAP) modifies the generation of hydrogen peroxide (H_2O_2) and oxidative strains in bursting forth *Brassica campestris* L. ssp. (Wang et al. 2020). The researchers discovered that higher oxygen levels may have an undesirable effect on edible floral quality by changing the response of the immune system and modifying the amount of reactive oxygen species, which affects the chemical composition of cell walls. MAP has been shown in previous research to be valuable in extending the cauliflower's expected shelf life. In the above studies, packaging that incorporated an oxygen-based mixture (92% N2, 5% CO_2, and 3% O_2) has been evaluated in comparison with conventional air packaging (Cai et al. 2022). The research on cauliflower covered in polyethylene packaging under MAP indicated that its shelf life might extend up to 30 days. Another study that used image processing and artificial neural networks suggested that it might endure up to 50 days. Packaging and preservation conditions of edible flowers have a significant impact on their nutritional value and metabolic activities. Although MAP shows positive outcomes for fresh preservation, more infrastructure will be needed to accommodate larger amounts (Paulauskienė et al. 2021). Edible flowers might now be transported through the latest developments in sensing technology, including flowers that use cloud-based technologies and the Internet of Things (IOT), which have facilitated dealing with supply chains and packed product inspection. The MAP technique modifies the internal surroundings of packaging to lower respiration rates, prevent water loss, and extend the life span of food goods. It is normal to increase carbon dioxide concentration while decreasing oxygen concentration for perishable food products, such as fresh flowers. By increasing CO_2 concentration, microbial activity-induced spoilage might be reduced successfully and mold development is largely avoided (Leena et al. 2020). The current technology of MAP includes preserving the quality and freshness of a variety of perishable food products, including fresh fruits and vegetables. For freshly cut flowers with long stems, may not constitute the most efficient option, but it has shown a lot of potential for other kinds of food preparation. Research has shown that MAP significantly extends the life span of flowers such as carnations, lilies, and roses. For example, research on how MAP affected specific flower bulbs and perennial herbivorous plants showed a significant increase in shelf life, extending from 3–4 weeks to 2–3 months (Mlcek et al. 2021). In practical applications, currently picked flowers may have been placed on trays with permeable material, which may include 0.5 µL/L of 1-Methylcyclopropene (1-MCP). 1-MCP interacts with ethylene, which is a receptor that is responsible and increases its duration of storage and functions as an undetectable ethanol modulator. According to research, integrating MAP for packing considerably reduces the diminution in oxygen levels while also reducing the accumulated amounts of carbon monoxide and ammonia in flowers like carnations and snapdragons. Snapdragon flowers' just emerging appear and high standards are maintained while their overall retention time is increased with the addition of 1-MCP, which effectively prevents petal disintegration. Samples packed as AP1 (polyethylene packing with ordinary air, AP2 (polypropylene bags with ordinary air), and MAP 2 (polypropylene packing with gas mixture) had storage existence of 30, 20, and 25 days, respectively, according to research considering different packaging alternates (Cardwell et al. 2023). However, the research specimen that was packed in MAP 1 (polyethylene gas with gas combination) maintained a shelf life that was greater than 30 days. Artificial neural network results indicated that samples containing MAP 1 had a promotion perspective of up to 50 days in terms of entire color modification, verifying that it is suitable to extend the shelf life of cauliflower.

13.11.2 High Hydrostatic Pressure Technology

Improved safety and nutritional content in food are emerging as becoming increasingly demanded by customers, especially for goods that provide certain health advantages. High hydrostatic pressure (HHP) technology is becoming an important contributor to the food manufacturing sector's future due to its benefits during processing and alternatives for food preservation (Bedoya et al. 2022). HHP works in a pressure span ranging from around 200 to 800 MPa, with particular processing parameters designed for different uses. The inactivation of enzymes, microbial decrease in unprocessed along with fresh industrial products, preservation of the product's nutritional and functional qualities, along the creation of naturally grown goods missing of preservatives are some examples of this application. HHP interventions generally remain uniformly distributed, regardless of the product architecture and magnitude, use little chemical additions, and are ecologically sustainable. This method extends the shelf life of food while maintaining its fundamental freshness, stability, and nutritional profile. HHP technology is widely applied to processing an extensive assortment of animal-derived fresh meat products, including hams and minced pork, as well as plant-based goods including fruit juices, purees, and smoothies (Aboonajmi et al. 2022). In addition to offering additional nutritional advantages, HHP is a non-thermal packaging procedure that assists in preserving edible flowers. However, it is extremely important to assess how this therapy affects both the sensory components and the stability of the structure.

Research indicates that pansies and *Centaurea* are more tolerant to harsh dealing than particular flowers, such as borage and camellia. Although *Centaurea* has great resistance to scrutiny, its shelf life was not longer than that of pansies. This was probably because pansy petals feature superhydrophobic papillae. The microbial burdens in pansies have been reported to be considerably decreased by short-duration, low-pressure treatments. Regardless, with a conclusion of the duration of storage, the chemically treated flowering pansies were found to have reduced microbial count than their ineffective competitors, showing that HHP might be a worthwhile preservation approach for this particular kind of flower (Anusiya et al. 2021). More study into modifying processing conditions is required to properly preserve edible flowers. Furthermore, studies on broccoli sprouts administered pressure ranging from 100 to 600 Mpa for 3 minutes at 30 °C revealed an adverse impact on the sprouts' physical appearance as well as enzymes in a solution responsible for browning. Increased cellular dysfunction and glucosinolate fluid leakage were the causes of this negative effect. It is recommended to use softer techniques for treatment because significant HHP therapy may have adverse effects on specific flowers. However, processed meals like cocktails and juices might still benefit from insignificant severe treatments. For example, the Juicy Line-Fruity line in the Netherlands uses HHP technology for manufacturing smoothies and juices from edible flowers like broccoli and cauliflower (Bartkiene et al. 2023).

Although delicate HHP treatments are advised to reduce harm to the sensation of taste, spoiling could be accomplished because enzymes and microbes are not fully eliminated during storage. In such circumstances, using a variety of technologies can have an advantageous impact, cutting down on treatment time and maintenance along with operating expenditure. The antibacterial effects of cauliflower waste products are noteworthy; they may reduce *Salmonella typhimurium* by 5 logarithm units at 10 and 37 °C over 10 h. When HHP is added, these avoid the amount improves significantly more; at 37 °C, a five-log decrease may be acquired during 6 hours. Such unconventional methods will be implemented to keep improved fresh-like texture while improving the degree of intensity of the individual interventions. The development of "non-thermal" technologies that optimize nutrient retention has been provoked by shifts in consumer demand. Despite having the prospect of significant advantages in thermal energy production, the temperatures attained are not comparable to the outcomes of conventional heating operations and could be reduced using cooling systems. Currently, one of the most sophisticated and widely used preservation techniques is high hydrostatic pressure (HHP) (Liu et al. 2020). The foods made with the above method successfully deactivate enzymes and deactivate pathogens while preserving their nutrient content and perceptual attractiveness. During a particular amount of time, HHP processes packed items at pressures between 100 and 1000 MPa utilizing groundwater or other fluids as an atmospheric pressure transfer medium. While effectively eliminating germs at room temperature, this method modulates the macromolecular composition of the food, resulting in changes including protein denaturation, starch gelatinization, and enzyme deactivation, which helps preserve food.

The appearance, taste, and nutritional significance of food can also be preserved especially because tiny molecules like vitamins and polyphenols are mainly unaltered. While a lot of investigation has been done on HHP implements in broccoli and cauliflower, while customers often neglect to consider flowers, less is known about other edible flowers. Flower samples kept at 4 °C were subjected to post-treatment examination that concentrated on their microbiological content, biological activity, and appearance (Barani et al. 2022). After rehabilitation at 100 MPa for 5 min, *Centaurea* exhibited a decent look, but its life expectancy remained the same. The consequences ranging from 75 to 450 MPa had been employed for holding durations of 1, 5, and 10 min. Pansies treated at 75 MPa for 5 or 10 min, on the other hand, sustained their fresh look and had a 20-day shelf life, as opposed to simply 6 days for untreated pansies kept at 4 °C. The creation of bioactive compounds in pansies was another consequence of HHP treatment. Ultimately, HHP treatment at 75 Mpa for 5–10 min is an efficient technique for prolonging pansy sample persistence. It is important to remain cognizant that each variety of flowers might respond differently to hydrochloric acid treatment.

13.11.3 Technology for Edible Coating

One of the techniques with demonstrated interest in maintaining the shelf life of food products is edible coating technology. Different indispensable attributes from renewable sources, such as proteins, lipids, and polysaccharides, are used distinctly or in combination to make appropriate coatings for food. The freshness and shelf life of fresh agricultural commodities might be greatly enhanced by this procedure. Both accidental harm reduction and an efficient barrier disregarding gasses, water vapor, and soluble chemicals are provided by edible coatings. This feature facilitates in reduction of the internally generated oxygen partial pressure of newly produced goods, which lowers deterioration and metabolic activity and helps to preserve volatile ingredients (Aayush et al. 2022). Edible coating arises as a direct response to this requirement, providing an easy, eco-friendly choice that claims to increase the standard of food preservation. Among the various types of food preservation technologies available, edible coating applications are currently gaining a major interest due to their significant opportunities to enhance the shelf life of agricultural and processed food products (Erceg et al. 2023). Besides their nutritional value, edible coatings are significant because they are directly concerned with the idea of reliability, which is gaining importance in present food industries. This new technology shows the industry's preservance to implement procedures that focus on both the sustainability and shelf life of food products, opening the possibilities for a detailed analysis of edible coatings as a major success in regulation (Hernández-Carrillo et al. 2021).

According to Vimala Bharathi et al. (2020), edible coatings and packaging substances mostly on natural ingredients like proteins, lipids, and polysaccharides are increasing their popularity in food industries (Vimala Bharathi et al. 2020). Fresh-cut fruit and legumes are regularly coated using dipping or spraying techniques that limit processing steps; however, recently, electrospray technology has been used to ensure uniform coating distribution. The use of edible coverings on flower-like foods like artichokes, brassica, broccoli, and cauliform remains to be researched. Alginate has become one of the most accepted biopolymers used for coating fruit and vegetables. Research studies showed that alginate-covered flowers stay fresh for 4 more days (in comparison to untouched flowers). Then flowers are encapsulated with sodium alginate 0.5% and conjugation occurs with 1% $CaCL_2$ solution. At 4 °C, alginate-coated flowers may resist deterioration for 14 days (Jurić et al. 2023). In broccoli garner, alginate coatings are added with essential oils and antibacterial drugs such as natamycin enhanced radiotherapy treatment and offer a huge advantage. Edible coatings which are applied directly on a wide range of agriculture-based products can enhance product quality, food safety, and bioavailability and reduce the cost of manufacturing. Edible coating's primary function is as a barrier to gas exchange and moisture loss while also shielding the inner color and aromatic volatile compounds of fresh fruits and vegetables.

Edible coatings also prevent food from taking in unpleasant orders, prolong the oxidation of enzymes, and protect fresh fruit from browning and shrinking while being transported (Kazemeini et al. 2021). They also help to lessen the harm that disturbances and mechanical pressure may induce. Additionally, edible coatings act as a barrier for useful substances including tastes, nutrients, antioxidants, and antibacterial agents, improving the quality and enhancing the duration of storage of fresh products. Edible coatings increase flexibility of the processed frozen meals while also reducing the decomposition and moisture absorption. Despite coating techniques that have been widely researched and used to preserve fresh-cut fruits and vegetables after harvest, minimum information is known about their use in edible flowers. In one research, Night-Fragrant blooms were coated with a mixture of sodium alginate, chitosan, and protein from soybean concentrate (Ahmadimaram et al. 2021). Results showed that the coatings of the composite surface successfully sustained sensory effectiveness, prevented shedding, lessened weight loss and the deterioration of vitamin C content, and suppressed polyphenol oxidase (PPO) activity. An aqueous solution including 3% soy protein isolate, 0.25% sodium alginate, and 0.7% glucosamine is known to be the best-performing formulation for increasing the freshness of *Night-Fragrant* flowers (Kiraman and Yusof 2022). This finding has Improved Night-Fragrant flower production, sales, and storage spaces.

Similarly, it has been demonstrated that these coatings enhance the concentrations of hydrolyzable tannins, anthocyanins, flavonoids, and antioxidant activities in pansy flowers in comparison with flowers that are not coated. Compared to unprocessed flowers, coated flowers had a 7-day shelf life, lasting 14 days. Upon storage for 14 days, the alginate-coated pansies indicated reduced levels of fungal and bacteria compared to their uncoated peers (Pinton et al. 2020). According to studies on the impacts of packing and alginate coating on pansy samples, there were several positive impacts on visual appearance, change in color, weight reduction, moisture activity, and morphological adjustments. Alginate coatings significantly preserved the visually appealing integrity of flowers in a chilled setting at 5 °C, prolonging the degradation time from 3–4 days to 14 days, in comparison to unprotected treatments. In the final analysis, spraying edible coatings on flowers while they are in the refrigerator may effectively extend their shelf life, making this a straightforward and useful preservation method. To ensure sustainability through the process, it makes sense to emphasize the use of culinary and biodegradable substances. Furthermore, safety studies need to be the initial priority when utilizing active edible coatings comprised of nanomaterials (Chen et al. 2022).

13.11.4 Irradiation Technology

An efficient financially as well acceptable alternative for enhancing the shelf life and safety of perishable foods depends on the incorporation of radiation technology. Irradiation techniques have been successfully implemented in food processing industries for many different kinds of objectives, such as microbiological decontamination, sprout oppression, insect disinfestation, and ripening delay. Delays in cellular death, which show up as changes in the physicochemical and physiological aspects of edible flowers, are responsible for their increased shelf life. Oxidative damage imposed through an excessive accumulation of peroxides and reactive oxygen species (ROS) can frequently result in these modifications (Chongtham et al. 2022). Food is exposed to ionizing radiation during the irradiation process, predominantly in the form of gamma radiation or electron-beam radiation. Many investigations have examined the consequences of irradiation on the biochemical components of edible flowers for preserving flowers. Studies using electron beam radiation show that the irradiation procedure has no noticeable effect on the nutritional makeup of flowers. Furthermore, the blooms' biological function did not diminish (Vimala Bharathi et al. 2020). Furthermore, it has been determined that irradiation employs the phenolic compounds more easily extracted, most likely because of the capacity they have to interrupt chemical bonds. Thus, it is potential to determine that edible flowers may be preserved via electron beam irradiation with minimal to no loss of quality (Bigi et al. 2023). Food manufacturers could benefit through the use of irradiation technology, which provides fortification and sterilizing without impacting the product's nutritional content, organoleptic compounds properties, or bioactive ingredients. Present research findings undoubtedly demonstrated that irradiation may improve the general phenol concentrations and antioxidant levels while still maintaining biological properties. For stronger marketplace acceptableness, it is important to teach and persuade people about the positive health benefits and advantages of incorporating irradiation is still and important component. Irradiation is an excellent non-chemical treatment for enhancing shelf life by eliminating disturbance while improving the sanitation and safety of edible flowers (Bisht et al. 2021).

Previous investigation has proven that irradiation has repellent effects on flowers; for example, a dosage of 300 Gy is thought to be adequate for accomplishing insect and parasite sterilization at every phase of development. Besides, low-dose irradiation has a neutralizing effect. But a dosage of a small amount as 200 Gy can cause the death of dangerous insects, it may not fully cause post-harvest fungal infections in fresh flowers. A multitude of variables which involve as pretreatment medication, the level of moisture, chemical makeup, and flower storage conditions, influence the consequences of radiation exposure in food. The research examined the result of electron beam radiation of 0.5 kGy, 0.8 kGy, and 1 kGy on chemical and nutritional value, also cytotoxicity, antioxidant, and anti-inflammatory activities of *Bauhinia variegata L. var. candida alba Buch.* -Ham flowers (Castillejo et al. 2021). The electron beam irradiation effect at doses of 0.5 kGy, 0.8 kGy, and 1 kGy on the nutritional and chemical characteristics, as well as the cytotoxicity, antioxidant, and anti-inflammatory activities of *Bauhinia variegata* L. var. *candida alba* Buch. -Ham flowers (Castillejo et al. 2021). The findings demonstrated a little reduction in anti-inflammatory effectiveness but no modification of the irradiation concentration on the nutritional status or cytotoxicity. Interestingly, antioxidant activity rose at a dose of 0.5 kGy, suggesting a rise in phenolic compounds. Overall, the radiation treatment demonstrated no noticeable impact on antioxidant activity or phenolic component enthusiastic, allowing *V. tricolor* samples to be preserved for a longer duration of time and provide an affordable answer for this peculiar food category (da Rocha and Noreña 2020). It is essential to remain mindful that the radiation dosage necessary to achieve fungal control could have a detrimental influence on the quality of the blooms. To successfully avoid the establishment of postharvest fungal diseases, the primary objective should be to minimize radiation doses or research other alternatives. A possible choice is to combine low doses of gamma rays with other restoration treatments, like physical therapy and ecologically friendly chemicals. A research investigation demonstrated that amalgamating gamma and UV-C irradiation enhanced effectiveness over solo treatments while reducing negative consequences on raw materials (Chen et al. 2020).

13.12 Conventional Method

Edible flowers are highly perishable and delicate, which can result in edible flower quality degradation which can cause discoloration, bad smell, harmed petals, and wilting, that's why they must be preserved using different methods that both preserve but also enhance quality. Edible flowers are traditionally processed via conventional drying. The drying process involves keeping the bloom in the sun, however, sun drying takes a long time to finish the process (Milião et al. 2022). Edible flower exposure for a long time in any drying process can cause changes in their physical and nutritional value, which reduces the quality of edible flowers (Awasthi et al. 2023). Non-thermal technologies are an excellent substitute for traditional drying techniques in thermal processing. Edible flowers are processed using low or mild temperatures in non-thermal methods. When it comes to processing and preserving edible flowers, they show promising outcomes (Feng et al. 2023).

13.13 Novel Drying Technology

This method of drying innovative food ingredients is an essential beginning to expanding their potential in the food industry. Drying simplifies the degradation of novel food commodities into segments and powdered substances, which have a longer shelf life and are more straightforward to handle and manage. Additionally, through the sterilization of novel food products from radiating and ionization impact, the drying method diminishes the danger of pathogens and allergies (Acharya et al. 2023). Drying techniques for commonly encountered fresh foods including meat, fish, fruits, and fresh produce have been the subject of several research investigations, but comprehensive analyses of drying techniques used for exceptional food resources remain displaced. In consideration of their various nutritional attributes, an in-depth investigation of the current research on drying technology for new food sources is imperative. The review explains the development of drying technology for new food resources such as microbes, microscopic algae, edible mushrooms, and whole grains (Demasi et al. 2021). There was further discussion to understand how dried new food resources could potentially be used in the preparation of foods, functional foods, and 3D-printed foods. To assemble an illustration that will promote the distribution and marketing of new food resources in future food advancement, this review would assist with comprehending the current advancements and barriers of drying methods used in the processing of novel food resources. Drying is essential to get adequate cutting-edge food resources and extracts (protein, polysaccharides, and so on), which are commonly used as food components in food preparation as chunks, slices, and powders (Barber et al. 2020). The nutrient composition and sensory characteristics of accomplished meals are greatly influenced by the permeation quality of new food products. Although they allow effective enough and substantial nutritional resources together in order with enhanced sensory qualities with intelligent management, dried novel foods are an essential supplement to future food development. Innovative foods provide superior benefits and future food processing opportunities, but the obstacles are still extremely serious at that moment. More investigation into revolutionary drying techniques must be conducted to maximize the duration of drying effectiveness and excellence of hydrated products (Bisht et al. 2021).

13.13.1 Freeze Drying Technology

Temperature plays a crucial role in maintaining the hardness and freshness of agricultural foods. Lowering the operational temperature enables decreased drainage and respiration, which can dramatically decrease the aging process of flowers. Additionally, reduced temperatures slow down bacteria development and spoiling by lowering their activity in enzymes. Refrigeration and cold storage may greatly improve the shelf life of flowers. However, it is crucial to recognize that cold damage might influence the overall quality of certain flowers (Aung et al. 2020). To execute an extensive investigation, researchers maintained numerous flower kinds in tightly sealed containers and inspected them every few days. The findings were intriguing—all flower types fared well in temperatures below freezing over the week! The most interesting were the viola, pansy, and nasturtium, which could be kept fresh for a maximum of a week at temperatures between 0 and 2.5 °C (Bilbao-Sainz et al. 2020). Most edible flower species may be securely kept at 4 °C for 7 to 14 days, according to recent research that looked at 17 different species. Since refrigeration retains edible flowers looking at and inhaling fresh while stopping germs from propagating this is a wonderful development for marketing vegetables at accessibility shops where refrigeration is frequent.

In a related study, scientists examined the behavior of Listeria monocytogenes on broccoli preserved at different temperatures. They perceived that, even though refrigeration was beneficial, broccoli may promote the growth of this bacteria at higher temperatures (Gao et al. 2020). It was surprising when it was observed that the bacteria could persist for up to 168 days at low manipulates. Therefore, it is highly advised to keep storage temperatures at 4 °C or −18 °C to lower the risk of germs. Although freezing is one of the easiest ways to preserve edible flowers, maintaining a consistent cold chain across the supply chain remains hazardous. Freezing may also work! Because frozen particles form, they not only prevent natural aging processes but also destabilize microbes. While freezing preserves the important nutrients in flowers, it can also damage some microbes. Keeping in mind that freezing is unlikely to eliminate the function of enzymes, so pre-treating with anything like being blanched can help deactivate them instead of freezing. Of course, blanching may alter the nutritional and taste composition of the flowers slightly, but creative procedures can reduce these losses. Fernandes et al. (2020) found that neither manual freezing nor freezing flowers in ice cubes were efficient for preserving four specific varieties: fresh white borage (*B. officinalis*), heartsease (*V. tricolor*), yellow kalanchoe (*K. blossfeldiana*), and dandelion (*T. officinale*). Both techniques Both strategies effectively controlled microbial development, though the ice-cube strategy performed much more effectively at maintaining an innovative appearance. However, several minerals and antioxidants degraded all during the preservation analysis, whereas straight freezing maintained good quality, with some activity as antioxidants rising after a month of storage.

Researchers have also attempted to incorporate freezing with non-thermal processing methods that include ultrasonic and high-pressure processing. Such techniques can encourage super-cooling and aid in controlling the process of freezing crystal formation, which is extremely important for the quality of end product. For example, Zhu et al. (2020) discovered that artichoke (Cynara scolymus) flowers are kept at various temperatures had variable levels of phenolic and vitamin C contents (Zhu et al. 2020). Compared to fresh flowers or flowers preserved at other temperatures, flowers memorized at 5 °C comprised more phenolics. In an unusual broccoli study, researchers studied two types of cauliflower (white and green) to compare regular and modified freezing technologies. By blanching some and injecting citric acid to others, they were able to assess how different approaches changed antioxidants, beta-carotene, and vitamin C over time. Vitamin C levels diminished by about 27–29% whenever white cauliflower was cooked, but after a year of storage, green cauliflower showed significantly greater decrease. The green variant was surprisingly evaluated higher for sensory properties (Halder and Khaled 2021). The pros and cons associated with creative preservation techniques have been demonstrated by the higher antioxidant activity and improved vitamin C retention of cauliflowers prepared using customized methods as opposed to those frozen using standard procedures. In conclusion, strict temperature control must be implemented to maintain the shelf life of fruits, vegetables, and herbs. These days, cold storage methods including freezing and refrigeration, and even hot-air drying, are widely used in the edible flower industry. Chilling delays aging and maintains quality by reducing respiration, tissue degradation, and water loss. Though each kind may respond strongly to temperature changes, it's vital to keep in mind that edible flowers haven't been thoroughly tested at temperatures lower than −2.5 °C yet. For example, certain develops, like scarlet runner beans, may be kept for 2 weeks at conditions among −2.5 and 20 °C. Our understanding of how particular temperatures impact the quality and duration of storage for various flowers will improve with more research (Bortolini et al. 2022).

13.13.2 Microwave Drying

Vacuum-microwave drying has already been demonstrated to be far superior then typical air drying in keeping the purple color of coneflower and amount of caffeic acid alternatives. Their iron-chelating abilities and high general phenolic content of black locust flowers that were vacuum-microwave dried demonstrated that this process is more cost-effective than freeze drying since it just takes minutes like hours or days, which makes it enhances innovative to think more about it in future. Far-infrared radiation combines with hot-air convection drying which protect the color of marigolds and retains the higher possible level of phenolic component, lutein, and lycopene as compared to fresh flowers (Huang et al. 2021).

13.14 Bioactive Compounds in Edible Flowers

Edible flowers are rich source of bioactive compounds i.e. phenolic compounds, flavonoids, antioxidant compounds, anthocyanins, carotenoids, saponins, xanthophylls, vitamin-C, and flavones (Pereira et al. 2020). Flowers have a high content of natural antioxidants such as anthocyanin, flavonoids and various other phenolic components (Rao and Poonia 2024a). Due to the presence of a variety of bioactive compounds the edible flowers used in nutraceutical foods provide many health benefits. These bioactive compounds act as anti-obesity, anti-cancer, anti-proliferative, and anti-inflammatory (Takahashi et al. 2020). As Fig. 13.2 Phenolic acids,

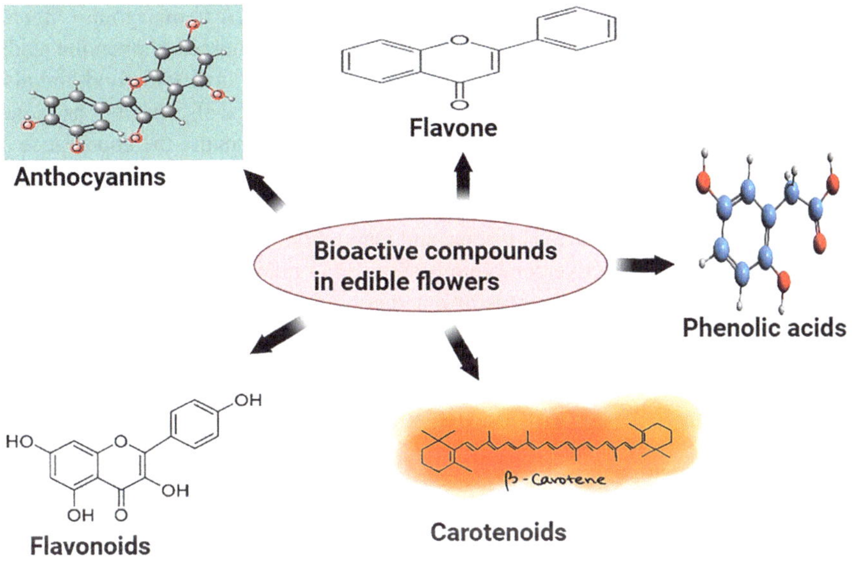

Fig. 13.2 Phytochemicals in edible flowers

stilbenes, anthocyanins, flavanols, flavanones, and flavonols are included in phenolic compounds present in edible flowers (Kritsi et al. 2022). The edible flowers named mini rose have good amount of phenolic compounds (Simin et al. 2023). Based on oxygen radical absorbance capacity (ORAC) the total phenolic compounds, carotenoids, and antioxidant properties of edible flowers can be determined (Prabawati et al. 2021). One of the best sources of polyphenols and antioxidants is the edible flower named cosmic yellow (Kandylis 2022). Some edible flowers are good origins of pro-vitamins i.e. β-carotene and β-cryptoxanthin (Purohit et al. 2021) Tagetestenui folia contains a high content of L. ascorbic acid and carotenoids. Xynthophylls are mostly present as carotenoids in edible flowers. In feijoa flower the flavonoids that are present are quercetin, flavone, isoquercetrin and hyperoside (Gil et al. 2023). As a phenolic compound p -hydroxybenzoic acid, p-coumaric, quercetin-3-rutinose, and ferulic acid, caffeic acid are present in apple blossom edible flower. Gastrointestinal tract problems are treated by phenolic compounds and flavonoids which are present in chamomile flowers (Sarkar and Chawla 2022).

From poly-phenolic compounds, flavonoids are an important part of edible flowers. The color of the flowers and plants is due to it (Kuljarusnont et al. 2024). They have antioxidant properties and also have lowering the risk of cancer and heart disease. Edible flowers like Hibiscus, marigolds, blue pea flowers, chamomile, and roses are wealthy sources of flavonoids (Teixeira et al. 2023). Flavonoids include flavones, flavanones, isoflavones, Anthocyanins. Hydroxybenzoic acid and hydroxycinnamic acid are two classifications of phenolic acids (Al Mamari 2021). Gallic acid and vanillic acid are included in hydroxybenzoic acid and caffeic acid, ferulic acid, p-coumaric acid, and chlorogenic acid are included in hydroxybenzoic acid (da Silva et al. 2023). In the edible flowers roses (*Rosa species*), marigolds (*Tagetes species*), and hibiscus (*Hibiscus sabdariffa*) are excellent sources of phenolic acids. These compounds when added to food enhance the functionality of food and also impart therapeutic effects on human health. Carotenoids are also present as bioactive compounds in edible flowers (Pires Jr et al. 2021). Normally the appearance of an edible flower is due to carotenoids but it contains a huge amount of phenolic and antioxidant compounds. The edible flowers that are rich in carotenoids are Canlendula, Saffron, Pansies, Hibiscus, and Dandelion. These edible flowers which are a rich source of bioactive compounds are used in many food products like salad dressings, used in herbal teas, infusions, and cooking oils (Duggirala et al. 2024).

13.15 Extraction Methods of Bioactive Compounds

There are following methods for the extraction of bioactive compounds from edible flowers are discussed below

1. Microwave-assisted extraction method (MAE)
2. Supercritical fluid extraction (SFE)
3. Ultrasonic -assisted extraction (UAE)

13.15.1 Microwave-Assisted Extraction (MAE)

An erudite, environmentally welcoming extraction technique called microwave-assisted extraction (MAE) is regularly used to distinct bioactive materials from a diversity of natural foundations, including edible flowers (Nonglait and Gokhale 2024). Likened to traditional approaches, this approach has the benefit of shorter removal times, less solvent use, and higher goal compound produces. Numerous traditional extraction techniques, including maceration, distillation, and Soxhlet extraction, have been used to extract bioactive mechanisms from saffron floral by-products (Ozkan et al. 2021). The expansion of novel processing techniques that can be used on an industrial scale has been encouraged by the requirement to use environmentally satisfactory technology to circumvent harmful solvents. When associated with traditional methods, new green energy-saving technologies like microwave-assisted extraction (MAE) offer several welfares. These comprise improved extraction kinetics, an earlier extraction time and temperature increase, increased effectiveness and extraction yield, and concentrated energy consumption and cost (Álvarez et al. 2021). MAE is based on joining conventional solid-liquid extraction with microwave heating. Breaks down plant cell walls and relieves bioactive chemicals by rapidly heating the solvent and sample using microwave energy.

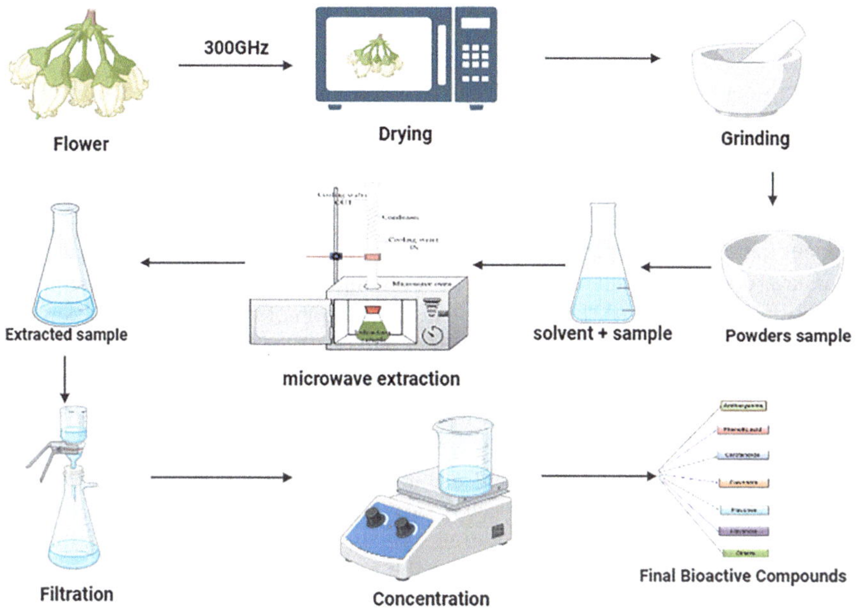

Fig. 13.3 Microwave-assisted extraction (MAE) method

This method is established on energy transfer mechanisms such as ionic transmission and dipole rotation. Cells are broken by the radiation, permitting the solvent to pass through the plant matrix. Contrariwise, plant matter arrives in the solution from outside the cells (Ozkan et al. 2021). The kind of solvent, temperature, extraction time, and power are some of the main variables that possibly impact the MAE process. Because it saves money, time, and energy, MAE is observed as a green technology (Cerdá-Bernad et al. 2022). Figure 13.3 shows the process of extraction of bioactive compounds from edible flowers.

To advance the extraction process from food and agro-industrial by-products, the method makes use of the communication between the target material and microwave radiation, which yields electromagnetic waves with a frequency that usually, ranges from 300 MHz to 300 GHz. When the material engrosses microwave energy, polar molecules are banished, causing dipole rotation. This forces the particles to align with the current electric field, producing internal heat and contravention chemical bonds that make extraction easier (Vinatoru et al. 2017). MAE habitually only needs a slight amount of solvent; in fact, the extraction process can be approved without any solvent at all, using the sample's natural water to act as a solvent and endorse cell lysis. Nevertheless, its benefits, not all sample types can benefit from microwave-assisted extraction, and the optimum outcomes for each specific application depend on the extraction parameters being improved (Cannavacciuolo et al. 2024).

13.15.2 Supercritical Fluid Extraction (SFE)

Supercritical Fluid extraction (SFE) is an alternative to solvent extraction of the soluble compounds from the plant matrix, an eco-friendly extraction process based on supercritical fluids. We can define it as a substance that provides a quick mass transfer rate having low viscosity and high diffusion rate. The extraction of compound solubility can be increased by increasing the pressure which may related to solvent density. The supercritical fluid extraction depends on the temperature and pressure, due to pressure enhancement the temperature increases the compound solubility (Chai et al. 2021). For industrial production of flower extract, supercritical CO_2 most often used extraction techniques that has been used (Lopez-Hortas et al. 2022). Under the observation of human Health and the environment with the high-level recovery of bioactive extract from edible plant tissues, supercritical solvent extraction is the most popular and well-established method that has been used. Due to its low polarity, the supercritical fluid extraction that is non-polar is difficult for anthocyanin extraction which is an ionic compound. For the extraction of anthocyanin, the synergistic solvent addition can significantly increase the extraction ability of supercritical fluid by changing the non-polar nature of the solvent. The supercritical CO_2 efficacy should be increased when used in combination with other extraction methods (Chai et al. 2021). Supercritical fluid extraction is a superior technology to traditional methods; by using these unique properties this extraction method facilitates the transport properties and the extraction of bioactive compounds from the vegetable matrix. Solvent selectivity should be enhanced by changing the solvent strength and the supercritical fluid has higher properties other than liquid and gasses (Nunes et al. 2022). The SFE is a most effective and efficient technique but when we compare it with the traditional method the main disadvantage is it's expensive. In this extraction CO_2 is the most commonly used solvent because of its superficial properties that include low cost, high diffusivity, non-toxic, and inert. The solvent has beneficial properties no residues will left in the product because it can easily be removed from the extract (Lefebvre et al. 2021).

13.15.3 Ultrasound-Assisted Extraction (UAE)

Ultrasound-assisted extraction is the most widely used method for the extraction of bio-active substances (Carrera et al. 2021). This technique has been s uggested for mass transfer and solvent extraction because it can effectively extract the bioactive compounds and a minimum quantity of solvent is required. Due to low cost and high working principle UAE has been used in industrial applications as an efficient method to separate and identify compounds from the tissue matrix (Saqib Farooq et al. 2020). Different researchers preferred UAE over traditional extraction methods for the extraction of dissolved compounds (Dash et al. 2021). In this method, ultrasound waves are used to extract bioactive compounds. The range of waves is typically about 20 kHz to 1 MHz (Yeasmen et al. 2024). Ultrasonic waves attack the cell wall which makes it easier to extract bioactive compounds from edible flowers (Bolat et al. 2024). Compared to the other methods UAE is less energy user and provides extracted material in less time (Demesa et al. 2024). This Technique is especially good for sensitive to temperature bioactive substance and preserve their nutritional composition. Three steps are involved in this extraction i.e. mixing the sample with solvent, applying high-frequency ultrasonic waves, and then the extracted material should separate by centrifugation or filtration (Bhadange et al. 2024). Here Table 13.4 shows health benefits and bioactive compounds in edible flowers. Nanotechnology represents an emerging field of scientific exploration focused on the examination and control of matter at the smallest scales, encompassing molecules and atoms (Rao and Poonia 2024a, b).

13.16 Future Perspective

Edible flowers are frequently used as ingredients in both traditional and gourmet cuisine, and they have a lot of capacity to intensify the nutritional value and visual appeal of food items. The chemical makeup and major groupings of chemical compounds found in the flowers of commonly used species are usefully described in this article. Finding potentially harmful substances in some species that require more distant thought before initiating or recommending their culinary use is also beneficial. Since flowers are so common, there would seem to be a great chance to reevaluate their place in modern cuisine and to stop thinking of them only as decorative or for their fragrance qualities. The use of flowers in meals has a lot of potential. It will benefit the current generation as well as perhaps provide food for future generations.

13.17 Conclusion

Edible flowers appear to be a desirable alternative for new food trends, with possible health benefits around the globe, despite cultural differences and consumption patterns. Due to their low fat content and high fiber or even protein content, flowers have been linked to a favorable nutritional profile that satisfies a variety of dietary requirements, including those of vegetarians and vegans. However, given the large number of species that are often used as edible flowers and the fact that only an insignificant portion of them have had their chemical composition and safety for human consumption thoroughly studied, more research is required to initiate regulations regarding the safe utilization of both traditional species where there is still a knowledge gap and novel species that sometimes introduced in culinary arts. Given this, the market niche for edible flowers is quite attractive to businesspeople in the agriculture and food industries.

Table 13.4 Edible flower, Health benefits, Bioactive compounds, and their mechanism

Scientific name	Figure	Common name	Advantages	Bioactive substance	Action mechanism	References
Bauhinia variegata		Orchid tree	Antidiabetic, and anti-oxidant	Polyphenol	Bioactive substance that has anti-diabetic activities	Tripathi et al. (2019)
Catharanthus roseus		Madagascar periwinkle, old maid	Antioxidant	Phenolic compound	Phenolic compound that gives it antioxidant properties	Ojulari et al. (2019)
Hibiscus sabdariffa		Roselle, Jamaica Sorrel	Anticancer activity, Anti-inflammatory, Anti-obesity	Flavonoids, phenolic compounds, and anthocyanins	Flavonoids found in Roselle petals have antioxidant properties which are crucial in the pathophysiology of cancer.	Abdelhafez et al. (2018)
Hibiscus malvaviscus		Wax Mallow	Hepatoprotective activity	β-Reso cyclic caffeic, 4-hydroxyphenylacetic acids, two flavonoids	Synergistic antioxidant and liver-protecting properties of phenolic compounds	Loizzo et al. (2016)
Malva sylvestris		Watercress	Anti-oxidant activity, Hypoglycemic effects	Quercetin	High level of radical scavenging activity	Chaudhary et al. (2018)

Rosa spp.		Rose	Anti-cancer activity	Phenolics and flavonoids	High anthocyanin content is thought to be the cause of the high antioxidant activity.	Moliner et al. (2018)
Tagetes erecta		African marigold	Antioxidant activity, Neuroprotective, Anticancer	Myricetin, Quercetin	Myricetin in T. erecta extracts reduces the synthesis of β amyloid protein in neurons.	Varahi Vedam et al. (2019)
Woodfordia fructicosa		Fire flame bush, Hypoglycemic effects	Hypoglycemic effects	Quercetin and quinin acid	Woodfordia fructicosa flowers contain flavonoids and phenolics that increase blood insulin and C-peptide levels.	Arya et al. (2015)

References

Aayush K, McClements DJ, Sharma S, Sharma R, Singh GP, Sharma K, Oberoi K (2022) Innovations in the development and application of edible coatings for fresh and minimally processed Apple. Food Control 141:109188

Abbet C, Mayor R, Roguet D, Spichiger R, Hamburger M, Potterat O (2014) Ethnobotanical survey on wild alpine food plants in lower and Central Valais (Switzerland). J Ethnopharmacol 151(1):624–634

Abdelhafez OH, Fawzy MA, Fahim JR, Desoukey SY, Krischke M, Mueller MJ, Abdelmohsen UR (2018) Hepatoprotective potential of Malvaviscus arboreus against carbon tetrachloride-induced liver injury in rats. PLoS One 13(8):e0202362

Abdel-Haleem SA, Ibrahim AY, Ismail RF, Shaffie NM, Hendawy SF, Omer EA (2017) In-vivo hypoglycemic and hypolipidemic properties of Tagetes lucida alcoholic extract in streptozotocin-induced hyperglycemic Wistar albino rats. Ann Agric Sci 62(2):169–181

Abdullah ATM, Rahman MM, Sharif M, Khan TA, Islam SN, Karim KMR (2024) Edible flowers efficiency to boost the thermal oxidation stability of soybean oil: polyphenolic and antioxidant insights. Future Foods 9:100338

Aboonajmi M, Ganjdoost M, Mirsaeedghazi H, Asefpour Vakilian K (2022) Effect of power ultrasound treatment on the shelf life of edible mushroom. J Food Res 32(2):139–152

Acharya B, Behera A, Sahu PK, Dilnawaz F, Behera S, Chowdhury B, Mishra DP (2023) Bamboo shoots: an exploration into its culinary heritage in India and its nutraceutical potential. J Ethnic Foods 10(1):22

Ahmadimaram F, Mostaghim T, Shahriari S (2021) Evaluation of alginate/collagen edible coatings with betanin and cumin to improve the shelf life of lighvan cheese. J Food Biosci Technol 11(2):69–84

Al Mamari HH (2021) Phenolic compounds: classification, chemistry, and updated techniques of analysis and synthesis. In: Phenolic compounds-chemistry, synthesis, diversity, non-conventional industrial, pharmaceutical and therapeutic applications, vol 10: Published by Springer Nature and Nutrition Journal (Nutr J)

Alshehri MM, Quispe C, Herrera-Bravo J, Sharifi-Rad J, Tutuncu S, Aydar EF et al (2022) A review of recent studies on the antioxidant and anti-infectious properties of Senna plants. Oxid Med Cell Longev 2022(1):6025900

Álvarez A, Terreros S, Cocero MJ, Mato RB (2021) Microwave pretreatment for the extraction of anthocyanins from saffron flowers: assessment of product quality. Antioxidants 10(7):1054

Alvarez-Suarez JM, Giampieri F, Tejera E, Battino M (2023) Malvidin: advances in the resources, biosynthesis pathway, bioavailability, bioactivity, and pharmacology. In: Handbook of dietary flavonoids. Springer, Cham, pp 1–35

Anusiya G, Gowthama Prabu U, Yamini NV, Sivarajasekar N, Rambabu K, Bharath G, Banat F (2021) A review of the therapeutic and biological effects of edible and wild mushrooms. Bioengineered 12(2):11239–11268

Arslaner A, Salik MA, Bakirci I (2021) The effects of adding Hibiscus sabdariffa L. flowers marmalade on some quality properties, mineral content and antioxidant activities of yogurt. J Food Sci Technol 58:223–233

Arya A, Al-Obaidi MMJ, Karim RB, Taha H, Khan AK, Shahid N et al (2015) Extract of Woodfordia fruticosa flowers ameliorates hyperglycemia, oxidative stress and improves β-cell function in streptozotocin–nicotinamide induced diabetic rats. J Ethnopharmacol 175:229–240

Aung TT, Myat YY, Mar MM, Kyu KK (2020) Nutritional compositions, elemental compositions and antinutrient factor in different varieties of water lily. 3rd Myanmar Korea Conf Res J 3(5):1917–1922

Awasthi MK, Kumar V, Hellwig C, Wikandari R, Harirchi S, Sar T et al (2023) Filamentous fungi for sustainable vegan food production systems within a circular economy: present status and future prospects. Food Res Int 164:112318

Baessa M, Rodrigues MJ, Pereira C, Santos T, da Rosa Neng N, Nogueira JMF et al (2019) A comparative study of the in vitro enzyme inhibitory and antioxidant activities of Butea monosperma (Lam.) Taub. and Sesbania grandiflora (L.) Poiret from Pakistan: new sources of natural products for public health problems. S Afr J Bot 120:146–156
Bahrami A, Baboli ZM, Schimmel K, Jafari SM, Williams L (2020) Efficiency of novel processing technologies for the control of Listeria monocytogenes in food products. Trends Food Sci Technol 96:61–78
Bao X, Min R, Zhou K, Traffano-Schiffo MV, Dong Q, Luo W (2023) Effects of vacuum drying assisted with condensation on drying characteristics and quality of apple slices. J Food Eng 340:111286
Barani YH, Zhang M, Mujumdar AS, Chang L (2022) Preservation of color and nutrients in anthocyanin-rich edible flowers: progress of new extraction and processing techniques. J Food Process Preserv 46(9):e16474
Baranović G, Šegota S (2018) Infrared spectroscopy of flavones and flavonols. Reexamination of the hydroxyl and carbonyl vibrations in relation to the interactions of flavonoids with membrane lipids. Spectrochim Acta A Mol Biomol Spectrosc 192:473–486
Barber TM, Kabisch S, Pfeiffer AF, Weickert MO (2020) The health benefits of dietary fibre. Nutrients 12(10):3209
Bartkiene E, Zarovaite P, Starkute V, Mockus E, Zokaityte E, Zokaityte G et al (2023) Changes in lacto-fermented Agaricus Bisporus (White and Brown varieties) Mushroom characteristics, including biogenic amine and volatile compound formation. Foods 12(13):2441
Bedoya MG, Montoya DR, Tabilo-Munizaga G, Pérez-Won M, Lemus-Mondaca R (2022) Promising perspectives on novel protein food sources combining artificial intelligence and 3D food printing for food industry. Trends Food Sci Technol 128:38–52
Benvenuti S, Bortolotti E, Maggini R (2016) Antioxidant power, anthocyanin content and organoleptic performance of edible flowers. Sci Hortic 199:170–177
Bhadange YA, Carpenter J, Saharan VK (2024) A comprehensive review on advanced extraction techniques for retrieving bioactive components from natural sources. ACS omega 9(29):31274–31297
Bigi F, Maurizzi E, Quartieri A, De Leo R, Gullo M, Pulvirenti A (2023) Non-thermal techniques and the "hurdle" approach: how is food technology evolving? Trends Food Sci Technol 132:11–39
Bilbao-Sainz C, Sinrod AJ, Williams T, Wood D, Chiou BS, Bridges DF et al (2020) Preservation of tilapia (Oreochromis aureus) fillet by isochoric (constant volume) freezing. J Aqua Food Prod Technol 29(7):629–640
Bisht B, Bhatnagar P, Gururani P, Kumar V, Tomar MS, Sinhmar R et al (2021) Food irradiation: effect of ionizing and non-ionizing radiations on preservation of fruits and vegetables–a review. Trends Food Sci Technol 114:372–385
Bolat E, Sarıtaş S, Duman H, Eker F, Akdaşçi E, Karav S, Witkowska AM (2024) Polyphenols: secondary metabolites with a biological impression. Nutrients 16(15):2550
Bordewijk M, Schifferstein HN (2020) The specifics of food design: insights from professional design practice. Int J Food Design 4(2):101–138
Bortolini DG, Barros L, Maciel GM, Brugnari T, Modkovski TA, Fachi MM et al (2022) Bioactive profile of edible nasturtium and rose flowers during simulated gastrointestinal digestion. Food Chem 381:132267
Cai ZX, Chen MY, Lu YP, Guo ZJ, Zeng ZH, Liao JH, Zeng H (2022) Metabolomics and transcriptomics unravel the mechanism of browning resistance in Agaricus bisporus. PLoS One 17(3):e0255765
Cannavacciuolo C, Pagliari S, Celano R, Campone L, Rastrelli L (2024) Critical analysis of green extraction techniques used for botanicals: trends, priorities, and optimization strategies-a review. TrAC Trends Anal Chem 173:117627

Cardwell G, Bornman JF, James AP, Daly A, Strobel N, Jakobsen J, Black LJ (2023) Effect of air-drying on the generation of vitamin D2 and 25-hydroxyvitamin D2 by pulsed UV irradiation in button mushroom (Agaricus bisporus). J Food Compos Anal 115:105034

Carrera C, Aliaño-González MJ, Valaityte M, Ferreiro-González M, Barbero GF, Palma M (2021) A novel ultrasound-assisted extraction method for the analysis of anthocyanins in potatoes (*Solanum tuberosum* L.). Antioxidants 10:1375

Castillejo N, Martínez-Zamora L, Artés-Hernández F (2021) Periodical UV-B radiation hormesis in biosynthesis of kale sprouts nutraceuticals. Plant Physiol Biochem 165:274–285

Cerdá-Bernad D, Baixinho JP, Fernández N, Frutos MJ (2022) Evaluation of microwave-assisted extraction as a potential green technology for the isolation of bioactive compounds from saffron (Crocus sativus L.) floral by-products. Foods 11(15):2335

Chai YH, Yusup S, Kadir WNA, Wong CY, Rosli SS, Ruslan MSH, Chin BLF, Yiin CL (2021) Valorization of tropical biomass waste by supercritical fluid extraction technology. Sustainability 13:233

Chaudhary SACHIN, Hisham HAZAR, Mohamed DOHA (2018) A review on phytochemical and pharmacological potential of watercress plant. Asian J Pharm Clin Res 11(12):102–107

Chen GL, Chen SG, Xiao Y, Fu NL (2018) Antioxidant capacities and total phenolic contents of 30 flowers. Ind Crop Prod 111:430–445

Chen F, Zhang M, Yang CH (2020) Application of ultrasound technology in processing of ready-to-eat fresh food: a review. Ultrason Sonochem 63:104953

Chen C, Tang T, Shi Q, Zhou Z, Fan J (2022) The potential and challenge of microalgae as promising future food sources. Trends Food Sci Technol 126:99–112

Chiou SY, Sung JM, Huang PW, Lin SD (2017) Antioxidant, antidiabetic, and antihypertensive properties of Echinacea purpurea flower extract and caffeic acid derivatives using in vitro models. J Med Food 20(2):171–179

Chongtham N, Bisht MS, Premlata T, Bajwa HK, Sharma V, Santosh O (2022) Quality improvement of bamboo shoots by removal of antinutrients using different processing techniques: a review. J Food Sci Technol 59:1–11

Coyago-Cruz E, Moya M, Méndez G, Villacís M, Rojas-Silva P, Corell M et al (2023) Exploring plants with flowers: from therapeutic nutritional benefits to innovative sustainable uses. Foods 12(22):4066

Coyago-Cruz E, Alarcón A, Guachamin A, Méndez G, Osorio E, Heredia-Moya J et al (2024) Functional, antioxidant, antibacterial, and antifungal activity of edible flowers. Antioxidants 13(11):1297

Dash DR, Pathak SS, Pradhan RC (2021) Extraction of oil from Terminalia chebula kernel by using ultrasound technology: influence of process parameters on extraction kinetics. Ind Crop Prod 171:11389

Dastidar AG, Chakraborty S, Saha A, Das S, Zaman QF (2023) Edible flowers: a new source of minerals. Int J Res Appl Sci Eng Technol 11:38–54

Demasi S, Mellano MG, Falla NM, Caser M, Scariot V (2021) Sensory profile, shelf life, and dynamics of bioactive compounds during cold storage of 17 edible flowers. Horticulturae 7(7):166

Demesa AG, Saavala S, Pöysä M, Koiranen T (2024) Overview and toxicity assessment of ultrasound-assisted extraction of natural ingredients from plants. Foods 13(19):3066

Denoya GI, Colletti AC, Vaudagna SR, Polenta GA (2021) Application of non-thermal technologies as a stress factor to increase the content of health-promoting compounds of minimally processed fruits and vegetables. Curr Opin Food Sci 42:224–236

Dhakad AK, Ikram M, Sharma S, Khan S, Pandey VV, Singh A (2019) Biological, nutritional, and therapeutic significance of Moringa oleifera Lam. Phytother Res 33(11):2870–2903

Dogra S (2024) Helianthus annuus L. In: Edible flowers. Academic, pp 105–126

Dourson D, Dourson J (2019) Wild yet tasty: a guide to edible plants of Eastern Kentucky. University Press of Kentucky

Duggirala K, Mummaleti G, Kong F, Roy A, Mohan A (2024) Edible flowers: a sustainable source of natural food ingredients. J Food Bioactiv 27:1–14

Erceg T, Šovljanski O, Stupar A, Ugarković J, Aćimović M, Pezo L et al (2023) A comprehensive approach to chitosan-gelatine edible coating with β-cyclodextrin/lemongrass essential oil inclusion complex—characterization and food application. Int J Biol Macromol 228:400–410

Espejel EAR, Alvarez OC, Muñoz JMM, Mateos MDRG, León MTBC, Damián MTM (2019) Physicochemical quality, antioxidant capacity and nutritional value of edible flowers of some wild dahlia species. Folia Hortic 31(2):331–342

Faisal Z, Saeed F, Afzaal M, Akram N, Shah YA, Islam F, Ateeq H (2022) Phytochemical profile and food applications of edible flowers: a comprehensive treatise. J Food Process Preserv 46(11):e17061

Farooq S, Shah MA, Siddiqui MW, Dar BN, Mir SA, Ali A (2020) Recent trends in extraction techniques of anthocyanins from plant materials. J Food Meas Charact 14:3508–3519

Feng M, Xu B, Islam MN, Zhou C, Wei B, Wang B et al (2023) Individual and synergistic effect of multi-frequency ultrasound and electro-infrared pretreatments on polyphenol accumulation and drying characteristics of edible roses. Food Res Int 163:112120

Fernandes L, Casal S, Pereira JA, Saraiva JA, Ramalhosa E (2017) Edible flowers: a review of the nutritional, antioxidant, antimicrobial properties and effects on human health. J Food Compos Anal 60:38–50

Fernandes L, Casal S, Pereira JA, Malheiro R, Rodrigues N, Saraiva JA, Ramalhosa E (2019a) Borage, calendula, cosmos, Johnny Jump up, and pansy flowers: volatiles, bioactive compounds, and sensory perception. Eur Food Res Technol 245:593–606

Fernandes L, Casal S, Pereira JA, Pereira EL, Saraiva JA, Ramalhosa E (2019b) Physicochemical, antioxidant, and microbial properties of crystallized pansies (Viola× wittrockiana) during storage. Food Sci Technol Int 25(6):472–479

Fernandes L, Saraiva JA, Pereira JA, Casal S, Ramalhosa E (2019c) Post-harvest technologies applied to edible flowers: a review: edible flowers preservation. Food Rev Intl 35(2):132–154

Fernandes L, Casal S, Pereira JA, Saraiva JA, Ramalhosa E (2020) An overview of the market of edible flowers. Food Rev Intl 36(3):258–275

Gao J, Sun Y, Li L, Zhou Q, Wang M (2020) The antiglycative effect of apple flowers in fructose/glucose-BSA models and cookies. Food Chem 330:127170

Ghavam M, Afzali A, Manconi M, Bacchetta G, Manca ML (2021) Variability in chemical composition and antimicrobial activity of essential oil of Rosa× damascena Herrm. from mountainous regions of Iran. Chem Biol Technol Agric 8:1–16

Gil KA, Jokić S, Cikoš AM, Banožić M, Jakovljević Kovač M, Fais A, Tuberoso CIG (2023) Comparison of different green extraction techniques used for the extraction of targeted flavonoids from edible feijoa (Acca sellowiana (O. Berg) Burret) flowers. Plants 12(7):1461

Gomes S, Belo AT, Alvarenga N, Dias J, Lage P, Pinheiro C et al (2019) Characterization of Cynara cardunculus L. flower from Alentejo as a coagulant agent for cheesemaking. Int Dairy J 91:178–184

Gostin AI, Waisundara VY (2019) Edible flowers as functional food: a review on artichoke (Cynara cardunculus L.). Trends Food Sci Technol 86:381–391

Guiné PF, Florença SG, Villalobos Moya K, Anjos O (2020) Edible flowers, old tradition or new gastronomic trend: a first look at consumption in Portugal versus Costa Rica. Foods 9(8):977

Guo S, Qin N, Wang X, Zuo Z, Li Q, Wang Y (2023) Freeze-dried powder of daylily bud improves bromocriptine-induced lactation disorder in rats via JAK2/STAT5 pathway. J Ethnopharmacol 313:116536

Halder S, Khaled KL (2021) Anti-nutritional profiling from the edible flowers of Allium cepa, Cucurbita maxima and Carica papaya and its comparison with other commonly consumed flowers. Int J Herb Med 9(5):55–61

Hamedani MP, Kashi MA, Goodarzi S, Shokouhi H, Ardestani MS, Hadjiakhoondi A et al (2023) Elucidation of flavonoids from potent Iranian Scutellaria species against Influenza A (H1N1) virus. Iran J Basic Med Sci 26(1):76

Harmayani E, Anal AK, Wichienchot S, Bhat R, Gardjito M, Santoso U et al (2019) Healthy food traditions of Asia: exploratory case studies from Indonesia, Thailand, Malaysia, and Nepal. J Ethnic Foods 6(1):1–18

Hernández-Carrillo JG, Orta-Zavalza E, González-Rodríguez SE, Montoya-Torres C, Sepúlveda-Ahumada DR, Ortiz-Rivera Y (2021) Evaluation of the effectivity of reuterin in pectin edible coatings to extend the shelf-life of strawberries during cold storage. Food Packag Shelf Life 30:100760

Huang D, Yang P, Tang X, Luo L, Sunden B (2021) Application of infrared radiation in the drying of food products. Trends Food Sci Technol 110:765–777

Hubert J, Kotland A, Henes B, Poigny S, Wandrey F (2022) Deciphering the phytochemical profile of an alpine rose (Rhododendron ferrugineum L.) Leaf extract for a better understanding of its senolytic and skin-rejuvenation effects. Cosmetics 9(2):37

Jadhav HB, Badwaik LS, Annapure U, Casanova F, Alaskar K (2023) A review on the journey of edible flowers from farm to consumer's plate. Appl Food Res 3(2):100312

Jurić M, Donsi F, Bandić LM, Jurić S (2023) Natural-based electrospun nanofibers: challenges and potential applications in agri-food sector. Food Biosci 56:103372

Kandylis P (2022) Phytochemicals and antioxidant properties of edible flowers. Appl Sci 12(19):9937

Kaur K, Asthir B, Verma DK (2021) Biosynthesis, bioavailability, and metabolism of plant polyphenols: biological activities and their potential benefits in human health. In: Phytochemicals in food and health. Apple Academic Press, pp 231–255

Kazemeini H, Azizian A, Adib H (2021) Inhibition of Listeria monocytogenes growth in Turkey fillets by alginate edible coating with Trachyspermum ammi essential oil nano-emulsion. Int J Food Microbiol 344:109104

Keservani RK, Kesharwani RK, Emerald M, Sharma AK (eds) (2023) Nutraceutical fruits and foods for neurodegenerative disorders. Elsevier

Kim BR, Paudel SB, Nam JW, Jin CH, Lee IS, Han AR (2020) Constituents of Coreopsis lanceolata flower and their dipeptidyl peptidase IV inhibitory effects. Molecules 25(19):4370

Kiraman NAS, Yusof H (2022) A mini review on the methods for the extraction, isolation, and determination of P. odorata's bioactive compounds. Healthscope 5(2):38–44

Kritsi E, Tsiaka T, Ioannou AG, Mantanika V, Strati IF, Panderi I et al (2022) In vitro and in silico studies to assess edible flowers' antioxidant activities. Appl Sci 12(14):7331

Kuljarusnont S, Iwakami S, Iwashina T, Tungmunnithum D (2024) Flavonoids and other phenolic compounds for physiological roles, plant species delimitation, and medical benefits: a promising view. Molecules 29(22):5351

Kumari P, Bhargava B (2021) Phytochemicals from edible flowers: opening a new arena for healthy lifestyle. J Funct Foods 78:104375

Kumari P, Raju DVS, Singh KP, Prasad KV, Panwar S (2018) Characterization of phenolic compounds in petal extracts of rose. Indian J Hortic 75(2):349–351

Kumari P, Raju DVS, Prasad KV, Saha S, Panwar S, Paul S et al (2022) Characterization of anthocyanins and their antioxidant activities in Indian rose varieties (Rosa× hybrida) using HPLC. Antioxidants 11(10):2032

Le TN, Luong HQ, Li HP, Chiu CH, Hsieh PC (2019) Broccoli (Brassica oleracea L. var. italica) sprouts as the potential food source for bioactive properties: a comprehensive study on in vitro disease models. Foods 8(11):532

Lee J, Park G, Chang YH (2019) Nutraceuticals and antioxidant properties of Lonicera japonica Thunb. as affected by heating time. Int J Food Prop 22(1):630–645

Leena MM, Yoha KS, Moses JA, Anandharamakrishnan CJFB (2020) Edible coating with resveratrol loaded electrospun zein nanofibers with enhanced bioaccessibility. Food Biosci 36:100669

Lefebvre T, Destandau E, Lesellier E (2021) Selective extraction of bioactive compounds from plants using recent extraction techniques: a review. J Chromatogr A 1635:461770 de Lima Franzen F, Rodríguez de Oliveira MS, Lidório HF, Farias Menegaes J, Martins Fries LL (2019) Composición química de pétalos de flores de rosa, girasol y caléndula para su uso en la alimentación humana. Ciencia y Tecnología Agropecuaria 20(1):149–168

Liu W, Zhang J, Zhang Q, Shan Y (2018) Effects of postharvest chilling and heating treatments on the sensory quality and antioxidant system of daylily flowers. Hortic Environ Biotechnol 59:671–685

Liu Y, Cheng H, Wu D (2020) Preparation of the orange flavoured "Boba" Ball in milk tea and its shelf-life. Appl Sci 11(1):200

Liu J, Ye T, Yang S, Zhong X, He W, Xu M et al (2022) Antidepressant-like activity, active components and related mechanism of Hemerocallis citrina Baroni extracts. Front Pharmacol 13:967670

Liu A, Wan Q, Li J, Li Q, Hu K, Ao X et al (2023) Rose bud extract as a natural antimicrobial agent against Staphylococcus aureus: mechanisms and application in maintaining pork safety. Lwt 176:114527

Loizzo MR, Pugliese A, Bonesi M, Tenuta MC, Menichini F, Xiao J, Tundis R (2016) Edible flowers: a rich source of phytochemicals with antioxidant and hypoglycemic properties. J Agric Food Chem 64(12):2467–2474

Lopez-Hortas L, Rodriguez P, Diaz-Reinoso B, Gaspar MC, de Sousa HC, Braga ME, Dominguez H (2022) Supercritical fluid extraction as a suitable technology to recover bioactive compounds from flowers. J Supercrit Fluids 188:105652

López-Romero D, Izquierdo-Vega JA, Morales-González JA, Madrigal-Bujaidar E, Chamorro-Cevallos G, Sánchez-Gutiérrez M et al (2018) Evidence of some natural products with antigenotoxic effects. Part 2: plants, vegetables, and natural resin. Nutrients 10(12):1954

Łysiak GP (2022) Ornamental flowers grown in human surroundings as a source of anthocyanins with high anti-inflammatory properties. Foods 11(7):948

McCabe MC, Hill RC, Calderone K, Cui Y, Yan Y, Quan T et al (2020) Alterations in extracellular matrix composition during aging and photoaging of the skin. Matrix biology plus 8:100041

Meenakshi SC, Beldal BS, Londonkar RL (2019) Evaluation of methanolic extract of Bombax ceiba bark for in-vitro antioxidant and anti-inflammatory activities. J Pharmacogn Phytochem 8(1):1504–1507

Milião GL, de Oliveira APH, de Souza Soares L, Arruda TR, Vieira ÉNR, Junior BRDCL (2022) Unconventional food plants: nutritional aspects and perspectives for industrial applications. Future Foods 5:100124

Mlcek J, Rop O (2011) Fresh edible flowers of ornamental plants–a new source of nutraceutical foods. Trends Food Sci Technol 22(10):561–569

Mlcek J, Plaskova A, Jurikova T, Sochor J, Baron M, Ercisli S (2021) Chemical, nutritional and sensory characteristics of six ornamental edible flowers species. Foods 10(9):2053

Moliner C, Barros L, Dias MI, López V, Langa E, Ferreira IC, Gómez-Rincón C (2018) Edible flowers of Tagetes erecta L. as functional ingredients: phenolic composition, antioxidant and protective effects on Caenorhabditis elegans. Nutrients 10(12):2002

Mulík S, Ozuna C (2020) Mexican edible flowers: cultural background, traditional culinary uses, and potential health benefits. Int J Gastron Food Sci 21:100235

Navarro-González I, González-Barrio R, García-Valverde V, Bautista-Ortín AB, Periago MJ (2015) Nutritional composition and antioxidant capacity in edible flowers: characterisation of phenolic compounds by HPLC-DAD-ESI/MSn. Int J Mol Sci 16(1):805–822

Nonglait DL, Gokhale JS (2024) Review insights on the demand for natural pigments and their recovery by emerging microwave-assisted extraction (MAE). Food Bioprocess Technol 17(7):1681–1705

Nunes AN, Borges A, Matias AA, Bronze MR, Oliveira J (2022) Alternative extraction and downstream purification processes for anthocyanins. Molecules 27(2):368

Ojulari OV, Lee SG, Nam JO (2019) Beneficial effects of natural bioactive compounds from Hibiscus sabdariffa L. on obesity. Molecules 24(1):210

Otles S, Bakirci GT (2021) Phytochemicals and health: an update. In: Phytopharmaceuticals: potential therapeutic applications, pp 437–454: Published by Springer Nature and Nutrition Journal (Nutr J)
Ozkan K, Bayram Y, Karasu S, Karadag A, Sagdic O (2021) Extraction of bioactive compounds from saffron species. In: Saffron. Elsevier, pp 99–141
Panda SK, Mohanta YK, Padhi L, Luyten W (2019) Antimicrobial activity of select edible plants from Odisha India against food-borne pathogens. Lwt 113:108246
Panda D, Behera RK, Behera PK, Padhi SK, Nayak JK (2024) Nutritional and nutraceutical potential of underutilised wild flowers used by tribal people of Koraput for health benefit. Proc Natl Acad Sci India Sect B Biol Sci 94(1):201–207
Park G, Baek S, Kim JE, Lim TG, Lee CC, Yang H et al (2015) Flt 3 is a target of coumestrol in protecting against UVB-induced skin photoaging. Biochem Pharmacol 98(3):473–483
Paulauskienė A, Tarasevičienė Ž, Laukagalis V (2021) Influence of harvesting time on the chemical composition of wild stinging nettle (Urtica dioica L.). Plants 10(4):686
Pereira AG, Fraga-Corral M, García-Oliveira P, Jimenez-Lopez C, Lourenço-Lopes C, Carpena M et al (2020) Culinary and nutritional value of edible wild plants from northern Spain rich in phenolic compounds with potential health benefits. Food Funct 11(10):8493–8515
Pereira AG, Silva A, Grosso C, Echave J, Chamorro F, Seyyedi-Mansour S et al (2023) Antimicrobial activity screening of Camellia japonica flowers (var. Carolyn Tuttle) for potential drug development. Eng Proc 56(1):314
Perić I, Costina V, Gass P, Findeisen P, Filipović D (2021) Hippocampal synaptoproteomic changes of susceptibility and resilience of male rats to chronic social isolation. Brain Res Bull 166:128–141
Pinton SC, Bardsley CA, Marik CM, Boyer RR, Strawn LK (2020) Fate of Listeria monocytogenes on broccoli and cauliflower at different storage temperatures. J Food Prot 83(5):858–864
Pires EDO Jr, Di Gioia F, Rouphael Y, Ferreira IC, Caleja C, Barros L, Petropoulos SA (2021) The compositional aspects of edible flowers as an emerging horticultural product. Molecules 26(22):6940
Prabawati NB, Oktavirina V, Palma M, Setyaningsih W (2021) Edible flowers: antioxidant compounds and their functional properties. Horticulturae 7(4):66
Primitivo MJ, Neves M, Pires CL, Cruz PF, Brito C, Rodrigues AC et al (2022) Edible flowers of Helichrysum italicum: composition, nutritive value, and bioactivities. Food Res Int 157:111399
Purohit SR, Rana SS, Idrishi R, Sharma V, Ghosh P (2021) A review on nutritional, bioactive, toxicological properties and preservation of edible flowers. Future Foods 4:100078
Rachkeeree A, Kantadoung K, Suksathan R, Puangpradab R, Page PA, Sommano SR (2018) Nutritional compositions and phytochemical properties of the edible flowers from selected Zingiberaceae found in Thailand. Front Nutr 5:3
Rameshwar V, Kishor D, Tushar G, Siddharth G, Sudarshan G (2014) A pharmacognostic and pharmacological overview on Bombax ceiba. Sch Acad J Pharm 3(2):100–107
Rao V, Poonia A (2024a) Microbial biosynthesis of selenium nanoparticles using probiotic strain and its characterization. J Food Meas Charact 18(7):5477–5488. https://doi.org/10.1007/s11694-024-02581-z
Rao V, Poonia A (2024b) Bioactive compounds, nanoparticle synthesis and potential utilization of edible flowers for development of functional dairy products: a review. J Food Sci Technol 61(6):1053–1068. https://doi.org/10.1007/s13197-023-05853-6
Riaz M, Zia-Ul-Haq M, Saad B (2016) Anthocyanins and human health: biomolecular and therapeutic aspects, vol 10. Springer, pp 978–973
da Rocha CB, Noreña CPZ (2020) Microwave-assisted extraction and ultrasound-assisted extraction of bioactive compounds from grape pomace. Int J Food Eng 16(1–2):20190191
Rodrigues H, Spence C (2023) Looking to the future, by studying the history of edible flowers. Int J Gastron Food Sci 34:100805
Rodriguez-Amaya DB (2016) Natural food pigments and colorants. Curr Opin Food Sci 7:20–26
Ryu J, Nam B, Kim BR, Kim SH, Jo YD, Ahn JW et al (2019) Comparative analysis of phytochemical composition of gamma-irradiated mutant cultivars of Chrysanthemum morifolium. Molecules 24(16):3003

Salekzamani S, Ebrahimi-Mameghani M, Rezazadeh K (2019) The antioxidant activity of artichoke (Cynara scolymus): a systematic review and meta-analysis of animal studies. Phytother Res 33(1):55–71

Sarkar M, Chawla P (2022) Curative properties of chamomile in gastrointestinal disorders. In: Herbs, spices, and medicinal plants for human gastrointestinal disorders. Apple Academic Press, pp 125–140

Sarla Saklani, S. S., & Subhash Chandra, S. C. (2015). Evaluation of in vitro antimicrobial activity, nutritional profile and phytochemical screening of Rhododendron arboreum

Segneanu AE (ed) (2024) Natural products and medicinal properties of Carpathian (Romanian) plants. CRC Press

da Silva APG, Sganzerla WG, John OD, Marchiosi R (2023) A comprehensive review of the classification, sources, biosynthesis, and biological properties of hydroxybenzoic and hydroxycinnamic acids. Phytochem Rev 24:1–30

Simin N, Lesjak M, Živanović N, Božanić Tanjga B, Orčić D, Ljubojević M (2023) Morphological characters, phytochemical profile and biological activities of novel garden roses edible cultivars. Horticulturae 9(10):1082

Singh S, Bhatt P, Kumar V, Singh NP (2024) Phytonutrients, anthocyanidins, and anthocyanins: dietary and medicinal pigments with possible health benefits. In: Advances in flavonoids for human health and prevention of diseases. Apple Academic Press, pp 23–46

Skrajda-Brdak M, Dąbrowski G, Konopka I (2020) Edible flowers, a source of valuable phytonutrients and their pro-healthy effects – a review. Trends Food Sci Technol 103:179–199

Sood Y, Lal M, Kalia A, Verma S (2024) Edible flowers: super foods with potential health benefits. Int J Plant Soil Sci 36(3):213–221

Takahashi JA, Rezende FAGG, Moura MAF, Dominguete LCB, Sande D (2020) Edible flowers: bioactive profile and its potential to be used in food development. Food Res Int 129:108868

Teixeira M, Tao W, Fernandes A, Faria A, Ferreira IM, He J et al (2023) Anthocyanin-rich edible flowers, current understanding of a potential new trend in dietary patterns. Trends Food Sci Technol 138

Tembo-Phiri C (2019) Edible fynbos plants: a soil types and irrigation regime investigation on Tetragonia decumbens and Mesembryanthemum crystallinum. Doctoral dissertation, Stellenbosch University, Stellenbosch

Tripathi AK, Gupta PS, Singh SK (2019) Antidiabetic, anti-hyperlipidemic and antioxidant activities of Bauhinia variegata flower extract. Biocatal Agric Biotechnol 19:101142

Uniyal A, Kumar A (2024) Minor flowers of European and American countries. In: Edible flowers. Academic, pp 385–430

Varahi Vedam VA, Xavier AS, David DC (2019) In-vitro evaluation of antifungal and anticancer properties of Tagetes Erecta petal extract. Biomed Pharmacol J 12(2):815–823

Villa-Rodriguez JA, Kerimi A, Abranko L, Tumova S, Ford L, Blackburn RS et al (2018) Acute metabolic actions of the major polyphenols in chamomile: an in vitro mechanistic study on their potential to attenuate postprandial hyperglycaemia. Sci Rep 8(1):5471

Vimala Bharathi SK, Maria Leena M, Moses JA, Anandharamakrishnan C (2020) Zein-based anti-browning cling wraps for fresh-cut apple slices. Int J Food Sci Technol 55(3):1238–1245

Vinatoru M, Mason T, Calinescu I (2017) Ultrasonically assisted extraction (UAE) and microwave assisted extraction (MAE) of functional compounds from plant materials. TrAC Trends Anal Chem 97:159–178

Wan H, Yu C, Han Y, Guo X, Luo L, Pan H et al (2019) Determination of flavonoids and carotenoids and their contributions to various colors of rose cultivars (Rosa spp.). Front Plant Sci 10:123

Wang N, Zhang C, Bian S, Chang P, Xuan L, Fan L et al (2019) Flavonoid components of different color Magnolia flowers and their relationship to cultivar selections. HortScience 54(3):404–408

Wang L, Wen M, Chen F, Luo Z, Yin J, Chen Y, Huang H (2020) High oxygen atmospheric packaging (HOAP) reduces H2O2 production by regulating the accumulation of oxidative stress-related proteins in Chinese flowering cabbage. Postharvest Biol Technol 165:111183

Wu X, Zhao Y, Haytowitz DB, Chen P, Pehrsson PR (2019) Effects of domestic cooking on flavonoids in broccoli and calculation of retention factors. Heliyon 5(3):e01310

Yang Y, Ling W (2024) Health benefits and future research of phytochemicals: a literature review. J Nutr 155:87

Yang H, Shin Y (2017) Antioxidant compounds and activities of edible roses (Rosa hybrida spp.) from different cultivars grown in Korea. Appl Biol Chem 60:129–136

Yeasmen N, Sharma N, Bhuiyan MHR, Orsat V (2024) Ultrasound as a techno-functional modifier in food frying and bioactive compounds extraction. Food Rev Intl 40(3):974–995

Yeo I, Lee YJ, Song K, Jin HS, Lee JE, Kim D et al (2018) Low-molecular weight keratins with anti-skin aging activity produced by anaerobic digestion of poultry feathers with Fervidobacterium islandicum AW-1. J Biotechnol 271:17–25

Zheng J, Meenu M, Xu B (2019) A systematic investigation on free phenolic acids and flavonoids profiles of commonly consumed edible flowers in China. J Pharm Biomed Anal 172:268–277

Zhu Z, Zhang P, Sun DW (2020) Effects of multi-frequency ultrasound on freezing rates and quality attributes of potatoes. Ultrason Sonochem 60:104733

Chapter 14
Valorization of Edible Flowers for Development of Novel Food and Dairy Products

Lizet Aguirre Güitrón, Amrita Poonia, Fernando González-Chávez, Bimlesh Mann, Rekha Yadav, and Priyanka

14.1 Introduction

Edible flowers have been gaining popularity in recent years because they are a potential source of antioxidants, essential oils, proteins, and vitamins; particularly when consumed in a raw or lightly processed form. These compounds, known as bioactives, are found not only in flowers but also in other parts of plants, providing a wide range of adaptive advantages (Gupta et al. 2024). Phenolic compounds are essential in plants and contribute to pigmentation, growth, pathogen resistance, and the maintenance of cellular redox balance (Hajam et al. 2023; Noreen et al. 2017). In addition to their vital functions in plant physiology, these compounds have been widely recognized for their antioxidant properties, anti-inflammatory effects, and capacity to mitigate oxidative stress. Furthermore, they may play a role in cancer prevention (Cerdá-Bernad et al. 2022; Clemente-Villalba et al. 2023).

L. A. Güitrón
Universidad Tecnológica de Nayarit, Carretera Federal, Xalisco, Nayarit, Mexico

A. Poonia (✉)
Department of Dairy Science and Food Technology, Institute of Agricultural Sciences, Banaras Hindu University, Varanasi, Uttar Pradesh, India
e-mail: amrita12@bhu.ac.in

F. González-Chávez
Universidad Politécnica del Estado de Nayarit, Tepic, Mexico

B. Mann
Indian Council of Agricultural Research, New Delhi, India

R. Yadav · Priyanka
Department of Food and Nutrition, I.C College of Community Science, Chaudhary Charan Singh Haryana Agricultural University, Hisar, Haryana, India

A. Poonia et al. (eds.), *Edible Flowers: Source of Phytonutrients, Valorization and Technological Advancements*, https://doi.org/10.1007/978-3-032-00667-7_14

Edible flowers typically have a high-water content, ranging from 70% to 95%. Water is an essential element of flowers, constituting over 80% of their total composition (Skrajda 2017). Flowers are recognized for their nutritional and medicinal properties. The nutrient and phytochemical makeup of these flowers varies based on the specific part of the flower being considered. Pollen, for example, is a rich source of proteins, carbohydrates, both saturated and unsaturated fats, carotenoids, and flavonoids. Nectar, on the other hand, is composed of amino acids, free sugars, proteins, inorganic ions, lipids, organic acids, phenolic compounds, alkaloids, and terpenoids, along with other bioactive components. Petals and other floral parts are known to be particularly rich in essential vitamins, minerals, and antioxidants (Fernandes et al. 2017; Mlcek and Rop 2011). Among edible flowers, carbohydrates are the primary macronutrient. For instance, flowers such as *Centaurea cyanus* L. and roses (*Rosa* spp.) show carbohydrate contents of 88.39 and 86.12 g per 100 g of dry weight (DW), respectively (Pires et al. 2017). The fiber content in these flowers can vary greatly, ranging from 6.1 to 55.4 g per 100 g of dry weight, depending on the species, such as *Allium schoenoprasum* L., *Spilanthes oleracea* L., and *Tagetes erecta* L. (Fernandes et al. 2017).

14.2 Composition and Functional Properties of Edible Flowers

Edible flowers are also rich in important minerals such as potassium, phosphorus, calcium, and magnesium. Potassium is especially prevalent, and it is associated with various cardiovascular health benefits (Fernandes et al. 2017; Rop et al. 2012). Historically, flowers have been used in child snacks in places like Poland, where species such as *Trifolium spp.*, *Lamium album*, and *Robinia pseudacacia* are incorporated (Luczaj et al. 2012). The growing demand for healthy, low-calorie food options has led to the creation of functional foods using natural ingredients like edible flowers. In one study, Pires et al. (2018a, b) explored the addition of edible flower extracts to yogurt, improving its color and nutritional content. These flowers are an excellent source of antioxidants, vitamins, and minerals. The market for nutraceuticals is driven primarily by increased consumer awareness, improved purchasing power, the popularisation of the beneficial effects of phytochemicals on the human body, consumer readiness to purchase functional foods, and a strong desire to maintain a healthy lifestyle. It is anticipated that dairy-based beverages will propel the market expansion for functional foods (Rao and Poonia 2024a, b).

The sunflower (*Helianthus annuus*) is another example of an edible flower with a rich nutritional profile. The flower, composed of thousands of individual florets, contains significant levels of potassium, calcium, and phosphorus, with dry weight concentrations of 1983.83 to 2006.76 mg/100 g, 622.81 to 748.26 mg/100 g, and 127.81 to 335.67 mg/100 g, respectively (Liang et al. 2013). Sunflowers are also a good source of iron, making them beneficial for those with anemia, and provide vitamin E and vitamin C in substantial amounts. Despite its nutritional benefits, the sunflower flower is often discarded as a byproduct of oil extraction. The disc florets are particularly rich in essential amino acids like valine, leucine, and isoleucine, leading some researchers to suggest their use as a supplement for athletes (Liang et al. 2013).

Another notable flower with high nutritional value is the chive flower (*Allium schoenoprasum* L.), known for its use in cooking and gardening. These pale purple flowers contain significant levels of unsaturated fatty acids, around 13% of their total composition, primarily linoleic and oleic acids (Grzeszczuk et al. 2011). Other flowers, such as *Calendula officinalis* (marigold) and *Primula spp.* (primrose), also exhibit noteworthy levels of fatty acids (Pires et al. 2017; (Chensom et al. 2019). Marigold flowers, in particular, stand out with an impressive 59.3% unsaturated fatty acid content. Although marigolds contain fewer fatty acids than chicken breast, they still surpass many plant-based foods, including banana, broccoli, and cauliflower (USDA 2018a, b, c, d, e, f, g).

Protein content in edible flowers is generally low compared to animal-based products but can be significant in certain varieties. For instance, the banana flower (*Musa sp. var. elakki Bale*) contains 12.5% protein by dry weight (Bhaskar et al. 2012), and sunflowers (*Helianthus annuus*) also provide notable protein levels (Liang et al. 2013). Flowers like *Rosa damascena*, *Calendula officinalis*, and *Centaurea cyanus* have been found to contain moderate amounts of protein, higher than certain vegetables like banana, broccoli, and cauliflower (Pires et al. 2017; USDA 2018a, b, c, d, e, f, g).

Carbohydrates are another major macronutrient in edible flowers. For instance, *Centaurea* petals contain a carbohydrate content of 88.39%, higher than that of rose petals, which are also used for their sweetness in culinary applications (Pires et al. 2017). In comparison, hibiscus petals and whole wheat flour contain lower carbohydrate levels (USDA 2018a, b, c, d, e, f, g). While the carbohydrate content in flowers may be high, they are typically used in small quantities in recipes, so they are generally suitable for diets that limit carbohydrate intake. In addition to petals, flower components like nectar and pollen, are rich in a variety of essential nutrients (Fernandes et al. 2017). Nectar contains a variety of carbohydrates, including fructose, sucrose, and glucose, along with proteins, lipids, organic acids, and more. Pollen is rich in carbohydrates and proteins and also provides bioactive compounds such as carotenoids and flavonoids, as well as fatty acids. Pollen is particularly abundant in carotenoids, flavonoids, carbohydrates, proteins, and both saturated and unsaturated fatty acids (Mlcek and Rop 2011). Additionally, petals and other floral components are excellent sources of antioxidants, minerals, and vitamins, all of which are crucial in supporting human health and nutrition.

Fiber content in edible flowers also varies by species. For example, the banana flower has a remarkable fiber content, exceeding 61% by dry weight (Begum and Deka 2019), while sunflower disc florets contain about 59% fiber by dry weight (Liang et al. 2013). These values far surpass the fiber content found in soybeans, a commonly recognized fiber source (USDA 2018a, b, c, d, e, f, g). Chive flowers, on the other hand, have lower fiber content, around 6.1% by dry weight (Grzeszczuk et al. 2011). Incorporating such flowers into the diet can help meet the recommended fiber intake, which is approximately 14 g per 1000 kcal (U.S. Department of Health and Human Services and U.S. Department of Agriculture 2015). Numerous studies worldwide have identified bioactive compounds in edible flowers. Table 14.1 shows the main phytochemical compounds found in various edible flowers.

Interest in edible flowers has grown in recent years, driven by their health benefits, unique flavors, and culinary versatility (Benvenuti and Mazzoncini 2021). Benvenuti et al. (2016) analyzed the sensory profile of different edible flowers, with panel tests revealing not only the general appreciation of certain flower species but also their distinct organoleptic properties, such as aroma, bitterness, spiciness, sweetness, and texture. Additionally, these tests highlight how closely some flowers resemble familiar vegetables and spices. The most highly valued flowers were *Tropaeolum majus*, which is known for its spicy, radish-like flavor, and *Ageratum houstonianum*, which exhibits a taste reminiscent of carrots. For its part, *Begonia semperflorens* garnered significant interest for its distinctly sour flavor, which is comparable to that of lemon juice. Another study by Simoni et al. (2018) explored the potential of incorporating irradiated edible flowers into a variety of salad dishes. Sensory analysis revealed that participants were generally open to the idea of consuming edible flowers, showing a particular preference for those with a sweet flavor.

14.3 Edible Flowers in Cultural Contexts

Across different cultures, edible flowers have been used in diverse ways, influenced by local customs, beliefs, and traditions (Singh et al. 2020). These traditional uses often reflect the availability of local resources and the environmental conditions of the area.

Aesthetic Uses In many societies, edible flowers are incorporated into personal adornments, body art, and decorative garlands (González-Barrio et al. 2018).

Natural Dyes Flowers are also used to produce natural dyes, which enhance the aesthetic quality of artworks, textiles, and clothing by imparting vibrant colors.

Garnishing and Decoration In various cultural celebrations and feasts, edible flowers play a key role in beautifying dishes and table settings, contributing to visually striking food presentations (Kelley et al. 2003).

Fragrances and Aromatics Flowers are often employed in the creation of scented products such as fragrant waters, oils, and perfumes, improving the ambiance of living spaces and contributing to personal grooming routines.

Food Additives Edible flowers are also used as flavor enhancers and preservatives, commonly added to pickled items, vinegar, and oils to elevate taste and aroma.

Table 14.1 Bioactive compounds of several edible flowers

Edible flower	Total phenolic content	Compounds	References
Anchusa azurea	33.6 ± 0.9 (mg CAE/g of extract)	Kampherol, myricetin, anthocyanins, carotenoids	Loizzo et al. (2016)
Chrysanthemum × *hybrid* (orange)	155.25 ± 0.73 (mg GAE/g DW)	Gallic acid, 4-Hidroxy benzoic acid, rutin, quercetin glucoside, α-carotene, β-carotene, Lutein	
Helianthus annuus (yellow)	77.42 ± 0.97 (mg GAE/g DW)	α-carotene, β-caroten, gallic acid, ρ-cumaric acid, m-cumaric acid, gallic acid, vanillic acid, caffeic acid, ferulic acid, kampherol, lutei, quercetin glucoside, violaxanthin, zeaxanthin	Coyago-Cruz et al. (2024)
Hibiscus acetosella	568.8 ± 0.8 (mg GAE/g)	Cinnamic acid, gallic acid, cyanidin 3-O-glucoside, protocatechuic acid, cyanidin, delphinidin, caffeic acid, sinapic acid, myricetin, quercitrin, luteolin	dos Santos Silva et al. (2023)
Mimulus × *hybridus* L. 'Magic Red'	4.33 ± 0.19 (mg of GAE/ g FW)	Chlorophyll, carotenoids, anthocyanins	Stefaniak and Grzeszczuk (2019)
Mimulus × *hybridus* L. 'Magic Yellow'	2.21 ± 0.18 (mg of GAE/g FW)	Chlorophyll, carotenoids, anthocyanins	
Monarda didyma L.	10.57 ± 0.88 (mg of GAE/g FW)	Chlorophyll, carotenoids, anthocyanins	
Nepeta x faassenii'Six	5.11 ± 0.21 (mg GAE/g FW)	Eucaliptol, β-elemene, isodihydronepetalactone, β-caryophyllene, germacrene D, β-sinensal, pentylcurcumene	Marchioni et al. (2020)
Paeonia suffruticosa	235.5 (mg CAE/g DW)	Gallic acid, chlorogenic acid, 3-hydroxy-4-methoxybenzaldehyde thiosemicarbazone, ferulic acid, rutin quercetin, apigenin kampherol	Xiong et al. (2014)
Rosa chinensis	205 (mg CAE/g DW)	Gallic acid, chlorogenic acid, *p*-hydroxybenzoic acid, caffeic acid, 3-hydroxy-4-methoxybenzaldehyde thiosemicarbazone, *p*-coumaric acid,	
Rosa rugosa	56.18 ± 0.49 (mg GAE/g DW)	Gallic acid, caffeic acid, *p*-coumaric acid	Socha et al. (2021)
Sambucus nigra	228.5 ± 3.2 (mg of CAE/g of extract	Kampherol, myricetin, quercetin, rutin	Loizzo et al. (2016)
Tagetes patula (orange)	194.84 ± 15.99 (mg GAE/g DW)	Kampherol, quercetin glucoside, quercetin, α-carotene, β-carotene, β-cryptoxanthin, violaxanthin, lutein	Coyago-Cruz et al. (2024)

DW dry weight, *FW* fresh weight, *CAE* chlorogenic acid equivalents, *GAE* gallic acid equivalents

Cultural and Historical Significance Throughout history, edible flowers have been incorporated into diverse cultural practices, culinary traditions, and everyday life.

Symbolism and Rituals In various cultures, flowers are symbolic, representing themes like renewal, fertility, and purity in religious and ceremonial settings.

Medicinal Uses Many edible flowers are valued for their therapeutic properties, commonly used in traditional medicine to treat various health conditions and promote general well-being.

Culinary Contributions Edible flowers have long been a part of traditional cuisines, enhancing the flavor, color, and visual appeal of dishes. They are commonly featured in desserts, teas, salads, and soups, enriching the overall dining experience (Takahashi et al. 2020).

Edible flowers are increasingly being incorporated into various industries, including cosmetics, perfumery, food, and pharmaceuticals. As natural food colorants, they enhance the visual appeal of dishes by adding vibrant, eye-catching hues. Moreover, edible flowers are valued as food additives due to their health benefits and safety, making them suitable for a variety of culinary applications. In the pharmaceutical field, the phenolic compounds found in edible flowers are highly regarded for their potential therapeutic effects. In the confectionery industry, these flowers are used to improve the aesthetics of cakes, cookies, and other sweet treats. They are also a popular addition to everyday meals, enhancing the sensory experience by contributing to the flavor, aroma, and texture of dishes. Additionally, edible flowers are utilized in the beverage industry to enrich the flavors of wines, sauces, and herbal teas (dos Santos et al. 2018).

Beyond culinary and medicinal uses, edible flowers play a significant role in the production of cosmetics, natural dyes, and perfumes. Jasmine flowers, a member of the Oleaceae family, are particularly prized in the fragrance industry for their complex scent profile. Furthermore, edible flowers are essential in the extraction of essential oils (Zhao et al. 2019).

14.4 Functional Food Products of Edible Flowers

The potential of flowers as a food source is often overlooked, despite their increasing use in enhancing the taste, color, and aesthetic value of dishes. Utilizing edible ornamental flowers for food production presents an eco-friendly alternative to the cut flower industry, which discards vast quantities of flowers daily, despite their nutritional and functional benefits. These flowers are often discarded because they lack visual appeal (Rezende et al. 2019). The growing concerns surrounding synthetic food additives have led to greater interest in natural alternatives. Natural preservatives, particularly phytochemicals found in flowers with antioxidant and antimicrobial properties, offer a promising solution to extend food shelf life and reduce chemical degradation. These natural compounds also help manage the risks posed by foodborne pathogens, enhancing food safety.

Natural colorants, derived from flowers, not only contribute vibrant hues but also offer various pharmacological benefits. The pigments found in flowers, such as anthocyanins and carotenoids, can play a role in combating nutritional deficiencies and addressing health issues. Research has shown that foods rich in bioactive compounds, particularly flavonoids and carotenoids, can help reduce the risk of numerous diseases (Abdel-Aal et al. 2013; Scripsema et al. 2015). Researchers have explored the use of flowers like *Madhuca indica* and *Rhododendron arboreum* in creating ready-to-serve beverages and squash (Mishra et al. 2013; Solanke et al. 2016). Rhododendron flowers, in particular, are packed with antioxidants such as anthocyanins, phenols, and ascorbic acid, and have been shown to improve the antioxidant profile of beverages like squash, especially when sweetened with honey rather than sugar (Bhatt et al. 2017).

Dash (2017) developed a technology to produce a variety of value-added products from dried *Madhuca longifolia* flowers, including candies, cakes, toffees, squashes, and ladoos. According to Ramakrishnan (2018), *Delonix regia* (Red Gulmohar) flowers are an excellent source of anthocyanin pigments, particularly cyanidin-3-O rutinoside, which adds significant value to their use. *Delonix regia* flower extract has also been utilized to enhance the nutritional value of dairy products like curd (Chhabra and Gupta 2015).

Other studies have explored the use of flower extracts in functional foods. For example, Nisa et al. (2018) developed candies enriched with lutein from *Tagetes erecta* extract, while Gulkand, a traditional product made from rose petals known for its digestive benefits, has been enhanced by combining it with fennel seeds (*Foeniculum vulgare*) to create a stomachic formulation (Sardari et al. 2018). In yogurt, extracts from *Rosa damascena* petals have been successfully integrated to improve flavor and nutritional content (Chanukya and Rastogi 2016). Functional uses and applications and nutraceutical properties of edible flowers are listed in Tables 14.2 and 14.3.

Additionally, research by Pires et al. (2018a, b) examined the potential of edible flower extracts as natural alternatives to commercial food colorants such as E163. The study found that rose extract was the most suitable substitute among various flower extracts, including those from *dahlia* and *cornflower*. According to Bahuguna et al. (2018), fresh red *Hibiscus rosa-sinensis* flowers are rich in proximate content, vitamin C, and β-carotene, and have been incorporated into products like gulkand and juices, where their high acceptability was noted at a 10% concentration for gulkand and 50% for juice. Moreover, the addition of rose syrup (10%) and marigold powder (0.5%) significantly improved the nutritional profile, shelf life, and consumer appeal of nutritious cereals like finger millet and oats, as well as milk-based functional beverages (Kumar et al. 2020a, b).

Table 14.2 Functional uses and applications of edible flowers

Aspect	Flowers	Usage	References
Vivid hues	Marigold, Pansy, Nasturtium	Introduces bold and striking colors to dishes like salads, beverages, and sweets, boosting visual appeal.	Shantamma et al. (2021)
Distinct flavors	Hibiscus, Lavender, Chive Blossom, Rose, Jasmine, Chamomile, Dandelion, Violet, Fennel Blossom	Adds a spectrum of flavors, from tangy (Hibiscus) to floral (Lavender) and savory (Chive), enriching teas, desserts, and sauces.	Asif et al. (2023), Fernandes et al. (2017)
Visual appeal	Borage, Calendula, Orchid	Utilized for their ornamental beauty, these flowers make cakes, cocktails, and salads more attractive with their intricate designs and bright tones.	Rivas-García et al. (2021)
Textural diversity	Squash Blossom, Rose Petal, Tulip Petal	Offers a range of textures—from delicate softness (Rose petals) to satisfying crunch (Squash blossoms)—perfect for varying dishes.	Mulík and Ozuna (2020)
Culinary flexibility	Marigold, Lavender, Basil Blossom	These flowers shine in diverse culinary settings, whether raw in salads or cooked in soups, sauces, and desserts, enhancing both flavor and presentation.	Chen et al. (2020a, b)
Nutritional benefits	Hibiscus, Dandelion, Sunflower Petal	Packed with antioxidants and essential nutrients like Vitamin C (found abundantly in Hibiscus), they are commonly incorporated into health-boosting beverages.	Fernandes et al. (2020)

14.5 Importance of Sustainable Flower Sourcing in Product Development

Sustainable flower sourcing is critical in modern product development, particularly in the food and dairy industries. Ethical and organic sourcing practices not only ensure environmental conservation but also support local economies and enhance product quality. This paper examines the importance of sustainable sourcing, its benefits, and challenges, with recommendations for implementing effective practices. Sourcing practices play a pivotal role in the lifecycle of flower-based products. Ethical and organic sourcing not only aligns with consumer preferences for sustainability but also addresses broader environmental and social concerns.

Table 14.3 Flowers and their nutraceutical properties

Flowers	Nutraceutical properties	References
Rose	Rose hips are rich in vitamin C, flavonoids, and tannins. Rose tea, made from rose petals and rose hips, is utilized to address stomach issues, including gallbladder infections and diarrhea. Rose petals contain anthocyanins that exhibit strong antioxidant, anticancer, and antiallergic properties.	Sharma et al. (2011), Lee et al. (2011)
Marigold	Carotenoid (lutein esters) derived from marigold demonstrate efficacy in inhibiting free radical generation, as well as in the prevention of age-related macular degeneration, cataracts, cancer, and coronary heart diseases. Dried marigold petals and concentrates serve as feed to enhance the pigmentation of poultry skin and the eggs produced by laying hens.	Piccaglia et al. (1998)
Cotton Rose (Hibiscus mutabilis)	Leaves possess properties that serve as anodyne, antidote, demulcent, and expectorant, attributed to the presence of rutin alkaloid. Decoction is utilized in the management of respiratory conditions.	Sharma et al. (2011)
Carnation	Caryophyllene, a terpene, is identified as the compound responsible for its anti-inflammatory properties. Flowers are recognized for their properties as antispasmodic, cardiotonic, diaphoretic, and nervine tonic.	Lyra et al. (2008), Chevallier (1996)
Chrysanthemum	Chrysanthemum exhibits significant antimicrobial, anti-inflammatory, and anticancer properties attributed to the presence of triterpenes such as arnidiol, faradiol, and heliantriol.	Ukiya et al. (2002)
Calendula officinalis	Petals and pollen comprise triterpenoid esters, which serve as anti-inflammatories, and carotenoids, recognized for their antioxidant properties. This substance exhibits properties that are both antispasmodic and antiseptic.	Sharma et al. (2011)
Nasturtium (Tropaeolum majus)	It contains glucosinolates, which exhibit antibiotic and anti-tumor properties. It serves as a beneficial source of vitamin C and spilanthol oxalic acid	Sharma et al. (2011)
Periwinkle (Catharanthus roseus)	Root alkaloids ajamalicine and serpentine are utilized in the treatment of hypertension. Vincristine and vinblastine, derived from leaf alkaloids, are utilized in the treatment of blood cancer.	Sharma et al. (2011)

14.5.1 Environmental Impact

Unsustainable flower farming often leads to soil degradation, water pollution, and loss of biodiversity. By adopting organic practices, producers can minimize these impacts and promote environmental sustainability (Halliwell et al. 2015).

14.5.2 Social and Economic Benefits

Ethical sourcing ensures fair wages and working conditions for farmers, particularly in developing regions. This approach strengthens local economies and fosters community development (Khan. 2023).

14.5.3 Consumer Trust and Market Trends

Consumers are increasingly favoring products with transparent sourcing practices. Sustainable sourcing enhances brand reputation and meets the growing demand for eco-friendly products (Mintel 2024).

14.5.4 Organic Sourcing Practices

Certification and Standards Certifications such as fair trade and USDA organic ensure that flowers are cultivated without harmful chemicals and adhere to ethical labor practices.

Biodiversity Conservation Encouraging biodiversity through crop rotation and polyculture farming enhances soil health and reduces reliance on synthetic inputs (Halliwell et al. 2015).

14.5.5 Ethical Sourcing Practices

Fair trade initiatives Fair trade practices guarantee equitable compensation for farmers, fostering sustainable livelihoods and ethical trade (Onyekwelu et al. 2024).

Reducing Carbon Footprints Sourcing locally minimizes transportation emissions and supports regional agriculture.

14.6 Consumption of Edible Flowers

Edible flowers originate from a variety of plant species, but only a limited selection of these flowers is appropriate for eating. Accurate identification of these flowers is essential, given that numerous species may possess toxic properties. Edible flowers are distributed among 97 families, 100 genera, and 180 species worldwide. The culinary traditions of various cultures worldwide incorporate these flowers (Lu et al. 2016).

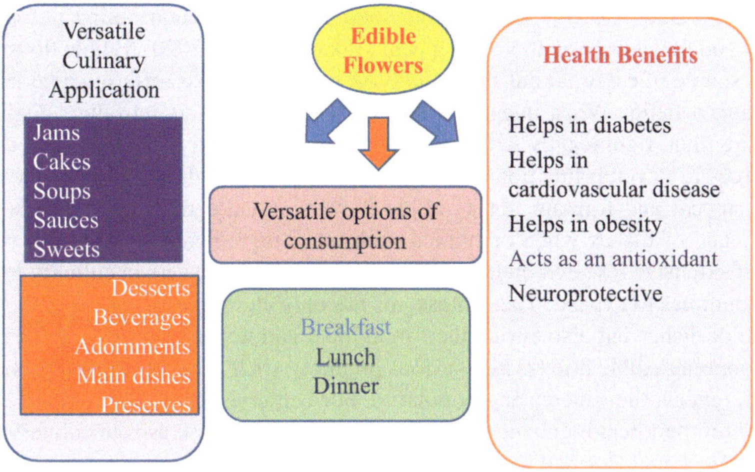

Fig. 14.1 Health benefits and food applications of edible flowers

Edible flowers are most frequently consumed fresh; however, they may also be dried, incorporated into cocktails (often frozen in ice cubes), canned with sugar, or preserved in distillates (Mlcek and Rop 2011). Flowers are commonly utilized in culinary applications to improve color, aroma, and taste, frequently featured in salads, soups, main dishes, desserts, and beverages (Mlcek and Rop 2011; Rop et al. 2012; Kelley et al. 2001a). The exploration of edible flowers has captured the interest of both researchers and chefs, indicating significant opportunities for broadening their culinary applications and enhancing understanding of their various uses (Basso and Papalia 2014). The investigation of edible flowers presents avenues to improve both the culinary experience and the nutritional benefits of food items. Edible flowers are generally eaten whole; however, the particular parts of the flower that are consumed may differ based on the species. In species such as Tulipa, Chrysanthemum, and Rosa spp., it is common for only the petals to be ingested. Conversely, in species like Bellis perennis (daisies) and Tropaeolum majus (garden nasturtium), the flower buds are the parts that are consumed. Furthermore, specific sections of certain flowers need to be eliminated before consumption because of undesirable characteristics, including the bitterness present in the white areas at the base of rose and chrysanthemum petals. The acceptability of edible flowers among consumers and professional chefs, noting that preferences differed based on flower species and sensory attributes (Fig. 14.1).

14.7 Applications of Edible Flowers in Food Products

Historically, flowers were consumed in ancient Rome, Greece, and China, where they served as both alternative medicines and essential components of traditional culinary practices. Alongside fruits, seeds, leaves, and vegetable roots, flowers played a significant role in enhancing food preparation, contributing not only to aroma and flavor but also to visual appeal (Takahashi et al. 2020). Edible flowers are a rich source of essential nutrients and bioactive compounds, offering a wide array of health benefits. When integrated into meals, these flowers introduce distinctive qualities that significantly contribute to the food industry, tapping into natural resources to develop functional, health-focused products. In addition to their aesthetic appeal and fragrant notes, edible flowers enhance the texture and sensory experience of dishes when combined with other ingredients. Despite the current lack of extensive research on the topic, the use of edible flowers in culinary innovation continues to expand. These blossoms not only elevate the color and nutritional profile of dishes but also enrich their mouthfeel and texture. The growing trend of incorporating edible flowers into various products, such as meats, drinks, and baked goods, reflects their increasing popularity. For centuries, edible flowers have been valued for their health-enhancing properties and their diverse uses in culinary traditions (Mlcek and Rop 2011).

Flowers are more than decorative embellishments for meals and desserts; they create a unique sensory experience while also enriching the nutritional content of the dishes. They can be enjoyed fresh, such as with marigold flowers, in salads, and are versatile enough to complement savory dishes featuring meat or fish. Flowers are also used in soups, beverages like wine and beer, desserts, candies, and jellies, and serve as natural spices and colorants, adding flavor and visual appeal (Takahashi et al. 2020).

Edible flowers have traditionally been used in infusions or teas. However, recent research by Madurangi and Maduwanthi (2024) has highlighted their distinctive flavors, aromas, and vibrant colors, making them valuable ingredients for innovative culinary applications. Their sensory qualities are largely influenced by bioactive and volatile compounds. For optimal flavor, these flowers are best harvested in the early morning when they are in full bloom. The taste profiles of edible flowers vary significantly across species, each offering a unique intensity of flavor (Fernandes et al. 2019). Flowers generally contain a large number of intricate bioactive compounds, like aliphatics, benzenoids, phenylpropanoids, and terpenes, which contribute to their distinct aromas. The use of different processing methods on fresh flowers can alter their pigmentation and flavor. Gaining a deeper understanding of the flavor profiles of edible flowers is essential for their value addition, integration into functional food products, and their role in the aroma industry (Chetia et al. 2024).

The multitude of chemical compounds present in flowers results in flavors and aromas that are reminiscent of other food ingredients. For example, *Anthriscus cerefolium* (chervil) has a flavor and aroma similar to parsley with a hint of citrus and tarragon. Similarly, *Ageratum houstonianum* shares flavor characteristics with carrots while *Hibiscus rosa-sinensis* presents a flavor profile that combines elements of citrus and cranberry (Chetia et al. 2024).

Recently, incorporating edible flowers into food products has gained relevance as a result of heightened demand for healthier foods and the search for natural alternatives to synthetic ingredients. From baked goods and beverages to sauces and garnishes, the applications are diverse and also reflect cultural traditions. However, its commercial exploitation faces challenges such as its high perishability and the lack of specific regulations, opening a wide field for research and technological development.

The use of flowers in a wide variety of foods is part of many cultures around the world. Recently, in the context of the pursuit of healthier foods and the substitution of synthetic ingredients, flowers have emerged as a potential option to meet these objectives. In particular, attention has been given to the use of flowers in baked goods, such as cookies, biscuits and bread.

Applications in foods such as sauces and condiments have also been explored, although in these cases the scientific literature offers limited descriptions, the most common details being found in cooking texts or in the common knowledge of different regions of the world (Bunting and Facer 2023; Newman and Kirker 2016). To mention a few examples, squash blossoms are used in the preparation of creamy sauces, especially in Mexican cuisine, where these flowers are combined with ingredients such as cream or cheese to enhance their delicate flavor. Similarly, in Asian cuisine, jasmine flowers are used to flavor light sauces, providing a distinctive floral nuance, while, in Mediterranean regions, calendula petals are used as a natural condiment, often replacing saffron, adding color and a slight spicy flavor to preparations.

In the case of baked goods, the addition of flowers is mainly aimed at incorporating compounds with antioxidant activity. Products made with refined flour tend to lose some of the components originally present in the grains, such as fiber and antioxidants. Therefore, the inclusion of flowers in different forms can contribute to the development of functional foods, i.e., those that not only provide nutrition, but also offer additional health benefits.

In some cases, the objective of adding flowers is not limited to enhance nutritional properties, but also to take advantage of their benefits in food stability. For example, Podgórska-Kryszczuk and Pankiewicz (2023) evaluated the fungistatic properties of an aqueous extract of *Calendula officinalis* (marigold) when it was used to partially replace water in bread formulation. The results indicated that the extract increased the total content of phenolic compounds by 77% and 95%, when using concentrations of 10% and 15%, respectively. In addition, it improved the microbiological quality of the bread, highlighting its potential to extending the storage life of baked goods.

Edible flowers have also found an important place in the elaboration of beverages, where their incorporation aims at objectives similar to those of their use in baked foods or dairy products, i.e. to improve both sensory and nutritional properties (Hutabarat and Irene 2021), particularly to increase the polyphenol content and to increase antioxidant activity. Lakshan et al. (2019) described the formulation of a beverage based on Blue Pea Flower extract, stevia, and lime. Through sensory evaluations, they determined that the optimal ratio between these ingredients was 983.25:1.75:15, achieving not only high acceptability, but also outstanding antioxidant activity. Similarly, Saati et al. (2020) reported the development of a beverage made with an extract obtained by maceration of rose flowers, to which they added natural flavors such as mint leaves, ginger and lemon juice in concentrations of

1.5%, 3% and 4% (v/v). Among the formulations evaluated, the combination with 3% mint leaves was the most outstanding, presenting an antioxidant activity of 84.44% and an anthocyanin content of 12.14 mg/L.

It is evident that the use of flowers in food is a topic that needs to be explored. According to the results, it has allowed the development of products with improved sensory and nutritional characteristics. Further exploration is needed in order to evaluate other effects that will allow maintaining the stability of these foods, giving them a longer shelf life, which, in short, could have an economic impact, as well as an impact on the health of consumers.

The use of flowers in garnishes around the world is a practice that has been done for a long time, however, sometimes it goes unnoticed by diners. Some flower-based preparations are also common, which are "ready to eat", or require minimal processing to be consumed. Examples include widely consumed flowers, such as artichoke (*Cynara scolymus*), chamomile (*Matricaria chamomilla*), which usually undergoes a drying process to produce, tea, and vegetables like broccoli and cauliflower, available fresh or in packaged, ready-to-eat presentations (ready to eat).

In some cultures, the use of edible flowers is a deeply rooted practice. An example of this is the squash blossom (*Cucurbita* spp.), which is used in a variety of preparations, including salads, creams and various traditional dishes in regions of the Americas, mainly in central Mexico and also in Asia. Another significant example is the calyx of *Hibiscus sabdariffa*, commonly marketed after a drying process. It can be used in salads as well as in the preparation of infusions. In addition, flowers such as azahar, from citrus fruits, are frequently used in confectionery and bakery as decorative elements. Calendula (*Calendula officinalis*) is used to decorate and add color to pastry preparations.

The diversity of applications of flowers as garnishes or toppings is remarkably wide and is documented mainly in recipe books, cookbooks and in the traditional knowledge of different cultures. However, much of this culinary heritage remains unrecorded, posing a risk to its preservation (Hegde et al. 2022; Miche Bacher 2013; Mulík et al. 2024; dos Santos and Reis 2021).

In Mexican cuisine, it is common to use flowers in food preparation: yucca (*Yucca* spp.), the American coral tree (*Erythrina americana*), the funeral tree (*Quararibea funebris*), the dahlia (*Dahlia* spp.) and the agave (*Agave* spp.), among many others. The ways of preparing these flowers have historical roots dating back to pre-Hispanic times. Although current recipes are often modern reinterpretations, they retain essential elements of the original traditions to meet the demands and preferences of contemporary generations. These flowers are used as an accompaniment to dishes such as salads, tacos or tamales, providing color, texture and a distinctive character to the preparations (Mulík and Ozuna 2020).

A flower, which is used in many regions around the world, *Hibiscus sabdariffa* L. In Mexico, where it is predominantly used to prepare infusions as fresh water, using the dried calyx of the flowers. However, its culinary application extends to salads, garnishes or sauces. Internationally, its use is documented in countries such as Nigeria and Ethiopia, among others, noted for its unique flavor, comparable to cranberries, and for its nutritional properties and health benefits. Scientific studies have shown promising effects of *Hibiscus sabdariffa* in and preventing diverse health issues, such as: anticancer activity, anti-inflammatory effects, antiparasitic activity and antioxidant properties (Edo et al. 2023).

Examples of successful products made from edible flowers already exist in different parts of the world. These include saffron, broccoli and cauliflower, although they are often not identified as edible flowers due to their categorization as condiments or vegetables. In line with the above, it is recognized that in many flowers are consumed as integral ingredients in traditional culinary preparations (Figueredo-Urbina et al. 2021). It is also possible to consume them in gourmet dishes, usually at a high price, since they are used as exclusive ingredients. Despite their high consumption, processed products that incorporate flowers, such as breads, beverages, sauces or condiments, are still scarce and their commercialization on a large scale is still limited (Jadhav et al. 2023).

In Mexico, a large number of edible flowers are consumed, however, there are very few value-added products from these flowers. Being the *Hibiscus sabdariffa* L. flower so popular in that country, it has been possible to commercialize concentrates of this flower to prepare "agua fresca", however, there are few cases that could be similar and data regarding this market are scarce.

14.7.1 Bakery Products

The addition of apple flower powder was found to reduce the formation of methylglyoxal and fluorescence in cookies (Gao et al. 2020). This could offer benefits for individuals with diabetes, particularly in relation to heat-processed foods like cookies. Hnin et al. (2021) observed that incorporating rose flower powder into cookie recipes improved not only the color but also the bioactive compounds and antioxidant properties of the cookies. The study tested various concentrations of rose flower powder (3%, 5%, 7%, and 10%) in cookie formulations. The results showed that both the bulk density and hardness of the cookies increased with higher levels of flower powder. Cookies containing 5% rose flower powder achieved the best color rating and were considered the most favorable in sensory assessments. Over a storage period of up to 8 weeks, the cookies with flower powder retained higher levels of total phenolic content, anthocyanins, and antioxidant activity compared to the control cookies made with only wheat flour. This suggests that adding edible rose flower powder can enhance the nutritional profile of cookies while extending their shelf life. Flower extracts are also used as natural colorants. For instance, lyophilized or spray-dried extracts from *Gomphrena globosa* L. flowers can impart a pink color to cookies, thanks to the presence of betacyanins. These extracts help preserve the chemical integrity of the product. The study found that cookies made with spray-dried extract displayed the most vibrant pink hue, while cookies made with lyophilized extract showed a more stable color over time (Roriz et al. 2020).

Edible flowers have been incorporated into foods to enhance their nutritional properties. For example, Nimakoa et al. (2024) studied the impact of fortifying a cereal bar with 5% edible flowers, including elderberry, lavender, rose, and marigold. The study found that bars containing elderberry and marigold flowers had higher total phenolic acid content. Additionally, those fortified with marigold flowers exhibited increased ash content compared to the control bar without flowers. In sensory evaluations, the bar enriched with rose flowers was rated highest for its organoleptic qualities. Sik et al. (2024) developed biscuits with added nutritional value from lavender, marigold, and rose flowers. Their study found that biscuits enriched with rose petals exhibited the highest antioxidant properties and received the greatest consumer acceptance compared to the other variants. For his part, Grzelczyk et al. (2025) investigated the enrichment of bamboo flour biscuits by incorporating edible flowers such as wild mallow and peony officinalis. Their findings revealed that biscuits made with bamboo flour and flowers contained twice the ash content of those made with wheat flour. Moreover, the addition of edible flowers significantly enhanced the polyphenol content, particularly flavonoids, increasing it by approximately three to four times. Another study evaluated the effects on the chemical and physical properties, bioactive effectiveness, sensory quality, and shelf life of nutricereals and a milk-derived functional beverage when enhaced with rose syrup and marigold powder. The addition of flower extracts to the functional beverage (FB) produced a significant elevation in phytochemical levels, including bioactive compounds like anthocyanins, total phenols, and β-carotene. The rose-flavored functional beverage demonstrated higher overall acceptability compared to functional beverage and marigold-flavored functional beverage (MFB). After 40 days of refrigerated storage, FB and MFB showed a notable drop in pH and an increase in acidity, whereas significant changes in pH, acidity, and overall acceptability were observed in RFB after 50 days. To ensure better consumer acceptance, a best-before period of 50 days is recommended for RFB. Finally, the authors conclude that the optimal concentrations for rose syrup and marigold powder were found to be 10% and 0.50%, respectively (Kumar et al. 2021).

Xu et al. (2022) analyzed the effects of replacing wheat flour with edible chrysanthemum powder at concentrations ranging from 3% to 15% w/w. The study examined its impact on dough structure, physicochemical properties, rheological behavior, and the aroma profile of the resulting bread. According to the authors, incorporating chrysanthemum powder reduced the relative crystallinity and structural order of starch, as evidenced by X-ray diffraction and Fourier transform infrared spectroscopy analyses. Furthermore, steady-state shear and oscillation tests revealed that chrysanthemum enhanced the dough's viscoelastic properties, indicating increased resistance to flow and deformation. Consequently, this led to a decrease in the specific volume and softness of the bread compared to bread made without chrysanthemum. Furthermore, the formulations with 3% and 6% of chrysanthemum powder not only contributed to the enhancement of viscoelastic and structural characteristics of the but also enriched with aromatic qualities of the bread relative to the reference sample. Sensory evaluation revealed that bread with 6% chrysanthemum obtained the best scores for appearance, texture, and general acceptance highlighting its strong sensory appeal to consumers.

The replacement of artificial dyes with natural colorants is a key priority for the food industry to meet evolving consumer preferences, emphasizing the importance of identifying new natural sources for these compounds. Therefore, Pires et al. (2018a, b) assessed hydrophilic extracts from edible flowers, including rose, cornflower, and dahlia, as potential alternatives to E163 (anthocyanin extract). They evaluated the colouring capacity and its effect on the chemical composition of different yogurts. Yogurts infused with edible floral extracts demonstrated a nutritional value and free sugar composition similar to those made with E163 dye and the control yogurt (without additives). Among the extracts tested, rose extract emerged as the most suitable alternative to E163, as the two yogurt groups shared similar nutritional composition, free sugar and fatty acid content, and achieved comparable scores in color parameters.

14.7.2 Beverages

Tamimi et al. (2020) found that adding roselle flowers to a date palm-based beverage enhanced its vitamin content and antioxidant properties. Additionally, combining 1% date palm pollen and 1% roselle flowers improved the beverage's protein, fat, carbohydrate levels, and certain trace minerals, although it negatively impacted the sensory characteristics of the drink. The process of removing alcohol from wine can often result in a loss of aroma. To restore some of this fragrance, extracts from edible flowers can be added. For example, the inclusion of peach, rose, and lily flower extracts in reconstituted dealcoholized Merlot red wine significantly enhanced the aroma, with no noticeable difference in the wine's chemical composition when compared to the original. Sam et al. (2023) reported this enhancement. Moreover, a nutritious spiced drink made with rhododendron (Rhododendron arboreum Sm.) flowers can be stored for up to 6 months in both glass and PET bottles, whether at room temperature (15–25 °C) or refrigerated (4–7 °C). The highest-rated version of this beverage contained 35% extract, 40°B total soluble solids (TSS), and 1.20% acidity. Additionally, it included a spice mix (cardamom, cumin, black pepper, salt, mint juice, and ginger juice) in specific proportions to boost its flavor profile. In southern China, the flowers of dragon fruit (*Hylocereus undatus*) are incorporated into soups and beverages to enhance cholesterol levels (Yi et al. 2012).

Cynara flowers in Europe are acknowledged for their milk-clotting capabilities, primarily utilized as rennet in cheese production, attributed to enzymes such as cardosin A and B. These enzymes operate in a manner akin to chymosin and pepsin, which are essential for the cheese-making process (Ciancolini et al. 2012).

14.7.3 Applications of Edible Flowers in Dairy Products

In certain regions of Spain and Portugal, Cynara flowers have long been used as a vegetable rennet in the production of sheep's cheese. The process begins with harvesting the flowers when they are in full bloom, followed by shade drying and storage in a dry, cool environment. Before use, the flowers are soaked in water for several hours to release their proteolytic compounds. Although salt can be added, it has little effect on the efficiency of this extraction process. The resulting liquid, which takes on a brownish hue, is then mixed into warm milk (35–40 °C), causing the milk to curdle. This technique, which dates back to earlier traditions, is based on the milk-clotting abilities of Cynara flowers, which are attributed to two proteolytic enzymes-cardosin A and B-that function similarly to chymosin and pepsin (Ciancolini et al. 2012). Research by Jacob et al. (2011) has shown that Cynara extracts provide a higher cheese yield than other vegetable coagulants like papain from papaya leaves and bromelain from pineapple. These extracts are now employed in industrial cheese production.

The potential of edible flowers extends beyond traditional uses, as they are increasingly being incorporated into a variety of food products. For instance, in 2013, a group of Romanian students created a hard cheese roll featuring dandelion flowers and chia seeds, which was presented at the Ecotrophelia Europe competition held during the International Trade Fair ANUGA in Cologne, Germany.

For those who don't gather edible flowers themselves, these blooms are available for purchase at local markets and supermarkets. Retailers such as Wegmans in the U.S. and M&S and Waitrose in the UK are increasingly offering edible flowers as part of their product lines. By adopting targeted marketing strategies, these stores are not only boosting sales but also making it easier for consumers to access these natural products.

The use of edible flowers in dairy products represents a convergence between culinary innovation and applied science, which offers opportunities to enrich both the sensory and nutritional properties of these foods (Fig. 14.2). Historically linked to cultural traditions and empirical uses, edible flowers have begun to receive scientific attention due to their potential functional benefits. From their use as natural colorants to their integration as sources of bioactive compounds, these flowers transform dairy matrices such as yogurt, cheese and ice cream into products that transcend their basic function. Recent research on edible flower applications in dairy products is explored below, emphasizing their sensory, technological and nutritional implications.

Dairy products, due to their versatility, offer an ideal matrix for the incorporation of edible flowers in various forms: extracts, powders, or even as visible pieces. These additions can serve different purposes, ranging from enhancing visual and organoleptic appeal to enriching products with nutrients such as fiber, minerals, or bioactive compounds. Below, specific studies on incorporating of edible flowers in milk, yogurt, cheese, and ice cream are described.

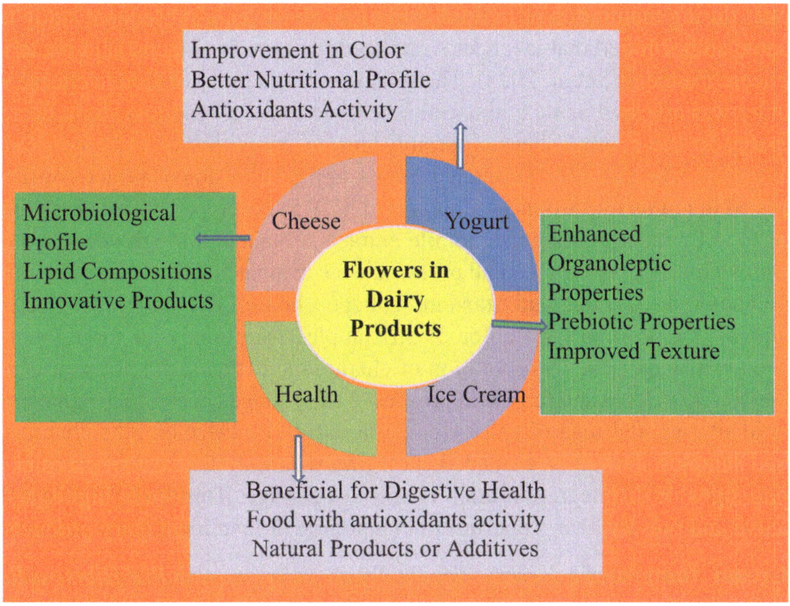

Fig. 14.2 Applications of edible flowers in dairy products and their beneficial impact on organoleptic & nutritional properties

Milk The addition of edible flowers to milk has been explored in studies based on traditional recipes. For instance, *Alcea rosea* flowers have been traditionally used in infusions due to their medicinal properties. In a recent study by Mortas (2023), a powdered product using spray drying techniques was developed, its physicochemical and nutritional properties were evaluated. It was determined that an optimal concentration of 5% flower in milk enhances its mineral content without compromising its organoleptic characteristics. These findings demonstrate the feasibility of producing powdered dairy products enriched with flowers, offering significant nutritional benefits.

Yogurt The incorporation of edible flowers into yogurt has been studied with multiple objectives, including improvements in color, nutritional profile, and antioxidant capacity. Sutakwa et al. (2021) analyzed the use of blue pea flower extracts in yogurts made with different types of milk, finding a significant increase in antioxidant capacity. Recently, the use of freeze-dried flowers, such as cornflower, was explored to enrich yogurt with fiber, vitamin C, and phenolic compounds. The results indicated notable increases in fiber (0.56 g/100 g), protein, and antioxidant activity, positioning edible flowers as functional ingredients in fermented foods (Król et al. 2024).

Cheese The use of flowers in cheese has mainly focused on evaluating their effects on lipid composition, oxidation, and microbiological profile. Semeniuc et al. (2022) studied the addition of lavender powder in Gouda cheese, observing no significant changes during maturation or in lipid oxidation. However, a subsequent study from the same author found that lavender stimulated the growth of lactic acid bacteria in the cheese (Semeniuc et al. 2023). The sensory acceptability of Gouda cheese with lavender was assessed using a nine-point hedonic scale, revealing a slight decrease compared to the control cheese, possibly due to unfamiliarity with the product. Nevertheless, the curiosity generated by its appearance could attract consumers interested in innovative products. Caleja et al. (2015) developed a dairy-based functional food by incorporating chamomile extracts, obtained through decoction, into goat-milk cottage cheese. The final products were evaluated against samples infused with chamomile powder and standard cottage cheese. The antioxidant activity, nutritional composition, and color of all samples were analyzed throughout the duration of storage. The incorporation of chamomile extracts, did not significantly modify the nutritional and fatty acid profiles of the cottage cheese, they enhanced its antioxidant potential regardless of storage duration. Furthermore, the chamomile-infused samples showed an extended shelf life, with only the control samples displaying signs of disintegration after 14 days of storage. This study highlights the bioactive and preservative properties of chamomile in functional dairy products.

Ice Cream formulated a dessert, specifically an ice cream with *Red Hibiscus Rosa*, analyzing its organoleptic and nutritional properties. The results highlighted the need to adjust ingredient proportions, such as Greek yogurt, honey, and skim milk, to optimize product acceptability. In another study, Thilakarathna et al. (2023) combined blue pea flower with banana flour to develop an ice cream with prebiotic properties. The best formulation contained 20% banana flour and 0.67% flower. The ratio between the ingredients significantly improving the product's appearance, texture, and flavor. These examples underscore the potential of edible flowers to diversify and enhance dairy products both sensorially and functionally.

Pairing dairy foods with flowers represents an innovative strategy to develop functional products with high sensory value. Incorporating flowers in dairy matrices, such as yogurts, cheeses, ice creams and other derivatives, has proven to be an effective technique to enrich the physicochemical, nutritional, organoleptic and sensory properties of these foods.

Several studies in this field have shown that the integration of infusions, extracts or parts of flowers in dairy products improves sensory profiles, as well as textural and visual properties such as color (Kumar et al. 2021; Liu and Lv 2019; AI-Musawi et al. 2018). For example, research has been conducted on the interaction between flowers or their extracts with different amounts of other ingredients in the dairy matrix. These studies make it possible to optimize the proportions to achieve combinations that enhance the sensory characteristics of foods. These strategies enrich the properties of the products, and also respond to the demands of consumers who are looking for nutritious, healthy options with a minimum of synthetic components. In 2021, Qiu et al. reported that the addition of a 0.1% *Rosa rugosa* cv *Plena* extract in yogurt significantly improved sensory characteristics, as well as viscoelastic, flavor and rheological properties compared to a yogurt without the flower extract. Among the different concentrations evaluated, 0.1% showed the greatest potential for developing a functional yogurt with improved sensory characteristics. On the other hand, not all impacts are evident in terms of visual or taste characteristics. For example, Limsuwan et al. (2014) evaluated the characteristics of ice cream made with 1%, 3% and 5% fresh flower petals. Although no significant differences were observed in most of the characteristics, the 3% ice cream, obtained the best results in terms of acceptability by the participants in charge of evaluating its sensory characteristics, even surpassing the 5% ice cream. This result underlines the importance of conducting studies to determine optimal proportions that ensure an adequate interaction of the extracts or flowers with the rest of the ingredients, maximizing the sensory properties.

An interesting case of synergy between components was documented in the analysis by Nyulas-Zeke et al. in 2024, in which the formulation of an ice cream using buttermilk and blue pea flower extract was explored. This work demonstrated that it is possible to obtain ice cream by using buttermilk, while the extract acts as a natural colorant. Furthermore, it was observed that a higher concentration of buttermilk intensified the blue hue of the color provided by the extract, while the fermentation process attenuated it. These findings highlight how the interaction between ingredients and production processes can influence key characteristics, such as color, contributing to the improvement of these products.

Some of the studies focus on floral species that, although commonly used in medicinal applications, have unexplored potential in food. For example, Pontes et al. (2024) evaluated the use of flowers and leaves of *Malvaviscus arboreus* in the preparation of yogurt based on goat milk. Although both flowers and leaf extracts were studied the best sensory analysis results were achieved with the flower extract. In particular, a concentration of 2% significantly improved sensory characteristics such as color and flavor, positioning it as the most suitable formulation (Table 14.4).

One of the most notable characteristics of edible flowers is their recognition as healthy foods, driven by the general perception that natural products or additives are beneficial to health due to their natural origin (Chen and Wei 2017; Martins et al. 2016). This concept has led to increasing research into the use of natural flower extracts as substitutes for artificial colorants, which are often considered less healthy. However, it is important to note that being natural does not necessarily guarantee safety or health benefits. Some of the most potent toxic substances come from natural sources, meaning that certain flowers, can be toxic and pose significant health risks (de Pires et al. 2021).

In the specific case of edible flowers, their consumption has been supported by empirical knowledge transmitted through traditions that, in some cases, date back hundreds of years. This cultural background has prompted modern research that goes beyond the organoleptic aspects of foods containing flowers. Currently, studies are focused on evaluating their nutritional properties, such as the addition of minerals, increased fiber content, beneficial for digestive health, and their antioxidant activity. A representative example is the use of blue pea flower; in addition to providing a distinctive blue coloration to foods, this flower contributes to increased polyphenol levels and antioxidant activity, which enhances their nutritional value (Sutakwa et al. 2021; Thilakarathna et al. 2023). Thus, research into the applications

Table 14.4 Different food products developed using edible flowers

Flower	Dairy products	Bakery products	Key findings	References
Rose (*Rosa* spp.)	Rose-flavored yogurt and milk	Rose-infused cake and cookies	These are rich in antioxidants, enhances flavor and aroma, antimicrobial properties	Sharma et al. (2020)
Hibiscus (*Hibiscus sabdariffa*)	Hibiscus yogurt and ice cream	Hibiscus muffins and bread	It helps in blood pressure-lowering. It contains high in vitamin C, natural colorant, potential effects	Kumar and Patel (2018)
Lavender (*Lavandula* spp.)	Lavender cheese and butter	Lavender biscuits, and scones	Very calming aroma, antibacterial properties, improves sensory appeal	Smith and Brown (2019)
Marigold (*Tagetes* spp.)	Marigold extract in dairy drinks	Marigold-flavored bread	Natural yellow-orange colorant and have antioxidant properties	Jones et al. (2021)
Jasmine (*Jasminum* spp.)	Jasmine-flavored milk and yogurt	Jasmine cupcakes and pastries	Helps in stress-relieving, enhances taste and aroma	Wang and Li (2022)
Chamomile (*Matricaria chamomilla*)	Chamomile-infused yogurt and milk	Chamomile tea cakes and cookies	Soothing effects, anti-inflammatory, mild sedative properties	Garcia and Lopez (2020)

of edible flowers allows the design or formulation of foods with improved functional properties, combining a traditional perspective with evidence supported by scientific research. Similarly, Liu and Lv described in (2019), the development of a yogurt to which blueberry flower pulp was added. The results showed a significant improvement in the antioxidant activity of the product. Hence, the incorporation of edible flowers enriches the organoleptic characteristics of foods, and improves their nutritional profile, positioning them as more attractive options for health-conscious consumers.

14.8 Selection and Processing of Edible Flowers for Food Use

Edible flowers, like all food products, should be safe for consumption. However, there are no clear legal regulations governing their use, including approved varieties, cultivation requirements, harvest timelines, storage conditions, or permissible protective substances. To minimize microbial contamination and chemical residues, edible flowers should be carefully cultivated, harvested, transported, and stored under strict safety standards. No international organization, including the Food and Drug Administration, European Food Safety Authority or Food and Agriculture Organization of the United Nations, has issued an official list of edible flowers or established recommended daily intake limits for each species (Demasi et al. 2021b; Matyjaszczyk and Śmiechowska 2019).

The determination of which flowers are safe for consumption was established by our ancestors many years ago; with their daring experiments, flowers have been organized into two different classifications: "edible" and "non-edible." The distinction between these categories lies in their alkaloid content, which are chemicals that serve as a defense mechanism for the plants. Flowers that contain alkaloids with pharmacological activity are generally edible and are often classified as medicinal plants. In contrast, non-edible flowers typically contain alkaloids with psychotropic or stimulant effects. While these flowers are hallucinogenic in nature, they have found application in entheogenic practices and are also used as recreational substances. Additionally, some flowers contain toxic alkaloids, rendering them poisonous. Since alkaloids are unevenly distributed throughout a plant or flower and are typically bitter, it is essential to determine which flowers and which parts are safe to eat. Some guides for edible flowers have been created to assist consumers in making the right choices, and this information is now available online. These guide materials should definitely feature the scientific names of flowers along with accurate drawings or photographs to aid in identification (Nicolau and Gostin 2016). For example, phytate can inhibit mineral absorption in the body, while other antinutrients may also degrade essential nutrients. The toxicity of these compounds is typically dose-dependent. Although saponins are recognized for their antinutritional properties, they also exhibit beneficial effects, including immunostimulatory, anticarcinogenic, and hypocholesterolemic activities (Morton 1992). Traditionally, various pretreatment methods, such as cooking and discarding the broth, were used to remove toxic compounds from certain edible flowers (Sotelo et al. 2007).

Beyond the physicochemical properties of flowers, several factors can render them inedible. These include toxics agents, including pesticides and sulfites. Additionally, microbiological hazards, including pathogenic bacterias and biohazards such as insect contamination, can further compromise their safety for consumption (Matyjaszczyk and Śmiechowska 2019).

Flowers have a high water content, making them highly perishable. As the most delicate part of a plant, they are genetically programmed for a short lifespan, for example, ephemeral flowers last only a day. Their rapid respiration rates indicate a higher metabolic activity than other plant organs, further accelerating deterioration. To maintain the quality and freshness of edible flowers, they should be stored and transported at low temperatures, ideally below 5 °C (Demasi et al. 2021a). For optimal freshness, they should be harvested and used as soon as possible. Keeping them in water while preparing a meal can help maintain their quality. Since flowers are seasonal and in demand year-round, various preservation methods have been developed. However, preservation can affect their sensory and nutritional properties. These methods are broadly categorized as traditional or modern. Traditional techniques, such as dehydration and osmosis, have long been used, while modern approaches, including refrigeration, freezing, freeze-drying, and other methods, which provide improved preservation options (Nicolau and Gostin 2016).

Drying has been a widely used preservation method for ages. This method has proven to be the most effective way to maintain flower freshness by inhibiting decay caused by the microorganisms and enzymes. Edible flowers have traditionally been dried using methods like sun drying, solar drying, and shade drying. However, advancements in technology have introduced various modern drying techniques including forced air drying, lyophilization, vacuum dehydration, and microwave drying, which integrates multiple drying approaches for enhanced efficiency (Zhao et al. 2019).

Freezing is another method for preserving the freshness of flowers. While they maintain their appearance when frozen, they tend to become limp and lose their color upon thawing. Despite this, frozen edible flowers can still be used in baked goods and ice creams. Another creative approach is to freeze them into ice cubes, adding a decorative touch to beverages ((Nicolau and Gostin 2016).

Candying or crystallizing flowers is a popular preservation method that enhances their texture and sweetness. The process involves coating the petals with a mixture of egg white and water (2:1), then dusting them with superfine sugar. The coated petals are then placed on wire racks or wax paper-lined trays in a warm, dry environment until they become crisp, allowing moisture to escape through osmosis. Flowers such as rose and geraniums, are well-suited for this technique. Once candied, they should be stored in airtight tins or plastic containers to maintain freshness. Often used to decorate cakes and confections, this method is also widely applied in industrial production, as it preserves the flowers' shape, color, and flavor (Nicolau and Gostin 2016).

The primary function of edible coatings is to form a protective barrier that regulates moisture, controls gas exchange, preserves color and flavor, and reduces enzymatic oxidation in horticultural products. Studies on the application of edible coatings in fresh fruits and vegetables suggest that this method could also be effective for preserving edible flowers. However, research in this area remains limited (Purohit et al. 2021). Similarly, ultrasound technology has proven to be an effective method for drying flowers. It helps retain essential compounds such as phenolics and carotenoids while preserving their antioxidant properties (Benimana et al. 2024).

14.9 Novel Preservation Techniques of Edible Flowers

14.9.1 Low-Temperature Storage

Low-temperature storage involves maintaining reduced storage temperatures to manage senescence. This is achieved by decreasing cellular respiration and minimizing water loss, both of which are critical factors affecting flower shelf stability and postharvest metabolic processes. Demasi et al. (2021a) assessed the shelf stability over time and the variations in photochemical profile during chilled storage (4 °C) of 17 edible blossoms, finding no adverse effects on the plant bioactive substances (total phenolic and anthocyanin content) over a period of 14 days.

14.9.2 Controlled Atmosphere Storage

This commercial system involves the regulation of gaseous concentrations of O_2, N_2, and CO_2, along with the temperature and humidity of the storage environment. Paradis et al. 1996 evaluated the organoleptic qualities (chromaticity, rigidity, deterioration) and antioxidant properties (vitamin C, β-carotene) of broccoli florets in atmospheric air and a controlled atmosphere (CA) consisting of 2% O_2 and 6% CO_2 during storage at 4 °C. Controlled atmosphere extended shelf life, maintaining greenish shades and chlorophyll pigments more effectively, while accelerating the loss of stem turgor. Samples stored in air exhibited decay within 4 weeks, whereas CA samples demonstrated minimal decay after 5 weeks, though they experienced browning on the stem surfaces.

Cefola et al. (2016) assessed the effects of controlled atmospheric packaging on the postharvest quality and marketability of fresh zucchini flowers (*Cucurbita pepo* L.) stored at 50 °C. The evaluation was performed under four different atmospheric compositions: 3% O_2 + 10% CO_2 in air, 3% O_2 in nitrogen, 10% CO_2 in air, and air. The study showed that 3% O_2 in nitrogen decreased mass loss and respiration rate while maintaining visual quality, sensory and nutritional properties, and ascorbic acid levels throughout the experiment. However, the quality of flowers stored in the air declined rapidly, restricting marketability after 5 days of storage. Flowers displayed signs of spoilage, such as water soaking, wilting, and diminished brightness. The severity of these symptoms was lessened under atmospheric conditions with lowered O_2 levels. A slight enhancement in shelf life was achieved through the incorporation of CO_2.

Aquino-Bolaños et al. (2013) conducted an experiment to investigate the effects of three distinct gaseous compositions in the storage environment on edible Squash (Cucurbita pepo) flowers. Four atmospheric conditions were tested: N_2 + 5% O_2 (GC_1); 10% CO_2 + 5% O_2 + N_2 (GC2); 10% CO_2 + air (GC_3); and air as the control. Flowers were stored at 5 °C. The flowers stored in the controlled atmosphere demonstrated enhanced preservation of pH, total sugars, titratable acidity, total soluble solids, and exhibited a reduction in mass compared to the control sample. The study determined that GC_2 and GC_3 treatments have the potential to extend the postharvest lifespan of squash flowers.

14.9.3 Modified Atmosphere Packaging

Modified atmosphere packaging involves the packing and sealing of edible flowers that are actively respiring within plastic pouches. This process utilizes adjusted CO_2 and O_2 levels inside the designated packaging. Aquino-Bolaños et al. (2013) examined the impact of controlled release of 1-methylcyclopropene and modified atmosphere packaging on enhancing the shelf life of freshly harvested snapdragon (*Antirrhinum majus* L.) and carnation (*Dianthus caryophyllus* L.) flowers. Flowers were arranged in containers that either contained 0.5 μl/l of 1-methylcyclopropene or did not, sealed with a gas-permeable membrane, and maintained at 50 °C. The study found that treatment with 1-methylcyclopropene effectively reduced ethylene activity, which in turn decreased dehydration and enhanced organoleptic properties compared to flowers packaged in standard retail packaging.

14.9.4 Pulsed Electric Field Technology (PEF)

Pulsed Electric Field is a preservation method that involves applying pulses of elevated electric fields to botanical ingredients situated between electrodes in the reaction cell. This process occurs over duration of a few microseconds to milliseconds, with intensity levels varying from 10 to 80 kV/cm (Nowosad et al. 2021). When subjected to high-voltage impulses, microbial cells experience swelling and disintegration, leading to a significant increase in the porosity of the cell membrane. As a result, it becomes permeable to smaller molecules, leading to the drainage of cell constituents and ultimately resulting in cell death Rowan et al. (2000). The experiment conducted by (Aguiló-Aguayo et al. 2015) demonstrates that PEF techniques can be effectively applied to optimize the availability of phytochemicals in broccoli flowers. Glucosinolates present in broccoli (*Brassica oleracea italica*) are recognized for their potential health benefits, including antimutagenic effects and the prevention of cardiovascular diseases. The application of PEF treatment at conditions of 4 kV cm^{-1} for durations of 525 and 1000 μs led to the highest levels of glucosinolates in broccoli flowers.

14.9.5 Ultrasound Processing

Ultrasound processing involves the application of ultrasound treatment to inactivate microbes. This is achieved by disrupting their cell membranes and generating free radicals. Localized hot spots result from cavitation, while shock waves generate free radicals. This process can damage cell walls and membranes, ultimately leading to cell death (Chavan et al. 2022).

Xin et al. (2014) investigated the effects of ultrasound-assisted freezing on the freezing duration and quality of broccoli (*Brassica oleracea* L. var. *botrytis* L.) florets during the immersion freezing process. The microstructure and firmness of broccoli tissues were preserved at an acceptable level through ultrasound-assisted freezing, with minimal drip loss observed in comparison to conventional immersion freezing. The use of ultrasound-assisted freezing at power levels of 150 W (30 kHz) or 175 W (20 kHz) resulted in a reduction of the overall freezing duration.

14.9.6 Edible Coating

Edible coating produced from biodegradable biopolymers, including polysaccharides, proteins, and lipids, is recognized as a sustainable packaging strategy. Edible coatings can be applied through either dipping or spraying methods. These biopolymeric coatings offer protective properties against adverse environmental conditions. They can impede physiological processes, including the breakdown of cell walls through enzymatic actions, while the accumulation of ethylene contributes to reducing microbial contamination, thereby enhancing shelf life (Garcia and Davidov-Pardo 2021).

Borage (Borago officinalis) is a plant known for its therapeutic benefits, commonly found in Mediterranean regions. It has a flavor reminiscent of cucumbers, with a mild sweetness derived from its stamens, making it suitable for use in summer fruit drinks. Nonetheless, the use of the flower in functional foods and culinary applications is limited due to its short shelf life, which is approximately 1 day. The post-harvest quality of the flowers deteriorated quickly, resulting in dryness, shrinkage, and darkening. Fernandes et al. (2020) applied a 0.5% alginate solution as an edible coating on borage flowers. The findings indicated that flowers coated with alginate attained a post-harvest shelf life of 5 days, surpassing the uncoated flowers by 4 days. In a separate investigation carried out by Fernandes et al. (2020) it was observed that alginate treatment extended the degradation period of pansy flowers (Viola × wittrockiana) from 3–4 days to 14 days during storage at refrigeration temperatures (4 °C).

14.10 Assessment of the Safety Profile of Edible Flowers

The safety of edible flowers for human consumption has been investigated by several studies, including those by Matyjaszczyk and Śmiechowska (2019) and Nicolau and Gostin (2016). Several factors i.e. internal and external can affect the safety of these flowers. Internal factors include the naturally occurring compounds within the plants that may be harmful to humans. On the other hand, external factors involve contamination by microorganisms, the presence of insect repellents, or residues from pesticides and insecticides. While certain edible flowers like *Convallaria majalis* and *Anemone ranunculoides* are known for their nutritional value and health-promoting properties, they can also contain potentially toxic substances such as sulfites, dimethoate, and harmful bacteria like Salmonella.

Edible flowers are commonly grouped with ornamental plants such as hibiscus, begonia, calendula, and daylilies, while flowers from fruits like bananas, citrus, and various vegetables fall into a different category. Unfortunately, there are no standardized guidelines for identifying safe edible flowers, and those typically sold at flower shops often intended for ornamental purposes can carry risks. Many of these flowers are grown using pesticides, which may leave harmful residues that could trigger allergic reactions, asthma, or more severe health issues.

While edible flowers are cultivated across the globe, a formal classification system based on their toxicity remains absent. Some flowers contain compounds like erucic acid, and excessive consumption can lead to health complications. For instance, consuming more than 39.5 g of *Tropaeolum majus* may exceed the safe daily intake of erucic acid. Similarly, consuming over 18 g of *Achillea millefolium* or 7 g of *Galium odoratum* can surpass the safe limits for thujone and coumarin, respectively (Egebjerg et al. 2018).

Therefore, further research is needed to assess the safety of edible flowers, especially when used as primary ingredients in food. When used as decorative elements in dishes like ice cream, cakes, or desserts, it is crucial that their consumption remains within safe limits to avoid any potential risks to health.

14.11 Marketing

Flowers contribute significantly to the economy in various ways, as they not only provide edible blooms but also support food production through their vital role in pollination. The growing demand for edible flowers in industries such as nutraceuticals, food, and pharmaceuticals is driving the expansion of the global edible flowers market. This increase in demand is boosting the production of these flowers, as they are considered both safe and nutritious by chefs and consumers alike. Edible flowers are used in a variety of products, including food, beverages, garnishes, baked goods, and salads (Lauderdale and Bradley 2014). Several countries engage in the commercial cultivation of edible flowers, often growing them alongside cut flowers, medicinal herbs, and salad crops. This practice is often motivated by the relatively low income it generates, as edible flowers are seasonal crops (Anca et al. 2013). The cultivation costs for edible flowers are significantly higher than those of other crops, mainly due to the practice of organic farming and the expense of importing certified seeds. Furthermore, there is limited information available regarding their exportation, though countries like Ethiopia and China have been known to export 7% and 8% of their edible flowers, respectively, for medicinal and industrial uses (Fernandes et al. 2020). Transporting edible flowers requires extra time and careful handling, and packaging these delicate flowers is labour-intensive. Given their short shelf life, fresh edible flowers must be quickly consumed once they reach the market (Keidar 2015). The global edible flower market is segmented by region, flower type, and application. The primary regions for this market are Europe, North America, Asia-Pacific, and the rest of the world. Europe's market includes countries such as Germany, Italy, Spain, the UK, and France, while North America is further broken down into Canada, Mexico, and the United States. The Asia-Pacific market encompasses countries like Australia, China, Japan, India, and New Zealand. Other regions, including the Middle East, South America, and Africa, are categorized under "the rest of the world" (https://www.marketresearchfuture.com/reports/

edibleflowers-market-6634). Europe is expected to lead the global edible flower market due to its significant consumption of these flowers, particularly for medicinal purposes. The distinct flavor and taste of edible flowers are key factors driving growth in this region. Portugal has recently joined the ranks of producers of edible flowers, alongside Japan, France, and Italy, with flowers like borage, roses, cornflower, pansy, courgette, nasturtium, and marigold available in the market. The price of fresh edible flowers ranges from 2.40 to 39.0 euros for a 20-g quantity (Fernandes et al. 2020).

The Asia-Pacific region has experienced the highest compound annual growth rate for edible flowers, driven by trends such as flower and petal garnishing, as well as the use of edible flowers in snacks and beverages. The growth of the edible flower market in India and China is particularly notable, though accurate data on the Indian market is lacking, as most edible flowers are sourced from the wild. There is a need to expand the market in developing economies to address this gap.

Key players in the global edible flower market include Cottage Botanicals (Canada), Herbs Unlimited (UK), J&P Park Acquisitions, Inc. (US), Maddocks Farm Organics (UK), Agriver Ltd. (Israel), Nurtured in Norfolk (UK), Petite Ingredient (Australia), LLC (US), Nouvelle Fresh (South Africa), and Edible Flower (UK) (https://www.marketresearchfuture.com/reports/edible-flowers-market-6634). According to Fernandes et al. (2020), there is a significant lack of reliable data on the global production and area dedicated to edible flower cultivation. However, the edible flower markets in India and China are experiencing substantial growth, signalling the need for more development in these regions.

The attention for edible flowers has been increasing in recent years, which is reflected in the increase of scientific publications and culinary books that explore their applications in food (Bunting and Facer 2023; Jacobs Miriam 1999; Bacher 2013). These studies mainly highlight their nutritional, sensory and functional properties, with emphasis on their antioxidant, mineral and fiber content, as well as acceptability studies, mainly with the objective of designing products that can be introduced to the market.

The potential of edible flowers in the market is clear, but there are still challenges that need to be overcome. To improve their commercialization, it will be crucial to improve the management of postharvest activities, given their fragility and the fact that they are highly perishable; without treatments or the application of postharvest technology, the shelf life is only a few days. In addition to the above, it is necessary to guarantee their safety through sustainable agricultural practices and proper management that limits the use of pesticides and fertilizers (Jadhav et al. 2023); this is even more relevant given the fact that in many countries there is no legislation regarding their management. The implementation of advanced postharvest technology (Machado et al. 2023; Pêgo et al. 2022; Shantamma et al. 2021) and further research on their chemical composition, nutritional properties and preservation techniques could optimize their commercialization, reduce costs and ensure the quality of these products for human consumption.

14.12 Future Trends and Expectations

The upcoming developments in the utilization of edible flowers will primarily be influenced by consumer preferences, alongside the reactions of researchers, food manufacturers, and distributors to these demands. Current research focuses on identifying more effective techniques for preserving edible flowers. Among these, freeze-drying has emerged as the most efficient method for retaining the flowers' original attributes, including color, appearance, and gloss. This technique is costly and presents significant technical challenges (Mlcek and Rop 2011). A study conducted by Aquino-Bolaños et al. (2013) investigated the impact of storing squash flowers (*Cucurbita pepo*) in controlled atmospheres (CA). The study indicated that controlled atmosphere storage, specifically utilizing a mixture of 5% O_2 and 10% CO_2, notably prolonged the shelf life of the flowers while maintaining essential compounds such as ascorbic acid, polyphenols, and carotenoids. With the growing consumer demand for healthier, natural foods, researchers are exploring various new varieties of edible flowers. The red-purple flowers of *Citrus medica* L. Cv. Diamante exhibit antimicrobial properties, demonstrating effectiveness against pathogenic bacteria such as *Pseudomonas aeruginosa, Staphylococcus aureus, and Escherichia coli.* Recent studies highlight the necessity of refreshing understanding regarding their composition and health advantages. Nanotechnology represents an emerging field of scientific exploration focused on the examination and control of matter at the smallest scales, encompassing molecules and atoms (Rao and Poonia 2024a; b).

A study conducted by Rop et al. (2012) investigated the antioxidant capacity of different edible flowers, revealing elevated levels of ascorbic acid and phenolic compounds in species such as Chrysanthemum, Dianthus, and Viola. The flowers also supplied significant minerals, including potassium, which was observed to be especially plentiful.

In nations such as Thailand, the incorporation of edible flowers into culinary practices has a rich history, and the Thai Health Promotion Foundation has actively promoted their nutritional advantages. Plants such as neem (Azadirachia indica), sesbania (Sesbania grandiflora), and cassia (Cassia siamea) possess culinary uses and exhibit health advantages, including antimicrobial effects and the capacity to counteract mutagens (Kaisoon et al. 2012; Wongwattanasathien et al. 2010). In Japan, the flowers of Chrysanthemum morifolium are utilized for their anti-inflammatory and potential cancer-protective effects (Ukiya et al. 2002). Research indicates that extracts from Cynara flowers produce a higher yield of cheese when compared to other vegetable coagulants (Jacob et al. 2011).

Individuals who do not cultivate their own edible flowers can find them more readily available for purchase in various markets and supermarkets, including major chains like Wegmans in the U.S. and M&S and Waitrose in the U.K. Retailers are implementing strategies to promote edible flowers, aiming to increase sales and enhance consumer access to these natural products.

14.13 Conclusion

Edible flowers represent an extraordinary fusion of aesthetic appeal, nutritional value, and therapeutic properties, making them an invaluable resource in the modern food and pharmaceutical industries. Rich in bioactive compounds such as phenolics, carotenoids, and flavonoids, these flowers exhibit significant health benefits, including anti-inflammatory, antioxidant, and anticancer properties. The integration of edible flowers into functional food development is driving innovation across multiple sectors, addressing malnutrition and diversifying agricultural practices. Despite limitations such as perishability and limited market data, ongoing research and technological advancements are enhancing their usability and market potential. The growing consumer demand for natural, health-oriented products underscores the importance of edible flowers in global markets. By fostering sustainable cultivation practices and exploring novel applications, edible flowers offer promising solutions to modern challenges in nutrition, agriculture, and environmental conservation.

References

Abdel-Aal ESM, Akhtar H, Zaheer K, Ali R (2013) Dietary sources of lutein and zeaxanthin carotenoids and their role in eye health. Nutrients 5(4):1169–1185. https://doi.org/10.3390/nu5041169

Aguiló-Aguayo I, Suarez M, Plaza L, Hossain MB, Brunton N, Lyng JG et al (2015) Optimization of pulsed electric field pre-treatments to enhance health-promoting glucosinolates in broccoli flowers and stalk. J Sci Food Agric 95(9):1868–1875

Al-Musawi AD, Fareed I, Abu-Almaaly RA (2018) Effect of rose demascena extracts in prolonging the duration of the preservation of yoghurt. Indian J Public Health Res Develop 9(10):923–931. https://doi.org/10.5958/0976-5506.2018.01259.7

Anca H, Cantor M, Buta E, Hort D (2013) Current trends of using ornamental plants in culinary arts. ProEnvironment 6:52–58

Aquino-Bolaños EN, Urrutia-Hernández TA, López Del Castillo-Lozano M, Chávez-Servia JL, Verdalet-Guzmán I (2013) Physicochemical parameters and antioxidant compounds in edible squash (*Cucurbita pepo*) flower stored under controlled atmospheres. J Food Qual 36(5):302–308

Asif M, Zahid T, Ahmad B, Noor S, Naqvi A, Yasmeen T, Imran M, Akhtar MU, Zaman M, Shafique H (2023) Therapeutics characteristics and application of *Aloe vera*: a review. RADS J Food Biosci 2(1):56

Bacher M (2013) Cooking with flowers: sweet and savory recipes with rose petals, lilacs, lavender, and other edible flowers. Quirk Books

Bahuguna A, Vijayalaxmi KG, Suvarna VC (2018) Formulation and evaluation of fresh red Hawaiian hibiscus (*Hibiscus rosa-sinensis*) incorporated value-added products. Int J Curr Microbiol App Sci 7(8):4282–4290

Begum YA, Deka SC (2019) Chemical profiling and functional properties of dietary fibre-rich inner and outer bracts of culinary banana flower. J Food Sci Technol 56:1–11

Benimana F, Kawata K, Kong F, Roy A, Mohan A (2024) Role of ultrasound application in the production, processing, and preservation of edible flowers: a review. J Food Process Eng 11:2–16. https://doi.org/10.1111/jfpe.70012

Benvenuti S, Mazzoncini M (2021) The biodiversity of edible flowers: discovering new tastes and new health benefits. In: Frontiers in plant science, vol 11, Frontiers Media. https://doi.org/10.3389/fpls.2020.569499

Benvenuti S, Bortolotti E, Maggini R (2016) Antioxidant power, anthocyanin content and organoleptic performance of edible flowers. Sci Hortic 199:170–177. https://doi.org/10.1016/j.scienta.2015.12.052

Bhaskar JJ, Mahadevamma S, Chilkunda ND, Salimath PV (2012) Banana (*Musa sp.* var. elakki bale) flower and pseudostem: dietary fiber and associated antioxidant capacity. J Agric Food Chem 60(1):427–432

Bhatt ID, Rawat S, Badhani A, Rawal RS (2017) Nutraceutical potential of selected wild edible fruits of the Indian Himalayan region. Food Chem 215:84–91

Basso C, Papalia J (2014) Aceitabilidade de flores comestíveis. Revista da Associação Brasileira de Nutrição 6:1391–1392

Bunting, E and Facer, J (2023) The edible flower: A modern guide to growing, cooking & eating edible flowers. Laurence King. ISBN-13: 978-0857829498

Caleja C, Barros L, Antonio AL, Ciric A, Barreira JCM, Sokovic M, Oliveira MBPP, Santos-Buelga C, Ferreira ICFR (2015) Development of a functional dairy food: exploring bioactive and preservation effects of chamomile (*Matricaria recutita* L.). J Funct Foods 16:114–124. https://doi.org/10.1016/j.jff.2015.04.033

Cefola M, Amodio ML, Colelli G (eds) (2016) Extending postharvest life of ready-to-use zucchini flowers: effects of the atmosphere composition. International Society for Horticultural Science (ISHS), Leuven

Cerdá-Bernad D, Valero-Cases E, Pastor J-J, Frutos MJ (2022) Saffron bioactives crocin, crocetin and safranal: effect on oxidative stress and mechanisms of action. Crit Rev Food Sci Nutr 62(12):3232–3249. https://doi.org/10.1080/10408398.2020.1864279

Chanukya BS, Rastogi NK (2016) A comparison of thermal processing, freeze drying, and forward osmosis for the downstream processing of anthocyanin from rose petals. J Food Process Preserv 40(6):1289–1296

Chavan P, Sharma P, Sharma SR, Mittal TC, Jaiswal AK (2022) Application of high-intensity ultrasound to improve food processing efficiency: a review. Foods 11(1)

Chen NH, Wei S (2017) Factors influencing consumers' attitudes towards the consumption of edible flowers. Food Qual Prefer 56:93–100. https://doi.org/10.1016/j.foodqual.2016.10.001

Chen L, Li M, Yang Z, Tao W, Wang P, Tian X, Li X, Wang W (2020a) *Gardenia jasminoides Ellis*: ethnopharmacology, phytochemistry, and pharmacological and industrial applications of an important traditional Chinese medicine. J Ethnopharmacol 257:112829

Chen Q, Xu B, Huang W, Amrouche AT, Maurizio B, Simal-Gandara J, Tundis R, Xiao J, Zou L, Lu B (2020b) Edible flowers as functional raw materials: a review on anti-aging properties. Trends Food Sci Technol 106:30–47

Chensom S, Okumura H, Mishima T (2019) Primary screening of antioxidant activity, total polyphenol content, carotenoid content, and nutritional composition of 13 edible flowers from Japan. Prevent Nutr Food Sci 24(2):171–178

Chetia I, Vijayakumar A, Badwaik LS (2024) Edible flowers' flavor, safety and their utilization as functional ingredients: a review. J Food Sci Technol 62. https://doi.org/10.1007/s13197-024-06071-4

Chevallier A (1996) The encyclopedia of medicinal plants. Dorling Kindersley, London, p 322

Chhabra D, Gupta RK (2015) Fortification of curd using *Delonix regia* flower petal extract and estimation of its phytochemical, antibacterial, and antioxidant activity. J Pharmacogn Phytochem 4(3):299

Ciancolini A, Rey NA, Pagnotta MA, Crinò P (2012) Characterization of Italian spring globe artichoke germplasm: morphological and molecular profiles. Euphytica 186:433–443

Clemente-Villalba J, Burló F, Hernández F, Carbonell-Barrachina ÁA (2023) Valorization of wild edible plants as food ingredients and their economic value. Foods 12(5):MDPI. https://doi.org/10.3390/foods12051012

Coyago-Cruz E, Alarcón A, Guachamin A, Méndez G, Osorio E, Heredia-Moya J, Zuñiga-Miranda J, Beltrán-Sinchiguano E, Vera E (2024) Functional, antioxidant, antibacterial, and antifungal activity of edible flowers. Antioxidants 13(11). https://doi.org/10.3390/antiox13111297

Dash SK (2017) Post-harvest management and value addition for tribal areas. Available at: http://zpd7icar.nic.in/Dr%20S%20K%20Das.pdf

Demasi S, Mellano MG, Falla NM, Caser M, Scariot V (2021a) Sensory profile, shelf life, and dynamics of bioactive compounds during cold storage of 17 edible flowers. Horticulturae 7(7):166

Demasi S, Caser M, Donno D, Enri SR, Lonati M, Scariot V (2021b) Exploring wild edible flowers as a source of bioactive compounds: new perspectives in horticulture. Folia Horticulturae 33(1):27–48. https://doi.org/10.2478/fhort-2021-0004

Edo GI, Samuel PO, Jikah AN, Oloni GO, Ifejika MN, Oghenegueke O, Ossai S, Ajokpaoghene MO, Asaah EU, Uloho PO, Akpoghelie PO, Ugbune U, Ezekiel GO, Onoharigho FO, Agbo JJ, Essaghah AEA (2023) Proximate composition and health benefit of Roselle leaf (Hibiscus sabdariffa). Insight on food and health benefits. In: Food chemistry advances, vol 3. Elsevier Ltd. https://doi.org/10.1016/j.focha.2023.100437

Egebjerg MM, Olesen PT, Eriksen FD, Ravn-Haren G, Bredsdorff L, Pilegaard K (2018) Are wild and cultivated flowers served in restaurants or sold by local producers in Denmark safe for the consumer? Food Chem Toxicol 120:129–142

Fernandes L, Casal S, Pereira JA, Saraiva JA, Ramalhosa E (2017) Edible flowers: a review of the nutritional, antioxidant, antimicrobial properties and effects on human health. J Food Compos Anal 60:38–50. https://doi.org/10.1016/j.jfca.2017.03.017

Fernandes L, Casal S, Pereira JA, Malheiro R, Rodrigues N, Saraiva JA, Ramalhosa E (2019) Borage, calendula, cosmos, Johnny Jump up, and pansy flowers: volatiles, bioactive compounds, and sensory perception. Eur Food Res Technol 245:593–606

Fernandes L, Casal S, Pereira JA, Pereira EL, Saraiva JA, Ramalhosa E (2020) Freezing of edible flowers: effect on microbial and antioxidant quality during storage. J Food Sci 85(4):1151–1159

Figueredo-Urbina CJ, Álvarez-Ríos GD, Zárraga LC (2021) Edible flowers commercialized in local markets of Pachuca de Soto, Hidalgo. Mex Bot Sci 100(1):120–138. https://doi.org/10.17129/BOTSCI.2831

Gao J, Sun Y, Li L, Zhou Q, Wang M (2020) The antiglycative effect of apple flowers in fructose/glucose-BSA models and cookies. Food Chem 330:127170

Garcia F, Davidov-Pardo G (2021) Recent advances in the use of edible coatings for preservation of avocados: a review. J Food Sci 86(1):6–15. 36

Garcia H, Lopez P (2020) Chamomile and its health benefits in functional foods. Nutr Res 19(3):98–112

González-Barrio R, Periago MJ, Luna-Recio C, Garcia-Alonso FJ, Navarro-González I (2018) Chemical composition of the edible flowers, pansy (Viola wittrockiana) and snapdragon (Antirrhinum majus) as new sources of bioactive compounds. Food Chem 252:373–380

Grzelczyk J, Drożdżyński P, Budryn G, Czarnecki A, Paprocka Z, Gałązka-Czarnecka I (2025) High-fiber cookies with bamboo flour and edible flowers: evaluation of structural properties, phenolic content, antioxidant activity and nutritional value. LWT 216. https://doi.org/10.1016/j.lwt.2025.117321

Grzeszczuk M, Wesolowska A, Jadczak D, Jakubowska B (2011) Nutritional value of chive edible flowers. Acta Sci Pol Hortorum Cultus 10(2):85–94

Gupta AK, Kumar V, Naik B, Mishra P (2024) Introduction. In: Edible flowers: health benefits, nutrition, processing, and applications. Elsevier, pp 1–8. https://doi.org/10.1016/B978-0-443-13769-3.00001-7

Hajam YA, Lone R, Kumar R (2023) Role of plant phenolics against reactive oxygen species (ROS) induced oxidative stress and biochemical alterations. In: Lone R, Khan S, Mohammed Al-Sadi A (eds) Plant phenolics in abiotic stress management. Springer Nature, pp 125–147. https://doi.org/10.1007/978-981-19-6426-8_7

Halliwell B (2015) Antioxidant and anti-inflammatory components of foods, p 34

Hegde AS, Gupta S, Sharma S, Srivatsan V, Kumari P (2022) Edible rose flowers: a doorway to gastronomic and nutraceutical research. In: Food research international, vol 162. Elsevier Ltd. https://doi.org/10.1016/j.foodres.2022.111977

Hnin KK, Zhang M, Ju R, Wang B (2021) A novel infrared pulse spouted freeze drying on the drying kinetics, energy consumption and quality of edible rose flowers. In Lwt. https://doi.org/10.1016/j.lwt.2020.110318

Hutabarat DJC, Irene (2021) Chemical and physical characteristics of fermented beverages from plant-based milk with the addition of butterfly pea flower (Clitoria ternatea L.) extracts. IOP Conf Ser Earth Environ Sci 794(1). https://doi.org/10.1088/1755-1315/794/1/012140

Jacobs M (1999) Cooking with edible flowers: storey country wisdom bulletin A-223, p 228

Jacob M, Jaros D, Rohm H (2011) Recent advances in milk clotting enzymes. Int J Dairy Technol 64:14–33

Jadhav HB, Badwaik LS, Annapure U, Casanova F, Alaskar K (2023) A review on the journey of edible flowers from farm to consumer's plate. Appl Food Res 3(2). https://doi.org/10.1016/j.afres.2023.100312

Jones M, Wilson K, Taylor D (2021) Marigold and its applications in food industry. Trends Food Sci Technol 32(4):210–223

Kaisoon O, Konczak I, Siriamornpun S (2012) Potential health enhancing properties of edible flowers from Thailand. Food Res Int 46(2):563–571

Keidar Y (2015) Edible flowers challenging to market. Hortidaily. Retrieved from: http://www.hortidaily.com/article/15049/edible-flowers-challenging-to-market

Kelley KM, Behe BK, Biernbaum JA, Poff KL (2001a) Consumer ratings of edible flower quality, mix, and color. HortTechnology 11(4):644–647. https://doi.org/10.21273/HORTTECH.11.4.644

Kelley KM, Cameron AC, Biernbaum JA, Poff KL (2003) Effect of storage temperature on the quality of edible flowers. Postharvest Biol Technol 27:341–344. https://doi.org/10.1016/S0925-5214(02)00096-0

Król J, Brodziak A, Ślusarczyk L, Matwijczuk A, Chwil M, Matraszek-Gawron R (2024) Yogurt with cornflower (Centaurea cyanus L.) petals as a source of antioxidant compounds and dietary fiber: physicochemical and spectroscopic research during storage. J Dairy Sci 108. https://doi.org/10.3168/jds.2024-25628

Kumar R, Patel D (2018) Hibiscus and its potential in food industry: a review. Int J Food Sci Nutr 45(2):78–89

Kumar A, Kaur A, Tomer V (2020a) Process optimization for the development of a synbiotic beverage based on lactic acid fermentation of nutricereals and milk-based beverage. LWT Food Sci Technol 131:Article 109774

Kumar A, Kaur A, Tomer V, Gupta K, Kaur K (2020b) Effect of rose syrup and marigold powder on the physicochemical, phytochemical, sensorial and storage properties of nutricereals and milk-based functional beverage. J Am Coll Nutr:1–8

Kumar A, Kaur A, Tomer V, Gupta K, Kaur K (2021) Effect of rose syrup and marigold powder on the physicochemical, phytochemical, sensorial and storage properties of nutricereals and milk-based functional beverage. J Am Coll Nutr 40(2):133–140. https://doi.org/10.1080/07315724.2020.1744487

Lakshan SAT, Jayanath NY, Abeysekera WPKM, Abeysekera WKSM (2019) A commercial potential blue pea (Clitoria ternatea L.) flower extract incorporated beverage having functional properties. Evid Based Complement Alternat Med. https://doi.org/10.1155/2019/2916914

Lauderdale C, Bradley L (2014) Choosing and using edible flowers: enjoy the flavor, color, and texture that flowers can bring to food. North Carolina Cooperative Extension Service

Lee JH, Lee HJ, Choung MG (2011) Anthocyanin compositions and biological activities from the red petals of Korean edible rose (*Rosa hybrida* cv. Noblered). Food Chem 129:272–278

Liang Q, Cui J, Li H, Liu J, Zhao G (2013) Florets of sunflower (*Helianthus annuus L.*): potential new sources of dietary fiber and phenolic acids. J Agric Food Chem 61(14):3435–3442

Limsuwan T, Paekul N, Ingsriwan L (2014) Asian journal of food and agro-industry effects of butterfly pea extract and flower petals on sensory, physical, chemical and microbiological characteristics of sugar-free ice cream. As J Food Ag-Ind 7(01):57–067. www.ajofai.info

Liu D, Lv XX (2019) Effect of blueberry flower pulp on sensory, physicochemical properties, lactic acid bacteria, and antioxidant activity of set-type yogurt during refrigeration. J Food Process Preserv 43(1). https://doi.org/10.1111/jfpp.13856

Loizzo MR, Pugliese A, Bonesi M, Tenuta MC, Menichini F, Xiao J, Tundis R (2016) Edible flowers: a rich source of phytochemicals with antioxidant and hypoglycemic properties. J Agric Food Chem 64(12):2467–2474. https://doi.org/10.1021/acs.jafc.5b03092

Lone R, Khan S, Al-Sadi AM (2023) Plant phenolics in abiotic stress management. Published by Springer Nature. https://doi.org/10.1007/978-981-19-6426-8

Lu B, Li M, Yin R (2016) Phytochemical content, health benefits, and toxicology of common edible flowers: a review (2000–2015) Crit Rev Food Sci Nutr 56(Suppl 1):130–148

Łuczaj L, Pieroni A, Tardío J, Santayana MP, Sõukand R, Svanberg I, Kalle R (2012) Wild food plant use in 21st century Europe: the disappearance of old traditions and the search for new cuisines involving wild edibles. Acta Societatis Botanicorum Poloniae 81(4):359–370. 5 5asbp.2012.031. https://doi.org/10.5586/asbp.2012.031

Lyra CCGV, Vieira RF, de Oliveira CBA, Santos SC, Seraphin JC, Ferri PH (2008) Infraspectric variability in the essential oil composition of *Lychnophora ericoides*. J Braz Chem Soc 19:842–848

Machado JS, Pieracci Y, Carmassi G, Ruffoni B, Copetta A, Pistelli L (2023) Effect of drying post-harvest on the nutritional compounds of edible flowers. Horticulturae 9(11). https://doi.org/10.3390/horticulturae9111248

Madurangi ADR, Maduwanthi SDT (2024) Edible flowers in herbal beverage development. Asian Food Sci J 23(8):1–21. https://doi.org/10.9734/afsj/2024/v23i8731

Marchioni I, Najar B, Ruffoni B, Copetta A, Pistelli L, Pistelli L (2020) Bioactive compounds and aroma profile of some lamiaceae edible flowers. Plants 9(6). https://doi.org/10.3390/plants9060691

Martins N, Roriz CL, Morales P, Barros L, Ferreira ICFR (2016) Food colorants: challenges, opportunities and current desires of agro-industries to ensure consumer expectations and regulatory practices. In: Trends in food science and technology, vol 52. Elsevier Ltd., pp 1–15. https://doi.org/10.1016/j.tifs.2016.03.009

Matyjaszczyk E, Śmiechowska M (2019) Edible flowers: benefits and risks pertaining to their consumption. Trends Food Sci Technol 91:670–674

Mintel (2024) Global food and drink trends. Retrieved from www.mintel.com

Mishra BK, Jain NK, Kumar S, Sharma KC (2013) Storage stability of ready-to-serve beverage from mahua (*Madhuca indica*) flowers. Indian J Hortic 70(3):459–462

Mlcek J, Rop O (2011) Fresh edible flowers of ornamental plants – a new source of nutraceutical foods. Trends Food Sci Technol 22(10):561–569

Mortas M (2023) Characterization of a new powdered, milk-based medicinal plant (Alcea rosea) drink product. Sustainability (Switzerland) 15(12). https://doi.org/10.3390/su15129320

Morton JF (1992) Widespread tannin intake via stimulants and masticatories, especially guarana, kola nut, betel vine, and accessories. Plant Polyphenols: Synthesis, Properties, Significance:739–765

Mulík S, Ozuna C (2020) Mexican edible flowers: cultural background, traditional culinary uses, and potential health benefits. Int J Gastron Food Sci 21:100235. AZTI-Tecnalia. https://doi.org/10.1016/j.ijgfs.2020.100235

Mulík S, Hernández-Carrión M, Pacheco-Pantoja SE, Ozuna C (2024) Endemic edible flowers in the Mexican diet: Understanding people's knowledge, consumption, and experience. Future Foods 9:100374

Newman M, Kirker CL (2016) Edible flowers a global history. Reaktion Books

Nicolau AI, Gostin AI (2016) Safety of edible flowers. In: Regulating safety of traditional and ethnic foods. Elsevier Inc., pp 395–419. https://doi.org/10.1016/B978-0-12-800605-4.00021-9

Nimakoa A, Ivanišová E, Didová M (2024) Nutritional, antioxidant and sensory properties of cereal bars fortified by edible flowers. J Process Energy Agric 28(2):53–60. https://doi.org/10.5937/jpea28-55159

Nisa A-u, Hina S, Mazhar S, Kalim I, Ijaz A, Zahra N, Masood S, Asif M (2018) Stability of lutein content in color extracted from marigold flower and its application in candies. Pak J Agric Res 31(1):15–23

Noreen H, Semmar N, Farman M, McCullagh JSO (2017) Measurement of total phenolic content and antioxidant activity of aerial parts of medicinal plant Coronopus didymus. Asian Pac J Trop Med 10(8):792–801. https://doi.org/10.1016/J.APJTM.2017.07.024

Nowosad K, Sujka M, Pankiewicz U, Kowalski R (2021) The application of PEF technology in food processing and human nutrition. J Food Sci Technol 58(2):397–411. https://doi.org/10.1007/s13197-020-04603-5

Nyulas-Zeke IC, Hidas KI, Pásztor-Huszár K, Dalmadi I, Szücs E, Pap R, Csurka T (2024) Analysis of techno-functional properties of fermented and non-fermented buttermilk-containing ice creams. Sustainability 16(17):7714. https://doi.org/10.3390/su16177714

Onyekwelu NP, Ezeafulukwe C, Owolabi OA et al. (2024) Ethics and corporate social responsibility in HR: a comprehensive review of policies and practices. Int J Sci Res Arch 11(01):1294–1303. https://doi.org/10.30574/ijsra.2024.11.1.0216

Paradis C, Castaigne F, Desrosiers T, Fortin J, Rodrigue N, Willemot C (1996) Sensory, nutrient, and chlorophyll changes in broccoli florets during controlled atmosphere storage. J Food Qual 19:303–316

Pêgo RG, Fiorini CVA, De Deco TA, Coneglian RCC, Xavier MCG, Ferreira WPC (2022) Postharvest of edible flowers. Pesq Agrop Brasileira 57. https://doi.org/10.1590/S1678-3921.PAB2022.V57.02953

Piccaglia R, Marotti M, Grandi S (1998) Lutein and lutein ester content in different types of *Tagetes patula* and *T. erecta*. Ind Crop Prod 8:45–51

Pires TCSP, Dias MI, Barros L, Ferreira ICFR (2017) Nutritional and chemical characterization of edible petals and corresponding infusions: valorization as new food ingredients. Food Chem 220:337–343. https://doi.org/10.1016/j.foodchem.2016.10.026

Pires TCSP, Dias MI, Barros L, Barreira JCM, Santos-Buelga C, Ferreira ICFR (2018a) Incorporation of natural colorants obtained from edible flowers in yogurts. LWT 97:668–675. https://doi.org/10.1016/j.lwt.2018.08.013

Pires TCSP, Dias MI, Barros L, Calhelha RC, Alves MJ, Oliveira MBPP et al (2018b) Edible flowers as sources of phenolic compounds with bioactive potential. Food Res Int 105:580–588. https://doi.org/10.1016/j.foodres.2017.11.014

de Pires EO, Di Gioia F, Rouphael Y, Ferreira ICFR, Caleja C, Barros L, Petropoulos SA (2021) The compositional aspects of edible flowers as an emerging horticultural product. Molecules 26(22):6940. https://doi.org/10.3390/molecules26226940

Podgórska-Kryszczuk I, Pankiewicz U (2023) Assessment of the fungistatic properties of Calendula officinalis L. water extract and the effect of its addition on the quality of wheat bread. Appl Sci (Switzerland) 13(12). https://doi.org/10.3390/app13127286

Pontes E, Viera V, Silva G, da Silva Neto M, Mendes B, Tome A, Almeida R, Santos NC, de Gusmão R, Lisboa H, Gusmão T (2024) Effect of malvaviscus arboreus flower and leaf extract on the functional, antioxidant, rheological, textural, and sensory properties of goat yogurt. Foods 13(23):3942. https://doi.org/10.3390/foods13233942

Purohit SR, Rana SS, Idrishi R, Sharma V, Ghosh P (2021) A review on nutritional, bioactive, toxicological properties and preservation of edible flowers. In: Future foods, vol 4, Elsevier B.V. https://doi.org/10.1016/j.fufo.2021.100078

Qiu L, Zhang M, Mujumdar AS, Chang L (2021) Effect of edible rose (Rosa rugosa cv. Plena) flower extract addition on the physicochemical, rheological, functional and sensory properties of set-type yogurt. Food Biosci 43. https://doi.org/10.1016/j.fbio.2021.101249

Ramakrishnan RB (2018) Evaluation of antioxidant and phytochemical activity in solvent extracts from *Delonix regia* flowers. Int J Green Pharm 12(3) https://doi.org/10.22377/ijgp.v12i03.2028

Rao V. and Poonia A (2024a) Microbial biosynthesis of selenium nanoparticles using probiotic strain and its characterization. Journal of Food Measurement and Characterization, 18(7):5477–5488. https://doi.org/10.1007/s11694-024-02581-z

Rao V. and Poonia A (2024b). Bioactive Compounds, nanoparticle synthesis and potential utilization of edible flowers for development of functional dairy products: a review. Journal of Food Science and Technology, 61(6):1053–1068. https://doi.org/10.1007/s13197-023-05853-6

Rezende F, Sande D, Coelho AC, Oliveira G, Boaventura MA, Takahashi JA (2019) Edible flowers as innovative ingredients for future food development: antialzheimer, antimicrobial, and antioxidant potential. Chem Eng Trans 75:337–342. https://doi.org/10.3303/CET1975057

Rivas-García L, Navarro-Hortal MD, Romero-Márquez JM, Forbes-Hernández TY, Varela-López A, Llopis J, Sánchez-González C, Quiles JL (2021) Edible flowers as a health promoter: an evidence-based review. Trends Food Sci Technol 117:46–59

Rop O, Mlcek J, Jurikova T, Neugebauerova J, Vabkova J (2012) Edible flowers—a new promising source of mineral elements in human nutrition. Molecules 17(12):6672–6683. https://doi.org/10.3390/molecules17066672

Roriz CL, Heleno SA, Carocho M, Rodrigues P, Pinela J, Dias MI et al (2020) Betacyanins from *Gomphrena globosa L.* flowers: incorporation in cookies as natural coloring agents. Food Chem 329:127178

Rowan NJ, MacGregor SJ, Anderson JG, Fouracre RA, Farish O (2000) Pulsed electric field inactivation of diarrhoeagenic *Bacillus cereus* through irreversible electroporation. Lett Appl Microbiol 31(2):110–114

Saati EA, Ramadhan AH, Lutfi M, Wahyudi VA, Manshur HA (2020) Utilization of rose flower extract as antioxidant rich-drink. IOP Con Ser Earth Environ Sci 458(1). https://doi.org/10.1088/1755-1315/458/1/012035

dos Santos IC, Reis SN (2021) Edible flowers: traditional and current use. Ornam Hortic 27(4):438–445. https://doi.org/10.1590/2447-536X.v27i4.2392

dos Santos Silva LY, da Silva Ramos A, Cavalcante DN, Kinupp VF, da Silva Rodrigues JV, Ventura BML, de Oliveira Mendes TA, Sanches EA, Campelo PH, de Araújo Bezerra J (2023) *Hibiscus acetosella*: an unconventional alternative edible flower rich in bioactive compounds. Molecules 28(12). https://doi.org/10.3390/molecules28124819

dos Santos AMP, Silva EFR, dos Santos WNL, da Silva EGP, dos Santos LO, Bruna BR, Maria MC, dos Santos WPC (2018) Evaluation of minerals, toxic elements and bioactive compounds in rose petals (*Rosa spp.*) using chemometric tools and artificial neural networks. Microchem J 138:98–108

Sam FE, Ma T, Wang J, Liang Y, Sheng W, Li J, Jiang Y, Zhang B (2023) Aroma improvement of dealcoholized Merlot red wine using edible flowers. Food Chem 404:134711. https://doi.org/10.1016/J.FOODCHEM.2022.134711

Sardari FA, Azadi A, Mohagheghzadeh A, Badr P (2018) Gulqand: a nutraceutical from sugared petals. Tradit Integr Med:180–185

Scripsema NK, Hu DN, Rosen RB (2015) Lutein, zeaxanthin, and mesozeaxanthin in the clinical management of eye disease. J Ophthalmol 2015. https://doi.org/10.1155/2015/865179

Semeniuc CA, Mandrioli M, Socaci BS, Socaciu MI, Fogarasi M, Podar AS, Michiu D, Jimborean AM, Mureşan V, Ionescu SR, Gallina Toschi T (2022) Changes in lipid composition and oxidative status during ripening of Gouda-type cheese as influenced by addition of lavender flower powder. Int Dairy J 133. https://doi.org/10.1016/j.idairyj.2022.105427

Semeniuc CA, Mandrioli M, Tura M, Socaci BS, Socaciu MI, Fogarasi M, Michiu D, Jimborean AM, Mureşan V, Ionescu SR, Rotar MA, Gallina Toschi T (2023) Impact of lavender flower powder as a flavoring ingredient on volatile composition and quality characteristics of gouda-type Cheese during ripening. Foods 12(8). https://doi.org/10.3390/foods12081703

Shantamma S, Vasikaran EM, Waghmare R, Nimbkar S, Moses JA, Anandharamakrishnan C (2021) Emerging techniques for the processing and preservation of edible flowers. Future Foods 4:100094

Sharma Y, Hegde RV, Venugopal CK (2011) Health and nutrition from ornamentals. In J Res Ayurveda Pharm 2(2):375–382

Sharma P, Verma R, Gupta S (2020) Floral extracts in dairy and bakery: a functional perspective. J Food Sci 58(3):120–134

Sik B, Ajtony Z, Lakatos E, Gál LH, Székelyhidi R (2024) Evaluation of the physical, antioxidant, and organoleptic properties of biscuits fortified with edible flower powders. Food Sci Nutr 12(5):3265–3272. https://doi.org/10.1002/fsn3.3993

Simoni NK, Santos FF, Andrade TA, Villavicencio ALCH, Pinto-e-Silva MEM (2018) The use of edible flowers in human food: sensory analysis of preparations. ETP Int J Food Eng:140–143. https://doi.org/10.18178/ijfe.4.2.140-143

Singh Y, Gupta A, Kannojia P (2020) *Tagetes erecta* (Marigold)—a review on its phytochemical and medicinal properties. Curr Medi Drug Res 4(1):201

Skrajda MN (2017) Phenolic compounds and antioxidant activity of edible flowers. J Educ Health Sport 7(8):946–956. https://doi.org/10.5281/zenodo.995637

Smith J, Brown L (2019) Lavender as a functional ingredient in dairy and bakery products. Food Chem 210:45–52

Socha R, Kałwik J, Juszczak L (2021) Phenolic profile and antioxidant activity of the selected edible flowers grown in Poland. Acta Univ Cibiniensis Ser E Food Technol 25(2):185–200. https://doi.org/10.2478/aucft-2021-0017

Solanke SN, Chopra CS, Sharma SK (2016) Expression of rhododendron aqueous extract & its use in preparation of RTS beverage. J Hill Agric 7(2):267–272

Sotelo A, López-García S, Basurto-Peña F (2007) Content of nutrient and antinutrient in edible flowers of wild plants in Mexico. Plant Foods Hum Nutr 62(3):133–138. https://doi.org/10.1007/s11130-007-0053-9

Stefaniak A, Grzeszczuk M (2019) Nutritional and biological value of five edible flower species. Notulae Botanicae Horti Agrobotanici Cluj-Napoca 47(1):128–134. https://doi.org/10.15835/nbha47111136

Sutakwa A, Nadia LS, Suharman S (2021) Addition of blue pea flower (Clitoria ternatea L.) extract increase antioxidant activity in yogurt from various types of milk. Jurnal Agercolere 3(1):31–37. https://doi.org/10.37195/jac.v3i1.123

Takahashi JA, Rezende FAGG, Moura MAF, Dominguete LCB, Sande D (2020) Edible flowers: bioactive profile and their potential to be used in food development. Food Res Int 129:108868. https://doi.org/10.1016/j.foodres.2019.108868

Tamimi AL, Alfaris JZ, Alghamdi NA, Abu-Hiamed FA, Albader HA, Almousa NA (2020) *Hibiscus sabdariffa L.* flower and date palm pollen fortification of date palm spathe beverage. Br Food J 122(7):2159–2170

Thilakarathna HIU, Jayarathna GN, Mudannayake DC (2023) Probiotic ice cream incorporated with blue pea flower (Clitoria ternatea) as a natural colorant and dehydrated banana flour as a potential prebiotic ingredient. Food Life 2023(2):67–75. https://doi.org/10.5851/fl.2023.e7

U.S. Department of Health and Human Services and U.S. Department of Agriculture (2015) 2015–2020 Dietary Guidelines for America*ns* (8th ed.) [e-book]. Retrieved from http://health.gov/dietaryguidelines/2015/guidelines/

Ukiya M, Akihisa T, Tokuda H, Hirano M, Oshikubo M, Nobukuni Y, Nishino H (2002) Inhibition of tumor-promoting effects by poricoic acids G and H and other lanostane-type triterpenes and cytotoxic activity of poricoic acids A and G from *Poria cocos*. J Nat Prod 65(4):462–465

USDA, United States Department of Agriculture (2018a) Standard reference database. Reference number 11740: Broccoli, flower clusters, raw. https://ndb.nal.usda.gov/ndb/foods

USDA, United States Department of Agriculture (2018b) Branded food products database. Reference number 45346720: Whole wheat flour. https://ndb.nal.usda.gov/ndb/foods. Accessed 22 February 2019

USDA, United States Department of Agriculture (2018c) Standard reference database. Reference number 11135: Cauliflower, raw. https://ndb.nal.usda.gov/ndb/foods

USDA, United States Department of Agriculture (2018d) Branded food products database. Reference number 16410: Soybeans, mature seeds, roasted, no salt added. https://ndb.nal.usda.gov/ndb/foods

USDA, United States Department of Agriculture (2018e) Branded food products database. Reference number 567848: Hibiscus flower. https://ndb.nal.usda.gov/ndb/foods. Accessed 10 June 2019

USDA, United States Department of Agriculture (2018f) Branded food products database. Reference number 45346720: Whole wheat flour. https://ndb.nal.usda.gov/ndb/foods. Accessed 22 February 2019

USDA, United States Department of Agriculture (2018g) Branded food products database. Reference number 16410: Soybeans, mature seeds, roasted, no salt added. https://ndb.nal.usda.gov/ndb/foods. Accessed 22 February 2019

Wang Y, Li Z (2022) The aromatic influence of jasmine in dairy and bakery products. J Sens Stud 65(1):35–48

Xin Y, Zhang M, Adhikari B (2014) Ultrasound-assisted immersion freezing of broccoli (*Brassica oleracea L. var. botrytis L.*). Ultrason Sonochem 21(5):1728–1735

Xiong L, Yang J, Jiang Y, Lu B, Hu Y, Zhou F, Mao S, Shen C (2014) Phenolic compounds and antioxidant capacities of 10 common edible flowers from China. J Food Sci 79(4). https://doi.org/10.1111/1750-3841.12404

Xu H, Wu M, Shang N, Wei W, Gao F (2022) Fortification of wheat bread with edible chrysanthemum (Chrysanthemum morifolium Ramat.): unraveling the mechanisms of dough rheology and bread quality changes. https://doi.org/10.21203/rs.3.rs-2114533/v1

Wongwattanasathien O, Kangsadalampai K, Tongyonk L (2010) Antimutagenicity of some flowers grown in Thailand. Food Chem Toxicol 48(4):1045–1051

Yi Y, Zhang Q, Li S, Wang Y, Ye W, Zhao J, Wang Y (2012) Simultaneous quantification of major flavonoids in "Bawanghua', the edible flower of Hylocereus undatus using pressurised liquid extraction and high performance liquid chromatography. Food Chem 135(2):528–533

Zhao L, Fan H, Zhang M, Chitrakar B, Bhandari B, Wang B (2019) Edible flowers: review of flower processing and extraction of bioactive compounds by novel technologies. Food Res Int 126. https://doi.org/10.1016/j.foodres.2019.108660

Chapter 15
Technological Advancements and Current Market Trends in the Edible Flowers Industry

D. E. Jimenez-Sánchez, E. J. Delgado-Núñez, A. Francisco-Palemon, S. A. Ortega-Acosta, and M. Luna-Flores

15.1 Introduction

The edible flower industry is experiencing unprecedented growth, fueled by a unique combination of technological innovations and changing market dynamics. Edible flowers being non-toxic, are considered healthy and safe for human consumption. They have been employed by humans since long time to beautify dishes, supply nutrients, and improve wellness (Rao and Poonia 2024a, b). As more consumers focus on health and seek new culinary experiences, the popularity of edible flowers has increased due to their visual charm and possible health advantages. This rising popularity has spurred advancements in various aspects of the industry, from cultivation and harvesting techniques to preservation and distribution methods. Technological innovations have fundamentally enhanced the quality and availability to flowers that can be consumed. Precision agriculture, such as, has transformed the way flowers are cultivated, ensuring optimal growing conditions and minimizing resource waste (Chaudhary et al. 2024). Hydroponic and aeroponic systems have enabled year-round cultivation, ensuring a consistent supply of fresh, high-quality flowers throughout the seasons (Brown and Green 2023). Furthermore,

D. E. Jimenez-Sánchez (✉)
Centro Regional de Educación Superior de la Zona Norte, Campus Taxco el viejo/ Universidad Autónoma de Guerrero, Taxco, Guerrero, Mexico
e-mail: 19248@uagro.mx

E. J. Delgado-Núñez · A. Francisco-Palemon · S. A. Ortega-Acosta
Departamento de Agronomía, Facultad de Ciencias Agropecuarias y Ambientales de la Universidad Autónoma de Guerrero, Iguala, Guerrero, Mexico

M. Luna-Flores
Departamento de Ingeniería en Procesos Bioalimentarios, Universidad Tecnológica del Centro de Veracruz, Veracruz, Mexico

A. Poonia et al. (eds.), *Edible Flowers: Source of Phytonutrients, Valorization and Technological Advancements*, https://doi.org/10.1007/978-3-032-00667-7_15

advancements in genetic engineering have resulted while developing fresh types of flowers varieties with enhanced flavors, colors, and nutritional profiles (Sarode et al. 2024a, b).

In the realm of preservation, innovative techniques like freeze-drying and vacuum sealing have extended the shelf life of edible flowers, making them more accessible to consumers and reducing post-harvest losses. These methods have also maintained the flowers' vibrant colors and delicate textures, ensuring that they remain visually appealing when used in culinary applications (Barani et al. 2022). Furthermore, packaging innovations have improved the protection and presentation of edible flowers, enhancing their marketability. Market trends within the edible flower industry have also evolved significantly. There is an increasing inclination towards organic and sustainably sourced products, driven by greater consumer awareness of environmental and health concerns. This trend has prompted producers to adopt eco-friendly practices and secure organic certifications, adding value to their products and attracting conscientious consumers (Fernandes et al. 2020). The rise of the farm-to-table movement and the popularity of artisanal and gourmet foods have further boosted the demand for edible flowers, with chefs and food enthusiasts eager to create visually stunning and unique dishes. Regarding market reach, the edible flower industry has expanded beyond traditional markets, finding new avenues through e-commerce and online retail platforms. This shift has democratized access to edible flowers.

The edible flowers industry sits at the crossroads of tradition and innovation, fueled by technological advancements and changing market preferences. As the industry continues to grow, it offers exciting opportunities for producers, chefs, and consumers to explore the culinary and health benefits of these delightful blooms. This section will explore the latest technological innovations and current market trends, highlighting key players, emerging technologies, and future growth opportunities in the edible flowers sector industry.

15.2 Production and Market of Edible Flowers

Edible flowers have become increasingly popular in global cuisine because of their visual appeal and potential health benefits. These flowers not only improve the presentation of dishes but also add unique flavors and essential nutrients. In recent years, there has been a notable rise in the cultivation of flowers fit for consumption, fueled by the rising demand for natural and organic ingredients. This essay aims to present an overview of the statistics on edible flower production, focusing on global trends, production by country, and the industry's obstacles and prospects.

Worldwide production related to consumable blooms has consistently grown. A document prepared by the FAO regions that global production arrived at approximately 1.2 million tons in 2023, with an yearly increase rate for 5% (FAO 2023). This increase is largely fueled by demand in developed markets such as Europe and North America, where consumers seek fresh and natural ingredients.

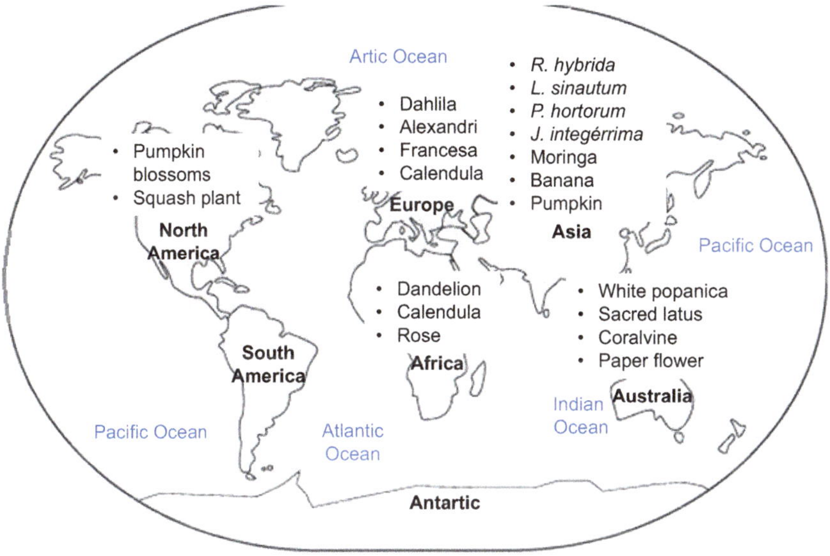

Fig. 15.1 Edible flowers accessible globally. (Purohit et al. 2021)

China remains one of the foremost producers of edible flowers, with an annual yield of approximately 300,000 tons. The application of flowers in Chinese cuisine boasts an extensive past, and the nation remains at the forefront of innovation and production (Rodrigues and Spence 2023). In India, the cultivation of edible flowers is also notable, producing around 150,000 tons per year, particularly in states such as Tamil Nadu and Karnataka (Chitrakar et al. 2019). In North America, the United States generates about 100,000 tons of edible flowers annually, with California and Florida being the primary producers. The Zuni tribe, for example, is widely admired for their enthusiasm and admiration for squash blooms (Murphy 2015). Mexico has seen an increase in production as well, with an annual output of 50,000 tons, mainly in regions like Michoacán and Jalisco. The distribution of widely consumed flowers worldwide is illustrated in Fig. 15.1.

Among the most produced edible flowers are marigold, rose, lavender, and nasturtium. Marigold is especially popular due to its citrusy flavor and natural coloring properties. In Mexico, the production of pumpkin flowers is particularly prominent, while in Europe, roses and violets are common in gourmet cuisine (Fernandes et al. 2017). The global market for consumable blossoms is valued at approximately $500 million, with growing demand in gourmet restaurants and consumers seeking unique and healthy ingredients. The popularity of farm-to-table movements and artisanal foods has further boosted the need for consumable flowers. The main customer markets include Europe, North America, and Asia (Fernandes et al. 2020). The production of consumable flowers significantly contributes to local economies, especially in rural areas. In countries like China and India, the cultivation and export of edible flowers employ thousands of farmers and workers. In the United States,

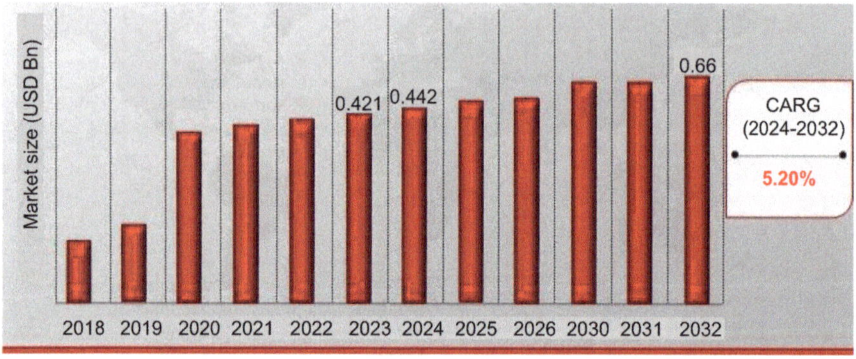

Fig. 15.2 Compound annual growth rate. (Source: Market Research Future (2025))

the industry generates substantial revenue, particularly in states like California, where production is intensive (Fortune Business Insights 2025). Although the edible flower industry has grown, it still encounters several challenges, such as climate change, pests, and diseases. Nevertheless, there are many opportunities, including creating new edible flower varieties through genetic engineering and expanding into new markets using e-commerce and digital platforms. In 2023, the market for edible flowers was worth $0.421 billion. It is anticipated that this industry will expand from $0.442 billion in 2024 to $0.66 billion by 2032, reflecting an annualized growth rate (CAGR) of 5.20% projected for the period between 2024 and 2032. The primary factors driving this market growth are the rising popularity of international cuisines and evolving consumer eating habits (Fig. 15.2).

The cultivation of consumable flowers is a growing industry with significant economic impact and potential for expansion. Global trends indicate an increasing demand for natural and organic ingredients, driving production in various regions worldwide. In 2023, the market for edible flowers was worth $0.421 billion. The primary factors driving this market growth are the rising popularity of international cuisines and evolving consumer eating habits. The popularity of edible flowers is on the rise post-pandemic as more people become health-conscious and learn about the wellness advantages offered by these flowers. Consumable blossoms serve multiple purposes directly or indirectly, improving both the aesthetic charm and dietary benefits of the end product. There is growing public support for a nutrient-rich diet, which has recently led to a surge in the popularity of edible flowers in food and beverages. For instance, in places like Indonesia, Etlingera elatior flowers are boiled and served hot with peanut sauce. These innovative uses are expanding the menus of global hotels, which predict a significant growth trajectory for the edible flower market during the forecast period. Consumable blossoms are utilized to enhance the taste of salads, desserts, and bakery items like cakes and cookies. The increased use of edible flowers for garnishing and flavoring food is expected to drive market growth. These flowers also help improve skin texture and treat skin conditions like acne, contributing to the growing demand for edible flowers. Additionally, the trend toward a healthier lifestyle is anticipated to boost the global edible flower market's growth, thereby increasing the market revenue (Market Research Future 2025).

For generations, consumable blossoms have contributed to culinary heritage, adding distinct flavors, vibrant colors, and unique textures to various dishes. Recently, technological advancements and changing market trends have brought significant attention to the edible flower industry. This chapter delves into these innovations and their effects on the industry (Sarode et al. 2024a, b).

Technological advancements have been essential in improving the production, quality, and marketability of edible flowers (Rabiya 2024). Some of the key advancements include.

15.3 Biotechnology

Advances in biotechnology, such as micropropagation and genetic engineering, have paved the way for improved cultivars with enhanced traits like resistance to pests, diseases, and environmental stress. These methods allow for the mass production of high-quality edible flowers with consistent characteristics (Sarode et al. 2024a, b). Nanotechnology represents an emerging field of scientific exploration focused on the examination and control of matter at the smallest scales, encompassing molecules and atoms (Rao and Poonia 2024a, b). Biotechnology has ushered in transformative changes across various agricultural sectors, including the burgeoning market for edible flowers. Through pioneering techniques like micropropagation and genetic engineering, enhanced varieties with superior characteristics like pest resistance, disease resistance, and environmental resilience have been developed. Such technological innovations enhance both the standard of edible flowers but also enable efficient mass production, catering to the increasing demand.

Micropropagation is a notable biotechnological technique that facilitates the rapid multiplication of plants using tissue cultures under controlled conditions. This method has revolutionized edible flower production for several reasons. Firstly, it allows the large-scale production of plants from a single mother plant, ensuring a steady availability of premium edible flowers year-round (Copetta et al. 2023). Secondly, it guarantees genetic consistency, meaning that all plants produced are exact genetic replicas of the mother plant, resulting in uniform characteristics (Cox 2018). Lastly, it aids in the creation of plant varieties with favorable characteristics, such as larger flower sizes, better flavors, and higher nutrient content (Ramawat et al. 2020).

Another significant advancement is genetic engineering, which involves the direct modification of plant genetic material to introduce specific desired traits. In the context of edible flowers, this technology has enabled several key developments. By introducing specific genes, scientists have created plants immune to prevalent pests and illnesses, minimizing the dependency on chemical pesticides and promoting sustainability (Wally and Punja 2010). Genetic engineering also enables plants to adapt to adverse environmental conditions, such as droughts or frosts, increasing production efficiency across various climates (Wani et al. 2023). Furthermore, it has enhanced the nutritional profiles of edible flowers, boosting their content of beneficial vitamins, antioxidants, and other health-promoting compounds (Sarode et al. 2024a, b).

The incorporation of these biotechnological advancements has had a profound impact on the edible flower industry. Farmers can now improve both the quality and quantity of their yield while meeting the growing demands of health-conscious and eco-aware consumers. The ability to consistently produce high-quality edible flowers has allowed producers to meet the global market demand (Market Research Future 2025). Reducing pesticide use and increasing environmental resilience foster more sustainable farming practices (Rabiya 2024). Moreover, these advancements in biotechnology reduce operational costs associated with pest and disease management, enhancing profitability for producers (Copetta et al. 2023).

Biotechnology has been instrumental in transforming the edible flower industry. Techniques such as micropropagation and genetic engineering have led to the development of improved cultivars that not only elevate the quality of edible flowers but also support efficient and sustainable mass production. The continued integration of these technologies is essential for future growth and sustainability of this thriving industry.

15.4 Precision Agriculture: Revolutionizing Edible Flower Cultivation

Precision agriculture methods, including the implementation of drones, sensors, and remote surveillance technologies, have transformed the cultivation of edible flowers. These technologies allow farmers to oversee crop health, efficiently utilize resources, and make informed decisions based on data, thereby enhancing yields and reducing losses (Wani et al. 2023).

Precision agriculture has proven to be a transformative approach in modern farming, significantly boosting the cultivation of edible flowers. This advanced agricultural practice leverages cutting-edge innovations like drones, sensors, and remote monitoring systems, enabling farmers to precisely track crop conditions, maximize resource efficiency, and base their decisions on robust data insights. These advancements enhance yields and reduce losses, thereby revolutionizing the edible flower industry.

The implementation of drones in precision agriculture brings numerous advantages. Drones fitted with high-definition imaging devices and broad-spectrum imaging sensors can capture detailed aerial views of crop fields, providing essential insights into plant vitality and developmental trends (Guebsi et al. 2024). This real-time data enables farmers to detect problems such as pest infestations, diseases, and nutrient deficiencies early on, allowing for prompt interventions. Furthermore, drones can serve to accurately administer nutrients and pest control substances, minimizing waste and environmental impact.

Sensors are integral to precision agriculture, gathering information on diverse environmental factors such as soil water content, thermal conditions, and atmospheric moisture.These devices are carefully positioned throughout the fields, continuously monitoring conditions and transmitting the data to a central system (Mohammed Aarif et al. 2025). This information helps farmers make informed decisions regarding irrigation schedules, ensuring optimal water usage and preventing issues related to overwatering or drought stress. Additionally, sensors can detect changes in soil nutrient levels, enabling precise fertilization that improves plant growth and productivity. Remote monitoring systems integrate data from drones and sensors, providing a comprehensive view of the entire crop field. Advanced software platforms analyze this data, generating actionable insights and recommendations for farmers (Laveglia et al. 2024). For instance, remote monitoring can identify areas of the field that require attention, such as zones with poor plant health or water deficits. By addressing these issues promptly, farmers can improve crop performance and reduce losses.

The impact of precision agriculture on the edible flower industry is profound. With these technologies, farmers can achieve higher yields and better-quality flowers, meeting the increasing need within the marketplace. The ability to monitor crop health accurately and respond to changes swiftly results in healthier plants and increased productivity (Hasna et al. 2023). Additionally, the prudent management of resources such as water and fertilizers not only boosts sustainability but also lowers production costs, supporting the economic feasibility of edible flower farming.

Precision agriculture also promotes environmental sustainability. The targeted application of inputs like agricultural supplements and herbicides minimizes the risk of substrate and water contamination, reducing the environmental footprint of farming activities (Sanyaolu and Sadowski 2024). Furthermore, the effective management of water supplies helps conserve this vital commodity, enhancing agriculture's adaptability to global warming and water scarcity issues.

Precision agriculture has transformed the edible flower industry by utilizing advanced technologies to oversee crop vitality, enhance resource efficiency, and base decisions on insightful data analysis. The incorporation of drones, sensors, and remote monitoring systems has allowed farmers to achieve higher yields, improve flower quality, and enhance sustainability. As these technologies continue to progress, the future of edible flower cultivation appears promising, with the potential for even greater efficiency and productivity.

15.5 Controlled Environment Agriculture

Controlled environment agriculture (CEA), such as hydroponics, vertical farming, and aeroponics, allows for the cultivation of edible flowers in a controlled environment (Rabiya 2024). This approach mitigates the challenges associated with seasonal variations and climate change, ensuring year-round production of high-quality edible flowers (Wani et al. 2023).

Controlled environment agriculture, or CEA, represents a transformative approach in modern farming that has significantly enhanced the cultivation of edible flowers. By employing innovative techniques such as hydroponics, vertical farming, and aeroponics, CEA enables farmers to grow plants in precisely controlled settings. This approach effectively mitigates the challenges associated with seasonal variations and climate change, ensuring year-round production of high-quality edible flowers.

Hydroponics, a method of farming without soil, enables plants to thrive in a nutrient-rich water solution. This approach has several benefits. First, it optimizes nutrient delivery by directly supplying nutrients to the plant roots, encouraging healthy growth and higher yields (Resh 2022). Moreover, hydroponic systems consume considerably less water than conventional land-based agriculture, rendering them a more eco-friendly alternative option, particularly in areas with limited water resources (Jones 2004). Furthermore, hydroponic systems can be set up indoors, allowing farmers to regulate conditions like illumination, thermal levels, and atmospheric moisture, ensuring consistent production regardless of external weather conditions (Velazquez-Gonzalez et al. 2022).

Vertical farming makes use of vertically stacked crop layers to optimize space and offer various benefits. This approach allows a high density of plants to be grown in compact areas, rendering it ideal for metropolitan environments settings and regions with restricted arable land (Despommier 2010). It also lowers transportation costs since vertical farms can be established near urban centers, thus reducing the distance between production sites and consumers and decreasing the carbon footprint (Zhang and Zha 2023). Moreover, vertical farming ensures a steady supply of edible flowers throughout the year by maintaining controlled environments that are not influenced by seasonal changes (Gurung et al. 2024).

Aeroponics, a different soilless farming method, grows Vegetation with roots hanging freely in the atmosphere, occasionally sprayed with a nutrient-infused mixture. This technique fosters accelerated development speeds and enhanced production outcomes due to the oxygen-rich environment for plant roots (Bubenheim 2024). Aeroponics efficiently uses resources, requiring less water and nutrients than traditional farming, making it a sustainable option (Weathers 2023). Additionally, without soil, the risk of earth-transmitted pathogens is minimized, leading to healthier plants and a reduced need for pesticides (Kozai et al. 2019).

The impact of CEA technologies on the edible flower industry is profound. These methods assure consistent production of high-quality edible flowers, meeting the rising market demand. By mitigating the risks associated with climate change and seasonal variations, CEA offers a dependable and eco-friendly approach to continuous cultivation throughout the year.

Controlled environment agriculture offers innovative solutions for the cultivation of edible flowers. Techniques such as hydroponics, vertical farming, and aeroponics enable the production of high-quality edible flowers in controlled settings. By addressing challenges related to seasonal variations and climate change, CEA ensures a continuous supply of edible flowers, supporting the industry's growth and sustainability.

15.6 Increasing Consumer Demand: A Rise in the Popularity of Edible Flowers

The consumable flowers market is experiencing notable growth, driven by several key trends. In recent years, consumer demand for distinctive and visually attractive food products has surged, resulting in the increasing popularity of edible flowers. This trend is evident in both home cooking and professional kitchens, where edible flowers are frequently employed to improve the presentation of meals, adding color, texture, and flavor. This essay explores the reasons behind this increase in demand and its impact on the edible flowers industry.

A key factor driving this demand is the increasing fascination with culinary experimentation and visual appeal. Consumers are increasingly seeking out innovative and visually striking food experiences. Edible flowers, with their vibrant colors and delicate textures, provide a organic way to elevate the aesthetic presentation of meals (Gupta et al. 2024). Whether sprinkled over salads, used to garnish desserts, or incorporated into beverages, edible flowers bring an element of artistry to the plate, making meals more attractive and enjoyable.

Another factor contributing to the rise sought after for culinary blossoms is the influence of social media and food bloggers. Platforms like Instagram, Pinterest, and TikTok have become hotspots for showcasing visually appealing culinary creations. As food presentation gains importance, chefs and home cooks alike turn to edible flowers to create Instagram-worthy dishes. The visual appeal of edible flowers resonates well with social media audiences, driving awareness and interest in these products (Cash et al. 2022).

Health and wellness trends also play a significant role in boosting consumer demand for edible flowers. Many edible flowers are packed with vitamins, antioxidants, and other bioactive compounds that contribute to good health (Sood et al. 2024). For example, marigolds are rich in lutein, which benefits eye health, and nasturtiums are noted for their high vitamin C content. The potential health advantages of edible flowers make them an appealing choice for health-conscious consumers.

The culinary industry's focus on sustainability and local sourcing has further fueled the popularity of edible flowers. Chefs and consumers are increasingly preferring ingredients that are sustainably grown and locally sourced. Edible flowers, often cultivated using environmentally friendly practices, align well with these preferences (Duggirala et al. 2024). The availability of diverse varieties grown by local farmers adds to the appeal of including edible flowers in dishes.

In addition, the multifunctionality of edible blooms in culinary uses is remarkable. They can be utilized fresh, dried, or crystallized, providing a broad array of applications in different dishes. Edible flowers are suitable for everything from appetizers and main courses to desserts and beverages, showcasing their extensive range of uses. This versatility allows chefs and home cooks to experiment with flavors and presentations, catering to diverse culinary preferences (Brown 2023).

The rising consumer interest in distinctive and visually attractive food products has greatly increased the popularity of edible flowers. Driven by culinary experimentation, social media influence, health and wellness trends, sustainability, and versatility, edible flowers have become a sought-after ingredient in both home kitchens and professional restaurants. This trend is likely to continue as consumers and chefs seek out new ways to create visually stunning and flavorful dishes.

15.7 Health and Wellness: The Nutritional Benefits of Edible Flowers

Vitamin-packed culinary blossoms, antioxidants, and other active biochemical agents, are increasingly popular among health-conscious consumers. They offer prospective wellness advantages, including supporting digestive processes and alleviating inflammation, which has fueled their appeal in the functional foods market (Market Research Future 2025).

Edible blossoms have attracted increasing interest because of their high levels of vitamins, antioxidants, and other bioactive compounds, making them a favored option for health-conscious consumers. These natural gems enhance the visual appeal of dishes and offer possible wellness advantages, including supporting digestion and soothing inflammation. This essay explores the nutritional benefits of edible flowers and their rising popularity in the functional foods market.

To start, let's explore the vitamin content of edible flowers. Numerous edible blossoms are abundant in essential vitamins that foster holistic wellness and vitality (Sood et al. 2024). For example, marigolds are abundant in vitamin A, crucial for supporting clear eyesight, a robust immune response, and radiant skin. Additionally, nasturtiums are excellent provider of vitamin C, a vital antioxidant that strengthens immunity and supports skin vitality (Nicknezhad et al. 2022). Including these flowers in one's diet can help fill nutritional gaps and promote overall wellness.

Antioxidants are another significant component of edible flowers (Fernandes et al. 2017). These compounds Here's a paraphrase with synonyms: assist in neutralizing free radicals within the body, mitigating oxidative damage, and decreasing the likelihood of long-term conditions like cardiovascular issues, malignancies, and metabolic disorders. The vibrant colors of flowers like pansies and violets are indicative of their high antioxidant content, specifically anthocyanins, which have been shown to protect against cardiovascular diseases and support brain health (Mattioli et al. 2020). Consistently eating edible flowers rich in antioxidants can lead to long-term health benefits.

Table 15.1 Bioactive and nutritional compounds edible flowers

Flower	Color	Nutrient concentration	Bioactive compounds	Reference
Nasturtium (*Tropaeolum majus*)	Orange	Vitamin C (100 mg/100 g), Calcium (70 mg/100 g)	Flavonoids, Carotenoids	Purohit et al. (2021)
Lavender (*Lavandula*)	Violet	Calcium (215 mg/100 g), Vitamin A (287 IU/100 g)	Linalool, Tannins	Purohit et al. (2021)
Rose (*Rosa*)	Pink	Vitamin C (60 mg/100 g), Vitamin E (0.87 mg/100 g)	Anthocyanins, Flavonoids	Mlcek and Rop (2011)
Marigold (*Tagetes*)	Yellow	Vitamin A (5610 IU/100 g), Vitamin E (1.89 mg/100 g)	Carotenoids, Flavonoids	Rodrigues and Spence (2023)
Hibiscus (*Hibiscus rosasinensis*)	Red	Vitamin C (43 mg/100 g), Magnesium (51 mg/100 g)	Organic acids, Anthocyanins	Chitrakar et al. (2019)
Pansy (*Viola tricolor*)	Various	Vitamin C (18 mg/100 g), Calcium (40 mg/100 g)	Flavonoids, Anthocyanins	Fernandes et al. (2017)
Chamomile (*Matricaria chamomilla*)	White/ Yellow	Magnesium (9 mg/100 g), Potassium (123 mg/100 g)	Apigenin, Luteolin	Murphy (2015)
Violet (*Viola odorata*)	Violet	Vitamin C (23 mg/100 g), Vitamin A (2000 IU/100 g)	Anthocyanins, Flavonoids	Fernandes et al. (2020)
Dandelion (*Taraxacum officinale*)	Yellow	Vitamin A (14,000 IU/100 g), Calcium (187 mg/100 g)	Polyphenols, Sesquiterpenes	Fortune Business Insights (2025)
Borage (*Borago officinalis*)	Blue	Vitamin C (35 mg/100 g), Magnesium (33 mg/100 g)	Gamma-linolenic acid, Flavonoids	Rodrigues and Spence (2023)

Besides vitamins and antioxidants, edible flowers possess various bioactive compounds that provide specific health advantages (Jakubczyk et al. 2021). For example, chamomile flowers are renowned for their anti-inflammatory and soothing effects, making them a common ingredient in teas and remedies for digestive problems and stress relief. Lavender, commonly used in both culinary and medicinal applications, contains compounds that promote relaxation and improve sleep quality (Patil and Waghmare 2024). These bioactive compounds enhance the functional properties of edible flowers, positioning them as valuable additions to a health-conscious diet (Table 15.1).

These flowers enhance the visual appeal of dishes and provide health benefits because they are abundant in nutrients, including vitamins, essential minerals, and active biological substances.

A key health advantage of edible flowers lies in their ability to support digestive health. Some flowers, like dandelions and calendulas, contain prebiotic fibers that aid gut health by encouraging the growth of beneficial bacteria in the digestive system (Clark 2023). This can improve nutrient absorption, decrease inflammation, and support overall gastrointestinal health. Adding these flowers to meals can be especially beneficial for those with digestive issues or anyone aiming to maintain a healthy gut microbiome (Rivas-García et al. 2021).

Furthermore, the anti-inflammatory properties of culinary blossoms have driven their popularity in the functional foods market (Prabawati et al. 2021). Persistent inflammation is linked to various medical conditions, such as joint disorders like arthritis and cardiovascular diseases and certain cancers. Flowers like hibiscus and roses contain anti-inflammatory compounds that can help lower inflammation and promote overall health. Consuming these flowers regularly can aid in managing inflammatory conditions and contribute to a balanced, health-promoting diet.

The rich nutritional profile of edible flowers, with their abundance of vitamins, antioxidants, and bioactive compounds, makes them a great option for health-conscious consumers. Their possible health benefits, such as enhanced digestion and decreased inflammation, have played a role in boosting their popularity in the functional foods market. As more people seek natural and nutritious food options, the demand for edible flowers is likely to continue rising, solidifying their place in modern culinary and wellness practices.

15.8 Sustainability: Advancing Practices in the Edible Flowers Industry

Eco-friendly agricultural methods, such as efficient water utilization and enhancing soil fertility and integrated pest management, are becoming increasingly important in the edible flowers industry (Rabiya 2024). Consumers are showing a preference for sustainably sourced products, driving demand for edible flowers grown using environmentally friendly methods (Market Research Future 2025).

Sustainable agricultural practices are increasingly becoming essential in the edible flowers industry. Such methods encompass efficient water usage, enhancing soil fertility, and employing comprehensive pest control strategies, are gaining importance as both producers and consumers recognize the need for eco-friendly cultivation methods. This essay explores the significance of these sustainable practices and their impact on consumer preferences and market trends in the edible flowers industry.

Effective hydrological planning is a key component of eco-conscious farming. As a precious resource, efficient use of water is crucial in agricultural practices, especially in areas enduring hydrological stress (Mpanga et al. 2022). Practices like drip irrigation and rainwater harvesting are being employed to reduce water wastage and ensure that edible flowers receive sufficient hydration without draining local water sources. Drip irrigation, for instance, directs water straight to the plant roots, minimizing evaporation and runoff (Guo and Li 2024). These water-saving methods help maintain the sustainability of edible flower cultivation.

Enhancing soil health is another essential component of eco-friendly farming. Healthy soil forms the basis of successful and sustainable farming, as it aids plant growth and resilience. Techniques including rotating crops, implementing cover crops, and applying organic soil enhancers are being employed to enhance soil richness and structure (Kama et al. 2025). Crop rotation helps prevent soil nutrient depletion, while protective crops shield the soil against erosion and enrich its organic material (Khan et al. 2024). Additionally, the utilization of organic fertilizers and composts enriches the soil with essential nutrients, promoting robust plant growth while diminishing the necessity for chemical inputs (Rastogi et al. 2023).

Integrated pest management (IPM) represents an integrated pest management strategy that minimizes reliance on chemical pesticides and focuses on natural and biological techniques (Headrick 2021). IPM strategies include introducing beneficial insects, using pheromone traps, and employing cultural practices that disrupt pest life cycles. These methods not only reduce the reliance on synthetic chemicals but also promote a balanced ecosystem within the agricultural environment (Muhammad Suhaib Ismayil et al. 2024). By adopting IPM, edible flower producers can ensure that their crops are protected from pests while minimizing damage to ecosystems and the well-being of people.

Consumers are increasingly showing a preference for sustainably produced products, including edible flowers. This change in consumer behavior is fueled by an increasing awareness of environmental issues and a commitment to supporting eco-friendly practices (Gabellini and Scaramuzzi 2024). Consequently, the interest in culinary blossoms has increased grown using sustainable methods has surged. Consumers are drawn to products that are not only visually appealing and nutritious but also align with their values of environmental stewardship and sustainability (Almusaed et al. 2023). This shift in demand is encouraging producers to adopt and expand sustainable practices in their cultivation processes.

The adoption of sustainable agricultural practices has several benefits for the edible flowers industry. Firstly, it enhances the long-term viability of farming by preserving natural resources and promoting biodiversity (Hobbs et al. 2008). Sustainable methods ensure that the soil remains fertile, water resources are conserved, and pest populations are managed in an environmentally friendly manner (Chhibber and Ravichandran 2024). Secondly, sustainable practices improve the marketability of edible flowers, as consumers are prepared to spend more on eco-friendly and ethically made goods (Fernandes et al. 2020). This increased market demand can lead to higher profits for producers who invest in sustainable methods.

Sustainable agricultural practices are becoming increasingly important in the edible flowers industry. Efficient use of water resources, enhancement of soil fertility, and comprehensive pest control strategies are essential components of sustainable farming that support eco-friendly cultivation. The growing consumer preference for sustainably produced products is driving the demand for edible flowers grown using these methods. As the industry continues to prioritize sustainability, it can look forward to long-term benefits with respect to preserving resources, enhancing product appeal, and reducing ecological footprint.

15.9 Conclusion

The edible flowers industry is experiencing remarkable growth driven by technological advancements and emerging market trends. Innovations in biotechnology, precision agriculture, and controlled environment agriculture have transformed production methods, significantly improving the efficiency and quality of edible flowers. At the same time, the increasing consumer interest in healthy eating, along with wellness and sustainability trends, is driving higher demand for these products. Social media influence also plays a crucial role, boosting the visibility and popularity of edible flowers among consumers. These developments not only optimize production but also increase the presence and variety of culinary blossoms available in the marketplace. Edible blossoms are finding their place in global gastronomy, enhancing dishes and offering nutritional benefits. Their inclusion in salads, desserts, and bakery products is on the rise. Moreover, culinary blossoms not only enhance taste and visual appeal as well as offer health benefits. The combination of technological advancements and changing consumer preferences is Laying the foundation for an eco-friendly and forward-thinking future in the edible flowers industry, ensuring its continuous growth and adaptability in an ever-evolving market. This boom promises new opportunities and an expanding market for the coming years.

References

Almusaed A, Almssad A, Alasadi A, Yitmen I, Al-Samaraee S (2023) Assessing the role and efficiency of thermal insulation by the "Bio-Green Panel" in enhancing sustainability in a built environment. Sustainability 15(13):10418. https://doi.org/10.3390/su151310418

Barani YH, Zhang M, Mujumdar AS, Chang L (2022) Preservation of color and nutrients in anthocyanin-rich edible flowers: progress of new extraction and processing techniques. J Food Process Preserv 46(9):e16474. https://doi.org/10.1111/jfpp.16474

Brown T (2023) Versatility of edible flowers in modern cuisine. Culinary Trends J

Brown L, Green R (2023) Innovations in hydroponic systems for edible flowers. Int J Hortic Sci 47(4):256–269

Bubenheim DL (2024) Aeroponic systems: future of high-tech plant production. Adv Agron 2(7):75–78

Cash SB, Alhabash S, Fretes G, Ma M (2022) Food, beverages and social media: trends and tools for economic research. In: Roosen J, Hobbs JE (eds) A modern guide to food economics. Edward Elgar Publishing, pp 297–326. https://doi.org/10.4337/9781800372054.00022

Chaudhary HH, Ahir MP, Chaudhary S, Mangroliya PM, Patel R, Chaudhari UC (2024) Blooming future: new emerging technologies in floriculture. Vigyan Varta 5(10):32–37. https://vigyanvarta.in/adminpanel/upload_doc/VV_1024_07.pdf

Chhibber A, Ravichandran S (2024) Eco friendly agriculture. Int J Anal Appl Chem 10(1):21–25p. https://doi.org/10.37628/IJAAC

Chitrakar B, Zhang M, Bhandari B (2019) Edible flowers with the common name "marigold": their therapeutic values and processing. Trends Food Sci Technol 89:76–87. https://doi.org/10.1016/J.TIFS.2019.05.008

Clark R (2023) Edible flowers and gut health: the role of prebiotics. Nutr Sci J. https://doi.org/10.3390/ijms25094834

Copetta A, Mussano P, Devi P, Lanteri A, Cassetti A, Mascarello C, Bisio A, Ruffoni B (2023) In vitro micropropagation, rooting and acclimatization of two Agastache species (A aurantiaca and A mexicana). Horticulturae 9(10):1065. https://doi.org/10.3390/horticulturae9101065

Cox DA (2018) Hartmann and Kester's plant propagation principles and practices. HortScience 53(5):741–741. https://doi.org/10.21273/HORTSCI535bkrev-17

Despommier D (2010) The vertical farm: feeding the world in the 21st century. Macmillan

Duggirala K, Mummaleti G, Kong F, Roy A, Mohan A (2024) Edible flowers: a sustainable source of natural food ingredient. J Food Bioact 27. https://doi.org/10.26599/JFB.2024.95027383

FAO (2023) Global production of edible flowers. Food and Agriculture Organization of the United Nations

Fernandes L, Casal S, Pereira JA, Ramalhosa E (2017) Edible flowers: a review of the nutritional, antioxidant, antimicrobial properties and effects on human health. J Food Sci 22(1):23–35. https://doi.org/10.1016/j.jfca.2017.03.017

Fernandes L, Casal S, Pereira JA, Saraiva JA, Ramalhosa E (2020) An overview on the market of edible flowers. Food Rev Intl 36(3):258–275. https://doi.org/10.1080/87559129.2019.1639727

Fortune Business Insights (2025) Edible flowers market size, share, and industry analysis, by type (Rose, Jasmine, Marigold, Hibiscus, Lavender, and others), by application (Food & Beverages, Baked Goods, Beverages, Dairy Products, Confectionery, and others), and regional forecast 2025–2032. Fortune Business Insights. https://www.fortunebusinessinsights.com/edible-flowers-market-110278

Gabellini S, Scaramuzzi S (2024) Evolving consumption trends and marketing strategies in ornamental horticulture–what are the new challenges? In: EHC2024: International Symposium on Ornamental Horticulture for the Service of Society, vol 1417, pp 255–278. https://doi.org/10.17660/ActaHortic.2025.1417.32

Guebsi R, Mami S, Chokmani K (2024) Drones in precision agriculture: a comprehensive review of applications, technologies, and challenges. Drones 8(11):686. https://doi.org/10.3390/drones8110686

Guo H, Li S (2024) A review of drip irrigation's effect on water, carbon fluxes, and crop growth in farmland. Water 16(15):2206. https://doi.org/10.3390/w16152206

Gupta A, Sharma S, Rajput D, Patil UK (2024) Medicinal and culinary importance of edible flowers of Indian origin: an in-depth review. Discov Food 4:148. https://doi.org/10.1007/s44187-024-00220-6

Gurung L, Rawal JS, Puspa RC, Joshi GR, Mandal A (2024) Vertical farming in urban agriculture: opportunities, challenges, and future directions. Big Data Agric 6:89–95. https://doi.org/10.26480/bda.02.2024.89.95

Hasna PM, Rafeekher M, Anaswara SJ (2023) Precision farming in commercial floriculture: a review. Biol Forum Int J 15(8a):473–479

Headrick D (2021) The future of organic insect pest management: be a better entomologist or pay for someone who is. Insects 12(2):140. https://doi.org/10.3390/insects12020140

Hobbs PR, Sayre K, Gupta R (2008) The role of conservation agriculture in sustainable agriculture. Philos Trans R Soc B Biol Sci 363(1491):543–555. https://doi.org/10.1098/rstb.2007.2169

Jakubczyk K, Łukomska A, Gutowska I, Kochman J, Janił J, Janda K (2021) Edible flowers extracts as a source of bioactive compounds with antioxidant properties—in vitro studies. Appl Sci 11(5):2120. https://doi.org/10.3390/app11052120

Jones JB (2004) Hydroponics: a practical guide for the soilless grower, 2nd edn. CRC Press. https://doi.org/10.1201/9780849331671

Kama R, He J, Nabi F, Aidara M, Faye B, Diatta S et al (2025) Crop rotation and green manure type enhance organic carbon fractions and reduce soil arsenic content. Agric Ecosyst Environ 378:109287. https://doi.org/10.1016/j.agee.2024.109287

Khan MT, Aleinikovienė J, Butkevičienė LM (2024) Innovative organic fertilizers and cover crops: perspectives for sustainable agriculture in the era of climate change and organic agriculture. Agronomy 14(12):2871. https://doi.org/10.3390/agronomy14122871

Kozai T, Niu G, Takagaki M (2019) Plant factory: an indoor vertical farming system for efficient quality food production, 2nd edn. Academic. https://doi.org/10.1016/C2014-0-01039-8

Laveglia S, Altieri G, Genovese F, Matera A, Di Renzo GC (2024) Advances in sustainable crop management: integrating precision agriculture and proximal sensing. AgriEngineering 6(3):3084–3120. https://doi.org/10.3390/agriengineering6030177

Market Research Future (2025) Edible flowers market trends – market research future. Edible Flowers Market Research Report. https://www.marketresearchfuture.com/reports/edible-flowers-market/market-trends

Mattioli R, Francioso A, Mosca L, Silva P (2020) Anthocyanins: a comprehensive review of their chemical properties and health effects on cardiovascular and neurodegenerative diseases. Molecules 25(17):3809. https://doi.org/10.3390/molecules25173809

Mlcek J, Rop O (2011) Fresh edible flowers of ornamental plants – a new source of nutraceutical foods. Trends Food Sci Technol 22(10):561–569. https://doi.org/10.1016/j.tifs.2011.04.006

Mohammed Aarif KO, Alam A, Hotak Y (2025) Smart sensor technologies shaping the future of precision agriculture: recent advances and future outlooks. J Sens 2025(1):2460098. https://doi.org/10.1155/js/2460098

Mpanga IK, Gaikpa DS, Koomson E, Dapaah HK (2022) Water management in agriculture: innovations for efficient irrigation. In: The Palgrave handbook of global sustainability, pp 1–23. https://doi.org/10.1007/978-3-030-38948-2_33-1

Muhammad Suhaib Ismayil M, Neha RK, Bisen S, Bisen D, Samreen (2024) Integrated pest management (IPM) in floriculture: strategies for sustainable pest control. Int J Res Agron 7(11):15–18. https://doi.org/10.33545/2618060X.2024.v7.i11a.1912

Murphy H (2015) Foods indigenous to the Western hemisphere. American Indian Health Diet Project; Available online: http://www.aihd.ku.edu/foods/squash.html. Accessed 18 Sept 2020

Nicknezhad S, Hashemabadi D, Allahyari MS (2022) Nutritional value of some flowers found in green spaces as new food sources. J Ornament Plants 12(2):143–155. https://dorl.net/dor/20.1001.1.22516433.2022.12.2.6.2

Patil H, Waghmare J (2024) Lavender oil: a comprehensive review of composition and applications. Asian J Res Chem 17(6):377–386. https://doi.org/10.52711/0974-4150.2024.00062

Prabawati NB, Oktavirina V, Palma M, Setyaningsih W (2021) Edible flowers: antioxidant compounds and their functional properties. Horticulturae 7(4):66. https://doi.org/10.3390/horticulturae7040066

Purohit SR, Rana SS, Idrishi R, Sharma V, Ghosh P (2021) A review on nutritional, bioactive, toxicological properties and preservation of edible flowers. Future Foods 4:100078. https://doi.org/10.1016/j.fufo.2021.100078

Rabiya UK (2024) Modern innovations and sustainability in floriculture: trends, technologies, and practices. J Divers 3(2):1–23. https://doi.org/10.51470/JOD.2024.03.02.01

Ramawat KG, Ekiert HM, Goyal S (eds) (2020) Plant cell and tissue differentiation and secondary metabolites: fundamentals and applications. Springer. https://doi.org/10.1007/978-3-030-11253-0

Rao V, Poonia A (2024a) Bioactive compounds, nanoparticle synthesis and potential utilization of edible flowers for development of functional dairy products: a review. J Food Sci Technol 61(6):1053–1068. https://doi.org/10.1007/s13197-023-05853-6

Rao V, Poonia A (2024b) Microbial biosynthesis of selenium nanoparticles using probiotic strain and its characterization. J Food Meas Charact 18(7):5477–5488. https://doi.org/10.1007/s11694-024-02581-z

Rastogi M, Verma S, Kumar S, Bharti S, Kumar G, Azam K, Singh V (2023) Soil health and sustainability in the age of organic amendments: a review. Int J Environ Clim Chang 13(10):2088–2102. https://doi.org/10.9734/ijecc/2023/v13i102870

Resh HM (2022) Hydroponic food production: a definitive guidebook for the advanced home gardener and the commercial hydroponic grower. CRC Press. https://doi.org/10.1201/9781003133254

Rivas-García L, Navarro-Hortal MD, Romero-Márquez JM, Forbes-Hernández TY, Varela-López A, Llopis J et al (2021) Edible flowers as a health promoter: an evidence-based review. Trends Food Sci Technol 117:46–59. https://doi.org/10.1016/j.tifs.2020.12.007

Rodrigues H, Spence C (2023) Looking to the future, by studying the history of edible flowers. Int J Gastron Food Sci 9:100805. https://doi.org/10.1016/j.ijgfs.2023.100805

Sanyaolu M, Sadowski A (2024) Sustainable farming practices: the role of precision agriculture. Sustainability 16(15):6668. https://doi.org/10.3390/su16156668

Sarode DK, Jadhav PR, Pagariya MC, Shinde NA, Devarumath RM, Shingote PR, Prasad KV, Kawar PG (2024a) Cutting-edge biotechnological progress in the world of edible flowers. Environ Anal Ecol Stud 12(1):779. https://doi.org/10.31031/EAES.2024.12.000779

Sarode DK, Pagariya MC, Jadhav PR, Patil SA, Devarumath RM, Shingote PR, Prasad KV, Jain SM, Penna S, Kawar PG (2024b) Edible flowers: biotechnological interventions for improving bioactives of food and health significance. J Food Compos Anal 134:Article 106506. https://doi.org/10.1016/j.jfca.2024.106506

Sood Y, Lal M, Kalia A, Verma S (2024) Edible flowers: super foods with potential health benefits. Int J Plant Soil Sci 36(3):213–221. https://doi.org/10.9734/ijpss/2024/v36i34417

Velazquez-Gonzalez RS, Garcia-Garcia AL, Ventura-Zapata E, Barceinas-Sanchez JDO, Sosa-Savedra JC (2022) A review on hydroponics and the technologies associated for medium- and small-scale operations. Agriculture 12(5):646. https://doi.org/10.3390/agriculture12050646

Wally O, Punja ZK (2010) Genetic engineering for increasing fungal and bacterial disease resistance in crop plants. GM Crops 1(4):199–206. https://doi.org/10.4161/gmcr.1.4.13225

Wani MA, Din A, Nazki IT, Rehman TU, Al-Khayri JM, Jain SM, Lone RA, Bhat ZA, Mushtaq M (2023) Navigating the future: exploring technological advancements and emerging trends in the sustainable ornamental industry. Front Environ Science 11:Article 1188643. https://doi.org/10.3389/fenvs.2023.1188643

Weathers PJ (2023) Innovations in aeroponics: enhancing plant growth and productivity. Plant Sci 38:1–10

Zhang J, Zha L (2023) Urban agriculture and vertical farming. In: Zhang Q (ed) Encyclopedia of digital agricultural technologies. Springer, Cham. https://doi.org/10.1007/978-3-031-24861-0_33

Index

A. Poonia et al. (eds.), *Edible Flowers: Source of Phytonutrients, Valorization and Technological Advancements*, https://doi.org/10.1007/978-3-032-00667-7

GPSR Compliance

The European Union's (EU) General Product Safety Regulation (GPSR) is a set of rules that requires consumer products to be safe and our obligations to ensure this.

If you have any concerns about our products, you can contact us on ProductSafety@springernature.com

In case Publisher is established outside the EU, the EU authorized representative is:

Springer Nature Customer Service Center GmbH
Europaplatz 3
69115 Heidelberg, Germany

Batch number: 09198611

Printed by Printforce, the Netherlands